X-RAYS, ELECTRONS, AND ANALYTICAL CHEMISTRY

Spectrochemical Analysis with X-Rays

X-RAYS, ELECTRONS, AND

Spectrochemical Analysis with X-Rays

ANALYTICAL CHEMISTRY

H. A. Liebhafsky General Electric Company 1934–1967
Texas A & M University 1967–
H. G. Pfeiffer General Electric Company 1948–

E. H. Winslow General Electric Company 1930–1962

P. D. Zemany General Electric Company 1945–

Collaborating Editor
Sybil Small Liebhafsky

NEW YORK · LONDON
SYDNEY · TORONTO

WILEY-INTERSCIENCE
a Division of
JOHN WILEY & SONS, INC.

Copyright © 1972, by John Wiley & Sons, Inc.

All rights reserved. Published simultaneously in Canada.

No part of this book may be reproduced by any means, nor transmitted, nor translated into a machine language without the written permission of the publisher.

Library of Congress Cataloging in Publication Data

Main entry under title:
X-Rays, electrons, and analytical chemistry.

  Bibliography: p.
  1. X-ray spectroscopy.   I. Liebhafsky, H. A.

QD95.X27   544'.66   78–39772
ISBN 0–471–53428–5

Printed in the United States of America.

10 9 8 7 6 5 4 3 2 1

To the memory of those physicists whose work on x-rays and electrons extended the horizons of analytical chemistry as one consequence of a scientific revolution.

*"O brave new world,
That has such people in't."*

Shakespeare, *The Tempest,*
Act V, Sc. 1, Line 183

Preface

This book was to be a revised edition of the 1960 *X-Ray Absorption and Emission in Analytical Chemistry*, but the "relevant happenings" of the last decade dictated a new book. During that time, electrons again became prominent for exciting x-rays, and emitted electrons became important sources of information about composition and chemical binding. Electrons thus deserve titular partnership. That is not all. X-ray diffraction in analytical chemistry warrants more than passing mention. X-ray absorptiometry continues to grow *relatively* less important. The approaching omnipresence of the computer cannot be ignored. Many topics—films, surfaces, light elements, process control, and new detectors as examples—need at least cursory discussion. And finally, a new book made it possible to extend the treatment of the classic early work on x-rays and electrons that now forms an important bridge from physics to analytical chemistry.

Our concern with the analytical chemist is clear from the reprinted preface that follows. Other readers, physicists included, should find the new book of interest. Much that is in it has twice been presented in a graduate course at Texas A&M University. Because such courses are not offered as widely as we think they should be, we have tried to write a book that will help a reader who has no instructor.

The book need not be read *in toto*. Especially for those interested in particular topics, we have provided cross references and comprehensive indices. Of the appendices, two deserve mention here. The tables of mass absorption coefficients include many newly calculated values. The last appendix will guide the analytical chemist element by element to the extensive practical literature that describes how they can be determined.

We thank all who have helped us, and our special thanks go to Mr. J. E.

Bigelow, Dr. L. K. Frevel, Dr. H. P. Klug, Dr. E. A. Meyers, Mr. M. L. Salmon, Dr. R. L. R. Towns, and Dr. R. L. Watson. Again we thank in advance all who inform us of our errors and shortcomings!

H. A. LIEBHAFSKY
H. G. PFEIFFER
E. H. WINSLOW
P. D. ZEMANY

College Station, Texas
Schenectady, New York
November 1971

Preface to X-ray Absorption and Emission in Analytical Chemistry

Like it or not, the chemistry is going out of analytical chemistry. For a long time indeed, Chaucer with his

> The lyf so short, the craft so long to lerne,
> Th' assay so hard, so sharp the conquering.
> *The Parlement of Foules*

proved a better prophet than he knew. But, nowadays, physics and electronics are in part being fused with analytical chemistry to make the assay easier and the conquering less painful.

This fusion of disciplines, though desirable and inevitable, complicates the writing of books in fields where it occurs. Spectrochemical analysis by means of x-rays is definitely such a field. For whom shall a book on this subject be written? Our answer is clear. This book was written for the analytical chemist who wants to use these x-ray methods and to understand them. We have striven for correctness in physics, electronics, and statistics; but we have tried first of all to help the analytical chemist in his work.

Nomenclature is bound to flourish in a field that rests on several disciplines. We have pruned the growth, judiciously we hope, after considering all the various shoots, even the more exotic. To keep the book within bounds, equipment for description and literature for reference had to be selected. The choice was made with the analytical chemist in mind. In general, references later than August 1957 could not be included.

We planned this book to be useful even if read only in part. The first chapter is a summary of what is known about x-rays that is pertinent to spectrochemical analysis, and it should receive at least cursory attention from all readers. Those interested primarily in absorptiometry may then turn to Chapter 3 or 5; in film thickness, to Chapter 6; and in x-ray emission spectrography, to Chapters 7 and 8. The remaining five chapters are ancillary and deal with special topics.

Again with the analytical chemist in mind, we have not treated all topics equally. The electronics expert is likely to feel we have skimped, especially in Chapter 2; Chapter 4 is oversimplified; statisticians will find much missing from Chapter 10; and other important developments could have been treated in Chapters 9 and 11.

Because this book rests on diverse disciplines, we have had to seek—and have received—help to an unusual degree. For their help, we gladly thank Dr. C. A. Bennett, Mr. J. E. Bigelow, Mr. H. C. Buchholtz, Mr. J. R. Churchill, Dr. R. L. Griffith, Dr. T. A. Hall, Mr. J. W. Kemp, Dr. H. P. Klug, Dr. W. F. Loranger, Dr. D. M. Miller, Dr. B. W. Roberts, Dr. J. Rouvina, and Mr. M. L. Salmon. We also thank the editors of *Analytical Chemistry*, who were kind enough to let us draw extensively upon the material we have published in that journal. And, finally, we shall be most happy to thank our readers for showing us mistakes and shortcomings that have escaped proofreading and revision!

H. A. LIEBHAFSKY
H. G. PFEIFFER
E. H. WINSLOW
P. D. ZEMANY

Schenectady, New York
April, 1960

Contents

Chapter			
	1.	Generation and Properties of X-Rays	1
	2.	The Measurement of X-Ray Intensity, X-Ray Detectors, and Detector Systems Energy Resolution	58
	3.	Absorptiometry with X-Rays	127
	4.	X-Ray Spectra	171
	5.	The Selection of X-Ray Wavelengths	198
	6.	X-Ray Diffraction in Chemical Analysis	232
	7.	Measurement of Film Thickness Simple Trace Determinations	281
	8.	Reliability of X-Ray Emission Spectrography Statistical Considerations	328
	9.	X-Ray Emission Spectrography. General	355
	10.	Equipment and Selected Applications	460
Appendix	I.	X-Ray Safety	501
	II.	The X-Ray Absorption Edges of the Elements	504
	III.	X-Ray Emission Spectra of the Elements	513
	IV.	Mass Absorption Coefficients of the Elements and of Selected Films	525

V.	Characteristic Photon Yields of the Elements and Absorption Jump Ratios	532
VI.	Determination of Elements by X-Ray Emission Spectrography A Guide to the Recent Literature (1964–1970)	535

Index 551

X-RAYS, ELECTRONS, AND ANALYTICAL CHEMISTRY

Spectrochemical Analysis with X-Rays

Chapter 1

Generation and Properties of X-Rays

Der Kürze halber möchte ich den Ausdruck 'Strahlen' und zwar zur Unterscheidung von anderen den Namen 'X-Strahlen' gebrauchen.

W. C. Roentgen [1]

INTRODUCTION

The x-ray story, one of the most brilliant in modern physics, is characterized by elegant simplicity, by a close relationship to quantum theory and to atomic structure, and by important applications that began almost immediately after x-rays were discovered. An important chapter on analytical chemistry is now added to the story. This chapter contains little that is new about the x-rays themselves; even the analytical methods were largely conceived before World War II. The scientific and technological activity that began during this war, and has expanded since then, accelerated the development of many kinds of equipment often intended for other purposes, but useful in connection with x-rays. Notably, methods of x-ray detection were improved until measurements of x-ray intensity became easy and highly precise. This improvement, which benefited from rapid progress in experimental nuclear physics, led to a rapidly growing appreciation of the great advantages of x-rays for analytical chemistry. Realized first on improvised or converted equipment—especially on converted diffractometers—these advantages soon led to the manufacture of equipment designed especially for laboratories of analytical chemistry and for the control of industrial processes. The conversion of an electron microscope to an electron microprobe by Castaing [2] was a particularly noteworthy event. During the last decade, great progress has been made in equipment using x-rays of long wavelength to make light elements accessible to determination by methods based on x-ray emission. In 1968 computers began to make themselves felt so that x-ray *systems* are multiplying for the *characterization* and *control* of materials, the mission of modern analytical chemistry [3]. The end is not in sight. However, it is likely that x-ray methods will become—if indeed they are not already—the most important single class of methods for the *determination* of *elements* with atomic numbers exceeding 7 (nitrogen). Meanwhile, progress continues unabated on x-ray methods that reveal structure and chemical state—methods in the main outside the scope of this book.

In this introductory chapter, we describe the important properties of x-rays, trace their discovery, and indicate their significance in quantum theory and in atomic structure—all to foreshadow their importance in analytical chemistry.

The great penetrating power of x-rays is often considered their most remarkable property. The essential simplicity of their spectra is of at least equal concern to us; after all, this leads to analytical methods in the results of which deviations from simplest behavior are often predictable. Both the great penetrating power (small chance of being absorbed) and the simplicity of x-ray spectra are traceable ultimately to the high energy of the rays, which also makes it possible to measure their intensity by counting quanta.

X-rays have high energies because of the way atoms are built. Their energies exceed those of any other spectra produced outside the atomic nucleus. This tells us that *most* x-ray absorption or emission spectra will result from displacements of the *inner* electrons of the atom; that is, the electrons between the valence electrons and the atomic nucleus. Needless to say, the simplicity of x-ray spectra is less marked for the light atoms (low x-ray energies), in which the inner electrons are few and the valence electrons closer to the nucleus. These matters need not concern us here, where the aim is to emphasize that neither x-ray absorption nor emission spectra reflect the periodic characteristics of Mendeleeff's table; if differences due to atomic *number* (not atomic weight) are ignored, the x-ray spectra (absorption or emission) of chlorine closely resemble those of argon, which in turn are much like those of potassium, and so on. It is precisely this independence of the chemical nature of elements that has made x-rays invaluable in modern analytical chemistry.

DISCOVERY OF X-RAYS

Ich komme deshalb zu dem Resultat, dass die
X-Strahlen nicht identisch sind mit den Kathodenstrahlen, dass sie aber von den Kathodenstrahlen
in der Glaswand des Entladungsapparates erzeugt
werden. [4]

1.1 Roentgen's Discovery

In October 1895, Roentgen began experiments on cathode rays (electron beams); these had been studied by many brilliant investigators, among them Lenard, who helped Roentgen begin his work. On November 8, 1895, which deserves to be remembered as the last day in the era of classical physics, Roentgen covered a Hittorf-Crookes tube (Figure 1.1–1) completely with black cardboard and passed through it a succession of electrical discharges at high voltage. In his darkened laboratory, he noticed a weak light on a bench about a yard from the tube.

Not believing this possible, he passed another series of discharges through the tube, and again the same fluorescence appeared, this time looking like faint green clouds moving in unison with the discharges ... [5]

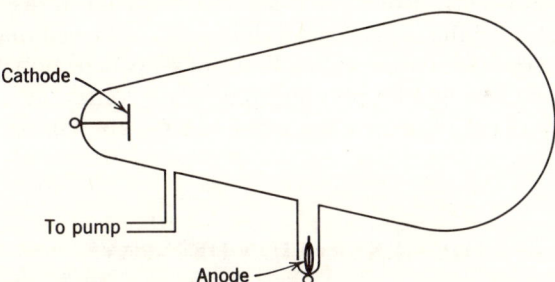

Fig. 1.1–1. Schematic diagram of a Hittorf-Crookes tube of the kind in which were generated the first x-rays to be discovered. With a high-voltage source of direct current connected to the anode and cathode of an air-filled Hittorf-Crookes tube, imagine the air pressure to be gradually lowered. The total electric strength of the air between anode and cathode will decrease. When this strength becomes less than the potential difference between these electrodes, electrical breakdown of air generates a succession of complex, striking phenomena. As the state of what Roentgen called "high evacuation" is approached, a bluish glow appears, which is produced by the beam of cathode rays (electrons) that leaves the cathode. As the pressure is further reduced and Roentgen's state of "high evacuation" is reached, the blue color fades; but the glass walls begin to glow a brilliant green where they are struck by electrons, mainly in regions opposite the cathode. X-rays accompany the green light. The walls have acted as a *target* for x-ray production by electron bombardment, and the tube has become an *x-ray source*.

On striking a match, he found that the fluorescent light came from a small barium platinocyanide screen. Today we might describe the experiment by saying that intermittent electron bombardment of the tube walls had generated x-ray *pulses* by *electron excitation*, and that the x-rays emanating from the tube as *x-ray source* had penetrated walls, cardboard, and air to produce synchronous pulses of visible light in a phosphor (the x-ray *detector*).

Roentgen classified his tubes as "hard" and "soft," terminology applied to x-rays today. The hard tubes had lower gas pressure than the soft did. The potential difference required to generate x-rays in the hard tube was higher, and these x-rays were higher in penetrating power. Hard tubes give hard x-rays, and the x-rays from soft tubes are soft. The origin of this terminology seems lost.

The x-ray work for which Roentgen in 1901 received the first Nobel Prize in physics is described in three great papers [1]. By 1927, further work on x-rays had merited five other Nobel Prizes: M. T. F. von Laue (1914), W. H. Bragg and W. L. Bragg (1915), C. G. Barkla (1917), K. M. G. Siegbahn (1924), and A. H. Compton (1927). In 1896, Becquerel discovered radioactivity while working with uranium to test the (erroneous) idea that radiant energy similar to x-rays could result from ordinary substances made phosphorescent by visible light. Roentgen thus helped to usher in the atomic age. X-ray diffraction played a key role in recent Nobel awards that concern the structure of proteins.

No single scientific discovery in the last century has had more important consequences. However, Roentgen's achievements far transcended mere discovery. He studied the properties of the new rays so well that he laid the foundations not only for important methods of x-ray detection (fluorescence of a phosphor, darkening of a photographic plate, ionization of a gas) and for radiography, but also for the application of x-ray absorption to analytical chemistry.

THE GENERATION OF X-RAYS

> The idea of using a hot cathode in a Roentgen tube was not new, but... the principle had never been successfully applied in a vacuum good enough so that positive ions did not play an essential role. [6]

1.2 Gas or Ion Tubes

Roentgen's "highly evacuated" electron tubes were the precursors of today's gas tubes—or ion tubes, which is a better name. In these, a potential difference usually exceeding 10 kV is placed across the electrodes to break

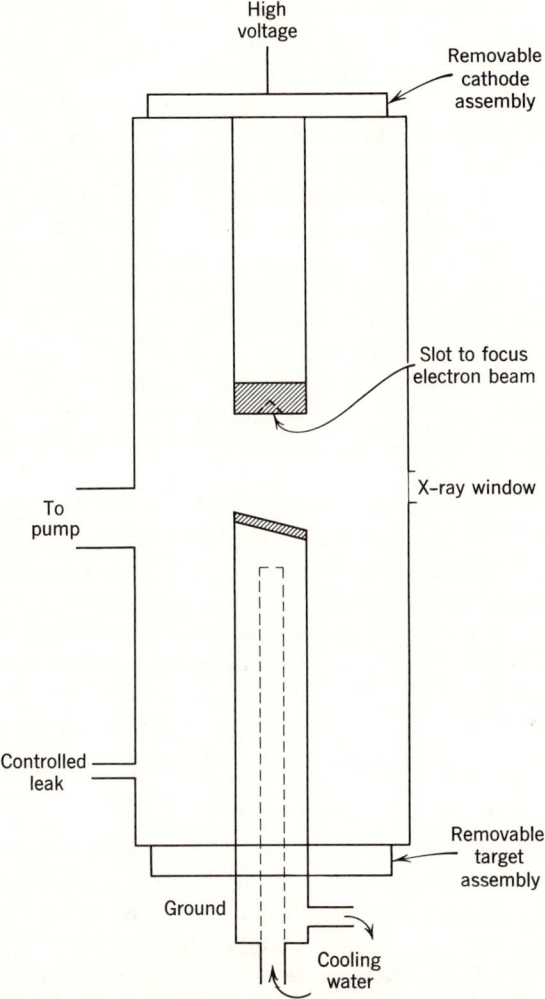

Fig. 1.2–1. Schematic diagram of demountable gas (ion) x-ray tube. For a tube of a given design, current and voltage are fixed by invariant continuous pumping against a controlled leak to maintain stable operating conditions. The whole target assembly and the target metal button are easily changed. The cathode button, usually slotted for focusing, is easily removable for cleaning and reslotting, which are needed to repair damage done by sputtering and ion bombardment. Now that Coolidge tubes give x-rays of spectral purity approaching that of the rays from gas tubes, these tubes have become less popular because they are more difficult to operate and control.

down electrically a gas at low pressure ($\sim 10^{-2}$ torr). The initial cause of breakdown can be a few electrons either pulled out of the cathode by field emission or generated by cosmic rays. Under the influence of the field, these few electrons gain enough energy to ionize the gas, and the positive ions thus produced strike the cathode to generate electrons that are accelerated by the field in their turn. When these electrons strike the target (anode), x-rays are generated. As mentioned above, the maximum energy of the x-rays increases with increasing potential difference between the electrodes, which increases with decreasing gas pressure. A constant and reproducible x-ray beam therefore requires careful stabilization of the gas pressure. Even Roentgen observed "self-evacuation" in his tubes on long continued operation. A "clean-up" of gas reduced the pressure and made tube and x-rays "harder." Stable gas pressures can be achieved by pumping against a controlled leak (see Figure 1.2–1), but such equipment is not readily portable. Portability and stability of operation tend to be mutually self-exclusive in these tubes.

Nevertheless, gas (ion) tubes have two advantages that can be overriding in special cases. First, in such tubes, provisions for ready interchangeability of targets are usually made so that x-ray spectra can be generated from one of a number of metals as needed. Second, spectral purity of the x-ray beam maintains itself in a gas tube properly operated because the risk of contaminating the target is then small; furthermore, contaminated targets are easily replaced.

1.3. The Coolidge Tube and Its Voltage Circuits

In the Coolidge tube (Figure 1.3–1), the vacuum is so good [6] that gases or positive ions derived therefrom do not influence the x-ray beam. Being sealed, the tube obviously requires no pump—a great advantage. The electrons needed to excite the x-ray beam "evaporate" from a heated tungsten filament according to the Richardson equation, which relates the increase of tube current to temperature. As the heating and electron-accelerating circuits are independent, the rate at which electrons strike the target (tube current) can be adjusted independently of their energy (tube voltage); consequently, intensity and wavelength of the x-ray beam are independently variable.

The foregoing characteristics of the Coolidge tube make it a highly convenient and flexible x-ray source, easily made portable. On the other hand, the high temperature of the cathode brings with it some risk of target contamination during life. The principal contaminant is tungsten evaporated from the cathode and the filament, and the loss of spectral purity is therefore most pronounced when the target is made of another metal. Target versatility in Coolidge tubes is becoming common, and targets of tungsten, molyb-

THE GENERATION OF X-RAYS 7

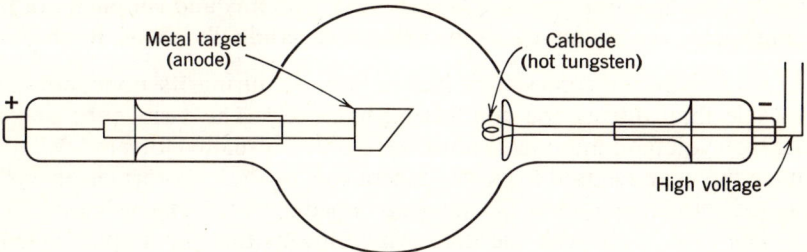

Fig. 1.3–1. Schematic diagram of the Coolidge (high-vacuum) x-ray tube. Coolidge tubes are widely used because they are stable, long-lived, and permit tube current and voltage to be controlled independently.

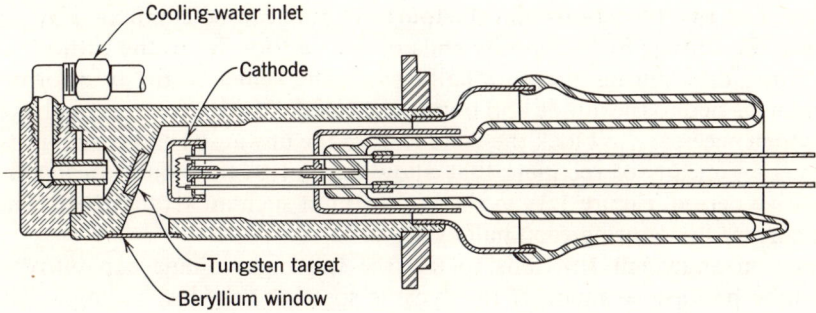

Fig. 1.3–2. A modern version of the Coolidge x-ray tube. The tube operates at high vacuum with storage battery (portable models) or step-up transformer as power source. Tube voltages range from a few hundred volts up to values (a million or more) beyond the needs of ordinary analytical chemistry. Because most of the energy fed into the tube is degraded to heat, effective cooling is of first importance.

denum, platinum, chromium, copper, silver, nickel, cobalt, iron, and even carbon are available; others could be made if needed. A modern version of the Coolidge tube appears in Figure 1.3–2.

The electrical engineering associated with x-ray tubes and their power supplies, though of first importance, is outside the scope of this book. However, most analytical chemists will wish to know the main features of the electrical system. When voltages below 50 kV were the rule and maximum intensity and stability were not necessary, Coolidge tubes were usually connected across the secondary of a high-voltage transformer. See Figure 1.3–3a. In this arrangement the tube conducts during the half-cycle; the filament is negative and thus produces x-rays only during this half-cycle. The anode (target) end is usually grounded to make working safe and to simplify cooling connections. In spite of the economy and simplicity of this arrangement, the following drawbacks have gradually made it obsolete.

1. A transformer comes to a higher voltage during its nonconducting half-cycle than during the conducting half-cycle. The x-ray tube and all other high voltage components must therefore be insulated for a much higher voltage than can be used for x-ray generation, or the transformer has to be designed with an extremely low internal impedance—an expensive solution.

2. The tube is idle half the time, which means that for a given average current the peak current must be doubled; this reduces the life of the tube.

3. The average voltage is 64% of the peak voltage and the x-ray production is thus below the capacity of the tube.

4. The detectors receive the counts in bursts and suffer from increased coincidence losses.

The first two objections and the fourth in part can be eliminated by applying the voltage from opposite ends of the transformer to the cathode of the x-ray tube during alternate half-cycles. The center of the transformer winding is tied to the anode and both are grounded for reasons given above. It is then necessary to block the voltage from the unwanted end of the transformer by means of rectifiers that allow current to flow only during the negative period. Figure 1.3–3b shows a form of such an arrangement with the current flow during each half-cycle.

This arrangement still fails to use the full high-voltage capability of the tube because so much of the cycle is spent at very low voltages. The rectifiers can be either solid-state or tube type, with the former gradually becoming more popular as the capabilities of solid-state diodes made of silicon are increased. Their main advantage over tube-type rectifiers is the complete absence of x-rays generated in the rectifiers themselves.

The third and fourth objections mentioned above are eliminated by direct-current operation, as with batteries. These have now been replaced by "constant-potential" circuits that accomplish "electronic filtering" by using a capacitor in conjunction with full-wave rectification. This circuit, shown in Figure 1.3–3c, has long been used for tube-type home radios. Specially designed transformers are often needed for such circuits owing to impedance considerations.

The "constant potential" is not truly constant. An a-c "ripple" is present because the capacitor suffers slight discharge as it tries to maintain voltage. This ripple is the quotient of charge transferred and capacitance. For Figure 1.3–3c it might be

$$V_{ripple} = \frac{0.050/120}{2.0\,(10^{-7})} = 2.1\,\text{kV}$$

with a current of 50 mA at 120 Hz and a capacitance of 0.2 μF.

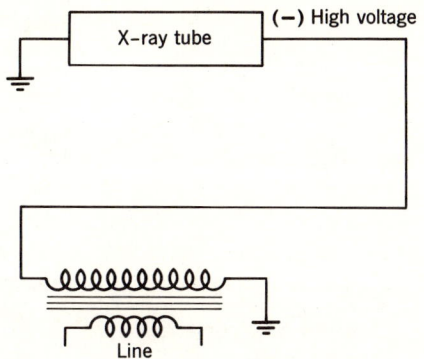

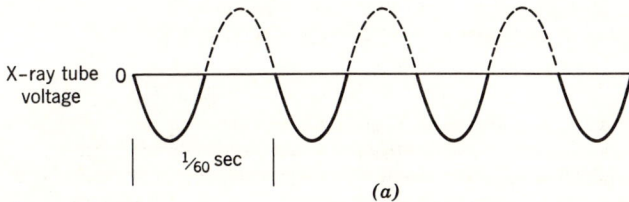

(a)

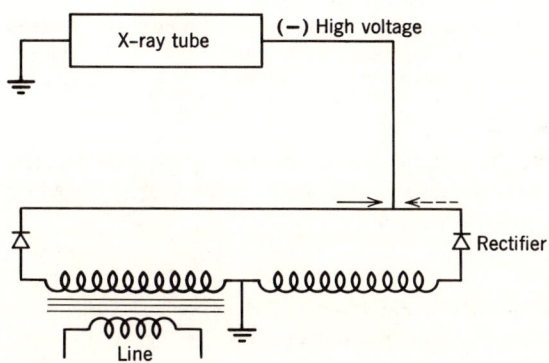

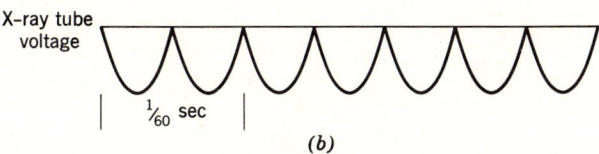

(b)

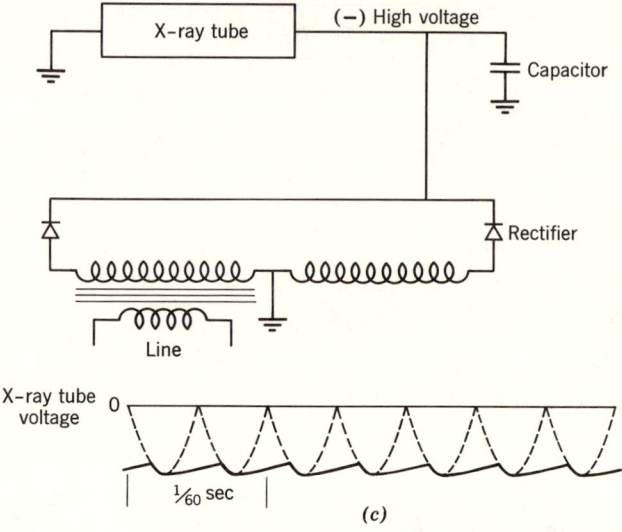

Fig. 1.3–3. Three ways of operating a Coolidge x-ray tube on standard (60 Hz) a-c. 1. During self-rectifying (or half-wave) operation (*a*) the tube is idle during each inverse half cycle. The average voltage during a pulse is only 63.6% the maximum so that useful x-ray generation will usually occur only near the center of each pulse. The average tube current is half the average current of a pulse. 2. With full-wave rectification, the pulse frequency is doubled, and the idle half-cycle eliminated, by the use of two rectifiers connected as in (*b*). In this circuit, electrons will flow as shown by the solid arrow when the left end of the transformer is negative, and in the *opposite* direction (broken arrow) during the other half-cycle. 3. Modern constant-voltage ("constant-potential") operation is achieved by having a capacitor act as shown in (*c*) to smooth the saw-tooth pattern of (*b*) until only an "a-c ripple" remains. (*a*) Self-rectifying operation. (*b*) Full-wave-rectifying operation. (*c*) Modern, constant-voltage operation.

The advantage of using constant potential for the excitation of the K spectra of representative elements is clear from Table 1.3–1. For further discussion, see Section 9.4.

Table 1.3–1. Time Effective as Percentage of Total for Excitation of Representative Elements with Two 100-kV Supplies

Element	Excitation Potential, kV	With Constant Potential, %	With Full-Wave Rectification, %
Bi	90.4	100	30
Pt	78.4	100	43
Ag	25.5	100	83
Cu	9.0	100	93

X-RAYS AND ELECTRONS

> Erzeugt ein Elementarvorgang einen
> anderen, so ist die Energie des letzteren nicht
> grösser als die des ersteren.   A. Einstein [7]

The interplay of electrons and x-rays is of first importance to anyone concerned with the importance of x-rays in analytical chemistry. Both have dual (wave-particle) character. Both are absorbed, emitted, scattered, and diffracted. Absorption of either can excite emission of the other. Electrons carry a negative charge while x-rays are uncharged and are much lower in mass. Yet collisions between the two (Compton effect) follow the laws of mechanics.

There are no experiments in physics more elegant and important than those we describe below to illustrate the interplay of electrons and x-rays [8–10]. They show why quantum theory is needed, and they do this more simply and directly than do the phenomena (anomalies in black-body radiation and in heat capacities) that gave rise (Planck, 1900; Einstein, 1905) to the theory. As the role of the interplay of electrons and x-rays in the characterization and control of materials continues to grow, so does the debt that analytical chemistry owes to physics.

1.4 The Photoelectric Effect

When a clean metal in vacuum absorbs light or x-rays sufficiently short in wavelength, electrons (called photoelectrons) are emitted, and the metal (if insulated) becomes positively charged. We distinguish between photoelectrons and *Auger* electrons, more of which later.

Classical physics, simply applied, would require that the kinetic energy of photoelectrons increase with the *intensity* of the irradiating beam. Lenard [8] discovered them to be independent of each other. Intenser irradiation produced *more* photoelectrons, but the kinetic energy of these electrons depended only on the energy (or wavelength) of the irradiating beam: the shorter this wavelength, the greater the kinetic energy. The time lag between irradiation and emission was below 10^{-8} sec even at the lowest intensities of irradiation. There was no indication that irradiation could be accompanied by an energy accumulation or storage mechanism that could make for high energy in the emitted photoelectrons.

In the experiment mentioned at the opening of this section, a steady state will be reached when the positive charge on the irradiated metal has become large enough so that electrons return to it at the rate they leave. Under these conditions the law of conservation of energy requires that

$$eV_e = \tfrac{1}{2} m_e v_e^2 \qquad (1.4\text{--}1)$$

where V_e is the potential difference between metal and ground, and e, m_e, and v_e are the charge, mass, and velocity of an electron. V_e is conveniently (though loosely) called the "voltage velocity" of the electrons. By substituting (approximate) values for the universal constants e and m_e, one can obtain

$$v_e \text{ (cm/sec)} = 6(10^7)\sqrt{V_e} \qquad (1.4\text{--}2)$$

Einstein's introduction of the quantum theory to explain the photoelectric effect amounts to writing

$$hv = \tfrac{1}{2} m_e v_e^2 + W_e = eV_e + W_e \qquad (1.4\text{--}3)$$

where W_e is the work required to get an electron out of the metal at zero velocity (the "work function"); h is Planck's constant and v the frequency of the irradiating light.

A test of (1.4–3) is thus a test of the quantum theory, which meets the test if V_e varies linearly with v, or if h/e is constant. Millikan [11], in work of high precision, found for sodium

$$h/e = 1.375(10^{-17}) \qquad (1.4\text{--}4)$$

and for lithium

$$h/e = 1.379(10^{-17}) \qquad (1.4\text{--}5)$$

and an average value $h = 6.557(10^{-27})$ erg-sec by use of $e = 4.774(10^{-10})$ esu, a value slightly too low as we shall see later.

1.5 The Excitation of X-Rays by Electrons

In the limit, electron excitation of x-rays may be regarded as a photoelectric effect in reverse. "In the limit" here means that the interaction of electrons and metal lattice (different in the two cases) is disregarded.

When the target of an x-ray tube is struck by electrons, these are retarded by the atoms of the target. The energy the electrons lose is radiated in a spectrum that ranges from the x-ray region into the infrared; we say this spectrum has been produced by electron excitation. The elementary processes involved in this energy transfer can take place also in a sample being analyzed by x-ray methods. In general, electrons do not penetrate much below the surface of such samples.

In most analyses by these methods, the x-ray source is likely to be a Coolidge tube. For this reason, we examine the spectra that have been observed when an "infinitely thick" tungsten target in such a tube is bombarded by electrons of various energies. In Figure 1.5–1, intensities (arbitrary units) are plotted against wavelength in angstrom units for these spectra.

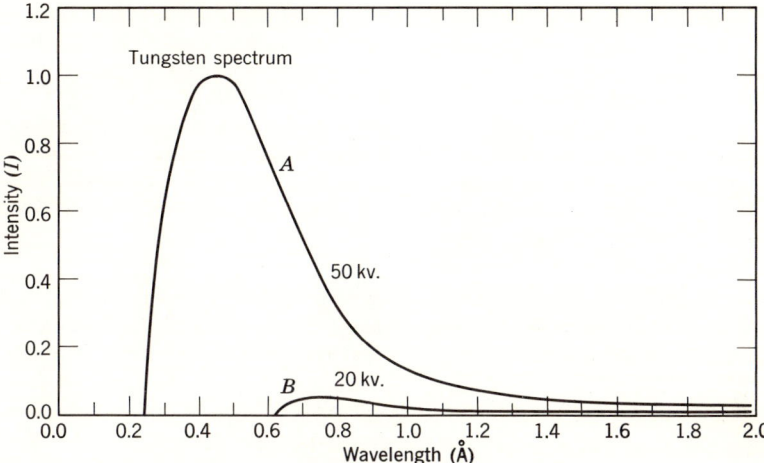

Fig. 1.5-1. The continuous x-ray spectrum. Note that the short-wavelength limit (1.6–1) is 0.248 Å for 50 kV and 0.620 Å for 20 kV. A spectrum from a rectified a-c tube would have the peak displaced to the right and for a given input energy would have less x-ray output. After Ulrey, *Phys. Rev.* [2], **11**, 401 (1918).

See *Notes* below. Three items are noteworthy.

1. All spectra begin abruptly at a short-wavelength limit that is always inversely proportional to the electron energy.

2. Each spectrum is a continuum that extends over a wide range of wavelengths. The curves are terminated near 2Å to call attention to the difficulty of working with long-wavelength x-rays.

3. The spectra in Figure 1.5–1 are observed spectra, obtained by an experimental method described later. The spectra as generated at the target differ from these in several important respects.

NOTES. 1. In most modern analytical methods that use x-rays, *intensity* is preferably measured by *counting* the quanta that reach a detector during a suitable time interval. The intensity I can then be expressed as counts per second. Sometimes intensities are high enough so that instantaneous values thereof can be read as electric currents after the x-rays have been converted to electrons.

2. In this book, intensity often serves as a *relative measure*, for example, of amount present of a substance sought. Many methods of concern here are *comparative*. Consequently, the interpretation of experimental results often involves *intensity ratios*, of which a well known example is I/I_0, where I is the intensity of a beam reduced by attenuation from I_0, the intensity of the source.

3. Ordinarily I will be used to represent I_λ, the intensity of a *monochromatic* beam of wavelength λ, or the integrated intensity, $I_{int} = \int_{\lambda_1}^{\lambda_2} I_\lambda \, d\lambda$, of a *polychromatic* beam in which I_λ is finite over the spectral region $\lambda_1 - \lambda_2$. When advisable for the sake of clearness, I_λ and I_{int} are used.

14   GENERATION AND PROPERTIES OF X-RAYS

4. In discussions of theory, the I's usually are considered independent of optics or of detecting systems.

5. The energy in a spectral region is the *radiant energy* of the region.

6. The total rate at which radiant energy flows is *radiant flux* or *power*.

7. In addition to the angstrom (Å), a slightly different unit of length is the X unit (XU), proposed by Siegbahn: 1000 XU = 1.00202 Å (angstroms). For further information on the origin of the X unit consult H. P. Klug and L. E. Alexander, *X-Ray Diffraction Procedures* (New York: John Wiley and Sons, 1954, p. 90); W. T. Sproull, *X-Rays in Practice* (New York: McGraw-Hill Book Co., 1946, p. 337).

8. A definite spectrum of the types shown in Figure 1.5–1 is produced for a given value of the d-c voltage applied. When rectified a-c is applied, the spectrum at each instant of the a-c wave is that produced by the voltage at that instant. Thus the spectrum of the x-ray tube under these conditions is the sum of the spectra produced by all voltages from zero to the peak value of the a-c voltage. The short-wavelength limit will be given by the peak voltage. See Figure 1.3–3. "Filtering" rectified a-c brings the x-ray spectrum closer to that from a d-c power supply.

1.6   The Continuous Spectrum

The continuous spectrum is thus characterized by a short-wavelength limit and an intensity distribution. Experiments on other target materials have shown that these characteristics are independent of the target material although the *integrated* intensity increases with atomic number. See (1.6–3). The continuous spectrum therefore results generally from the interaction of electrons with matter. Attempts (none completely successful) have been made to treat this interaction theoretically by both classical and quantum mechanics.

In Figure 1.5–1, the abrupt termination of continuous spectra at short-wavelength limits points clearly to a quantum phenomenon. The full significance of the short-wavelength limit appears when one regards it as an inverse photoelectric effect, which makes it another way of determining h.

Duane and Hunt [12] discovered the short-wavelength limit as described in the caption of Figure 1.6–1. If the limit truly represents the x-ray quantum of *maximum* energy obtainable at a given tube voltage, then the bombarding electrons must not lose energy to the target; the term that may be regarded as the analogue of W_e in (1.4–3) must here be zero. Under these conditions the short-wavelength limit λ_0 is governed by

$$hc/\lambda_0 = h\nu_0 = \tfrac{1}{2}mv^2 = Ve \tag{1.6–1}$$

which is to be compared with (1.4–3); here, c is the velocity of light in vacuum and the subscript in (1.4–3) has been omitted.

Provided the *relativistic increase* in m with v is allowed for, (1.6–1) stands experimentally verified [8] for the range of tube voltages from 4500 to 170,000 volts, and yields $h = (6.558 \pm 0.009)10^{-27}$ erg-sec for comparison

X-RAYS AND ELECTRONS

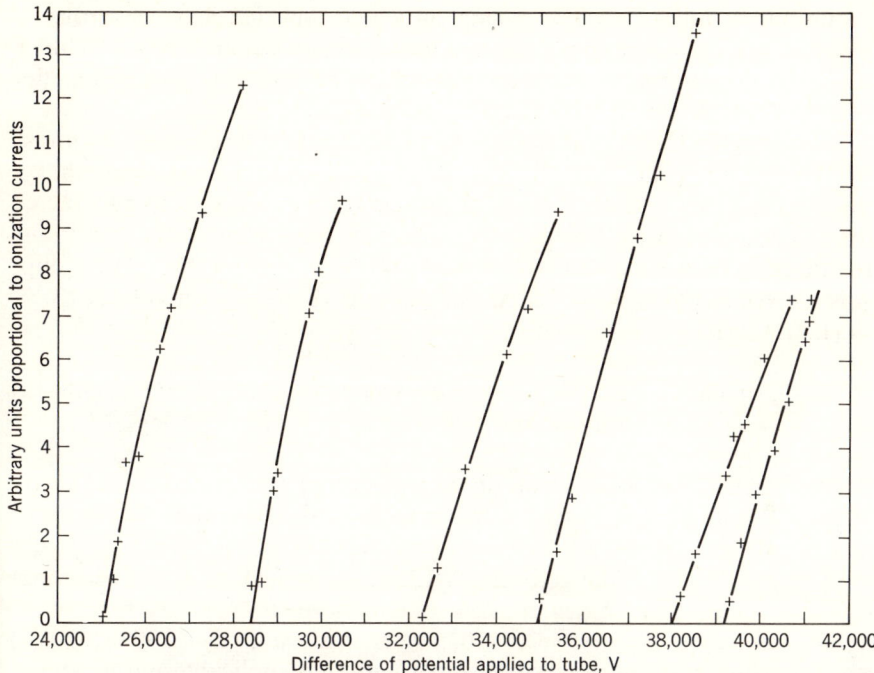

Fig. 1.6–1. Discovery of the short-wavelength limit by Duane and Hunt [12]. Monochromatic x-ray beams were obtained by Bragg reflection (which see). With the crystal set at the angle θ that corresponds to the desired wavelength, V was varied as indicated in the figure. The voltage V_0 at the short-wavelength limit λ_0 is the voltage at which the intensity first becomes zero. [See (1.6–1).] The six values of h obtained in this pioneering research ranged from 6.34 to 6.44 · $\cdot (10^{-27})$ erg-sec.

with $h = 6.557 \, (10^{-27})$ given above as having been found by (1.4–3). The comparison is gratifying indeed!

Because of this identity of h from two processes (photoelectric effect and electron excitation) that may be regarded as inverse, it is *ideally* possible to have electrons of "voltage velocity" V_e excite x-rays of the wavelength λ_0 at the short-wavelength limit in one x-ray tube, and to have these x-rays regenerate electrons of the same voltage velocity (or energy) by striking a suitable target in another. To explain this, a quantum (or corpuscular) theory is necessary; the classical wave theory of light is helpless.

No one has exposed this helplessness more graphically or more forcefully than W. H. Bragg, who in 1933 used this analogue in his *The Universe of Light* [13, p. 28]: "Suppose we drop a plank into the sea from ... say 100 feet; there is a splash and waves spread away over the surface of the water. They pass by boats and ships without any effect, and then after traveling thousands of miles they find a ship on which their effect is disastrous: a plank is torn out of the ship's side and hurled ninety feet into the air." Under the ideal conditions of the experiments in the previous paragraph, the "emitted" plank would have risen to 100 feet above the ship.

According to (1.6–1), all the energy of an electron accelerated from rest by voltage V appears ultimately in the x-ray quantum of wavelength λ_0; the equation thus restates the law of conservation of energy and may therefore be generalized by dropping the subscript. Furthermore, the important relationship between λ and V is implicit in (1.6–1). To find this relationship, it is convenient to express λ in angstroms and Ve in electron-volts (1 eV = = (1.60210) (10^{-12}) erg); that is,

$$hc/\lambda \text{ (angstroms)} \times 10^{-8} = Ve \text{ (electron-volts)} (1.60210)(10^{-12}) \quad (1.6\text{–}1\text{a})$$
$$\text{ergs} = \text{ergs}$$

Now, the *energy* of an electron (in electron volts) accelerated from rest by voltage V is *numerically equal* to that *voltage*. This convenient fact leads to

$$\lambda \text{ (angstroms)} = 12{,}398.1/V \text{(volts)} \quad (1.6\text{–}2)$$

upon insertion of $h = 6.6256(10^{-27})$ erg-sec and $c = 2.997925(10^{10})$ cm/sec. Thus (1.6–2) is the relationship sought; it is applicable as a statement of the conservation of energy to all wavelengths, not only to those in the x-ray region. (Note that the comparison following (1.6–1) remains valid even though the presently accepted value of h is higher than those based on older values of the electronic charge. Note also that (1.6–1) deals with excitation of an x-ray at the short-wavelength limit, which differs from the excitation of a characteristic line in that *no photoelectron is emitted*. See Section 1.12.)

According to a rough empirical rule, the wavelength of maximum intensity in a continuous spectrum is about $3\lambda_0/2$. One use for this rule is in the estimation of the *effective wavelength* of a *polychromatic beam*, this being the best single wavelength that can be selected to describe how the polychromatic beam behaves when it is absorbed.

The total power, or integrated intensity, of the x-ray beam, in watts, is the product of the (empirical) efficiency of x-ray production and cathode-ray power iV; or

$$I_{int} = 1.4\,(10^{-9})iZV^2 \quad (1.6\text{–}3)$$

where i is the electron current in amperes; Z the atomic number, and V the potential difference across the tube, in volts [13]. This estimate of the beam energy becomes less reliable as V decreases, but it emphasizes the inefficiency of x-ray production by electron bombardment. In typical cases, less than 1 % of the electron energy appears in the x-ray beam. Almost all the rest is degraded to heat, and special provisions for cooling the target are often necessary. Though an empirical equation, (1.6–3) is of great practical value; for example in the selecting of conditions for radiography. Note especially the dependence of the integrated x-ray intensity on the current, the atomic number of the target, and on the *square* of the tube voltage.

When electrons strike an "infinitely thick" target, they are quickly slowed by interaction with its atoms, which produces the "braking radiation," or "Bremsstrahlung," a continuous spectrum. The electrons do not penetrate at all deeply, and very thin foils must be used to study how the continuous spectrum is generated [14]. Furthermore, different electrons lose different amounts of energy as they interact with individual atoms of the target, so that an electron-energy spectrum is set up even in a target being struck only by electrons identical in energy. The actual excitation of x-rays is therefore brought about by electrons differing widely in energy, with the result that the emitted x-rays show a corresponding wavelength range. Even monoenergetic electrons will generate a polychromatic x-ray beam.

The absorption of x-rays on their way out of the target is also important. Because x-rays are usually absorbed more readily the longer their wavelength, the intensity distribution is changed by absorption, and the relative intensities of the longer wavelengths are decreased. The target thus tends to *filter* the longer wavelengths out of the beam.

The two effects just described determine the emergent x-ray distribution at the target. Before an x-ray beam strikes a sample to be analyzed, the beam is usually modified further. For example, there may be absorption (and filtering) by the window of the x-ray tube, by an air path between tube and sample, by the walls of a cell containing the sample, and finally by the sample itself. Analogous considerations govern the absorption (and, with polychromatic beams, the filtering) of the beam entering the detector from the sample.

The last two paragraphs explain why *observed* spectra, such as those in Figure 1.5–1, usually differ significantly from the spectra generated at the target of an x-ray tube.

THE ABSORPTION OF X-RAYS

Das ... zunächst Auffallende ist, dass durch die
schwarze Cartonhülse, welche keine sichtbaren oder ultra-
violetten Strahlen des Sonnen—oder des electrischen
Bogenlichtes durchlässt, ein Agens hindurchgeht, das im
Stande ist, lebhafte Fluorescenz zu erzeugen, und man wird

deshalb wohl zuerst untersuchen, ob auch andere Körper diese Eigenschaft besitzen.

Man findet bald, dass alle Körper für dasselbe durchlässig sind, aber in sehr verschiedenem Grade. [1]

1.7 Beer's Law Applied to X-Ray Absorption

Inasmuch as atoms (or ions derived from them) act as absorbing centers for x-rays, Beer's law might be expected to govern the absorption process [15, 16]. Owing to the high energies of x-rays, the electrons primarily involved in absorption will lie close to the nucleus; valence electrons respond to longer wavelengths. As a consequence, one might expect also that x-ray absorption coefficients will be more simply interrelated than are absorption coefficients for radiant energy of longer wavelengths—of wavelengths long enough to affect valence electrons, hence to make the absorption process subject to chemical influences. Both expectations are realized. In the main, x-ray absorption is a physical process governed by Beer's law.

For a narrow, parallel monochromatic x-ray beam incident perpendicularly upon an absorber of unit area, and of uniform thickness and composition as shown in Figure 1.7–1, Beer's law has been found to hold and may be written

$$\log (I_1/I_2) = k\Delta m = k_d \Delta d \qquad (1.7\text{–}1)$$

where m refers to mass and d to thickness of sample. The values of the proportionality constants k and k_d depend not only on the units in which the amount of sample is measured, but also on the wavelength and the elements

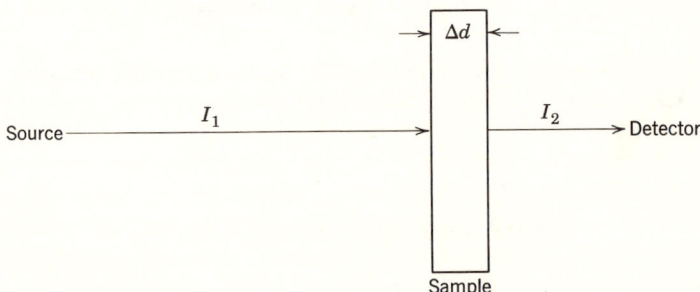

Fig. 1.7–1. Simple absorption experiment. Note three important elements of optical system. Two others, the collimator (to give narrow, parallel beam) and the monochromator, have been omitted. A *source* (x-ray, electron, radioactive, etc.), a *sample*, and a *detector* are the elements fundamental to most x-ray equipment used in analytical chemistry. Nature and arrangement of these three important elements can differ considerably from one equipment to the next.

THE ABSORPTION OF X-RAYS

in the sample as well. The area of the sample must equal or exceed that of the beam. For a sample of unit area, k_d is k times the density.

1.8 Polychromatic Beams. Filtering

Roentgen [1] and others soon discovered that k decreases markedly with increasing thickness of absorber when the x-ray beam is polychromatic. For example, a curve resembling those in Figure 1.8–1 will result when the

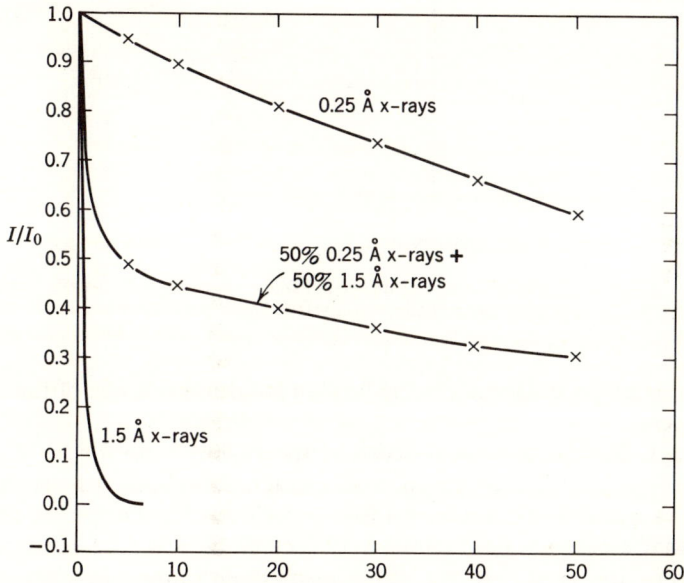

Fig. 1.8–1. Transmittance as a function of the thickness of absorber. Intensities are quanta per sec or counts per sec. The two curves for the transmittance of monochromatic x-rays are pure exponentials. The curve for the transmittance of mixed x-rays is the sum of two exponential curves. The figure illustrates the filtering of x-rays, so important with polychromatic beams. The experimental arrangement is shown in Figure 1.7–1.

experiment of Figure 1.7–1 is carried out with a polychromatic beam and different numbers of standard aluminum sheets. The more pronounced the curvature in Figure 1.8–1, the greater is the wavelength range in the incident beam. "Absorbability" thus determined was an important method for the characterization and classification of x-rays until wavelength determination by means of Bragg reflection (Section 1.16) was discovered.

1.9 Discovery of Characteristic Emission Spectra

Absorption measurements by Barkla revealed the existence of characteristic x-ray *emission* lines before x-ray wavelengths could be measured.

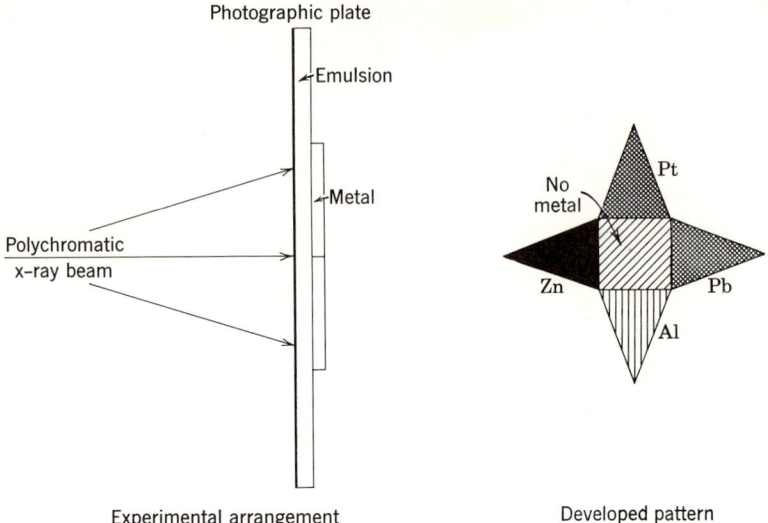

Fig. 1.9–1. Roentgen's production of secondary x-rays. The secondary x-rays from zinc are most efficiently absorbed by the emulsion and give the maximum darkening.

There is no better evidence for the fundamental importance of the absorption process.

Although Barkla deservedly received the Nobel Prize for his discovery [17], the first clue to the existence of characteristic lines is to be found in Roentgen's papers [1]. Roentgen found difficulty in reconciling his conclusion that x-rays are not "regularly reflected" by powders and by certain bodies with the results of the experiment diagrammed in Figure 1.9–1. He passed a polychromatic beam through a photographic plate the sensitive side of which was in contact with four metals about as shown. After development of the plate, Roentgen saw that the areas in contact with three of the metals were markedly darker than the center of the plate. Interposition of a thin film of aluminum between metals and emulsion had virtually no effect. Clearly, the three heaviest metals produced an unexpected result.

The modern interpretation of Roentgen's experiments is this. The (slight) darkening at the center is a background effect resulting from absorption of x-rays by the emulsion. All additional darkening is due to x-rays from the metals. The great darkening for the three heaviest metals resulted from absorption of their characteristic lines, which were excited by the x-ray beam. In the case of aluminum, however, any additional darkening must have been due to the absorption of *back-scattered* x-rays, for this metal absorbs x-rays poorly, and the excitation of its characteristic lines is inefficient.

Winkelmann and Straubel [18] carried out a similar experiment, but they

THE ABSORPTION OF X-RAYS

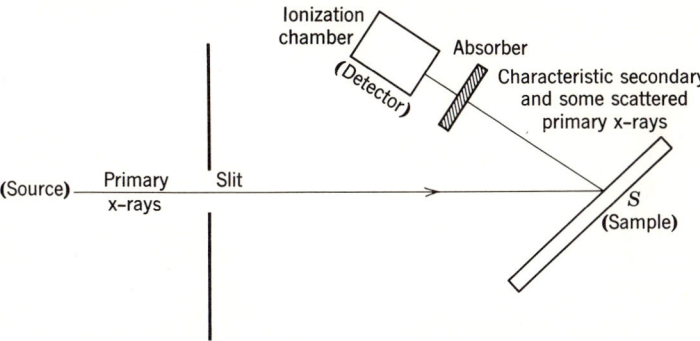

Fig. 1.9–2. Schematic diagram of Barkla's experiment. Barkla proved that the component of the x-ray beam reaching the ionization chamber is *independent* of angle and *characteristic* of the element used as secondary emitter or sample (except for carbon; see Figure 1.9–3).

were fortunate in using fluorspar (calcium fluoride) instead of Roentgen's metals. In their work, the darkening under the fluorspar was very pronounced, but it disappeared completely when paper was inserted as the absorbing barrier. They could consequently conclude—as Roentgen could not—that the specimen (fluorspar) could change x-rays into wavelengths much more readily absorbed than the original beam, and (consequently) a hundred times more active in the photographic emulsion. This intensifying technique is in use today [19].

Barkla, originally interested mainly in x-ray scattering, discovered characteristic x-rays by an experimental method similar in principle to that described above. His experimental arrangement (Figure 1.9–2) is reminiscent of that used today in studies of the Raman effect. By using sheets of a metal (e.g., aluminum) as absorber to analyze the "scattered" beam, he obtained results that clarified the experiments of Roentgen (Figure 1.9–1).

Figure 1.9–3 shows curves calculated with minor changes from Barkla's absorption-coefficient data. For carbon, the wavelength distribution is virtually unchanged from that of the incident polychromatic beam, mainly scattered x-rays being detected; the situation is reminiscent of Figure 1.8–1. The curve for calcium, on the other hand, begins with a straight line that shows the presence in the scattered beam of a relatively intense component for which k is large and sensibly constant as (1.7–1) requires. The curve for tin shows two such components. Barkla realized that these components are emitted, and he eventually called them **K** and **L** spectra [20]. He chose these letters because he anticipated that there might be some spectra less readily absorbed (e.g., **J** spectra) than the **K**, and others more readily absorbed (e.g., **M, N, O**) than the **L**. We know today that atoms are built in a way that precludes the existence of **J** spectra and requires the existence

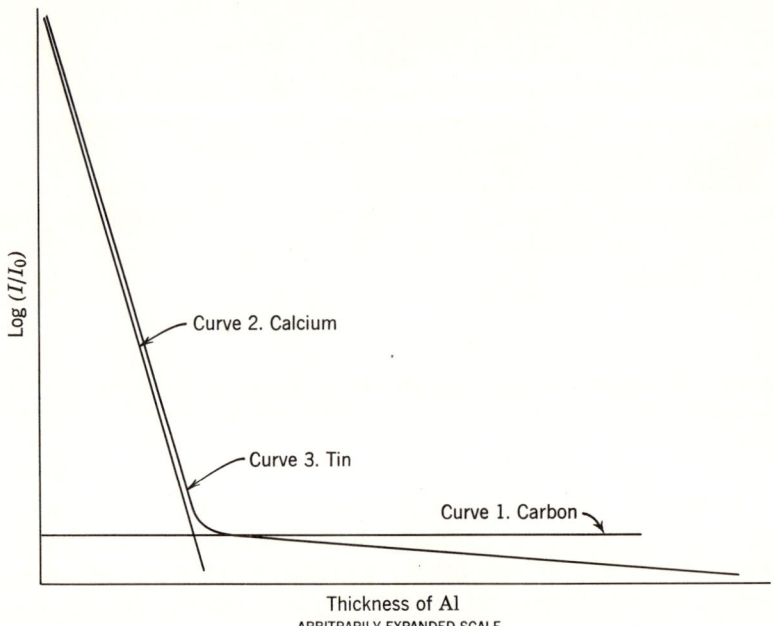

Fig. 1.9–3. Barkla's discovery of characteristic x-ray spectra by the absorptiometric method of Figure 1.9–2. The ordinate for all three curves is calculated with a single value for I_0. *Carbon* shows only the back-scattered primary polychromatic beam, naturally at greatly reduced intensity, which explains the low ordinate values for curve 1. The primary beam was hard enough so that no detectable filtering occurred: the curve is a straight line. *Calcium* shows only the **K** spectrum; note that the intensity of this spectrum exceeds that of curve 1 initially, but that the characteristic spectrum is more strongly absorbed than the back-scattered beam of curve 1. *Tin* shows both the **K** and **L** spectra. The steep part of curve 3 may be regarded as due to a filtering out of the **L** spectrum, which has a wavelength near that of the **K** spectrum of calcium. The **K** spectrum of tin is absorbed more strongly than the back-scattered beam but much less strongly than the other characteristic spectra in the figure. For fuller understanding, the reader may wish to return to this figure after he has read through Section 1.21, which describes the work of Moseley.

of the others. The **M**, **N**, and **O** spectra are of little importance in analytical chemistry at present. They are difficult to work with because their wavelengths are long and their intensities low. Usually the analytical information they give is more easily obtained from **K** or **L** spectra.

In x-ray language, optical atomic spectra due to transitions of valence electrons are *either* characteristic *emission* or characteristic *absorption* spectra: the **D** lines of sodium, for example, are both emitted and absorbed. As appears later, the nature of x-rays precludes the occurrence of characteristic x-ray lines in *absorption* spectra.

THE ABSORPTION OF X-RAYS

1.10 The Mass Absorption Coefficient

In the preceding discussion of Beer's law, it was argued that x-ray absorption is a simpler process than the absorption of ultraviolet, visible, and infrared wavelengths. This greater simplicity becomes particularly obvious when x-ray absorption coefficients are examined.

The general validity of the empirical (1.7–1) points to a mass absorption coefficient, μ, as being the most logical for x-ray absorption. If the absorption equation is written in terms of natural logarithms, and μ is introduced, the relation becomes

$$\ln(I_1/I_2) = \mu \Delta m = \mu \rho \Delta d \quad (1.10-1)$$

where Δm is the difference in the masses, m_2 and m_1 of two samples, expressed in grams per square centimeter of irradiated sample area, and ρ is the sample density in grams per cubic centimeter. If c.g.s. units are used also in (1.7–1), then $\mu/2.303$ is equal to k in (1.7–1). The Δd is the sample thickness in centimeters. The mass absorption coefficient (μ) is often written μ_m by others.

The use of (1.10–1) is obviously restricted to the area of the sample that is uniformly irradiated by the beam. It is also obvious that the part of the sample not in the beam, and the part of the beam that does not strike the sample, do not contribute to the absorption process.

Other absorption coefficients are also used, the most important of which is the *linear* absorption coefficient, μ_l, used in thickness measurements. In accordance with its definition, μ_l is given by the equation

$$\ln(I_1/I_2) = \mu_l \Delta d \quad (1.10-2)$$

where Δd is the thickness (in centimeters) of sample irradiated. Evidently,

$$\mu_l = \mu \rho \quad (1.10-3)$$

The *atomic* mass absorption coefficient is

$$\mu_a = \mu M/N \quad (1.10-4)$$

where M is the atomic weight of the absorber and N is Avogadro's number. The μ_a measures the "cross-section" of the atom for x-ray capture. This cross-section is smaller than the geometrical by several powers of ten.

There is convincing experimental evidence for the following important statement. To a degree of approximation satisfactory for most analytical work, the mass absorption coefficient of an element is independent of chemical or physical state. This means, for example, that an atom of bromine has the same chance of absorbing an x-ray quantum incident upon it in bromine vapor, completely or partially dissociated; in potassium bromide or sodium bromate; in liquid or solid bromide. X-ray absorption is predominantly an atomic property. This simplicity is without parellel in absorptiometry.

Equation 1.10–1 was written for a sample containing a single element on which monochromatic x-rays are incident. Insofar as x-ray absorption is an atomic property, the mass absorption coefficients for other samples are additive functions of the weight-fractions of the elements, free or combined, that are present; that is,

$$\mu_S = W_A\mu_A + W_B\mu_B + \cdots + W_M\mu_M \tag{1.10–5}$$

for a sample S containing M elements, each at its weight-fraction W with mass absorption coefficient μ.

Equation 1.10–5 may be considered always valid for absorptiometry with monochromatic x-rays, and valid for absorptiometry with polychromatic x-rays also when a satisfactory effective wavelength exists. Difficulties sometimes arise with polychromatic beams.

Relationships among the mass absorption coefficients for different elements and for different wavelengths, to be discussed later, further emphasize the simplicity of x-ray absorption.

1.11 Absorption Edges

The way in which mass absorption coefficient varies with wavelength is shown in Figure 1.11–1 for three metals differing widely in atomic number. For the present, we are concerned only with the sharp discontinuities, called *absorption edges*, each of which appears at its own *critical absorption wavelength*.

Barkla and Sadler [21] discovered absorption edges by an extension of the experimental method that led them to find the characteristic spectra before x-rays of known wavelength were available. It will be remembered that their earlier work led to the generation of characteristic lines by x-ray excitation of sample S in Figure 1.9–2 and to the identification of these lines by virtue of their absorption (for example) by aluminum. The extension of the method consisted (1) of using a large number (nine in Table 1.11–1) of elements as samples; and (2), for each sample, of measuring the absorbance in aluminum and in iron. The results are given in Table 1.11–1 and plotted in Figure 1.11–2.

Table 1.11–1. Experimental Results of Barkla and Sadler

Sample	Cr	Fe	Co	Ni	Cu	Zn	As	Se	Ag
μ_{Al} (abscissa)[a]	136.0	88.5	71.6	59.1	47.7	39.4	22.5	18.9	2.5
μ_{Fe} (ordinate)[a]	103.8	66.1	67.2	314.0	268.0	221.0	134.0	116.3	17.4

[a] Mass absorption coefficients with aluminum as absorber (Figure 1.9–2) are plotted as abscissas in Figure 1.11–2, and corresponding mass absorption coefficients with iron as absorber are plotted as ordinates.

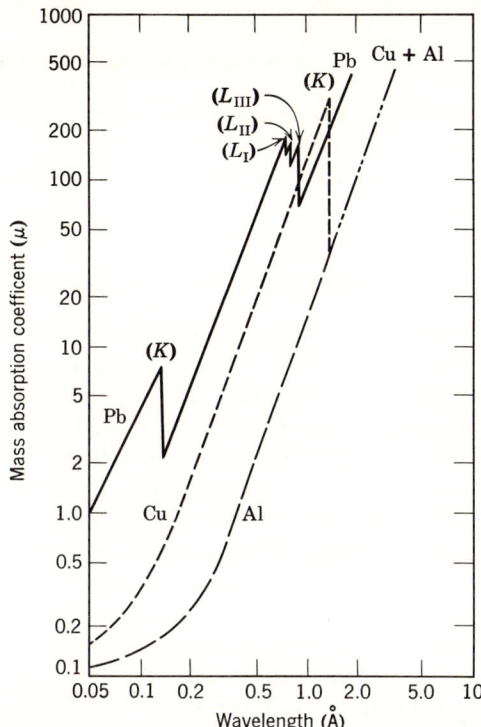

Fig. 1.11–1. Log-log plot showing mass absorption coefficient as a function of wavelength for three common metals. Note that the discontinuities locate the absorption edges, (**K**) and (**L**), each at its critical absorption wavelength. The coincidence of the copper and aluminum curves is not of fundamental significance. Note the absence of lines such as are found in optical absorption spectra.

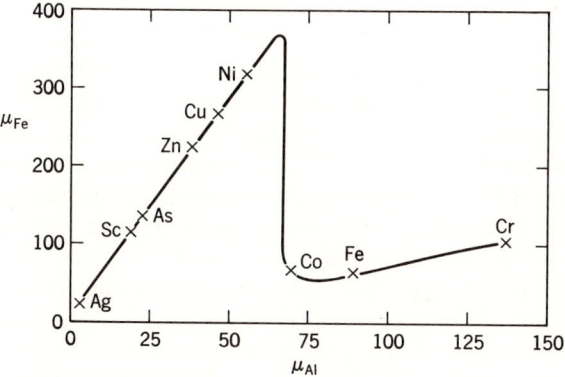

Fig. 1.11–2. Discovery of the absorption edge by Barkla and Sadler [21]. See Table 1.11–1 for their original numerical results. The curve shown is for the experiment of Figure 1.9–2 with iron and aluminum as the absorbers and elements of increasing atomic number as samples. In this region of the spectrum, absorption by aluminum increases uniformly with λ. Iron shows the sharp drop at its **K** edge.

The absorption edge has the earmarks of a quantum phenomenon, and the fact that the critical absorption wavelength is shorter (but not much shorter) than the wavelength of Fe **Kα** suggests that the edge in Figure 1.11–2 is the **K** edge of iron.

1.12 Absorption Edges and Quantum Theory

The analytical chemist usually deals with the absorption of light in and near the visible region. Here the absorption processes involve *valence* electrons; and electronic, vibrational, and rotational energy levels, the energies decreasing in that order. The quanta involved in transitions among these levels are relatively small; one deals here with electron volts and with fractions of electron volts—not with kiloelectron-volts as is usual with x-rays. Consequently, different kinds of absorbance curves result. The simple wavelength dependence and the absorption edges that characterize Figure 1.11–1 are absent. The similarity and regularity of the curves in Figure 1.11–1 do not exist for absorbance in the visible region.

The absorption edge is clearly associated with a quantum phenomenon; the critical absorption wavelength is that of the smallest quantum that can excite the characteristic x-ray lines with which it is associated. In addition, this absorption process is accompanied by the ejection of a photoelectron (and of Auger electrons, which we disregard here.) There are two important differences between this process and the simpler photoelectric effect of Section 1.4: here the energies involved are generally much higher, and *both* a photoelectron and a characteristic line are emitted in the simplest case. The Einstein relationship 1.4–3 should apply in some form.

That it does apply was elegantly demonstrated by use of a magnetic photoelectron spectrograph, used in primitive form by Robinson and Rawlinson in 1914, then by de Broglie, and in the form of Figure 1.12–1 by Robinson [22] in 1923 and as late as 1940 [23]. The operation of the spectrograph is outlined in the caption of Figure 1.12–1; its function is to measure the energy of photoelectrons emitted by a target that absorbs monochromatic x-rays. Focusing is only approximate, for the electrons that pass the slit S do not all travel the same circle. Using a compass, it is easy to show what also appears in the figure; namely, that circles approaching each other most closely at S will approach each other most closely again at L. As ultimately refined, the magnetic electron spectrograph became one of the most reliable instruments for the determination of physical constants. Recently, notably in the laboratory of Kai Siegbahn (see Section 10.10), it is being used to investigate chemical bonding.

The Einstein relation 1.4–3 in a form convenient for the magnetic electron spectrograph is

$$\tfrac{1}{2} m_e v_e^2 = hc(1/\lambda_I - 1/\lambda_C) \qquad (1.12\text{–}1)$$

THE ABSORPTION OF X-RAYS

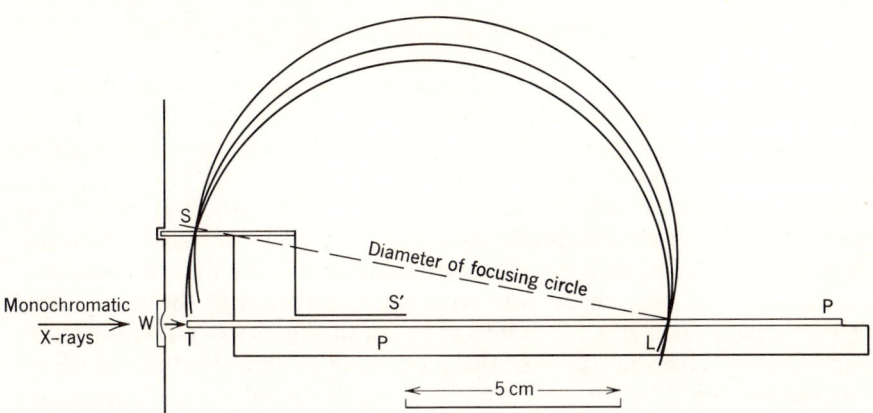

Fig. 1.12–1. The magnetic photoelectron spectrograph of Robinson [22]. Photoelectrons (and Auger electrons) emitted by the target in high vacuum pass through the narrow slit S (which may be regarded as an electron source) and are constrained to travel a circular path by a uniform magnetic field with lines of force perpendicular to the plane of the paper. A novel feature is that *exact* positioning of the target T is unnecessary; the target need only have its center *approximately* on the prolongation of the plate PP and be vertically below S. To compute r, the radius of curvature, one need know only the relative positions of S and S′ (a fiducial slit), and the distances S′L to the high-energy edge of a registered line. Exchanging targets according to the conditions given above does not change r for an electron of given energy traveling in an unchanged magnetic field.

in which λ_I is the wavelength of the incident monochromatic beam, and λ_C is the critical absorption wavelength for the edge with which the emitted photoelectron is identified. In the spectrograph, for an electron constrained to move in a circle,

$$Hev_e = m_e v_e^2 / r \quad \text{or} \quad Hr = m_e v_e / e \quad (1.12\text{–}2)$$

where H is the magnetic field strength in oersteds and r is the radius in centimeters. As in the case of (1.6–1), the relativistic value of m_e must be used; use of the value for the electron at rest could make the results in error by nearly 1% in extreme cases.

Robinson's experiments give the number of absorption edges and the value of λ_C for each (because there is an Hr for each v_e). In this work, *one* **K** edge, *three* **L** edges, *five* **M** edges, and (for uranium) *five* of the *seven* **N** edges were found; two N edges escaped resolution. The corresponding λ_C values agreed closely with those established by direct absorptiometry. This agreement shows incidentally that v_e in (1.12–1) is a *maximum* value; the photoelectrons lose negligible energy in being emitted. Further discussion appears in Chapter 4.

1.13 Detailed View of the Mass Absorption Coefficient

The actual history of an x-ray beam on its progress through matter is more complex than the foregoing treatment would lead one to believe. As Figure 1.13–1 shows, there is a transmitted beam, and there may be characteristic x-rays, both of which have been discussed. However, there are also scattering, modified and unmodified, and photoelectric absorption, the latter giving rise to photoelectrons in part emitted and in part absorbed. Figure 1.13–1 includes only processes significant in analytical chemistry.

Beer's law, (1.7–1), deals only with the absorbed and the transmitted beam; in particular, it is not concerned with the history of a quantum subsequent to absorption. Because the number of quanta disappearing from the incident beam by whatever process is proportional, other things equal, to the number of quanta present, we may group under "absorption" all the processes that reduce this number of quanta. This procedure is convenient, though not strictly rigorous. The transmitted beam, however, must include only quanta that have not been changed in passing through the sample. Consequently, it was necessary to specify in connection with (1.7–1) that the beam be so narrow as virtually to exclude scattered x-rays and photoelectrons from reaching the detector.

X-ray absorption processes are therefore of two general kinds: (1) *photoelectric absorption*, in which the entire energy of an incident x-ray quantum is transformed into the kinetic energy of a photoelectron and the potential energy of an excited atom that (*a*) subsequently emits a characteristic line, (*b*) emits Auger electrons (discussed later), or (*c*) has an even more complex history; and (2) *absorption leading to scattering*. Inasmuch as these two processes are differently affected by changes in wavelength and in the atomic number of the absorber, it is convenient to subdivide the (overall)

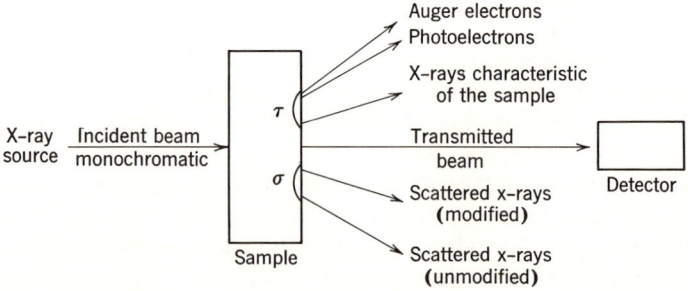

Fig. 1.13–1 Fate of an x-ray beam. Two types of events occur as an x-ray beam is absorbed in matter—more precisely, as x-ray quanta disappear from the beam. These types of events are photoelectric absorption, measured by τ, and scattering measured by σ. See (1.13–1).

THE ABSORPTION OF X-RAYS

mass absorption coefficient as in Figure 1.13–1, that is,

$$\mu = \tau + \sigma \tag{1.13-1}$$

This equation disregards "pair production" (the conversion of a photon of high energy into an electron and a positron), which requires more than a million electron volts and has no present importance in analytical chemistry.

The values of τ and σ show that photoelectric absorption generally makes the greater contribution to the mass absorption coefficient. Absorption leading to scattering, which is discussed in the next section, gains in *relative* importance as atomic number, Z, and wavelength, λ, decrease.

Between absorption edges, the photoelectric (true) mass absorption coefficient τ can be expressed as the following approximate empirical function of Z and λ:

$$\tau = \tau_a(N/A) = (CN/A)(Z^4\lambda^3) \simeq \mu \tag{1.13-2}$$

The equation, similar to one found by Bragg and Pierce in 1914, contains a proportionality constant C and the number of atoms per gram (quotient of Avogadro's number N by atomic weight A), the introduction of which is necessary to change from atomic photoelectric absorption coefficient τ_a to photoelectric mass absorption coefficient τ. In light of the above, it is clear that the values of Z and λ determine the degree of approximation involved in substituting μ for τ in (1.13–2), the substitution being more nearly justified as Z and λ increase.

An examination of Figure 1.11–1 will substantiate the statements just made. The following features are noteworthy. (1) Within the wavelength range shown, aluminum has no absorption edges; copper has only the **K** edge; lead has the **K** edge and the **L** edges, of which there are three, each associated with a different characteristic spectrum. (2) The slopes between absorption edges calculated from the logarithmic values approach the value 3 as Z and λ increase. (3) The decrease in slope at short wavelengths, which occurs mainly because μ (and not τ) has been plotted [see (1.13–1)], is most pronounced for aluminum and imperceptible for lead. (4) The x-ray absorption curves, except for the edges, show no structure—in marked contrast to optical spectra, which (as has been mentioned) can show lines in absorption.

In the discussion above, the fine structure of absorption edges has been neglected, and no mention has been made of the (very small) effect of chemical state on the wavelength of an edge. A complete understanding of these features, and of the form of (1.13–2), is not yet available. Such an understanding presupposes a detailed knowledge of how x-rays interact with the electrons near the nuclei of atoms to cause the ejection of photoelectrons. For our purposes, it is enough that (1.13–2) is empirically useful.

THE SCATTERING OF X-RAYS

> ... So kommt man zu Anschauung ... dass aber die Körper sich den X-Strahlen gegenüber ähnlich verhalten, wie die trüben Medien dem Licht gegenüber. [1]

> ... Thus on the quantum theory we should expect the wavelength of the scattered x-rays to be greater than that of the incident rays. [24]

1.14 Nature and Importance of Scattering

The importance of x-ray scattering is easy to underestimate, especially by an analytical chemist to whom it is generally a nuisance; however, on occasion it can be used as to provide a standard of reference for intensity measurements.

One cannot hope to understand x-rays without at least a little knowledge of how they are scattered [25]. Such scattering is the basis of x-ray reflection, diffraction, and refraction. Scattering is the only way in which x-rays can be polarized. Scattering has "proved" beyond doubt that x-rays are waves, and "proved" beyond doubt that x-rays are corpuscles. Scattering provided the first connection between x-rays and atomic number and continues to give us valuable information about the structure of matter.

Scattering is the interaction of electromagnetic waves with "scattering centers," which may be dust particles (Rayleigh scattering, which makes the sky look blue), molecules, atoms, ions, or electrons. We see by scattered light. No matter what kind of scattering center it is convenient to assume, the fundamental process is always the interaction of an electromagnetic wave with a single electron as a result of which the electron is accelerated and sent into oscillatory motion. Scattering by larger centers is the summation of scattering by their individual electrons; in general, the individual scattering does not proceed independently but is influenced by that of the other electrons in the center. According to Maxwell's electromagnetic theory of light, the fundamental process can be described by saying that the electromagnetic field forces the electron to oscillate in the direction of the electric vector of the wave with the result that a *linearly polarized scattered* wave is sent out by the electron in a direction 90° to that of the incident wave.

A quantitative discussion of scattering is outside the scope of this book. Figure 1.14–1 attempts to give an elementary qualitative picture. An excellent treatment of x-ray scattering is given by Semat [10, p. 174, *et seq.*].

An electron forced into oscillation by an electromagnetic wave will radiate energy in all three directions. The simple two-dimensional represen-

THE SCATTERING OF X-RAYS

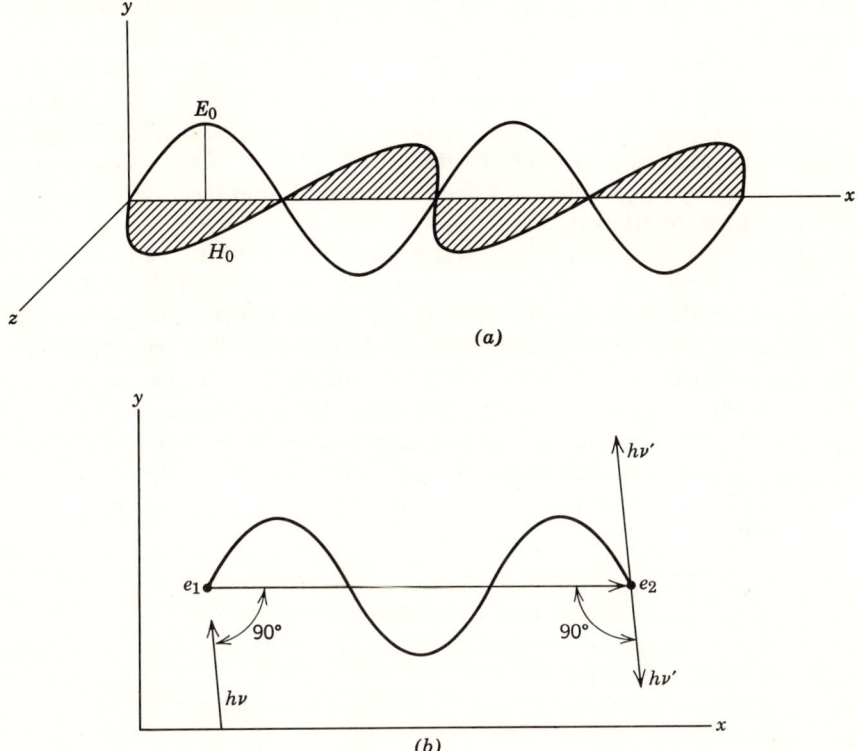

Fig. 1.14–1. Qualitative representation of classical scattering of a linearly polarized electromagnetic wave. (*a*) Electric field intensity, magnetic field intensity (both vectors), and direction of propagation are each perpendicular to the other. The intensities vary sinusoidally, but the squares of their amplitudes are always equal, which means that the magnetic field intensity may be dropped from consideration. (*b*) Interaction with the incident wave forces electron e_1 to oscillate. As this wave is linearly polarized (intensity vectors in perpendicular planes), scattering by electron e_2 must occur as shown if the fundamental perpendicularity conditions in (*a*) are to be met.

tation in Figure 1.14–1 is therefore inadequate. Mathematical treatment of the problem (first carried through by J. J. Thomson) for a single electron is possible on the basis that the electron, moving at much less than the velocity of light, is at the center of a sphere tangent to the surface of which are the electric and magnetic intensities [9, 10]. The treatment results in an equation that gives the variation with angle θ of the intensity of the x-ray beam scattered by the electrons, each acting independently, in a unit volume of sample. This amounts to a determination of *electron density* in matter on the simplest basis. Even though this simplicity is realized only under special conditions, studies of x-ray scattering have had far-reaching consequences.

Before we mention three such studies, we must characterize the kinds of x-ray scattering. They are *unmodified* (no change in wavelength); *modified* (wave length increased); *coherent* (x-rays in definite phase relationship to one another, after having been scattered by a number of electrons; that is, scattering according to classical electrodynamics); and *incoherent* (no such definite phase relationship).

Let us consider Thomson scattering first. By integrating the scattered intensity over all values of θ for a large sphere symmetrical about the scattering center (the sample), one obtains the fraction of the energy removed by scattering from the incident beam per centimeter of sample path. This fraction is called the *scattering coefficient*, and from its value the electron density of the sample can be calculated. That this is tantamount to a determination of atomic number was first appreciated by Barkla and successfully carried out for carbon by Barkla and Sadler [26]. The following numerical example for carbon is from Compton and Allison [9, p. 123]:

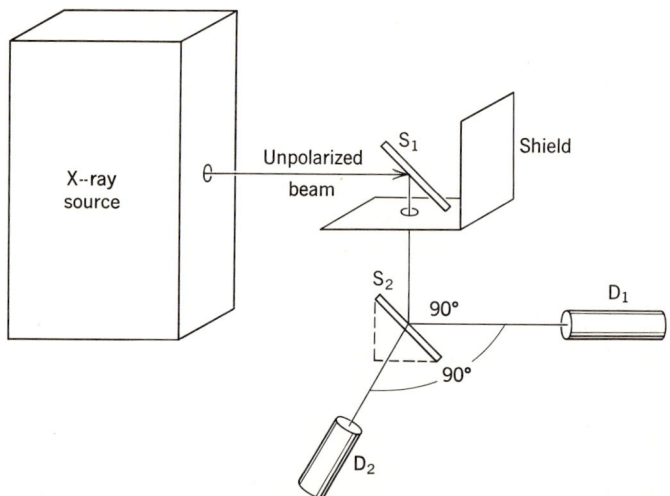

Fig. 1.14–2. Experimental arrangement of Compton and Hagenow for studying the polarization of an x-ray beam produced by scattering at 90° from S_1 (the "polarizer") and detected by S_2 (the "analyzer"). Note the similarity of this arrangement to instruments (e.g., a polarimeter) that use polarized visible light for purposes of chemical analysis. Note also that this arrangement is an elaboration of Barkla's (Figure 1.9–2), with which he discovered characteristic spectra. The ionization detectors D_1 and D_2 share a horizontal plane. The experiment showed that the intensity measured by D_1 greatly exceeded that measured by D_2. Compton and Hagenow found I_1 to be 90% of the incident intensity; Barkla had found 70%. These results are in accord with Figure 1.14–1b. Remember that here the direction of the electric vector is *perpendicular* to the plane of the paper.

THE SCATTERING OF X-RAYS

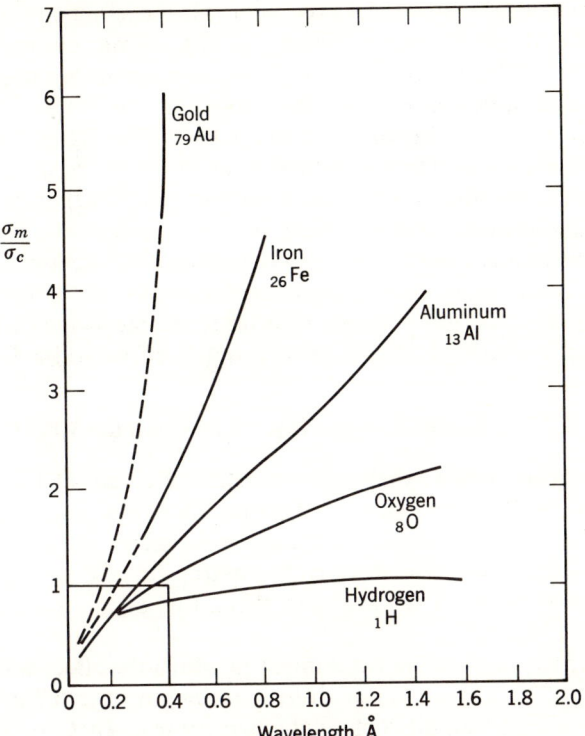

Fig. 1.14–3. Ratios of measured scattering coefficients to those calculated for (classical) Thomson scattering for five elements over a range of wavelengths. The elements are identified by name and symbol, with the atomic number as subscript preceding the symbol. This convention will be followed henceforth.

Experimental value of *mass* scattering coefficient: 0.2 (Hewlett)
From experimental value of scattering coefficient: electrons/g = $3(10^{23})$
Number of electrons that scatter per atom: Z (experimental) = 6

The result may be taken as proof that classical (independent) scattering by the electrons does occur in carbon for x-rays 0.71 Å in wavelength; as we shall see, the situation in general is more complex.

Second, we consider polarization by scattering as demonstrated first by Barkla in 1906, and in improved fashion by Compton and Hagenow in 1924 [9, pp. 120 and 121]. Visible light can be polarized by reflection, refraction, selective absorption, double refraction, and scattering. Only the last method serves for very short wavelengths: x-rays are in this class. Experiment (Figure 1.14–2) and expectation on the basis of Maxwell's electromagnetic wave theory (Figure 1.14–1) are in agreement.

Third, we consider how observed scattering coefficients compare with those calculated on Thomson's theory as the atomic number is changed. (Scattering to establish structure and electron distribution in gases, liquids, and crystalline solids is outside the scope of this book.) Experimental results [8, p. 122] are shown in Figure 1.14–3. The figure shows (1) that scattering *in excess* of Thomson scattering increases rapidly with atomic number for all but the shortest wavelengths, and (2) that at the shortest wavelengths (rectangle nearest origin of figure) scattering is *smaller* than Thomson scattering. The increased scattering results because, as the electron density increases with atomic number, electrons no longer scatter independently as individuals: interaction brings reinforcement. The decrease in *unmodified* scattering is due to the Compton effect, to be discussed next.

1.15 Modified Scattering. The Compton Effect

The "scattering deficit" that appears near the origin of Figure 1.14–3 is not a real deficit. What has happened is that the x-rays scattered under these conditions have been modified (had their wavelength increased) and therefore do not register at the wavelength setting for the unchanged monochromatic x-ray beam. The Compton effect, one of the most significant in physics, has occurred.

Figure 1.15–1 outlines the experiment in which the effect was discovered. When monochromatic x-rays of short wavelength (e.g., 0.7 and 0.2 Å) are scattered by a small sample S of a light element (e.g., Li, C, or Al), the wavelength of the scattered beam is increased by an amount that depends upon ϕ, the scattering angle, but is *independent* of the *element* and of the *wavelength*: see Figure 1.15–2. The increase is quantitatively in accord with the hypothesis that an x-ray and an electron collide *elastically* with energy and momentum conserved as when two billiard balls collide: the hypothesis leads to the equation

$$\Delta\lambda = \frac{h}{m_0 c}(1 - \cos\phi) \qquad (1.15\text{--}1)$$

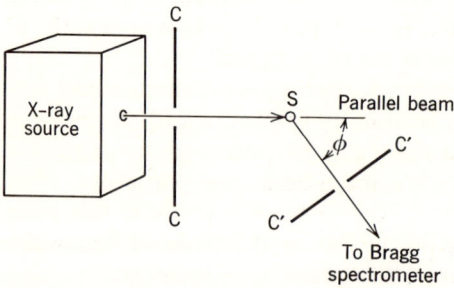

Fig. 1.15–1. The Compton effect: experimental. Collimators C C and C' C' make possible a precise measurement of x-ray intensity and wavelength as function of ϕ, the scattering angle. A Bragg spectrometer, Section 1.17, is used to make the measurements. The sample S is made small to reduce the chance of multiple scattering and to give a well-defined ϕ. The sample is an element light enough so that the electrons outside the nucleus scatter independently.

THE SCATTERING OF X-RAYS

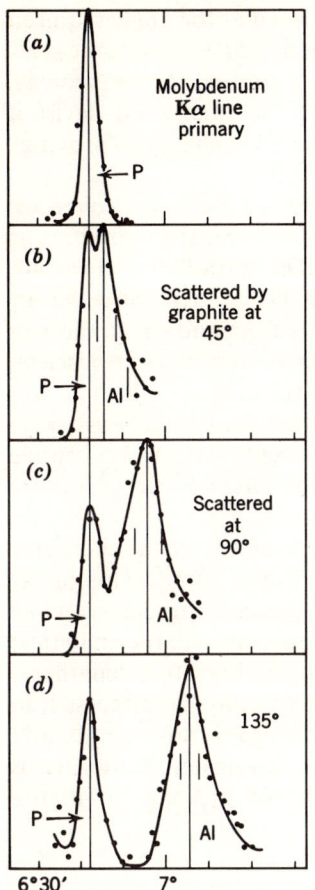

(a) Molybdenum Kα line primary

(b) Scattered by graphite at 45°

(c) Scattered at 90°

(d) 135°

Fig. 1.15–2. The Compton effect: results. The upper beam is the incident unmodified beam (no sample) in Figure 1.15–1. With the sample in place, a second intensity maximum appears, which is displaced by $\Delta\lambda$ toward longer wavelengths. The displacement increases with the scattering angle.

where the $\Delta\lambda$ is the *increase* in wavelength, m_0 the mass of the electron at rest, ϕ the scattering angle; h and c have their usual meanings. The factor multiplying the parenthesis is often called the *Compton wavelength* of the electron. See Section 9.20.

The electrons that interact according to (1.15–1), called *recoil electrons*, are ejected from the atom. The presence of λ_0, the unmodified wavelength, in the scattered beam shows that some electrons, the *bound electrons*, in the sample scatter x-rays without receiving either energy or momentum. The proportion of bound electrons increases with atomic number.

The simple mechanical picture "established" by (1.15–1) is fully supported by the results of Compton-scattering experiments with γ rays, which produce recoil electrons of high energy. By use of *coincidence circuits* (Chapter 2), it has been shown that the recoil electrons are emitted at the angle required by the mechanical picture, and that (to within about $1.5(10^{-8})$ sec) the scattered, modified photon and the recoil electron are emitted simultaneously, again as the picture requires [10, pp. 184–186]. No one could devise a simpler, more direct or elegant way than the Compton effect of "proving" that x-rays, "known" to be waves, are also corpuscles.

The discovery of the Compton effect has had far-reaching consequences. The discoverer of the Raman effect found it helpful to regard that effect as the optical analogue of the one we are discussing. The "proof" that electrons, "known" to be corpuscles, could be diffracted (hence were also waves) came four years after the discovery of the Compton effect had paved the way to the de Broglie concept of waves associated with all particles (corpuscles); and to the Heisenberg uncertainty principle, according to which it will never be possible to discover experimentally whether entities such as x-rays or electrons are in fact waves or corpuscles. The concepts wave and corpuscle have only operational significance, and we are free to use whichever is more convenient [27]. Further we cannot go.

It has become customary in the x-ray literature of analytical chemistry to regard *Compton* and *incoherent* scattering as synonymous. This juxtaposition is correct because Compton scattering is necessarily *uncorrelated in phase*. But it is not the whole truth. Of greater importance to the analytical chemist is the fact that Compton scattering also modifies (i.e., lengthens) x-ray wavelength. When x-ray emission spectrography is discussed in Chapter 9, we speak of Rayleigh scattering (no change in wavelength) and of Compton scattering (wavelength increased). Thomson scattering is Rayleigh scattering under conditions for which the ordinate in Figure 1.14–3 is unity.

THE DIFFRACTION OF X-RAYS BY CRYSTALS

> It has been shown by Herr Laue and his colleagues that the diffraction patterns which they obtain with x-rays and crystals are naturally explained by assuming the existence of very short electromagnetic waves.... The spots of the pattern represent interference maxima of waves diffracted by the regularly arranged atoms of the crystal... these waves ought to be regularly reflected by a (suitable) surface.... Such surfaces are provided by the cleavage planes of a crystal.... [28]

THE DIFFRACTION OF X-RAYS BY CRYSTALS

1.16 History

As one may infer from the quotation, W. L. Bragg realized that a crystal can act as an x-ray grating made up of equidistant parallel planes (Bragg planes) of atoms or ions from which unmodified scattering of x-rays can occur in such fashion that the waves from different planes are in phase and reinforce each other. When this happens, the x-rays are said to undergo *Bragg reflection* by the crystal and a *diffraction* pattern results.

Bragg "*reflection*" is thus diffraction that results from the *coherent unmodified scattering* of x-rays by electrons. Mrs. Lonsdale (*Crystals and X-Rays*, New York: Van Nostrand, p. 13) tells how discovery of the phenomenon made "reflection" seem a logical name. W. L. Bragg, in the belief that x-rays might be corpuscles, was investigating the possibility that deviations of corpuscles as they passed between atoms in a crystal might be responsible for the spotted Laue patterns. This explanation was false, but Bragg noticed that the diffracted beam producing any spot was slightly *convergent* as if the slightly *divergent* incident beam was being *reflected* from within the crystal. The matter was tested by slowly turning the crystal: as this was done, the spot moved slowly and synchronously as though it were an optical reflection from a rotating mirror. What better evidence of reflection could one want?

The conditions for Bragg reflection are diagrammed in Figure 1.16–1, where AA' and BB' are the traces of successive Bragg planes, d their distance apart, and θ the glancing angle of the incident x-ray beam on these planes. Bragg reflection occurs when a wave scattered at O' can reinforce an identical wave scattered at O, these being the points at which the incident beam meets the Bragg planes.

According to the laws that govern the reflection of light, the reinforcement in question will occur if the path length of a beam specularly reflected at O' exceeds by an integral number (n) of wavelengths the path length of a beam similarly reflected at O. This requires for Figure 1.16–1 that

$$\theta = \theta' \tag{1.16–1}$$

and

$$n\lambda = 2d \sin \theta \qquad \text{Bragg's law} \tag{1.16–2}$$

Although only two Bragg planes have been considered, it is clear that planes below BB' will also contribute to the reflected beam, though to an exponentially decreasing extent. The relationship between λ and $\sin \theta$ will evidently be important in determining the useful wavelength range of a particular crystal.

In (1.16–1), the integer n gives the order of the reflection, $n\lambda$ always being the path difference in Figure 1.16–1. A second-order beam, for example, of

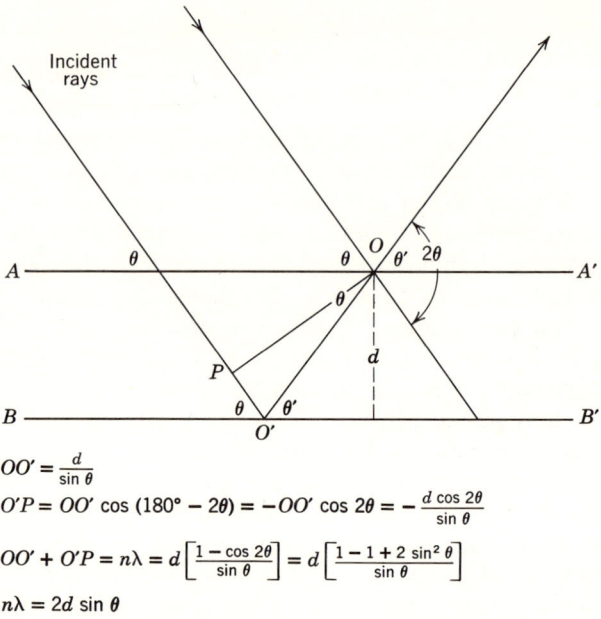

$OO' = \frac{d}{\sin \theta}$

$O'P = OO' \cos (180° - 2\theta) = -OO' \cos 2\theta = -\frac{d \cos 2\theta}{\sin \theta}$

$OO' + O'P = n\lambda = d\left[\frac{1 - \cos 2\theta}{\sin \theta}\right] = d\left[\frac{1 - 1 + 2 \sin^2 \theta}{\sin \theta}\right]$

$n\lambda = 2d \sin \theta$

Fig. 1.16–1. Simplified derivation of Bragg's law.

wavelength $\lambda/2$ will be reflected at the same angle as a first-order beam of wavelength λ, and a crystal cannot of itself distinguish between them. However, according to (1.13–2), these two beams will have greatly different mass absorption coefficients in a suitable absorber, that of the first-order beam being the larger. Insertion of such an absorber between crystal and detector will therefore attenuate the first-order beam to the greater extent, and this technique can be used to estimate the relative intensities of the two beams.

Bragg's law, (1.16–2), is obeyed so well that it is possible to use x-ray diffraction from crystals for highly precise determinations either of d or of λ. See, however, Section 1.19. The former type of determination is basic in establishing crystal structure.

The determination (or selection) of x-ray wavelengths by diffraction from crystals is highly precise ultimately because d is remarkably constant for the same Bragg planes in different crystals properly grown. When a polychromatic beam strikes such a crystal, only those wavelengths are detectable above the background for which (1.16–1) and (1.16–2) are obeyed. By positioning the detector to intercept a reflected beam that makes an angle 2θ with the beam incident on the crystal, it is thus possible to measure the combined intensities of the wavelengths for which these equations are satisfied with a given crystal. By varying 2θ (and, if necessary, by changing crystals), the intensity-wavelength distribution of a polychromatic beam can be obtained as in the case of Figure 1.5–1. In this way, also, a "monochromatic" beam of desired wavelength can be selected from a polychromatic beam. If an even purer beam is needed, two crystals in series (double monochromator) can be used. Clearly, the purer the beam, the lower will be its intensity.

1.17 The X-Ray Spectrometer

An x-ray *spectrometer*, as distinguished from an x-ray emission *spectrograph*, is an instrument primarily intended to measure x-ray wavelength and (usually) intensity as a function of wavelength, The elements of a spectrometer are the source, the collimator, the crystal monochromator, and the detector. The schematic diagram of Figure 1.17–1 applies both to the

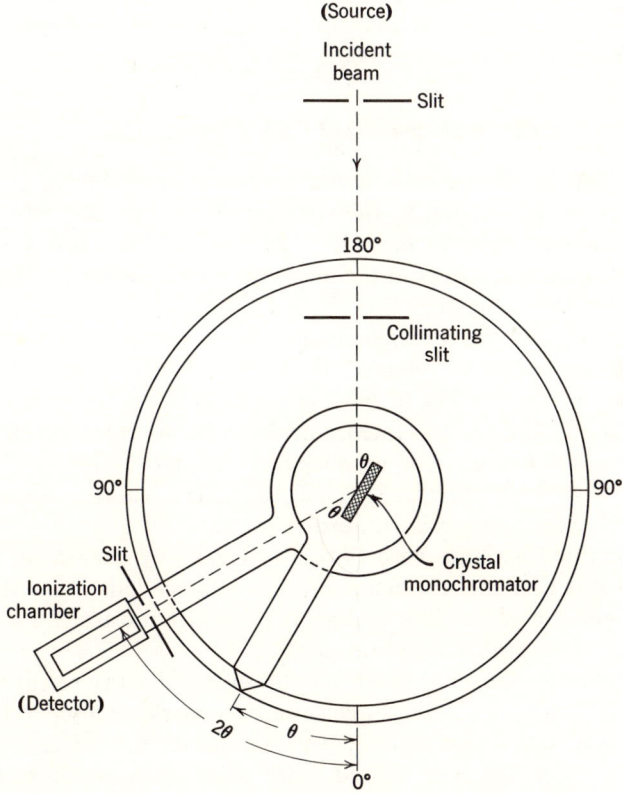

Fig. 1.17–1. Schematic diagram of the Bragg spectrometer. To scan a wavelength range, the crystal is rotated at angular velocity $\dot{\theta}$, and the detector at angular velocity $2\dot{\theta}$.

original Bragg spectrometer and to the modern instruments; the actual equipment today differs considerably from that of 40 years ago: the original spectrometer was a modified simple goniometer.

To obtain the information in Figure 1.18–1 with an x-ray spectrometer, we could proceed as follows. With electrons not too high in energy bombarding a tungsten target at constant rate, intensity readings are made on the detector for angular positions of the crystal monochromator. These positions will correspond to known wavelengths (Bragg's law) and be specified in terms of 2θ, the angle between incident and reflected beam, and twice the angle in (1.16–2). This angle must be maintained between a fixed source and movable detector if a ray is reflected from a crystal plane so as to satisfy (1.16–1) and (1.16–2). The intensity readings, normally uncorrected, are plotted against λ, or against 2θ, if one prefers not to reduce angular positions to wavelength.

Provision is sometimes made for varying θ by continuous rotation of the crystal. From what was said above, it follows that the detector must be rotated at twice the angular velocity of the crystal to obtain Bragg reflection of all wavelengths in the range being investigated.

1.18 Wavelengths of Characteristic Lines

The elder (W. H.) Bragg soon scored a great triumph on the basis of the x-ray "reflection" postulated by the younger (W. L.) [29, 30]. With the first x-ray spectrometer, which he made by replacing the collimator of an optical goniometer by a slit system and the telescope by an ionization chamber, he measured the wavelengths of the three **L** lines of platinum. By using a platinum target and the experimental method outlined in Section 1.17, he obtained the results in Figure 1.18–1. He correctly assumed the more intense group of three peaks to be characteristic "lines" of platinum (first-order reflections), and the less intense peaks to be the corresponding reflections of the second order. The wavelengths calculated from (1.16–2) (0.97, 1.13, and 1.32 Å) were remarkably close to modern values, although some of the "lines" were necessarily unresolved.

Bragg identified his lines with Barkla's L series by measuring their μ_{Al}, and by showing that this characteristic was independent of the diffracting crystal. He completed the link to Barkla's work by measuring the critical wavelengths of several absorption edges.

The discovery of the x-ray spectrometer changed x-ray research from a roughly quantitative to a highly precise activity. Barkla's discoveries, which owed so much to absorptiometry, were in the main confirmed, and the emphasis in x-ray research shifted from absorption to diffraction and emission.

X-ray diffraction makes most of its great contribution to science outside

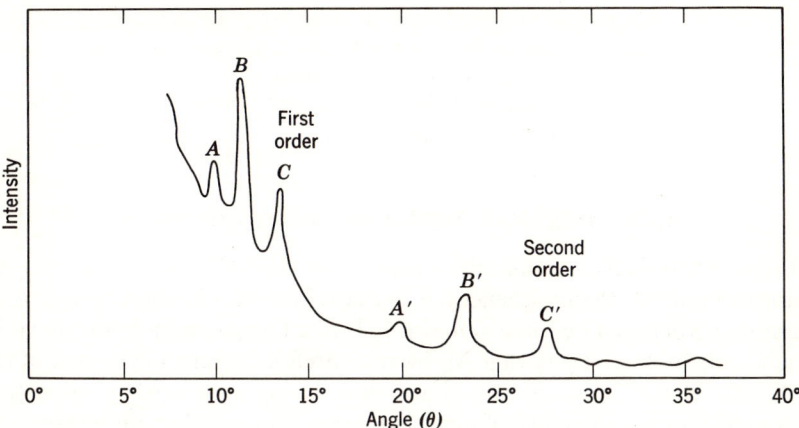

Fig. 1.18–1. Bragg's resolution of the platinum **L** spectrum. Each of his peaks contains a series of lines as shown in Figure 1.22–1. After Bragg and Bragg, *Proc. Roy. Soc. (London)*, **88A**, 428 (1913).

of analytical chemistry as we have defined it. Because x-ray diffraction originates with the scattering of x-rays by the individual electrons outside the nucleus, it offers through the measurement of *intensities* in Bragg reflection a way of learning how these electrons are distributed. The computer is enormously helpful in establishing these distributions, especially for very complex molecules such as those of the proteins. The locations of atoms, their radii, and the forces that bind them all become amenable to direct determination for solids that remain unaltered. These properties are established for the *unit cell* in a crystal, the unit cell being an entity that repeats itself in *three* dimensions. The role of x-ray diffraction in analytical chemistry is more prosaic. For this role, all that need generally be known is the *interplanar spacing d*, which is most often determined on crystalline powders. To identify a crystal and to determine the amount present in a sample, one need not know its structure. See Chapter 6.

THE REFRACTION OF X-RAYS

> At the present time the most precise, and I believe also the most accurate method of determining e is by means of the ratio of the Faraday to the Avogadro number....
>
> In fact, I believe that N is now one of the best established of the general physical constants.

> My present weighted average value, based on the results from *five* different varieties of crystal, is
> $N = (6.02338 \pm 0.00043) \times 10^{23}$ mole^{-1},
> on the chemical scale of atomic weights. (Subscript to N omitted.)   R. T. Birge [31]

1.19 Avogadro's Number and X-Ray Diffraction

If there was ever a seemingly insignificant scientific effect that can lay claim to important consequences, it is the refraction of x-rays, which is so slight that Roentgen's efforts to find it [1] were foredoomed. Yet it played a crucial role in uncovering that Millikan's carefully determined value of the electronic charge was about 1% too low owing to a wrong value (not Millikan's) for the viscosity of air. Thereupon x-ray diffraction became the best means for establishing Avogadro's number. The story is a scientific thriller [9, 10, 32] with authenticity as an added virtue, but it can only be sketched here.

If x-rays are refracted, Bragg's law requires correction because λ and θ will then be different inside the crystal from outside, where they are measured. That this situation exists was demonstrated by W. Stenström (Dissertation, Lund, 1919), who showed that the change in apparent wavelength of the Mo Lβ1 line with order n on reflection from a sugar crystal could be quantitatively explained on the basis that the index of refraction is *less* than unity by a *few parts per million*. To show how small the correction is we quote from Compton and Allison [9, p. 676] who give an excellent discussion of the matter. For Mo Kα1 [λ taken as 0.708 (10^{-8}) cm] they give the corrections required to the observed wavelength at five values of the order n:

Order (n)	1	2	3	4	5
Correction [cm(10^{14})]	-107	-29	-13	-7	-5

For the smallest correction to be significant, the wavelength would have to be known to better than 10^{-5} Å!

A parenthetical note: when the index of refraction is less than unity,

$$\mu_r = \text{refractive index} = c/v_p < 1 \qquad (1.19\text{--}1)$$

that is, the *phase* velocity v_p is greater in a solid or liquid than it is in vacuum. (The *group* velocity of the x-rays in the dense medium, which is the velocity at which energy is transmitted, does not exceed c, the velocity of light, so that there is no conflict with the theory of relativity.)

Elementary optics teaches that a difference in indices of refraction at an interface implies the possibility of *total reflection*; when the index of refraction is less than unity, as it is with x-rays, total reflection at a vacuum-solid interface will occur into the vacuum and not into the solid, as it does with

THE REFRACTION OF X-RAYS

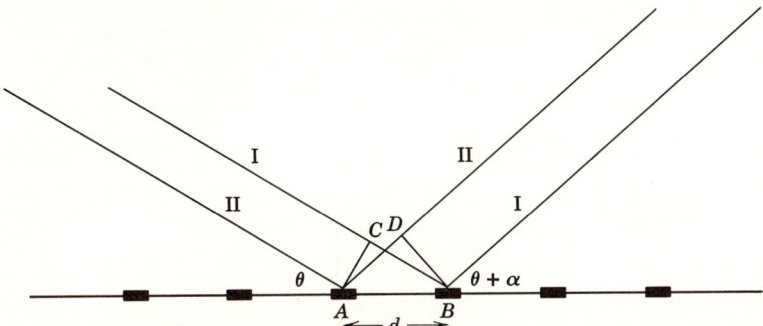

Fig. 1.19–1. Determination of x-ray wavelength by use of a ruled reflection grating. If $\theta < \theta_c$ (see text), intensity maxima are to be expected at values of $(\theta + \alpha)$ for which the difference in path between adjacent rays is a whole number of wavelengths; or, in the figure, when

$$CB - AD = d \cos \theta - d \cos (\theta + \alpha) = n\lambda$$

(Note that waves I and II are both incident at glancing angle θ and diffracted at glancing angle $(\theta + \alpha)$, and that both travel the same distance before they reach AC and after they leave BD.) Compare Figures 1.19–1 and 1.16–1.

ordinary light. Such total reflection of x-rays has been found. Reliable values for the refractive index have been obtained; for calcite and Mo Kα1, for which λ was taken as 0.708 Å,

$$\delta = 1 - \mu_r = 1.85(10^{-6}) \qquad (1.19-2)$$

The results are conveniently expressed in this way because μ_r is so near unity.

More importantly, the existence of total reflection into vacuum at a vacuum-solid interface makes possible the measurement of x-ray wavelength by use of a *ruled* reflection grating as Figure 1.19–1 shows. As in all such gratings, the spaces between the rulings act as diffracting centers that give by constructive interference the different orders of diffraction on either side of the reflected beam. Gratings for x-rays have from 50 to 600 lines per millimeter ruled on glass, speculum metal, silver, or gold. Total reflection is found for these gratings when the angle of incidence θ is less than θ_c, the critical angle, which can be shown theoretically [10, p. 128] to be equal to $\sqrt{2\delta}$ radians; θ_c is generally less than 1°. When $\theta > \theta_c$, x-rays are almost completely absorbed.

The connection between x-ray refraction and Avogadro's number came about because x-ray wavelengths from ruled reflection gratings were always about 0.25% greater than the wavelengths from Bragg reflection by crystals. It is important to remember that estimates of d, the interplanar spacing, could be made for simple crystals on the basis of crystallographic knowledge

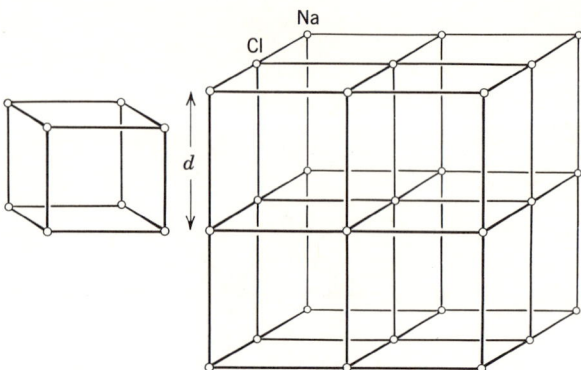

Fig. 1.19–2. The original Bragg computation of d from (1.19–3). The larger crystal shows the central chloride ion to be shared by eight of the small cubes.

that by far antedated x-ray diffraction. To show this, we repeat the original reasoning of Bragg [33]. In the sodium chloride crystal, which has the simplest of the several cubic structures, each smallest possible cube contains eight scattering centers, four being sodium ions, and four chloride; but each scattering center, being located on a corner, is shared by eight cubes. Overall then, one ion for one cube; see Figure 1.19–2. Each *molecule* of sodium chloride will then be associated with twice the volume of such a cube, or with $2d^3$. It follows that

$$d = \sqrt[3]{M/2\rho N} \qquad (1.19\text{–}3)$$

where M is the molecular weight, ρ the density, and N Avogadro's number. For crystals of less simple symmetry, a function of the angles in the crystal lattice must be included in (1.19–3). The d values obtained in this way were the ones that gave too small a value for the x-ray wavelength. These values were too small because the value of N was too large. The N was then determined as the quotient of the Faraday and the electronic charge. The trouble was eventually traced to Millikan's value of $4.774(10^{-10})$ esu for the latter. The quotation at the head of this section gives the final outcome of this, one of the most celebrated error hunts in science. Birge's value [31] of e is $(4.8021 \pm 0.0006)(10^{-10})$ esu, and Bearden [32] raised Millikan's value to $4.815(10^{-10})$ esu by recalculating Millikan's results with Bearden's value for the viscosity of air [32].

The five crystals to which Birge refers above are calcite, sodium chloride, diamond, lithium fluoride, and potassium chloride; the first of these is the generally accepted standard for x-ray diffraction. The high precision Birge quotes means that crystals of each kind can make admirable gratings. The triumphs of x-ray diffraction are a tribute to perfection in nature and to ingenuity in man!

X-RAY EMISSION

The secondary x-rays emitted by certain substances are remarkably homogeneous in character, though the primary radiation producing them is very heterogeneous. [21]

... The very close similarity between the x-ray spectra of the different elements shows that these radiations originate inside the atom, and have no direct connection with the complicated light-spectra and chemical properties which are governed by the structure of its surface. [36a, p. 1031]

The first problem we have to solve ... is therefore that of measuring up and analyzing the wave systems emitted by the atoms of 92 elements. [35]

1.20 Excitation of Characteristic Spectra

Even before Bragg measured the wavelength of the spectra produced by electron excitation of a platinum target, Kaye [34] had proved that electron excitation could be used to obtain the characteristic spectra of different metals serving as targets in an x-ray tube. X-ray excitation was used by Barkla in his discovery of these spectra. The high-energy particles of physics (α–particles and others more recently discovered) will also excite characteristic spectra. These methods of excitation are of growing interest in analytical chemistry.

Figure 1.20–1 illustrates an outstanding difference between electron and x-ray excitation of characteristic spectra. The former method conveniently yields spectra of high intensity, but (as might have been expected from Figure 1.5–1) it unfortunately leads to a much higher background, the continuous spectrum, than is obtained with x-ray excitation.

1.21 The Work of Moseley

It was thought that a careful study of the Röntgen radiation emitted by various elements, when used as anticathodes in discharge tubes, might be repaid by the discovery of some sort of relation between their *atomic weights* and the quantity and quality of the Röntgen rays given out and transmitted under various conditions. [34, pp. 123–124; italics supplied.]

The elements used as anticathodes were mounted in line on a car made of aluminum which ran along horizontal rails fastened by sealing

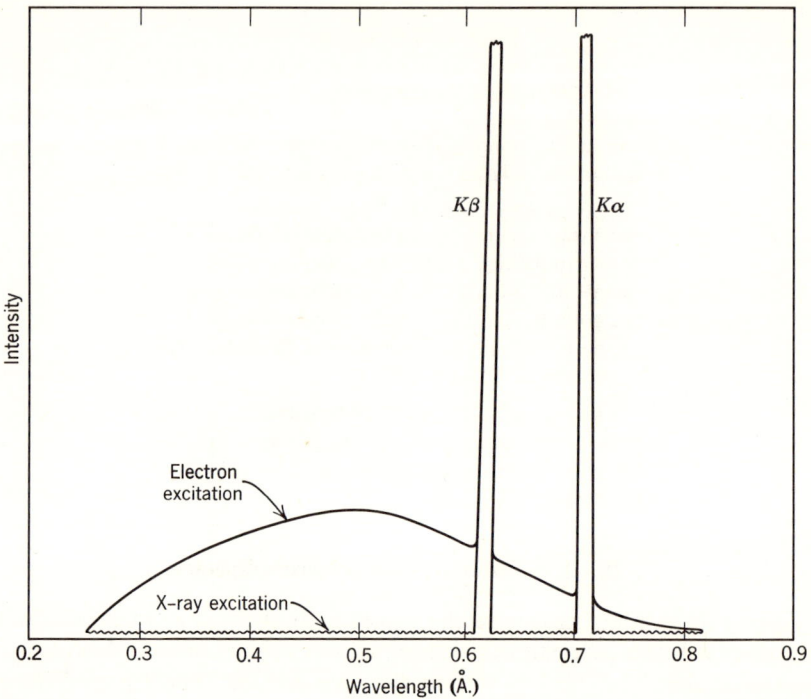

Fig. 1.20–1. The molybdenum spectrum excited by 50-kV electrons and by the polychromatic beam from a 50-kV x-ray tube. With x-ray excitation, most of the energy appears in the characteristic lines. With electron excitation, much of it is diverted to the continuous spectrum.

wax. . . . Underneath each axle of the car was fastened a piece of soft iron, and by means of a small electro-magnet outside, the car could be moved and any metal desired brought under the beam of cathode rays. [34, p. 125.]

More than 30 other elements have now been investigated (had their x-ray emission spectra photographed), and simple laws have been found which govern the results, and make it possible to predict with confidence the position of the principal lines in the spectrum of any element from aluminum to gold. The present contribution is a general preliminary survey, which claims neither to be complete nor very accurate.!! [36b, p. 705. Material in parentheses and emphasis supplied.]

These quotations show the progress from Kaye (1909) to Moseley. Moseley was 28 years old when killed at the Dardanelles in 1915 and tragically removed from consideration for the Nobel Prize. He had worked under Rutherford at Manchester, with Bohr for a time as contemporary [17, p. 231]. After Moseley and Darwin by using a fine slit had resolved the newly discovered Bragg platinum peaks B and C (Figure 1.18–1), Moseley

X-RAY EMISSION

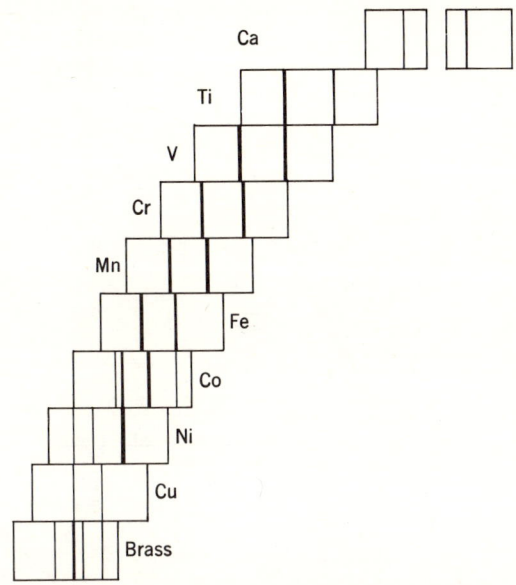

Fig. 1.21–1. Reproduction of Moseley's first published photograph of characteristic x-ray emission spectra. Each element gave two **K** lines, the stronger of which Moseley called α; the weaker, β. Brass, which "was substituted for zinc to avoid volatilization by the intense heat at a point struck by the cathode rays" [36a, p. 1029], shows lines for both copper and zinc. Moseley appreciated that impurities present could give their characteristic lines (see the Co spectrum), and that such lines could be used to determine the impurity elements.

modified Kaye's equipment [34] and photographed characteristic emission spectra produced by the electron bombardment of 42 elements [36]. The quotation that follows and Figure 1.21–1 establish him as the founder of x-ray emission spectrography. "The prevalence of lines due to impurities suggests that this may prove a powerful method of chemical analysis. Its advantages over ordinary spectroscopic methods lie in the simplicity of the spectra and the impossibility of one substance masking the radiation from another" [36a, p. 1030]. In addition, Moseley uncovered structure in the **K** and **L** spectra, established the atomic number as more fundamental than the atomic weight, and provided brilliant support for the Bohr theory of atomic structure. Later work, especially by Siegbahn [37], showed that **M**, **N**, and **O** spectra also exist.

In Figure 1.21–2, Moseley's data show that atomic number is clearly preferable, as a fundamental quantity, to atomic weight. The linear relationship between the frequency v of the **K**α line for element of atomic number Z is

$$v = 0.248 \times 10^{16} (Z - 1)^2 \qquad (1.21\text{–}1)$$

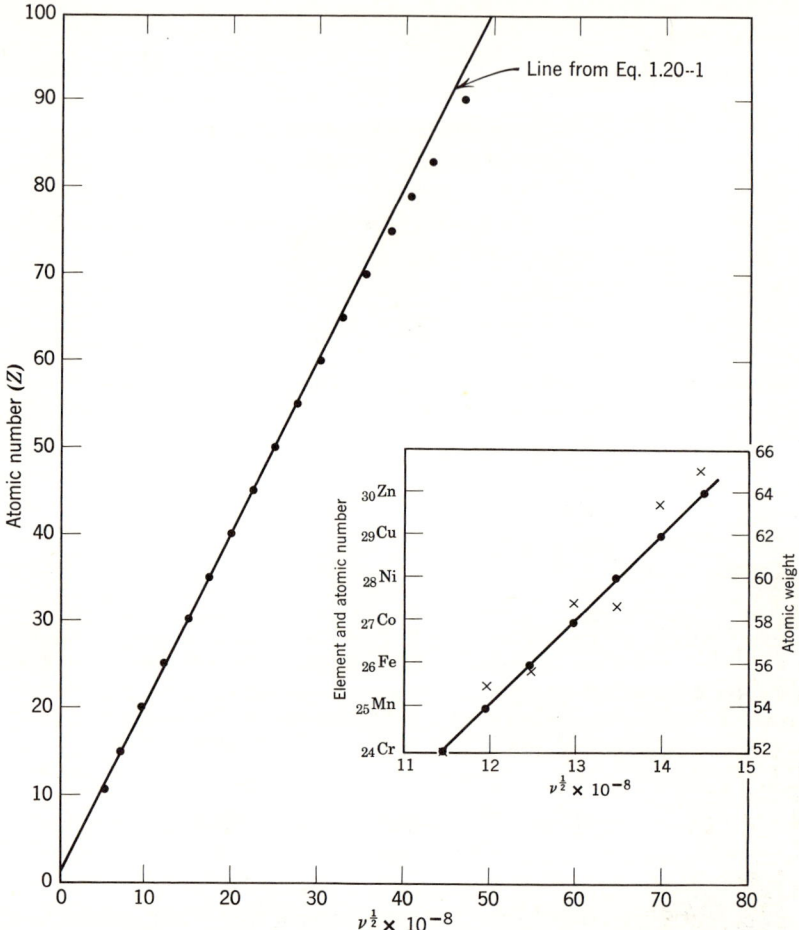

Fig. 1.21–2. Moseley plot for Kα2 lines. The curvature at high Z is due to a change in the effective nuclear charge (Z–1). In the insert, note that the crosses (atomic weights) do not agree as well with the line as the solid circles (atomic numbers). Atomic numbers are clearly more fundamental than atomic weights.

The value of the multiplicative constant differs little from that in (1.21–2) and was *theoretically* obtained by Bohr; that is,

$$\nu = 0.2467382 \times 10^{16} Z^2 \qquad \text{see (4.6–5)} \qquad (1.21\text{–}2)$$

This agreement provided powerful early support for the Bohr atom, and we return to this matter, and to the need for correcting Z in (1.21–1), in Chapter

X-RAY EMISSION

4. The validity of (1.21–1) proves that there are no isotope effects in x-ray emission spectra.

On the basis of Moseley's work, it became possible to correct and complete the periodic table, isotopes included. Thus x-ray emission spectrography could be used in returning misplaced elements to their proper places in the table, and to predict the existence of elements as yet unknown. Moseley himself had concluded that elements 43, 61, 72, 75, 85 and 87 remained to be discovered. Perhaps the most celebrated case among these is that of $_{72}$Hf, which was found by Coster and (von) Hevesy [38] on the basis of its six **L** lines. The story is a classic example of x-ray emission spectrography, in which the voltage across the x-ray tube was kept between 10 and 18 kV so as to remove the interference by Zr **K** lines (which cannot be excited below 18 kV) in the *second order* with certain Hf **L** lines in the *first order*.

1.22 Characteristic Lines and Absorption Edges

To complete the introduction of characteristic lines and absorption edges, it is necessary to consider how they are related. The Einstein quotation preceding Section 1.4 is a convenient point of departure. The best experimental approach is provided by the magnetic photoelectron spectrograph (Section 1.12). Einstein tells us that the *energy* of a characteristic line cannot *exceed* that of the associated absorption edge; via the magnetic photoelectron spectrograph (and in other ways as well), we learn that the *wavelength* of the line *exceeds* the critical absorption wavelength of the edge. Barkla (Figure 1.11–2) had already made this clear.

The experimental work (mainly by Robinson; see Section 1.12) on the magnetic photoelectron spectrograph proves that the ejection of a photoelectron from outside the nucleus of a target atom is associated with the emission of a characteristic line. The line is emitted because an electron, further removed from the nucleus than was the photoelectron, falls into and fills the hole that the photoelectron left behind. In x-ray language (which is opposed to optical), the normal atom has zero energy, and an atom with an electron hole anywhere in its structure has a positive energy. For example, an atom from which a **K** electron has been ejected is in the **K** state, which is a *higher* energy state than that of the normal atom. Suppose an **L** electron fills the hole in the **K** shell—the hole that characterizes the **K** state. The state of the atom will change from **K** to **L** and a **K** line will be emitted. Now, an atom in the **L** state still has an energy greater than zero on the x-ray scale, whence the energy of the **K** line must be less than that of the **K** absorption edge. (The reasoning here is valid for the **K** state no matter how produced—whether by electron bombardment or by x-ray excitation.)

With platinum as an example to illustrate nomenclature (other elements could serve as well):

$$\text{Energy } \mathbf{K} \text{ state} - \text{energy } \mathbf{L}_{III} \text{ state} = \text{energy Pt } \mathbf{K}\alpha 1 \qquad (1.22\text{--}1)$$

The energies of corresponding absorption edges are the *minimum* excitation energies for the states in (1.22–1); hence they equal the energies of the states. It therefore becomes possible to calculate the wavelength of the emitted line (e.g., Pt $\mathbf{K}\alpha 1$) from critical absorption wavelengths (here for the $\mathbf{K}$ and $\mathbf{L}_{III}$ edges of platinum) determined on a magnetic electron spectrograph. Comparison of the wavelength thus calculated with that directly measured constitutes an experimental test of (1.22–1), as Robinson has shown. The agreement is good enough for many elements (to 1 part in 6000 in the best cases) to prove that (1.22–1) is valid.

Relations of the type of (1.22–1) apply for the $\mathbf{L}$, $\mathbf{M}$, ... lines also. Whereever accessible to experiment, such relations have been verified. They are the only way of locating x-ray energy levels with respect to the normal state of the atom.

Figure 1.18–1 is one of the most important x-ray diffraction tracings ever made. The relationship of absorption edges and characteristic lines for this tracing is shown in Figure 1.22–1, and the relationship is in accord with the foregoing discussion. One other thing follows: It has been mentioned that x-ray absorption *lines* (edges are *not* lines) do not exist. Why not? Such a line would have to be generated by a process the inverse of the emission process exemplified by (1.22–1). Let us try to reverse this emission process by envisioning the absorption of Pt $\mathbf{K}\alpha 1$, as a consequence of which an electron would leave the $\mathbf{K}$ shell to find a home in the $\mathbf{L}$ shell. However, this shell under all ordinary circumstances is *completely occupied*, which makes the envisioned absorption process impossible. Experience shows that equations such as (1.22–1) cannot proceed from right to left.

1.23 The Characteristic-Photon (Fluorescence) Yield. The Auger Effect

There is unusually strong justification for the historical approach in presenting the important facts about x-rays to analytical chemists: the information these chemists need in their work is largely that discovered in the early researches. This is particularly clear in connection with absorption and emission spectra, in which more refined investigations with more powerful equipment later revealed important complexities that the analytical chemist may ignore until they become of demonstrated importance in analytical chemistry. Several such complexities are recorded below.

Let us consider the x-ray excitation of a characteristic line. For a given number of quanta per second in the exciting beam, the following factors will help determine the intensity of the emitted line: geometry, sample thickness, composition of sample, and wavelength distribution in the

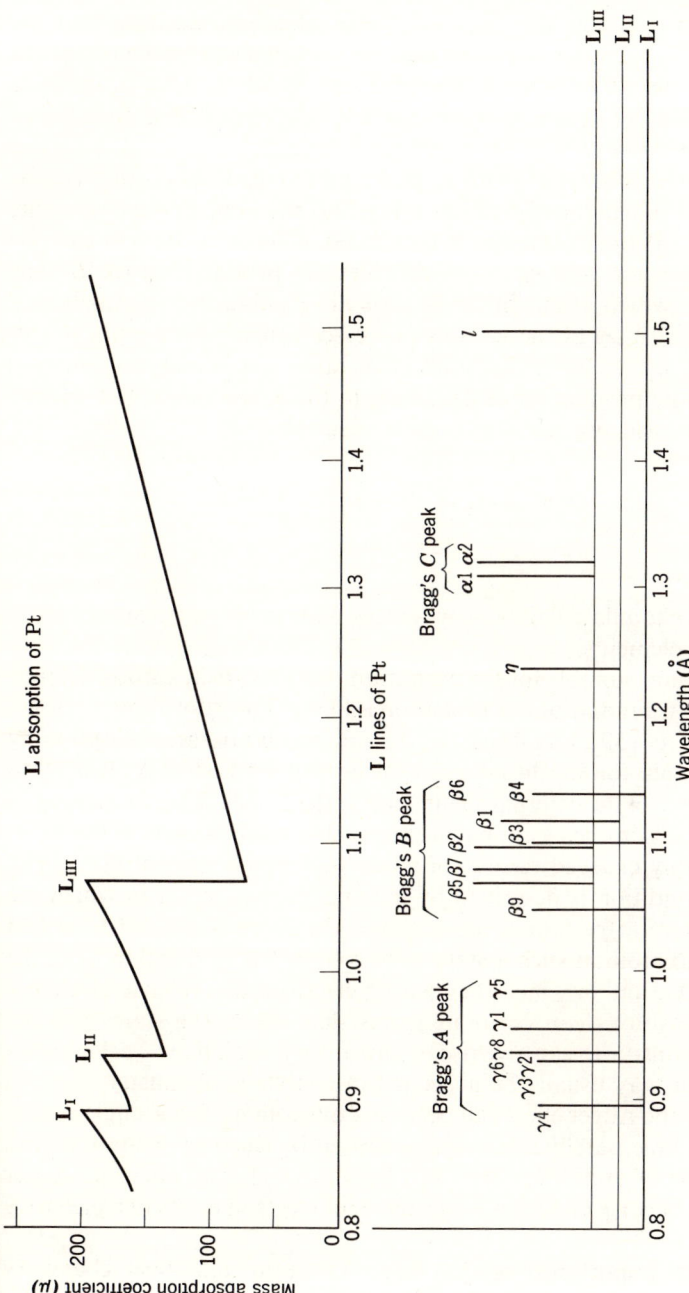

Fig. 1.22–1. Relationship of characteristic lines and absorption edges for the case of Figure 1.18–1. The **L** levels from which the lines originate can be identified by looking at the lower right-hand corner of the figure shown above. Note that the Einstein equivalence law is satisfied. Note also that some of the Bragg peaks contain lines from more than one **L** level. See Chapter 4.

exciting beam. These factors all influence the absorption of the exciting beam and of the characteristic line. None of these takes into account that not every quantum absorbed by an atom leads to the radiation by that atom of a quantum with the wavelength of the characteristic line. This situation is usually described by saying that the *fluorescence yield* is less than unity. *Characteristic-photon yield* is the more precise name.

To illustrate the concept of yield, we turn again to the **K** spectrum. Assume that an element is irradiated with an x-ray line energetic enough to excite the **K** spectrum. If the irradiation is continued, a steady state will soon be reached in which the rate n_K at which holes are produced in the **K** shell (i.e., the rate at which atoms in the **K** state are produced) is just balanced by the combined rates of the various processes causing such holes to disappear. Let $n_1, n_2, \ldots, n_f$ be the individual rates n_i at which the filling of holes leads to the production of the i lines in the **K** spectrum. The characteristic-photon yield, w_K, for this simple case is

$$w_K = \frac{n_1 + n_2 + \cdots + n_f}{n_K} = \frac{\sum_1^f n_i}{n_K} \qquad (1.23-1)$$

Values of w_K are tabulated in the Appendices. Values below 0.3 are common for the lighter elements.

If certain quanta suitable for the excitation of a line are absorbed without photon emission, a radiationless transition is likely. This transition is known as the Auger effect [39], and it may be thought to involve an absorption by the atom of the photon produced when the hole in the **K** shell is filled by an electron from one of the external shells such as the **L** shell. The absorption of this photon results in the ejection of a *second* electron from one of the shells to leave a doubly charged residue of what had been a normal atom. The atom in this condition is described by naming the two states in which the electron holes are to be found; for example, the atom is in the **LL** or **LM** or **LN** state. An atom in such a state is, of course, vastly different from the usual divalent cation, in which the missing electrons are valence electrons.

Electron transitions can occur in atoms that are in the states just described. Such transitions can give rise to lines—very weak lines—called satellite lines, which may usually be neglected in analytical chemistry.

Inasmuch as the Auger effect can occur in any atom having a single appropriate electron hole, satellite lines are produced by electron excitation also. With this mode of excitation, they will be produced also when the Auger effect does not occur provided a single incident electron ejects two electrons from the atom.

The growing importance of the Auger effect in analytical chemistry warrants its brief discussion later in the book.

CHEMICAL INFLUENCES IN X-RAY ABSORPTION AND EMISSION

1.24 General Considerations

Up to this point, our position has been that the elementary processes by which x-rays are absorbed and emitted are free of chemical influences because these processes involve energy levels nearer the nucleus than the levels in which valence electrons are to be found. This simplified position suffices for most x-ray applications in analytical chemistry. Nevertheless, chemical influences on both types of elementary processes have been demonstrated, but only at very high resolution—at much higher resolution than the analytical chemist usually requires.

As is to be expected from the way atoms are built, these chemical influences are more pronounced in the region of low atomic numbers, in which the valence shells are nearer the nucleus. Another fact to keep in mind is that the crowding of atoms, as in a compound or in the solid state, will tend to modify any x-ray process that involves the outer electron shells; the states of the valence electrons will usually be greatly changed from what they are in an isolated atom. In modern theories of the solid state, the outer electrons are regarded as belonging to the entire solid, and these electrons are thought to occupy bands of electronic states characteristic of the entire solid.

Chemical forces affect fine structure, cause shifts in wavelength, and thus change the intensity-wavelength distribution. Detailed discussion appears in a later chapter.

1.25 Effects of X-Rays on Samples

One of the attractive features of analytical methods using x-rays is that they can be applied without destroying samples ready for analysis. This generalization, though valid as written, must not be taken to imply that samples necessarily come through unchanged when analyzed by such methods.

In x-ray emission spectrography, for example, samples are exposed to a polychromatic x-ray beam that excites the characteristic lines of the elements present. Three samples exposed in such a spectrograph under standard analytical conditions were changed as follows. A glass slide suffered permanent darkening in 5 minutes. Potassium chloride changed from white to dark purple in 2 minutes; but this change was fleeting, for the F-centers responsible for the purple color had faded out almost completely in half an hour. A more serious change occurred in a dilute solution of tetraethyllead fluid in a hydrocarbon. Within 30 minutes, the solution turned yellow, and lead oxide appeared to precipitate. Methane and other hydrocarbons were formed also [40].

The effects of x-rays on samples being analyzed will almost never cause difficulty. Yet it is well to be on the lookout for such effects [41], an extreme example of which can occur when liquid hydrocarbon samples are analyzed for sulfur by x-ray methods. The irradiation causes the precipitation of solid sulfur, which will sink, leaving the upper layers too low in sulfur and making the bottom layers of the sample too high. If analysis is by x-ray emission, the results will be too low if irradiation is from above, and vice versa.

1.26 Concluding Guides

The x-ray story has been painted with a broad brush in this introductory chapter, partly to convince the reader that this—one of the most fascinating of scientific stories—has much to offer any one interested in the contributions of physics to chemistry. The overall value of the story, we believe, transcends the utilitarian, though that value is great enough. To guide readers new to x-rays through the specialized chapters that follow, we give here (1) a convenient classification of x-rays, (2) a crude correlation of x-ray spectra with others, and (3) a naive model to show how the correlation is reflected in atomic structure. The first two guides are in Tables 1.26–1 and 1.26–2 that have explanatory notes.

Table 1.26–1. A Convenient Classification of X-Rays

	Verbal Description		
	Hard	Soft	Ultrasoft
Wavelength range	Up to 1Å	From 1 to 10Å	From 10 to 200 Å
"Voltage" classification[a]	Above 10 kV	From 1 to 10 kV	Below 1kV

The "voltage" classification is practically useful and is meant to correspond to an energy classification in volt-electrons. See Table 1.26–2.

Figure 1.26–1 is even less quantitative than the tables. The "valence-electron" belt is enlarged and the radii of the three inner-electron shells do not represent their energies. However, the figure does emphasize that the inner electrons *generally* are of overriding importance as compared with the valence electrons in the absorption and emission of x-rays. Of course, as the wavelengths of x-rays increase to the point where their energies become comparable with those of valence-electron transitions, we must expect that these electrons will assume increasing importance in x-ray processes.

CHEMICAL INFLUENCES IN X-RAY ABSORPTION AND EMISSION

Table 1.26–2. A Crude Correlation of Various Spectra

Spectrum	Spectral Region	Representative λ, Å	Quantum energy, erg	eV, electron-volts
γ-ray	γ-ray	0.01	$2(10^{-6})$	$12.4(10^5)$
x-ray	x-ray	1	$2(10^{-8})$	$12.4(10^3)$
Electronic	Ultraviolet	1000	$2(10^{-11})$	12.4
Vibrational	Infrared (near)	10^4	$2(10^{-12})$	1.24
Rotational	Infrared (far)	10^6	$2(10^{-14})$	0.0124

Note. (1) It is assumed the atom is combined. Only molecules have vibrational and rotational spectra. (2) The wavelengths are chosen for convenience. The spectral regions are poorly defined; for example, the ultraviolet region shades into the visible. (3) One electron-volt is $1.6(10^{-12})$ erg. Often an electron with energy Ve electron-volts is called a V electron (see previous table). Kilovolts often replace volts in this designation.

The same expectation applies as we pass to elements of lower and lower atomic number: the "inner" electrons become fewer and fewer until they have disappeared at $_2$He, where the **K** electrons have become the "outer" electrons.

Gamma- and x-rays are unequivocally distinguished by different origins, but not always by different energies. Electrons exceeding $12,400\ (10^3)$ eV

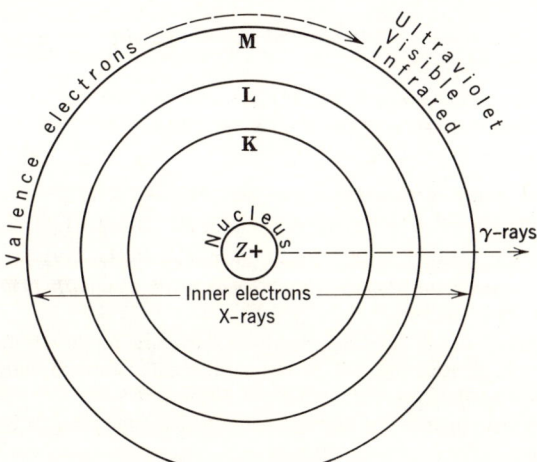

Fig. 1.26–1. Oversimplified model of any atom (combined or uncombined) between $_{11}$Na and $_{36}$Kr. The **K** and **L** shells are filled. The **M** shell and the valence band (or shell) need not be. Elements above $_{18}$Ar have electrons in the **N** shell (not shown). The **L** and **M** shells divide logically into subshells or energy levels (not shown, but see Figure 4.5–1). The different kinds of radiant energy are shown with origins indicated.

in energy can on bombardment of a target generate x-rays of wavelength shorter than 0.01 Å. These are still x-rays though their energies might exceed those of low-energy γ-rays.

Having arrived at this point, the reader who wishes should be ready for the chapters that deal with applications of x-ray methods if he is interested mainly in what results these methods can yield.

REFERENCES

1. W. C. Roentgen, *Ann. Physik. u. Chem.*, **64,** 1–37 (1898). These three great papers are more accessible here than in the journals of first publication. The quotation is on p. 2.
2. R. Castaing, *Application des Sondes Électroniques à une Méthode d'Analyse Ponctuelle Chimique et Crystallographique*, Thesis, University of Paris, 1951.
3. H. A. Liebhafsky, *Anal. Chem.*, **34**, 23A (1962).
4. Ref. 1, p. 9.
5. O. Glasser, *Dr. W. C. Roentgen*, C. C. Thomas, Springfield, Ill., 1945, p. 36. This version of Roentgen's discovery differs slightly from that in Glasser's *William Conrad Röntgen*, Bale, Sons & Danielson, London, 1933. Both books are worth reading.
6. W. D. Coolidge, *Phys. Rev.*, **2**, 412 (1913).
7. A. Einstein, *Physik. Z.*, **10**, 191 (1909).
8. C. Schaefer, *Einfuhrung in die theoretische Physik*, Vol. 3, No. 2, Walter de Gruyter, Berlin, 1937. The Lenard work is mentioned in Chapter 1.
9. A. H. Compton and S. K. Allison, *X-Rays in Theory and Experiment*, D. Van Nostrand, New York, 1935.
10. H. Semat, *Introduction to Atomic and Nuclear Physics*, 4th ed., Holt, Rinehart and Winston, New York, 1966.
11. R. A. Millikan, *Phys. Rev.*, **7**, 362 (1916).
12. W. Duane and F. L. Hunt, *Phys. Rev.*, **6**, 166 (1915).
13. W. T. Sproull, *X-Rays in Practice*, McGraw-Hill, New York, 1946, p. 14.
14. K. Harworth and P. Kirkpatrick, *Phys. Rev.*, **62**, 334 (1942).
15. H. G. Pfeiffer and H. A. Liebhafsky, *J. Chem. Ed.*, **28**, 123 (1951).
16. H. A. Liebhafsky and H. G. Pfeiffer, *J, Chem. Ed.*, **30**, 450 (1953).
17. N. H. de V. Heathcote, *Nobel Prize Winners in Physics 1901–1950*, Henry Schuman, New York, 1953, p. 141.
18. A. Winkelmann and R. Straubel, *Jenaisch. Z. Naturwiss.*, **30** (1896). This work is also discussed by F. K. Richtmyer, E. H. Kennard, and T. Lauritsen, *Introduction to Modern Physics*, 5th ed., McGraw-Hill, New York, 1955, p. 353.
19. H. Hirst, *X-Rays in Research and Industry*, Chemical Publishing Co., Brooklyn, N. Y., 1943, p. 108.
20. C. G. Barkla, *Phil. Mag.* [*6*], **22**, 406 (1911).
21. C. G. Barkla and C. A. Sadler, *Phil. Mag.* [*6*], **17**, 739 (1909).
22. H. R. Robinson, *Proc. Roy. Soc. (London)*, **104A**, 455 (1923).
23. C. J. B. Clews and H. R. Robinson, *Proc. Roy. Soc. (London)*, **176A**, 28 (1940).
24. A. H. Compton, *Phys. Rev.*, **21**, 485 (1923).

REFERENCES

25. See Ref. 9, especially pp. 12, 116, and 125.
26. Ref. 21. See also C. G. Barkla, *Phil. Mag.* [6], **21**, 648 (1911).
27. W. Heitler, *The Quantum Theory of Radiation,* Oxford University Press, London.
28. W. L. Bragg, *Nature,* **90**, 410 (1912).
29. W. H. Bragg and W. L. Bragg, *Proc. Roy. Soc. (London),* **88A**, 428 (1913).
30. W. H. Bragg, *Proc. Roy. Soc. (London),* **89A**, 246 (1914).
31. R. T. Birge, *Am. J. Phys.,* **13**, 63 (1945).
32. E. R. Cohen, K. M. Crowe, and J. W. M. Dumond, *The Fundamental Constants of Physics,* Interscience, New York, 1957. See J. A. Bearden, *Phys. Rev.,* **56**, 1023 (1939) for the last word on the viscosity of air, which was so important to the story.
33. W. H. Bragg and W. L. Bragg, *X-Rays and Crystal Structure,* Harcourt Brace & Co., New York, 1924, pp. 93 and 94.
34. G. W. C. Kaye, *Phil. Trans. Roy. Soc. (London),* **209A**, 123 (1909).
35. Siegbahn's Nobel Lecture, cited by N. H. de V. Heathcote in *Nobel Prize Winners in Physics 1901–1950,* p. 224.
36. (a) H. G. J. Moseley, *Phil. Mag.* [6], **26**, 1024 (1913); (b) *ibid.,* **27**, 703 (1914).
37. M. Siegbahn, *Spektroskopie der Röntgenstrahlen,* 2nd ed., Julius Springer, Berlin, 1931. An English translation of the first edition, *The Spectroscopy of X-Rays* (G. A. Lindsay, translator), was published by the Oxford University Press, London, 1925.
38. D. Coster and G. Hevesy, *Nature,* **111**, 79, 182 (1923).
39. P. Auger, *J. phys. radium,* **6**, 205 (1925); *Ann. phys.,* **6**, 183 (1926). E. H. S. Burhop, *The Auger Effect and Other Radiationless Transitions,* University Press, Cambridge, England, 1952.
40. Authors' unpublished results.
41. G. L. Clark, *Applied X-Rays,* 4th ed., McGraw-Hill, New York, 1955, Chap. 11.

Chapter 2

The Measurement of X-Ray Intensity. X-Ray Detectors and Detector Systems Energy Resolution

GENERAL INFORMATION

2.1 Introduction

The name "x-ray detector" came into use when observations on x-rays were predominantly qualitative. Nowadays, the emphasis is on high precision and efficiency so that most modern observations are measurements either of intensity or of dosage (x-ray quanta accumulated during exposure time). "X-ray detector" as a name has survived this change in emphasis although it does not describe the quantitative function of these devices.

X-rays cannot be detected until they have been absorbed. Chapter 1 has shown that x-rays and electrons are closely related, and that the absorption of x-rays affects the electrons in the absorber. Consequently, x-rays can be detected by observing and measuring the effects on electrons resulting directly or indirectly from x-ray absorption. Three of the four most important of these effects were discovered by Roentgen; to wit: (1) latent image formation in a photographic plate (chemical effect); (2) ionization in a gas (electrical effect); (3) excitation of a phosphor to yield visible light (optical effect). The fourth effect, of growing importance, is the separation of charge that can be made to occur when x-rays are absorbed in certain

GENERAL INFORMATION

crystals. The generation of heat, which always accompanies the absorption of x-rays, is not used in analytical chemistry to detect them; it can be used to measure x-ray dosage. The production of Čerenkov radiation, important though it is in nuclear physics, requires quanta of energy too high for analytical chemistry.

The electron effects of concern here are widely different: latent image formation results eventually in the *transfer* of electrons from bromide to silver ions; ionization in a gas *frees* an electron from an atom (or molecule), whereupon it may be collected or produce by collision other electrons before collection occurs; when visible light is instrumental in the detection process, the electrons resulting from x-ray absorption must eventually *excite* outer electrons of a phosphor or activated crystal so that these valence electrons can emit light when they return to their normal states; and conductivity in a crystal presupposes that x-ray absorption produces positive holes and negative electrons, both of which move freely in an electric field.

The effects in question are often translated into electric currents, pulsed or continuous. For the convenient reading or recording of these currents, complex electronic circuitry may be needed. Modern methods of measuring x-ray intensity are therefore primarily a concern of the experimental physicist. Nevertheless, the analytical chemist must know something about them because x-ray detectors are now among the tools of his trade. This chapter attempts to provide him with an *acceptable minimum* of knowledge about them.

X-ray detection is usually complex enough to require an *x-ray detection system*, which includes the ancillary electronics. Such systems often do more than count quanta. Some of them can sort quanta according to energy, and this leads to *energy resolution* of x-rays as a currently less precise alternative to *wavelength* (or Bragg) *resolution*. Other systems can perform even more sophisticated tasks. A detailed discussion follows.

An instantaneous detector measures I directly. For this to be possible the photon flux (quanta per second) must be high enough to give a steady-state current that can be read satisfactorily. Strictly speaking, there is no *constant* x-ray intensity; x-ray emission is itself random, and the detection process introduces additional fluctuations: therefore, I is a *rolling average*. An instantaneous detector is analogous to an *ammeter*.

When the I becomes too low to be read instantaneously with sufficient precision, an *accumulative* detector system is needed. In such a system, counts are accumulated for a fixed time, or the time is measured for the accumulation of a fixed number of counts. In either case, we have a *counting interval* Δt and a *total number* of counts N with $\bar{I} = N/\Delta t$. An accumulative detector system is analogous to a *coulometer*.

Intensity is often measured in counts per second, abbreviated cps.

2.2 More about X-Ray Absorption

X-ray absorption *by* the detector is obviously desirable as it produces the effect that is measured. Absorption *on the way* to the detector is undesirable as it reduces intensity. This section deals with detectors other than the photographic plate, which is discussed later.

Absorption problems fall under three headings: (1) attenuation along the beam path, (2) attenuation by the detector window, and (3) absorption by the detecting medium. Absorption by filters and blocking by collimators fall under the first heading, but we consider here only absorption by gas in

Table 2.2–1. X-Ray Transmission and Absorption in Spectrographs

	0.1 Å	0.3 Å	1.0 Å	3.0 Å	10 Å
Fraction *transmitted* by 35-cm[a] beam path (STP)					
Air	0.9935	0.9863	0.896	0.074	~ 0
He	0.9993	0.9991	0.9985	0.9912	0.740
Fraction *transmitted* by various windows					
Be 10 mil (0.025 cm)	0.994	0.992	0.972	0.625	~ 0
Be 1 mil (0.0025 cm)[b]	0.9994	0.9992	0.997	0.953	0.216
Mylar 0.25 mil (0.0006 cm)[c]	0.9999	0.9998	0.9985	0.963	0.287
Mylar 0.1 mil (0.00025 cm)[c,e]	0.9999	0.9999	0.9994	0.985	0.607
Al 0.2 mil (0.0005 cm)	0.9998	0.9993	0.980	0.606	0.487
Fraction *absorbed* by detector gas (3-cm path, STP)					
Argon[d]	0.0009	0.0065	0.176	0.9852	> 0.9999
Krypton	0.006	0.089	0.343	> 0.9999	> 0.9999
Xenon	0.025	0.320	0.831	> 0.9999	> 0.9999
Fraction *absorbed* by solid detectors					
Sodium iodide, 0.2 cm	0.58	> 0.9999	0.9999	> 0.9999	> 0.9999
Si 0.3 cm	0.11	0.35	0.9999	> 0.9999	> 0.9999
Ge 0.3 cm	0.47	> 0.9999	0.9999	> 0.9999	> 0.9999

[a] The length of beam path in a typical spectrometer is 35 cm.
[b] The 1 mil Be is liable to be porous.
[c] Gases used in detector and spectrometer permeate thin polymers readily.
[d] "Argon" used in detectors usually contains 10% CH_4. This mixture is known as P-10.
[e] Ultrathin polymer windows, 10^{-4} cm thick or less, with thin electrically conducting coatings, transmit to very long wavelengths (200 Å). They are extremely fragile and are often mounted on supporting screens, which of course reduce transmission.

GENERAL INFORMATION

the beam path. The results of illustrative absorption calculations are given in Table 2.2–1: where absorption is harmful (gas, window), the fraction *transmitted* is listed; where it is beneficial (detector), the fraction *absorbed* is listed.

To calculate the fraction of an x-ray beam that actually appears in the detector, one must consider geometry and the appropriate factors from Table 2.2–1, which will act multiplicatively; in addition, one must allow also for any reduction of intensity by filters and collimators or slits.

2.3 Functions of Electronic Circuitry

The circuits under discussion here are those in the detector system. Stable x-ray tube voltage and current, and a stable voltage across the detector are taken for granted.

Under the simplest conditions, the electrons produced in the detector system by the absorption of an x-ray quantum will appear as separate, well-defined pulses (bundles of electrons), one for each quantum. The pulses can then be counted as individuals, and each pulse (hence each x-ray quantum) will register as a unit. As the intensity increases, it becomes increasingly difficult to maintain this pulse individuality in the detector system. Individuality may be lost in the detector proper, in the associated circuitry, or in both.

The introduction of solid-state devices has revolutionized the circuits in detector systems, and the revolution has not ended. The introductory discussion here is followed by detailed description in later sections. These circuits are classified as *linear* or *logic* according to the way they handle signals. The primary function of a linear circuit is to amplify, the most common example being to increase the *amplitude* of a pulse so that the energy of an absorbed quantum can be measured. The logic circuit makes decisions about pulses; for example, it decides whether the energy of a pulse, which is proportional to its *amplitude*, exceeds an arbitrary minimum. To do this, the pulses entering the circuit must have a fixed *shape*. Both linear and logic circuits are required for the pulse-height selection and the counting of x-ray quanta.

We shall classify as follows the functions performed by electronic circuitry in detector systems:

1. Impedance matching (circuit coupling).
2. Amplification as required to give a useful signal.
3. Pulse shaping (time clipping included).
4. Pulse-height selection.
5. Counting and scaling.
6. Achievement of instantaneous readout.
7. Pulse rejection.

An introductory discussion of each function follows:

1. *Impedance matching.* This phrase usually expresses the idea that matched impedance of circuits is needed for maximum transfer of power. Here the emphasis is more on the transfer of a charge dq that measures the energy of an absorbed x-ray quantum. This charge must be transferred from the detector to the preamplifier without loss, and without loss of pulse individuality. The preamplifier input capacitance is the most important impedance component in this application and must be matched to the detector. Throughout the rest of the circuitry, proper matching of the output and input impedances is important. In vacuum-tube circuits, impedance matching was often accomplished by a "cathode follower," a device made obsolete by solid-state electronics.

2. *Amplification as required to give a useful signal.* The x-ray photons of interest in analytical chemistry have energies on the order of 10^{-15} W-sec. Upon absorption by the detector, not all this energy is available to produce ionization: some of it is inevitably degraded to heat. Considerably more energy than that directly available is usually needed before a detector system can operate to produce a useful readout. Consequently, amplification is needed, and this amplification is accomplished in different ways in the different detector systems: amplification can occur within the *detector* (internal); within the *preamplifier*; and within the *amplifier* (external).

The need for external amplification decreases as the internal amplification of a detector increases. A *Geiger detector* has enormous internal amplification so that an external amplifier is needed mainly for impedance matching. The *ionization detector* (ionization chamber), on the other hand, produces pulses so weak that the charge collected is near the noise level of the electronic components, and sophisticated amplifiers are consequently needed. The *proportional detector* and the doped *germanium* and *silicon detectors* give stronger pulses, but need an amplifier with a wide linear range of gain so that pulse heights can be recorded. See below.

3. *Pulse shaping (time clipping included).* These functions are often carried out by the *preamplifier* and the *amplifier*. The following oversimplified discussion uses an example (Figure 2.3-1) from Catalog No. 1001, ORTEC, Oak Ridge, Tenn., p. 80.

The pulses (bundles of electrons) produced by an electronic detector vary in amplitude and occur at random intervals because x-ray emission is itself a random process and because other random processes occur within the detector. The result of all this is that the output signal from the detector is very complex, with pulses of varying heights superimposed to varying extents and at varying intervals. Before such pulses can be selected (see below) and counted, they must be "clipped"—shortened in duration—to separate and give them individuality.

GENERAL INFORMATION

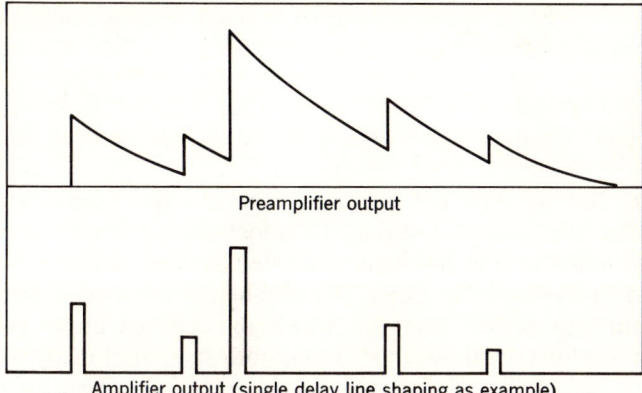

Fig. 2.3-1. An example by ORTEC of pulse clipping and shaping. ORTEC Catalog No. 1001, p. 80, Fig. 4. By permission.

Figure 2.3-1 illustrates pulse clipping and shaping. The complex output of the detector has been modified by the preamplifier to give the pulses shown in the upper half of the figure. These pulses rise as rapidly as the circuit permits (say, in 0.1 μsec) and they decay exponentially with a time constant ("clipping time") of 50 μsec, which means that they require four times that long to fall within 2% of the baseline for pulse-height selection (see below). Clipping and shaping in the amplifier give the (idealized) rectangular pulses shown in the lower half of the figure. Other shapes (for example, semi-Gaussian) are possible.

Modern solid-state circuits accomplish pulse clipping and shaping simultaneously. It is therefore correct and convenient to say that they *generate* new pulses of the desired height and shape—that is, pulses correctly representing the energy of the absorbed x-ray quantum and properly shaped for pulse-height selection.

4. *Pulse-height selection* accomplishes the energy resolution mentioned above. Considerable confusion results because the terms *pulse-height analysis, pulse-height discrimination,* and *pulse-height selection* are sometimes used loosely or interchangeably. We shall follow *The International Dictionary of Physics and Electronics* (Princeton, N.J.: Van Nostrand Company, 1956, pp. 715 and 716), where the instruments for these techniques are described as follows:

Pulse Height Analyzer. A device which records or counts a *pulse* only if the amplitude of the pulse falls within specified limits (single channel) or within specified sets of limits (multichannel). It thus yields the pulse height spectrum of a group of pulses. In many applications this gives the energy spectrum of nuclear radiations.

Pulse Height Discriminator. A circuit designed to select and pass voltage *pulses* of a certain minimum amplitude.

Pulse Height Selector. A circuit designed to select and pass voltage *pulses* in a certain range of amplitudes.

To mitigate confusion, we shall speak of pulse *height* not of pulse *amplitude*. Though "analysis" has so many meanings that we must minimize its use, we have no choice but to describe as "analyzers" the instruments just mentioned. But we shall not use "nondispersive spectrum analysis" and similar names for energy resolution. The most common function of a pulse-height discriminator is to eliminate unwanted background such as scattered x-rays and photomultiplier noise. A pulse-height selector or analyzer has one or more "windows" through which the selected pulses pass. Each window has a lower and an upper level, both measured in volts and each window marks a channel [e.g., 10 V (upper) − 3 V (lower) gives a 7-V window]. These voltages are settings on instruments and must not be confused with voltages that are sometimes used illogically and inexactly to denote the energies of x-rays. Perhaps it helps to envision a multichannel analyzer as an apartment house with windows in vertical rows (each row a channel) and with only one window open in each row, but with each open window one story higher than its neighbor to the left. Each open window is characterized by its voltage (see above) and its channel: the former fixes the range of pulse heights that can pass through the window; the latter establishes the mean of these heights—in the apartment house model, each channel is of a higher energy than its neighbor to the left. The pulses from a sample pass through these windows independently and simultaneously, each pulse through its proper window, for accumulation and storage prior to readout.

5. *Counting and scaling.* These functions accomplish the *accumulation* of pulses to give, as was mentioned in Section 2.1, either a number of counts over a fixed counting interval or the time required for a fixed number of counts; either path leads to an average intensity $\bar{I}$ measured in counts per second. The data are often printed out automatically on a paper strip; digital readout is also common. In the older accumulative detectors, these functions were separated.

Scaling is necessary because the number of counts is often too large for a counter to handle. A fixed number of counts (usually 2 or 10) is then accumulated into a larger unit; these units are accumulated into still larger units; and so on through as many stages as is necessary. This abacus-like process is called scaling: thus 64 counts are recorded as a unit in the seventh stage of a binary scaler. The process has much in common with digital computer technology, which rests on a binary foundation. By means of solid-state circuitry and "flip-flop" circuits, counting and scaling can be combined into a single operation more conveniently than was possible with vacuum tubes. With multichannel analyzers, computers are often used to store the sorted counts.

GENERAL INFORMATION 65

6. *Instantaneous readout*. A detector that yields discrete pulses can be used as an instantaneous detector if the pulses can be averaged to form a continuous electric current, as in a counting-rate meter. These devices can consist of a capacitor that is charged by the pulses and a resistance through which the capacitor is discharged. Their response time depends on the product of the capacitance and resistance, but their response time does not limit the counting rate. A further discussion is found later.

More sophisticated versions also exist. For example, only pulses of certain heights may be permitted to enter the meter. Those that do enter then trigger a monostable ("one-shot") multivibrator, driving this circuit into its quasi-stable state, where it remains for a predetermined time before returning to its stable state of readiness for the next trigger. This electronic prestidigitation is the "flip-flop" circuitry mentioned above. An important point: Counting-rate meters alone cannot give energy resolution because all pulses, no matter what their height, that trigger the counting mechanism appear identically in the readout of the meter. This statement applies also to the accumulation of counts by counting and scaling.

7. *Pulse rejection*. Pulse-height selection is achieved by rejecting pulses *lower* and *higher* in energy than those to be selected. The lower-energy pulses are rejected by a pulse-height discriminator (see point 4 above). The higher-energy pulses must be rejected in a more sophisticated way. They are isolated by a second pulse-height discriminator from those desired and then rejected by an *anticoincidence* unit, which is a device to block passage of pulses simultaneously surviving two discriminators. Details follow.

Instability in the electronic circuitry, which tends to become more serious as the counting circuits grow more complex, introduces errors in the intensity measurements that will combine with, and may overshadow, the irreducible statistical counting error. The instability may appear as a steady drift in the recorded detector output, or as fluctuations therein, or as the two effects superimposed. The drift is relatively easy to correct. The occurrence of fluctuations may make it necessary to shorten the counting interval, or to make more intensity measurements than would be required were the system stable. The risk of increasing instability, which can be reduced by adding special stabilizing circuits, must be weighed against the advantages that accrue from adding to or complicating the electronic circuitry used for making x-ray intensity measurements. Finally, at high counts per second, the scaling-and-counting circuit will lead to losses during counting if this circuit is considerably more sluggish than the detector.

The introduction of solid-state circuitry has greatly reduced the seriousness of the difficulties just mentioned. In addition, it has brought simplifying advantages such as the combining of counting and scaling and eliminated the cathode follower from detector circuits (see above). It has brought faster warmup of equipment, longer life, smaller generation of heat during operation (consequently lower thermal drift), lower electronic noise, miniaturization, easier servicing, and reduced capital investment.

PHOTOGRAPHIC X-RAY DETECTION

2.4 History and Status

The use of the photographic film for x-ray detection began with Roentgen and continues importantly today, especially in radiography and x-ray diffraction. Improvements in photographic emulsions are still being made [1]. The fine-grained emulsions now available make it possible to carry out x-ray absorptiometry on a very small scale [2]. The indispensability of photographic detection for the growth and development of x-ray emission spectrography is clear from Table 2.4–1, which records the results of a celebrated analysis by Hevesy discussed later; but the table is presented now to show the lines in an emission spectrum from a complex sample and to demonstrate the high precision attainable about 40 years ago in the photographic identification of x-ray lines.

In analytical chemistry today, x-ray diffraction is the last great stronghold of photographic detection, radiography excepted. In x-ray diffraction, it maintains its position because it provides—as most electrical methods do not—a *simultaneous record* of all significant x-ray lines from a sample (especially a powdered sample) upon proper exposure. See Chapter 6.

2.5 X-Rays and the Photographic Process

The way in which x-rays are detected photographically tells us a good deal about them. It is well known that the photographic process consists in the production of a *latent image* upon exposure of microcrystals of a silver halide (here, AgBr); that the presence of a latent image in a grain makes the grain susceptible of *development* by an organic reducing agent to produce enough metallic silver to blacken the grain; and that the developed image is subsequently *fixed* by leaching out the unchanged silver bromide.

Eggert and Noddack, having completed an investigation of the photographic process with visible (blue, 4600 Å) light [3a], did a similar investigation for x-rays (0.45 Å) on plates of one kind [3b], which was subsequently extended to plates of two other kinds [3c] with identical results. They set out to answer two questions: (1) How many atoms of silver can be recovered from a latent image formed by an x-ray quantum? About 10^3. (2) How many grains of silver bromide are affected by an x-ray quantum? One. The first question was answered by determining the silver from a plate *fixed* (but not developed) after known exposure. The second was answered by counting

Table 2.4–1. An Early Example of X-Ray Emission Spectrography with the Photographic Plate: The X-Ray Lines given by Thucolite

Element and Line		Wavelengths, XU		Element and Line		Wavelengths, XU	
		Measured	Siegbahn			Measured	Siegbahn
Th	$L\gamma6$	629.28	630.10	Th	$L\beta1$ (II)	1527.48	763.56
Th	$L\gamma1$	651.97	651.76	Cu	$K\alpha3$	1530.41	1530.70
Zr	$K\beta1$	700.31	699.98	Cu	$K\alpha1$	1537.45	1537.26
U	$L\beta1$	718.76	763.56	Cu	$K\alpha2$	1541.41	1541.16
Y	$K\beta2$	726.38	726.77	Er	$L\beta1$	1584.41	1583.44
Y	$K\beta1$	738.55	739.02	Lu	$L\alpha1$	1615.05	1615.51
Th	$L\beta3$	752.86	752.10	Dy	$L\beta2$	1620.15	1619.80
Th	$L\beta1$	763.44	763.56	Ho	$L\beta1$	1644.41	1643.52
Zr	$K\alpha1$	784.39	784.86	Y	$K\alpha1$ (II)	1653.05	827.03
Zr	$K\alpha2$	788.20	788.27	Y	$K\alpha2$ (II)	1662.00	831.21
Th	$L\beta2$	791.40	791.92	Yb	$L\alpha1$	1668.30	1667.79
Y	$K\alpha1$	827.00	827.03	Yb	$L\alpha2$	1678.89	1678.9
Y	$K\alpha2$	830.50	831.21	Dy	$L\beta1$	1707.00	1706.58
Sr	$K\alpha1$	873.00	873.28	Sm	$L\gamma1$	1722.49	1723.09
Sr	$K\alpha2$	876.30	877.45	Gd	$L\beta2$	1742.39	1741.90
U	$L\alpha1$	908.00	908.74	Sr	$K\alpha1$ (II)	1745.73	873.28
Br	K edge	917.00	917.00	Fe	$K\beta1$	1752.92	1752.72
Th	$L\alpha1$	952.05	954.05	Fe	$K\beta7$	1756.74	1756.46
Th	$L\alpha2$	966.10	965.24	Tb	$L\beta1$	1772.31	1772.68
Pb	$L\beta1$	981.04	980.83	Er	$L\alpha1$	1781.78	1780.40
W	$L\gamma3$	1060.34	1059.87	Er	$L\alpha2$	1791.28	1791.40
W	$L\gamma2$	1065.07	1065.88	Gd	$L\beta3$	1810.10	1810.90
W	$L\gamma1$	1096.98	1096.30	U	$L\alpha1$ (II)	1816.17	908.74
Th	Ll	1112.40	1112.41	Ho	$L\alpha1$	} 1842.00	1840.92
Pb	$L\alpha1$	} 1173.03	1172.58	Gd	$L\beta1$		1842.46
As	$K\alpha1$		1173.43	Sm	$L\beta7$	1851.90	1852.30
As	$K\alpha2$	1178.33	1177.40	Pr	$L\gamma2$	1875.26	1875.0
W	$L\beta2$	1242.40	1242.03	Sm	$L\beta2$	1878.98	1878.10
W	$L\beta3$	1259.03	1259.92	Ce	$L\gamma4$	1894.51	1895.20
W	$L\beta1$	1279.62	1279.17	Cu	$K\beta1$	1388.85	1389.33
W	$L\beta3$	1287.02	1287.00	Yb	$L\beta2$	1414.03	1412.80
W	$L\beta4$	1298.80	1298.79	Lu	$L\beta1$	1420.00	1420.70
Yb	$L\gamma5$	1302.45	1303.00	Zn	$K\alpha1$	1432.73	1432.06
Cu	$K\beta2$	1378.79	1378.00	Zn	$K\alpha2$	1435.04	1435.87
Yb	$L\beta3$	1449.51	1449.40	Fe	$K\alpha2$	1936.25	1936.51
W	$L\alpha1$	1473.70	1473.36	Ce	$L\gamma3$	1951.39	1950.90
W	$L\alpha2$	1484.66	1484.38	Ce	$L\gamma2$	1956.50	1955.90
Th	$L\beta2$(II)	1507.46	752.1	Tb	$L\alpha1$	1972.28	1971.49
Er	$L\beta2\beta14$	1513.76	1510.6	Sm	$L\beta1$	1993.29	1993.57

Notes. (1) This table appears in G. (von) Hevesy, *Chemical Analysis by X-Rays and its Applications*, McGraw-Hill, New York, 1932, pp. 99 and 100. (2) Wavelengths in XU 1.00202 (10^{-3}) = wavelengths in angstroms. (3) The Siegbahn values of wavelength were the best available in 1932. (4) The proper values of three wavelengths have been inserted. (5) The Roman numerals

Table 2.4–1 (continued)

Element and Line		Wavelengths, XU		Element and Line		Wavelengths, XU	
		Measured	Siegbahn			Measured	Siegbahn
Nd	Lβ2	2030.97	2031.40	Zn	Kα1 (II)	2864.16	1432.06
Gd	Lα1	} 2043.10	2041.93	W	Lα1 (II)	2949.91	1473.36
Ce	Lγ1		2044.33	W	Lα2 (II)	2968.60	1484.38
Ce	Lγ9	} 2051.27	2051.—	Th	M$_{IV}$O$_{IV}$	2999.37	2999.00
Gd	Lα2		2052.62	Cu	Kα3 (II)	3063.09	1531.00
Pr	Lβ2	2113.93	2114.80	Cu	Kα1 (II)	3076.37	1537.30
Nd	Lβ3	2121.26	2122.20	Cu	Kα2 (II)	3081.92	1541.16
La	Lγ1	2138.50	2137.20	Th	M$_V$O$_V$	3128.92	3127.00
Nd	Lβ1	2162.01	2162.21	Er	Lβ1 (II)	3168.20	1583.44
Ce	Lβ10	2192.15	2191.60	Dy	Lβ2 (II)	3242.00	1619.80
Sm	Lα1	2195.11	2195.01	W	Lγ1 (III)	3288.80	1096.3
Ce	Lβ2	2203.94	2204.1	Y	Kα1 (IV)	3304.40	827.03
Pr	Lβ1	2254.61	2253.90	Yb	Lα1 (II)	3337.14	1667.79
V	Kβ1	2279.40	2279.72	Ca	Kα1	3352.60	3351.69
Th	Lβ1 (III)	2290.30	763.	Ca	Kα2	3355.34	3354.95
La	Lβ2	2298.00	2298.00	Dy	Lβ1 (II)	3417.55	1706.58
Ce	Lβ3	2305.39	2305.90	Fe	Kβ1 (II)	3506.00	1752.72
Ce	Lβ4	2345.00	2344.20	Er	Lα1 (II)	3563.20	1780.40
Ce	Lβ1	2352.00	2351.00	Y	Kβ1 (V)	3408.80	741.00
Nd	Lα1	2365.90	2365.31	W	Lβ2 (III)	3727.50	1242.03
Nd	Lα2	2374.58	2375.63	Nd	Lγ1 (II)	3748.45	1873.83
La	Lβ3	2405.11	2405.30	Th	Lβ3 (V)	3755.60	752.1
Th	M$_V$O$_{III}$	2435.25	2437.—	W	Lβ3 (III)	3780.00	1259.92
La	Lβ1	2453.91	2453.30	Y	Kα1 (III)	2479.80	827.03
Pr	Lα1	2457.39	2457.70	W	Lβ2 (II)	2483.09	1242.03
Pr	Lα2	2467.61	2467.63	Y	Kα2 (III)	2492.60	831.21
Th	Lα1 (II)	} 1906.28	954.05	V	Kα1	2497.93	2498.35
Dy	Lα1		1904.60	V	Kα2	2502.05	2502.13
Dy	Lα2	1916.63	1915.64	W	Lβ3 (II)	2519.98	1259.92
Fe	Kα3	1923.51	1923.30	Ce	Lα1	2557.30	2556.00
Fe	Kα1	1931.88	1932.30	Ce	Lα2	2565.78	2565.11
Ru	Kα1 (IV) (??)	2570.38	651.54	Th	Lβ1 (V)	3814.00	763.00
Ru	Kα2 (IV) (??)	2584.48	645.88	Th	Lα1 (IV)	3814.92	966.00
W	Lβ4 (II)	2596.48	1279.17	W	Lβ1 (III)	3837.52	1279.17
Sr	Kα1 (III)	2618.98	873.28	Fe	Kα1 (II)	3864.00	1932.30
La	Lα1	2660.10	2659.68	Fe	Kα2 (II)	3869.30	1936.51
La	Lα2	2669.53	2668.93	U	M$_I$N$_I$	3900.02	3901.40
Cu	Kβ2 (II)	2756.48	1378.0	Th	M$_{II}$N$_{II}$	3932.08	3933.30
Cu	Kβ1 (II)	2778.88	1389.33	Sm	Lβ1 (II)	3988.50	1993.57

show the order (higher than the first) of Bragg reflection; thus ThLβ1 (III) represents a characteristic line of Th reflected in the third order. (6) This table will be needed in Chapter 10. (7) The line at 1530.41 differs from all the others in being produced by an atom, in this case of copper, with *two* missing electrons. (8) In a few cases, energy levels, not lines, are designated.

PHOTOGRAPHIC X-RAY DETECTION

the black grains in selected areas of a plate *developed and fixed* after known exposure. Table 2.5–1 compares the action of x-rays with that of visible light.

Table 2.5–1. Effects of X-Rays and Visible Light Compared [3b]

Process	Effect of One Visible Quantum, 4600 Å	Effect of One X-ray Quantum, 0.45 Å
Photolysis of AgBr	1 Ag atom produced	10^3 Ag atoms produced
Ionization[a]	1 ion pair produced	10^3 ion pairs produced
Excitation of fluorescence[b]	1 quantum visible light produced	30 quanta blue light produced

[a] See later discussion of gas-filled detectors.
[b] Datum for visible light based on Lenard's work with electrons. Calcium tungstate as phosphor for x-ray quantum.

In addition, the investigations showed that 300 absorbed quanta were needed to produce a black grain with visible light, and only *one* absorbed quantum with x-rays.

The experimental work makes clear that x-ray intensity is related to the blackening of the plate: development may be looked upon as *chemical amplification* that results from *electron transfer* to the silver ion. Although the relationship between intensity and blackening is complex in general, it has one valuable, simple feature: sometimes the photographic density measured by a photometer is identical for intensity A acting for time B and for intensity B acting for time A. This reciprocity law holds for x-rays, but not for visible light [4]. X-ray intensity has been established photographically by counting developed grains in arbitrarily chosen areas on the plate (prohibitively tedious when other means will serve); more usually, by visual comparison and photometering or microphotometering, nowadays less tedious than it was because the equipment has been improved.

An experiment by Roentgen (Figure 1.9–1) first illustrated an intensifying technique used today to increase blackening of photographic emulsion by x-ray lines. The following data (see Table 2.2–1) show why intensification is often needed: Fractions of x-ray beams (three wavelengths) absorbed by emulsion layer containing AgBr at thickness of 1 μ: 0.0002(0.1Å); 0.037(1.0) Å); 0.92(10 Å). Table 2.5–1 shows that an x-ray quantum (0.45 Å) absorbed by a calcium tungstate screen after penetrating a photographic emulsion returns to the emulsion 30 quanta of visible light. The energy of the x-ray quantum is that of 10^4 visible quanta. The process is not at all efficient under these conditions (though much better than nothing)—many of the visible quanta produced in the screen are obviously lost to the emulsion.

X-RAY INTENSITY MEASUREMENTS. ENERGY RESOLUTION

PHOTOELECTRIC DETECTORS

Photoelectric detectors are the modern way of measuring and recording what Roentgen saw when he discovered x-rays. The greatest improvement since his day is the substitution of the photomultiplier for the human eye. In the photomultiplier, visible light generated by x-rays ejects electrons from a photoelectric surface, and the name of the detectors derives from this fact.

2.6 The Photomultiplier

The human eye served as detector when Rutherford in 1908 used Sir William Crookes' sphinthariscope [5] as a scintillation detector for alpha particles. In the spinthariscope, photons or particles of high energy strike a phosphor to produce visible scintillations that are directed from a large solid angle by a magnifier to the dark-adapted eye of the observer. Observation is highly fatiguing; and, as intensity increases, the scintillations begin to merge so that individuality is lost. With the eye as detector, a practical observer might count four or five events per second for a brief time, an achievement that speaks highly for the human eye and for human endurance. A modern scintillation detector system, made possible by the advent of the photomultiplier, can count 10^5 or 10^6 times as rapidly and never feel fatigue. Such a system can handle x-rays, γ-rays, and nuclear particles.

The Radio Corporation of America first introduced the photomultiplier. A diagram of one of their early tubes (Type 931–A) appears in Figure 2.6–1, not only for historical reasons, but also because it shows clearly how photomultipliers work. Light produced by the phosphor enters as shown through the tube envelope to strike the photocathode, from which it ejects

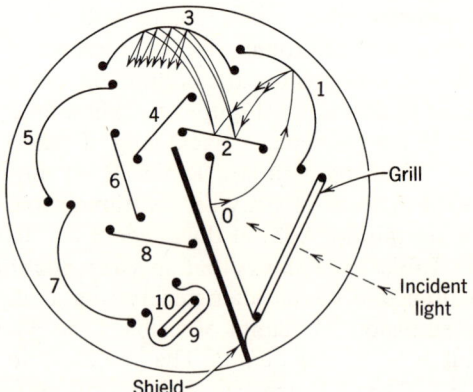

Fig. 2.6–1. Schematic diagram of No. 391-A photomultiplier. 0 = photocathode; 1–9 = dynodes; 10 = anode. Radio Corporation of America.

PHOTOELECTRIC DETECTORS

Fig. 2.6–2. RCA Photomultiplier tube type 8054. Compare with Figure 2.6–1. Above, the electrons travel along the axis of the tube. The last dynode (No. 10), which encloses the anode, is near the bottom of the tube. The dynodes have a Venetian-blind structure. By permission of RCA Electronic Components.

electrons that are focused electrostatically onto dynode 1. For each electron thus striking the dynode surface, there are produced a number of secondary electrons, which are focused electrostatically onto dynode 2. The process is described by saying there is a gain of so and so much in the first stage. The gain naturally increases with the electric field. The electrodes are called "*dy*nodes" because they perform *two* functions: that of collecting and that of multiplying the electron beam.

The multiplication process is repeated in each succeeding stage, the electrons from the (specially shaped) dynode 9 being collected by the anode 10. A multiplier phototube of this type is normally operated at 75 to 100 V/stage, and an overall gain of a million can be realized.

The response time of the multiplier phototube ($\sim 10^{-9}$ sec) is small enough for all practical counting rates and generally lower by several powers of ten than that of the phosphor used to convert the x-rays to visible light, or that of the scaling circuits. In sum, the photomultiplier has become indispensable because of its rapid response, its high gain, its small power requirements, and its low level of electronic noise.

Figure 2.6–2 is a photograph of a modern photomultiplier. Here, in contrast to Figure 2.6–1, the electrons travel along the axis of the tube. See Figure 2.8–1.

The use in vacuum of electron multipliers for the direct measurement of x-ray intensity is inherently attractive because it bypasses the conversion to visible light and eliminates the detector window. Beryllium-copper electron multipliers may be used in this way, but they have their best sensitivity for very soft x-rays (50 Å and more) and are thus of limited value in analytical chemistry.

2.7 The Phosphor-Photoelectric Detector

In the phosphor-photoelectric detector, the x-ray beam impinges upon a phosphor powder on the tube envelope. Because the phosphor layer is thin, high energy x-rays and nuclear particles will not be absorbed strongly enough for highly efficient detection. Although these detectors cannot compete with scintillation detectors (see below) on such assignments, they are historically important, simple, and still useful.

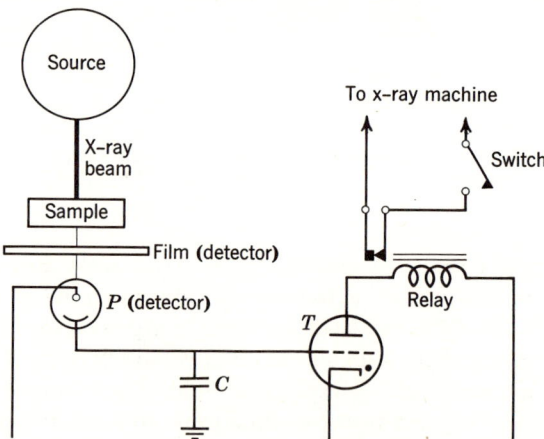

Fig. 2.7–1. Schematic diagram containing the essential components of Morgan's phototimer circuit. When enough x-ray quanta have been absorbed by the phosphor-photoelectric detector P to charge capacitor C to a predetermined voltage, the thyratron T fires and turns off the x-ray beam. In this way, the sample receives the x-ray dosage needed for a good exposure of the photographic detector (the film). After Morgan, *Am. J. Roentgenol. Radium Therapy*, **48,** 220 (1942).

Morgan [6], who had been investigating photoelectric cells for the control of roentgenographic exposures, was the first to discover the usefulness of photomultipliers in the detection of x-rays. Independently and somewhat later, Smith and Moriarty [7] made the same discovery in successfully completing a war assignment not fundamentally different from many problems in chemical control.

The phosphor-photoelectric detector is generally used with polychromatic beams the intensity of which is high enough to make the detector instantaneous. External amplification easily increases its output currents to values that can be read on a micro- or milliammeter. Output currents thus amplified could be used through servomechanisms to control operations such as blending.

In Morgan's work, this detector was made accumulative by having it charge a capacitor. The application is of interest because the charge accumulated in the capacitor was used to control the exposure time in roentgenography according to the schematic diagram of Figure 2.7-1. Here one detector is therefore used to enhance the value of another—the photographic plate.

2.8 The Scintillation Detector [8–10]

The detectors of Section 2.7 are optically inefficient and suffer further because the thin layer of powdered phosphor cannot absorb enough of the incident x-ray beam. Thickening the layer becomes self-defeating as the gain from increased x-ray absorption is largely offset by increased losses of visible light owing to scattering. Broser and Kallman [8] realized in 1947 that the need is for a large single crystal of an efficient phosphor transparent to the visible light it generates. They made from mothballs a large piece of transparent naphthalene, placed it on the window of a photomultiplier, and observed on an oscilloscope the flashes produced by the absorption of gamma rays as individual quanta. The modern scintillation detector had been born.

For x-ray detection, the decay time [8] of naphthalene $[\sim 8(10^{-9})$ sec] is satisfactory, but its mass absorption coefficient is low (Chapter 1). Thallium-activated sodium iodide—NaI(Tl)—is satisfactory on both counts. Its mass absorption coefficient is high enough so that all but the hardest x-rays will be sufficiently absorbed in a thickness of a few millimeters (Table 2.2-1). Its decay time, $250(10^{-9})$ sec [8], while longer than that of naphthalene, is shorter than those of inorganic phosphors. Most modern scintillation detectors use NaI(Tl) for x-rays.

Energy conversion within a phosphor can be complex. Pair (electron + positron) production may usually be ignored with x-rays (Section 1.13). Photoelectric absorption and Compton scattering must be considered. All that need be said here is this: An ideal phosphor transmits energy only in

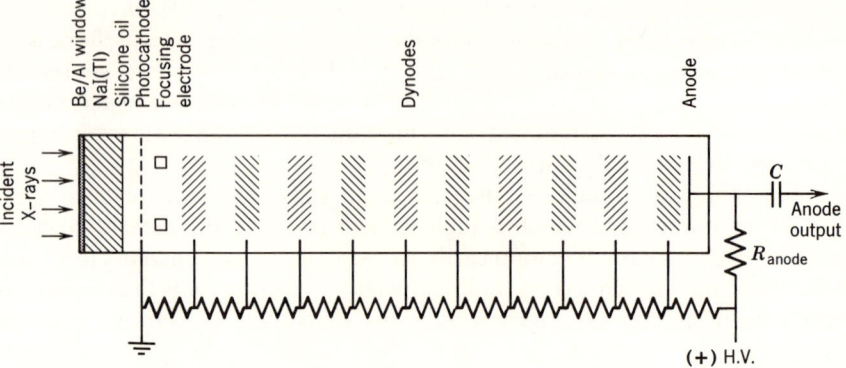

Fig. 2.8–1. Schematic diagram of a scintillation detector. The window is beryllium (0.2 mm thick) coated with aluminum (about 1 μ) to give opacity. Not shown: (1) outer glass envelope (2) glass seal between phosphor and silicone oil (3) crystalline MgO above and below phosphor to reflect visible light (4) voltage circuit for focusing electrode. As explained in Section 2.6, the phosphor converts the incident beam to visible light that generates electrons when it strikes the photocathode (Cs-Sb) in the photomultiplier. The dynodes, each acting in turn, collect and amplify the electron beam as it moves to the anode. Voltage control for the dynodes is provided within the detector assembly by the series of resistors shown. The bleeder current flowing provides a uniformly higher voltage at each successive dynode.

the form of visible light and will give the same pulse energy (pulse height) for every quantum of an incident monochromatic x-ray beam. NaI(Tl) comes close to meeting this requirement. So long as no energy is lost to the detector, multiple processes are desirable for the degradation of x-ray energy in the phosphor because efficiency then increases and statistical fluctuations are reduced. The excitation potential of NaI(Tl) being near 3 V, it produces— according to (1.6–2)—light of wavelength near 4100 Å (blue). The photomultiplier must of course respond satisfactorily to this wavelength, which it does if it has a cesium-antimony photocathode.

Energy can be lost to the detector as heat (may be ignored); as visible light (see below); as unabsorbed or scattered x-rays (negligible with a thick enough phosphor); and as characteristic x-ray lines of the phosphor (which can produce *escape peaks*). These peaks are governed by the law of conservation of energy in the form:

$$\text{Energy (incident x-ray)} - \text{energy (characteristic line)} = \text{energy (escape peak)} \quad (2.8-1)$$

With NaI(Tl), only the **K**α line of iodine is likely to escape at significant intensity.

The energy considerations outlined above and the occurrence of escape peaks are important also in other detectors. There will be further discussion.

GAS-FILLED DETECTORS

Fig. 2.8-2. Photograph of a Harshaw Integral Line Scintillation Detector. By permission of Crystal & Electronics Products Department, The Harshaw Chemical Co., Division of Kewanee Oil Company.

We repeat here that the complexity of the processes by which x-ray energy is degraded need not be considered so long as the energy of the x-ray quantum appears in the pulse. It makes no difference to us, for example, whether the ejection of a photoelectron is followed by the emission of a characteristic line or of Auger electrons. See Section 1.13.

The two principal parts of a scintillation detector, phosphor and photomultiplier, must be *optically coupled* to minimize loss of visible quanta. This coupling is accomplished, for example, by spreading a film of silicone oil between the optically similar faces of phosphor and photomultiplier. As NaI(Tl) is hygroscopic, it must be hermetically sealed. The detector window must transmit x-rays but not visible light. The device sketched in Figure 2.8-1 meets these requirements. Figure 2.8-2 is a photograph of a modern scintillation detector.

GAS-FILLED DETECTORS

Gas-filled detectors have lost ground in analytical chemistry to the scintillation detector, and they are now threatened by the lithium-doped germanium and silicon detectors, Ge(Li) and Si(Li), which have been highly successful in nuclear physics. They will nevertheless be given thorough treatment. They illustrate the interplay of x-rays and electrons, and a study of the proportional detector teaches much that promotes an understanding of the other two classes of detectors—especially in the matter of pulse height selection or energy resolution.

2.9 General Information [11–15]

Gas-filled detectors make use of the ionization, called here the initial ionization, that follows the absorption of x-rays by the filler gas. As in Section 2.8, we need not concern ourselves in detail with the processes by which the initial ionization is produced; whether the ejection of a photoelectron from a gas molecule is followed by the emission of a characteristic line that is reabsorbed or by the ejection of Auger electrons makes no difference; escape peaks must of course be considered. Under the simplest conditions, the energy of the x-ray quantum is distributed without loss among a number of ion pairs, each consisting of an electron and a relatively immobile positive ion. If the ion pairs do not recombine, the extent of initial ionization is determined by and measures the energy of the x-ray quantum.

However, the generation of positive and negative charges is not enough. Information about the absorbed x-rays will be forthcoming only if the charges are separated and collected. This requires an electric field. Such a field accelerates the (light) electrons much more effectively than it can accelerate the (heavy) positive ions. Electrons thus accelerated will collide with gas molecules, which they may ionize if the electron energy suffices. In this way, the electrons resulting from the absorption of an x-ray quantum can amplify (multiply) the initial ionization. The extent of amplification increases in a complex way with the electrical field, and it can be used to distinguish one gas-filled detector from another.

Three detectors are thus distinguishable. In the *ionization detector*, a nineteenth-century device, there is *no amplification*, and the electric field need only be high enough to keep electrons and positive ions from recombining. In the *proportional detector*, first used by Rutherford and Geiger in 1908, the amplification is *moderate*, as is the electric field. In the *Geiger detector*, developed soon thereafter and so greatly improved in 1928 by Geiger and Müller that it is often called by both names, the amplification and the field are both *high;* the field, in fact, is near the maximum that the gas can sustain without breaking down in the absence of x-rays.

The three detectors being related, we shall take the risks of oversimplification and use Figure 2.9–1 to represent them all. The figure, its caption, and Section 2.3 should be examined along with Figure 2.9–2, which shows how the change in high voltage V (taken as + 1000 V in Figure 2.9–1) affects

GAS-FILLED DETECTORS

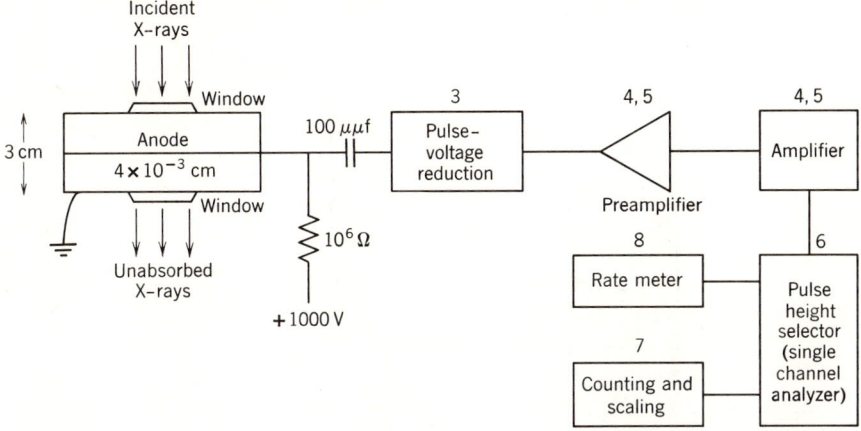

Fig. 2.9–1. Schematic diagram of gas-filled detector system. Numbers are those of relevant discussions in Section 2.3. System laid out for sealed proportional detector. Not every detector system would have all components; for example, functions 3, 4, 5, and 6 might be eliminated with Geiger detector. Reduction of pulse voltage, as by a cathode follower, is the first step in impedance matching. Ideally, the pulses passing through the system, though they may be greatly changed in shape and size, will retain the energy relationships existing among the x-ray quanta that produced them. To accomplish this there must be *linear* amplification. The matter is most important when only energy resolution is used in arriving at the intensities of characteristic lines.

the number of ion pairs collected for each ion pair in the initial ionization.

Figure 2.9–2 gains significance once we know the number of ion pairs initially produced when an x-ray quantum is absorbed. To calculate this number, we need the average energy required to produce an ion pair. Most filler gases are or contain a rare gas, so that the initial process producing ion pairs may be written

$$A_{RG} + h\nu_{x\text{-ray}} = A_{RG}^+ + e^- \qquad (2.9\text{–}1)$$

where A_{RG}^+ is a positive rare-gas ion. The average energy required in this reaction is always greater, and sometimes considerably greater, than that corresponding to the ionization potential of the rare gas. We shall take this average energy to be 30 eV, which means that 30 V is to be used along with the x-ray wavelength in (1.6–2) to calculate the number of ion pairs produced per absorbed x-ray quantum. For a quantum of wavelength 1 Å, the answer is 400 ion pairs; and for a quantum of wavelength 10 Å, it is 40.

Let us now use Figure 2.9–2 to see what might happen in the detector when 400 ion pairs are produced by the absorption of the quantum of wavelength 1 Å. For now, we shall concentrate on the electrons in these ion pairs, a justified approach because the positive ions move 1000 or more

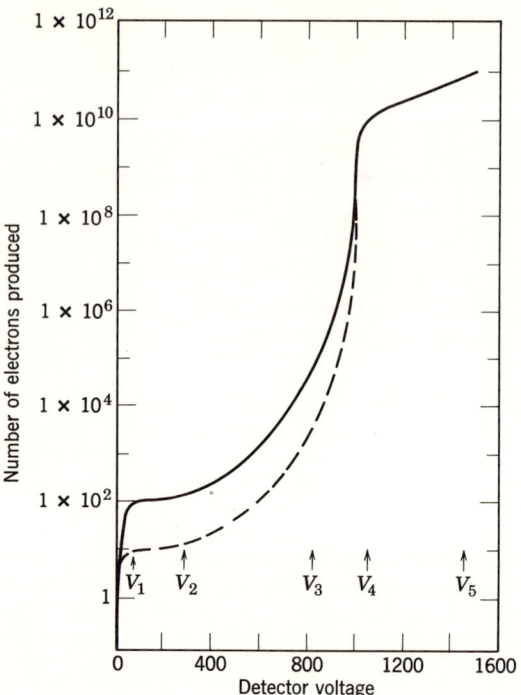

Fig. 2.9–2. Number of ion pairs collected in a gas-filled detector at different voltages when there are 40 ion pairs in the initial ionization resulting from the absorption of an x-ray quantum (broken curve). Same information for the case of 400 ion pairs (solid curve). After Wilkinson, *Ionization Chambers and Counters*, University Press, Cambridge.

times as slowly as the electrons and lead a more humdrum existence. The electrons can (1) be collected at the anode, (2) recombine with a positive ion, (3) become attached to a neutral molecule and lose heavily in mobility, and (4) with the help of the electric field, collide with one or more gas molecules to form new ion pairs (Townsend mechanism for formation of avalanches; these avalanches are the final result of ion-pair *multiplication*). Here is what will happen to our 400 electrons in the different voltage regions indicated in Figure 2.9–2:

1. Below V_1 volts. The pulse of electrons reaching the positively charged central wire will contain somewhat less than 400 electrons owing to some decrease of primary ionization through recombination, which occurs because the field is too low.

2. V_1 to V_2 volts. The ionization-detector region. The field is high enough to prevent recombination but not high enough to produce amplification. A pulse of 400 electrons reaches the wire, and the *pulse height* is propor-

GAS-FILLED DETECTORS

tional to the energy of the x-ray quantum. (The pulse height is fixed by the initial ionization, which is in turn fixed by the x-ray energy.)

3. V_2 to V_3 volts. The proportional-detector region. Here the field is high enough that, on the average, $A - 1$ electrons are produced by ionization for *each* of the 400 initial electrons. The amplification factor A increases with the field, but it is independent of the initial ionization. The pulse will contain $400\,A$ electrons. As in the ionization detector, the pulse height in the proportional detector will be proportional to x-ray energy. The value of the pulse height will be determined in part by the value of A, which can be fixed at a value lying between unity (ionization detector) and 10^4 or more.

4. V_3 to V_4. A transition region of no present importance in analytical chemistry.

5. V_4 to V_5. Geiger-detector region. The field is now so high that the enormous amplification factor (A is 10^9 or more) produces pulses whose size is independent of initial ionization, hence of x-ray energy—a single electron can trigger a catastrophe. The pulse size is fixed by the positive-ion space charge that remains near the positively charged central wire when the electrons, being much more mobile, have been swept out. The discharge spreads along the entire anode and is influenced by the visible photons produced.

6. Above V_5. The region of gaseous breakdown, where catastrophe requires no triggering because the field is higher than the gas can support. Obviously useless for purposes of x-ray detection.

What happens to the 40-electron pulse from the 10-Å x-ray in these voltage regions? The expected ratio of the relative number of electrons in the two pulses is: ideal, $R = 400/40 = 10$; below V_1, $R < 10$; V_1 to V_2, $R = 400\,A/40\,A = 10$; V_3 to V_4, $R < 10$; V_4 to V_5, $R = 1$; and above V_5, R has no meaning.

Important point: The pulse size prior to amplification (hence the pulse height for pulses properly shaped) is proportional to x-ray energy *because the average work required to produce an ion pair is constant.*

The relationship of detector voltage to number of *pulses* (not to be confused with number of *electrons* in a pulse) is of first importance. Unless conditions can be found under which the number of *pulses* counted measures the number of *x-ray quanta* absorbed over the counting interval, the detector is useless. These conditions exist if the detector has a *voltage plateau;* that is, a region within which detector voltage can vary without influencing the counts per second. Such a plateau exists for the Geiger detector in the $V_4 - V_5$ region of Figure 2.9–2 because within this region each quantum (of whatever energy) absorbed gives *one identical response.* Each quantum is counted once, and the degree of amplification does not enter the picture.

The proportional detector is a more complex case because the degree of

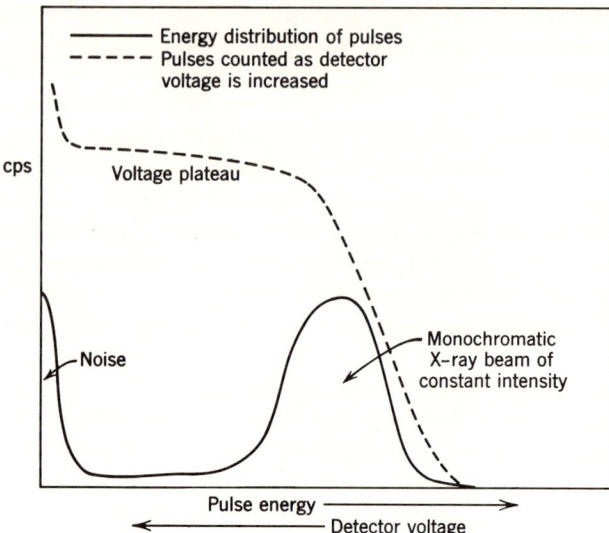

Fig. 2.9–3. Demonstration of voltage plateau for proportional detector (coordinate scales arbitrary). See text for details. Upper curve is integral of lower curve.

amplification, which depends upon detector voltage, must be considered, as must the fact that even a monochromatic x-ray beam generates detector pulses of various sizes (amplitudes); see Figure 2.20–1. That a voltage plateau is nevertheless possible follows from an experiment, the expected results of which appear in Figure 2.9–3. In a proportional-detector system, set the pulse-height discriminator so that no pulses (neither noise nor x-ray pulses) will be counted at a given detector voltage. Then increase the detector voltage. The x-ray pulses of highest amplitude will be the first to trigger the counting circuit and these will be followed by the pulses of lower amplitude produced by the monochromatic x-ray beam that we imagine is being counted at constant intensity. When *all these* x-ray quanta are being counted, the voltage plateau will have begun. It will terminate when the detector voltage has been increased enough so that pulses of much lower amplitude (noise, other x-ray, or whatever) begin to register. See Figure 2.9–3.

2.10 Gas-Filled Ionization Detectors

These detectors have been important in physics virtually since the discovery of x-rays, and they gained further importance with the advent of nuclear physics. They have not as yet become important in analytical chemistry, mainly because the electric currents they generate are so minute as to make them less convenient for precise measurements than others. They are often called ionization *chambers*.

GAS-FILLED DETECTORS

With appropriate circuitry, the ionization detector can serve as either a current-indicating (instantaneous) or a pulse-indicating (accumulative) device. In conjunction with the improved (external) amplifiers now available, the current-indicating detector deserves to be considered by the analytical chemist for measuring the intensity of strong polychromatic beams, and it is so used in the General Electric Raymike^R (Chapter 3).

Because ionization detectors are the simplest among the gas-filled types, it is worthwhile to examine how they work even though their importance in analytical chemistry is small. Let an x-ray photon 1 Å in wavelength be absorbed anywhere in the detector with the voltage between V_1 and V_2; 400 ion pairs form, and the 400 electrons reach the anode (central wire) in less than 1 μ sec. The 400 positive ions reach the cathode (outer detector wall) in about 1 msec. Under the simplest conditions (no effect of space charge—i.e., no interaction of electrons and positive ions), the voltage rise on the anode when all charged particles have been collected will be

$$dV = dq/C = (1.60)(400)(10^{-19})/10^{-12}\,C' = (1.60)(400)(10^{-7})/C' \quad (2.10-1)$$

where C' is the distributed capacity (in picofarads) of the central wire and anything attached electrically to it. The electronic charge is $1.6(10^{-19})$ coulomb, 400 is the number of electrons in the pulse, dq is the amount of electricity (coulombs) it contains, and dV the voltage increase its creation has produced.

To get an idea (again oversimplified) of how the pulse might be discharged, let us consider the part of the detector system within the broken lines in Figure 2.9–1 as a simple circuit with resistance and capacitance in series. For such a circuit, Kirchhoff's laws give

$$\varepsilon = q/C + Ri = q/C + R\,dq/dt \quad (2.10-2)$$

for the electromotive force when q is the charge on the capacitor and the current i is flowing. Integration gives

$$\ln(\varepsilon C - q) = (-t/RC) + \ln D \quad (2.10-3)$$

which for $q = 0$ at $t = 0$ gives $D = \varepsilon C$ as integration constant, and

$$q = \varepsilon C(1 - \exp - t/RC) \quad (2.10-4)$$

For the *discharge* of the capacitor, (2.10–4) is differentiated to give

$$dq/dt = i = (\varepsilon/R)\exp - t/RC \quad (2.10-5)$$

which shows that i at $t = RC$ will have decreased to $1/e$ times its maximum value. Hence RC is called the *time constant* of the circuit.

The time constant determines whether an ionization detector will be instantaneous or accumulative. If the readout is to be a *rolling average* current i, then RC must be large; say, 10 sec. If pulses are to be counted and accumulated, then RC must be small enough so that the individual pulses can be recovered from the detector output. See Figure 2.3–1.

The time constant RC is of further significance. As Figure 2.3–1 shows, the pulses are clipped (have their "tails" removed) in the preamplifier. One way of doing this, called RC clipping, uses a "clipping circuit," the time constant of which is known as the "clipping time" and is the value of RC for the circuit.

The foregoing brief excursion into electronic circuitry may give the analytical chemist unfamiliar with the field an inkling of what must go on to produce a counts-per-second (cps) reading that measures the intensity of an analytical x-ray line.

We turn now to the two other gas-filled detectors, which differ from the ionization detector in that they operate at detector voltage high enough to produce *internal amplification* by a *multiplication* of ion pairs (see Figure 2.9–2). Such amplification becomes possible because of the high voltage gradient in the neighborhood of the central (anode) wire. For example, in a cylindrical detector [of radius 1.5 cm with an anode of radius $4(10^{-3})$ cm], operated at 1000 V, electrons will drift *without multiplication* toward the anode at a velocity near 10^6 cm/sec. When they approach to within about $4(10^{-6})$ cm of the anode surface, which requires about $5(10^{-7})$ sec, the voltage gradient becomes high enough so that significant acceleration of electrons can occur between their collisions with gas molecules. Multiplication now begins. Under these conditions, collection of primary and secondary electrons by the anode can occur in less than 10^{-10} sec. The positive ions created move slowly toward the outer wall (cathode) and—owing to their sluggishness—set up a space charge that reduces detector output. Internal amplification thus produced has one great advantage: it does not contribute to electronic noise.

2.11 The Geiger (Geiger-Müller) Detector

The Geiger detector greatly accelerated the use of x-ray emission spectrography in analytical chemistry. In spite of its virtues, it is being superseded by the proportional and scintillation detectors operated separately or in tandem.

The virtues and faults of the Geiger detector stem from its enormous internal amplification ($A = 10^9$ or more), this being the highest possible without electrical breakdown of the filler gas (see Figure 2.9–2). The virtues are simple ancillary circuitry, good stability, trouble-free operation, powerful output signals (measured in volts), and sensitivity to soft x-rays. The faults need more detailed description.

The first fault is simple. The energy of the x-ray quantum does not influence the response of the Geiger detector.

The second fault is described by the terms *dead time, coincidence loss*, and *choking*, more serious with the Geiger than with other detectors. Linear

GAS-FILLED DETECTORS

detector response, obviously desirable, requires that the detector give one output pulse (and no more) for each x-ray quantum absorbed. For low intensities, when the rates of incidence of x-ray quanta upon the detector are low, the Geiger detector fulfills this condition. As this rate increases above (about) 500 cps, the number of pulses per second decreases progressively below the number of quanta absorbed per second. As in the case of the photographic plate and of photoelectric detectors, x-ray quanta may be said to lose their individuality as intensity increases. This decrease in counts per second occurs even with electronic circuits that can handle higher counting rates without appreciable losses.

The cause of this difficulty therefore resides within the detector itself. The difficulty is described by saying that the Geiger detector has a *dead time*, by which is meant the time interval after a pulse during which the detector cannot respond to a succeeding pulse. This interval, which is usually well below 0.5 msec, limits the useful maximum counting rate of the detector. The cause of the dead time is the slowness with which the positive-ion space charge leaves the central wire under the influence of the electric field. The number of positive ions increases with the amplification factor A; hence dead time is most serious with the Geiger detector, tolerable in the proportional detector (below), and negligible in the ionization detector.

The seriousness of the dead-time problem can be estimated roughly from the expression

$$(cps)_{true} = \frac{(cps)_{meas}}{1 - (cps)_{meas} \times [t_{dead}]} \quad (2.11\text{--}1)$$

Illustrative example:

$$(cps)_{true} = 1333 = \frac{1000}{1 - 1000\,[2.5(10^{-4})]} \quad (2.11\text{--}2)$$

$$\text{coincidence loss} = 1333 - 1000 = 333 \text{ cps} \quad (2.11\text{--}3)$$

that is, the *coincidence loss* is one-fourth of $(cps)_{true}$. As the coincidence loss increases, the Geiger detector eventually becomes inoperative and is said to *choke*. To summarize: the Geiger detector is on the way out in analytical chemistry because it gags on strong x-ray beams and is blind to energy differences among x-ray quanta.

The noble gases make good detector fillers because they are stable and the heavier ones (e.g., argon) have the high mass absorption coefficients desirable for efficient x-ray absorption. However, they break down easily at high detector voltage. Their excited atoms have long lifetimes. Emission of secondary electrons or photons from the cathode upon positive-ion bombardment also contributes to easy breakdown. The addition of a few per cent of a quenching gas (methane, butane, alcohol, or halogen) to the rare gas (argon, krypton, or xenon) improves Geiger-detector operation. The life of such a sealed detector is limited to about 10^9 total counts owing to decomposition of the quenching gas. Filler gases for proportional detectors (below) must contain no gas that forms negative ions (as do the halogens) by electron attachment, for energy resolution and detector efficiency suffer if this process occurs.

The detector in Figure 2.9–1 has side windows instead of the single end-window construction often used to enhance the absorption of the incident x-ray beam. The construction in Figure 2.9–1 allows x-rays unabsorbed by the gas to escape from the detector, thus eliminates the risk that such x-rays will disturb detector operation as by producing electrons and x-rays upon being absorbed by the walls. This gain must be balanced against the loss in detector output resulting from reduced x-ray absorption owing to a shorter gas path.

2.12 The Proportional Detector

The proportional detector strikes a happy mean between the other two gas-filled detectors. Internal amplification can be sufficient at constant A so that the possibility of energy resolution is not destroyed, yet moderate enough that dead time is negligible, and adequate counting rates (cps $\simeq 10^5$) are possible. Mixed methane and argon make a good filler gas. In contrast to the Geiger detector, the filler gas for a proportional detector must contain no gas that forms negative ions (as do the halogens) by electron attachment. If such gases are present, energy resolution becomes unreliable. Owing to the reduced destruction of the quenching gases, usually methane, the useful life of proportional detectors can be many times ($\sim 10^3$) that of the Geiger.

Obviously, if A (or what is the same thing, $A - 1$) ion-pairs are formed by collision for each electron that reaches the anode in a proportional detector, we have in place of (2.10–1)

$$dV = 1.60\,(10^{-7})\,An/C' \qquad (2.12\text{–}1)$$

for n electrons produced in the initial ionization corresponding to the absorption of one x-ray quantum; or, with illustrative numerical values for the 1-Å quantum inserted:

$$dV = (1.60)(10^{-7})(10^3)(400/100) = 6.4(10^{-4})\ \text{V} \qquad (2.12\text{–}2)$$

With a resistance of 1 MΩ, the circuit would have a time constant of $10^6 \times 100(10^{-12})$, or 10^{-4} sec. During an assumed time of collection of 10^{-10} sec, the average current would be $4(10^5) \times 1.6(10^{-19})/10^{-10}$, or $6.4(10^{-4})$ A. For the 10-Å quantum, the calculations would be valid with 40 electrons in

GAS-FILLED DETECTORS

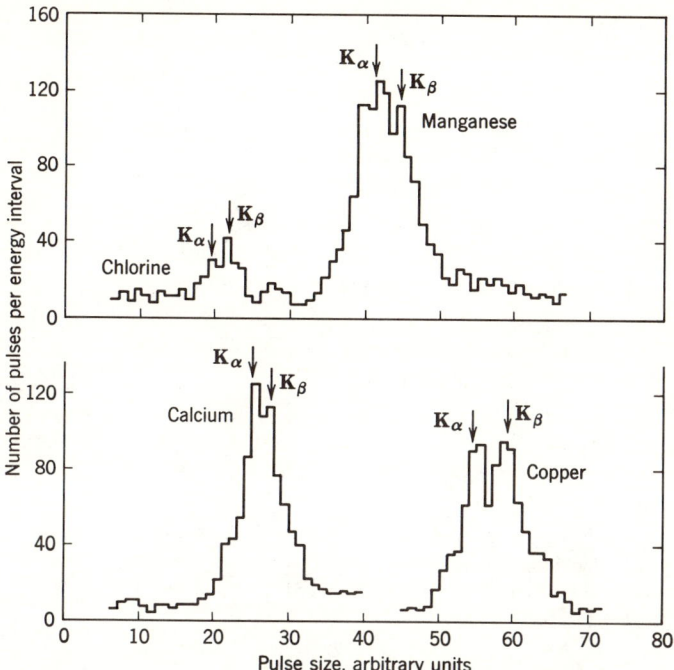

Fig. 2.12–1. Histograms for Kα and Kβ lines of chlorine and manganese (from manganese chloride), calcium, and copper. The abscissa may also be read as either *pulse height* or *pulse energy* or *x-ray-quantum energy*. See insert, Figure 2.12–2. After Curran, Angus, and Cockroft, Reference 16, Fig. 4.

place of 400. A smaller capacity increases dV and lowers RC, other things the same.

We owe the modern proportional detector to Curran, Angus, and Cockroft, who dealt with it in a series of important papers, in one of which [16] they say: "In the present papers it will be shown that proportional tubes used in conjunction with a strictly linear amplifier of high gain can be applied to the study of many types of radiation with considerable advantage over other techniques." As the uniformity of ion-pair multiplication, which is vital to proportionality, depends mainly on the condition of the electrical field near the anode, it is important that the central wire be very smooth, of uniform diameter, carefully aligned along the detector axis, and arranged to minimize end-effects.

The proportionality of their detector was established simply and directly on x-rays of known energies between 2.6 and 40 keV (kiloelectron volts). A weak polychromatic x-ray beam was used to excite the characteristic x-rays of thin foil samples in air, and these lines entered the detector at right

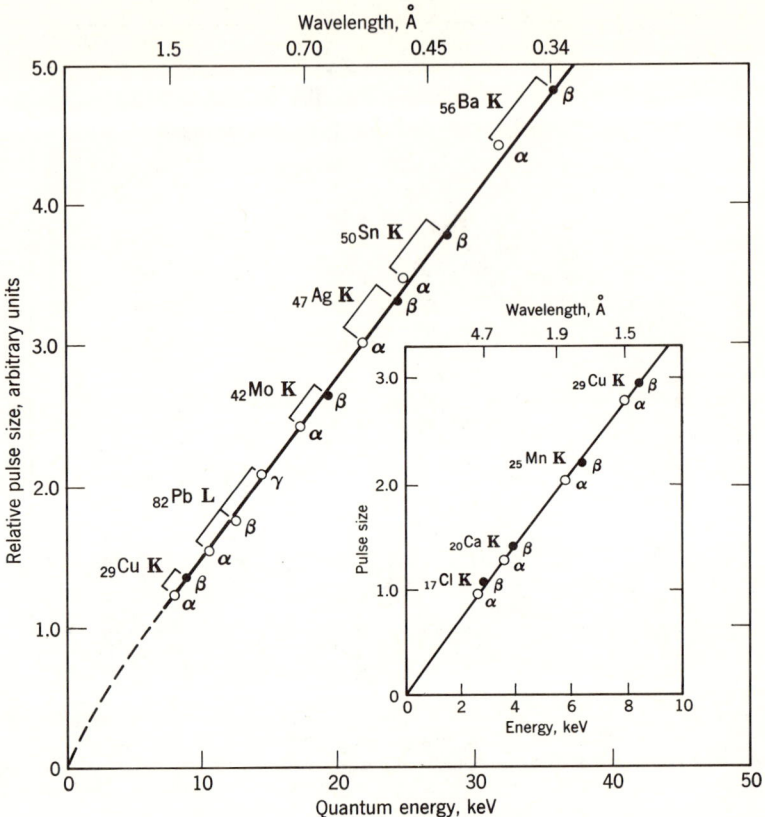

Fig. 2.12–2. Proportionality of pulse size and x-ray-quantum energy of characteristic lines for various elements. The filler gas in the proportional detector was argon (partial pressure, 60 cm Hg) and methane (partial pressure, 15 cm Hg). After Curran, Angus, and Cockroft, Reference 16, Fig. 5.

angles to the beam direction. The pulse output of the detector system was made visible on an oscilloscope and recorded on moving photographic film so that each pulse appeared as a vertical line, the height of which was proportional to pulse size. With the film moving at 2 in./sec, cps values up to about 160 could be recorded. The film was analyzed by projecting it upon a screen subdivided into 110 intervals each $\frac{1}{2}$ cm high. Histograms were then made of the results: these histograms recorded the number of pulses per energy interval (ordinate) as a function of pulse size (or pulse height, or pulse energy). It is important to remember that energy resolution alone was relied upon; there was no diffracting crystal. Detailed results for four elements are shown in Figures 2.12–1 and 2.12–2 as evidence that the following are true:

1. Remarkably good proportionality between x-ray-quantum energy and pulse size (or height) has been achieved.

2. Energy resolution unaided by Bragg resolution is worth investigating for analytical chemistry. Note that **K**α and **K**β lines of the same element, which differ but little in wavelength, gave surprisingly distinct peaks.

3. A *monochromatic* x-ray line will give pulses *varying rather widely in size*. See Figures 2.9–3 and 2.20–1.

Figure 2.12–3 shows the *identification* (not an intensity measurement, which is more difficult) of an x-ray line via unaided energy resolution. Copper has a radioactive isotope, $_{29}Cu^{64}$, that undergoes **K**-capture, being changed to $_{28}Ni^{64}$ and radiating the **K** lines of Ni. (A further discussion of **K**-capture appears in Section 3.2 and is followed by an example in which one isotope, $_{26}Fe^{55}$, of this important type is used as an x-ray source for analytical work.) The decay scheme for $_{29}Cu^{64}$ is

$$_{29}Cu^{64} \begin{cases} \rightarrow {_{28}Ni^{64}} + e^+ \\ \rightarrow {_{30}Zn^{64}} + e^- \\ \rightarrow {_{28}Ni^{64}} + \textbf{K-capture} \end{cases}$$

A sample of the isotope was covered with polyethylene thick enough to stop all electrons and positrons, which left only the **K** lines of nickel, some γ-rays, and the background to enter the proportional detector. Figure 2.12–3 shows that the expected was exactly realized: in the histogram from $_{29}Cu^{64}$, the **K** lines of nickel appeared above an enhanced background. The resolving power of the proportional detector is great enough to separate the **K** lines of copper and nickel, elements with neighboring atomic numbers, for purposes of identification—not necessarily for purposes of intensity measurement, which is complicated by the background.

Results similar to those in Figure 2.12–2 were obtained by Friedman, Birks, and Brooks [17] on equipment better adapted to analytical chemistry; these are shown in Figure 2.12–4.

Proportional detectors have proved most valuable in x-ray emission spectrography as an adjunct to Bragg resolution. They are useful in three ways: (1) as simple detectors, in the manner of Geiger detectors, especially for high counting rates, (2) with a pulse-height analyzer, as an adequate means of resolving an x-ray beam when the wavelengths of interest have no close neighbors, and (3) with a pulse-height selector or discriminator to obtain reduced background.

Escape peaks require special consideration when a proportional detector is used with a pulse-height selector. There are two reasons for this. Owing to the lower absorbance shown by proportional detectors with their gas filling, escape peaks are far more likely than they are with the solid scintilla-

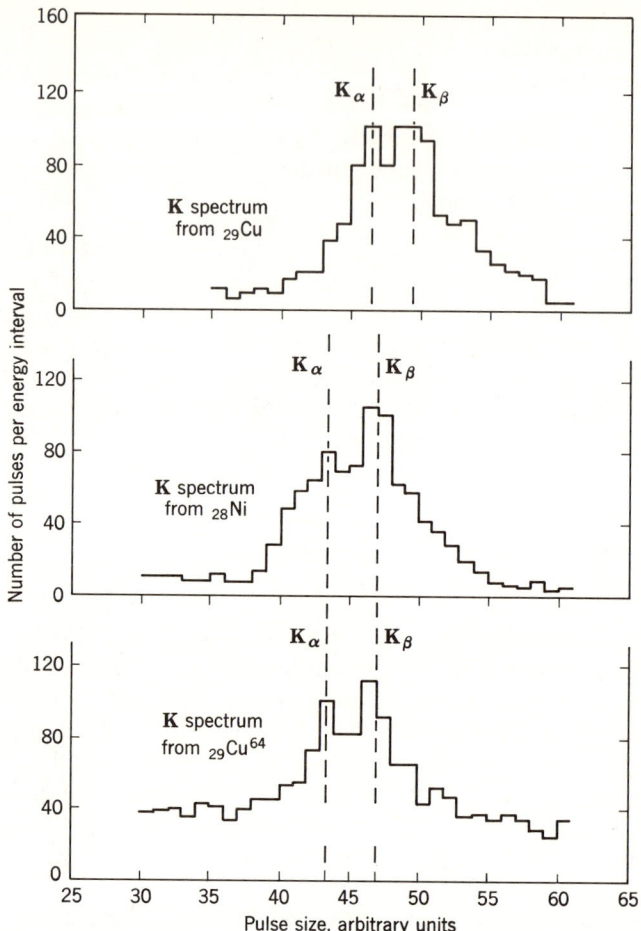

Fig. 2.12–3. Identification of characteristic x-ray lines by unaided energy resolution with a proportional detector. The lines are produced when $_{29}Cu^{64}$ undergoes **K**-capture and they must therefore be $_{28}Ni^{64}$ (**K**α and **K**β), which are identical with those of ordinary $_{28}Ni$. The figure shows that they were positively identified as such by use of a proportional detector. After Curran, Angus, and Cockroft, Reference 16, Fig. 9.

tion detectors. And with a pulse-height selector in the system, the escape peaks will go uncounted if they fail to pass the selector window. When the intensity of an analytical line is being measured, (2.8–1) may be written:

Energy (analytical line) − energy (characteristic line of filler gas) =
energy (escape peak) (2.12–3)

It must be remembered that the characteristic line of the filler gas escapes

GAS-FILLED DETECTORS

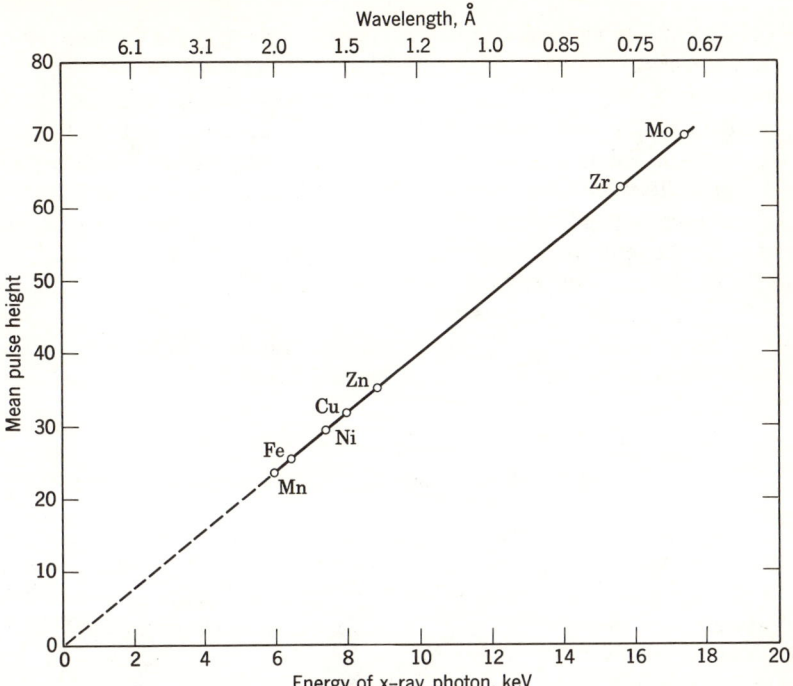

Fig. 2.12–4. Mean pulse height versus quantum energy, to illustrate pulse-height selection of characteristic lines. Side window tube, 3-in. diameter; 0.005-in. wire operating at 1275 V; filled with argon plus 10% ethylene; total pressure, 15 cm of mercury (proportional detector); gain, 3.5×10^4. After Friedman, Birks, and Brooks, *ASTM Spec. Tech. Publ.*, No. 157, page 3. Copyright 1954. American Society for Testing Materials.

the detector—the peak itself remains behind. The intensity at which the characteristic line escapes is difficult to calculate; but the following guides, based on Chapter 1, are useful. (1) Only wavelengths shorter than the critical absorption wavelength of the characteristic line can excite it. (2) The characteristic-photon yield (Section 1.23) of the filler gas is important; neon, with a yield below 0.1 for the K lines, is less likely (other things equal) to have its characteristic line leave the detector than is xenon, for which the same yield exceeds 0.7. (3) The mass-absorption coefficient for the analytical line and that for the filler-gas characteristic-line are both important; the higher they are, the less likely is the escape of the characteristic line. (4) The wavelength of the escape peak may be calculated according to the energy relationship given above. (5) The longer the path length in the detector for the analytical line, the lower (other things equal) will be the intensity of the escaping characteristic line, hence of the escape peak; in this connection, note the short path length in the detector in Figure 2.9–1. (6) As the wavelength of the analytical line decreases (energy increases), the relative separation of the escape peak and of the peak due to the analytical line decreases. Escape peaks at long wavelengths will merge with electronic noise. Escape peaks at short wavelengths will be near enough the peak for the analytical line so that both will be counted. To summarize: The problem of escape peaks is less serious than might appear at first sight. When it exists, it can be dealt with by changing experimental conditions, or it can be allowed for by calibration of the emission spectrograph against a suitable standard. An example of escape peaks will be given later in this chapter. Geiger detectors of course have no escape peaks.

A final caution. In work of high precision with proportional detectors, the dead-time correction should always be estimated according to (2.11–1) and applied when appropriate. See Section 2.25.

2.13 Flow Proportional Detectors

Sealed proportional detectors, the only kind so far discussed, are not adequate for the x-ray emission spectrography of light elements because the absorbance of their windows for the long-wavelength analytical lines is too high. As such emission spectrography grows in importance, *flow* proportional detectors [18] increasingly supplement the sealed, and may eventually displace them.

The flow detectors have thin and ultrathin windows, usually supported

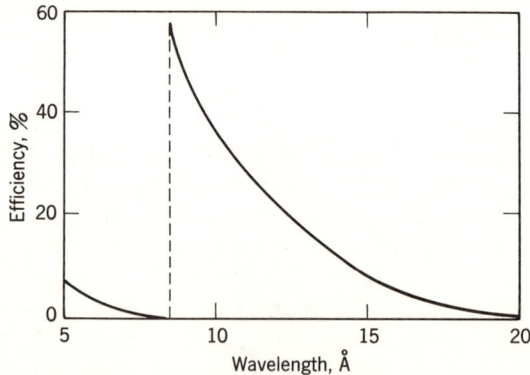

Fig. 2.13–1. The combined detector efficiency of a flow proportional detector in the determination of Al, Mg, and Na (wavelengths up to 12 Å). The detector is filled with P-10 (90% Ar–10% CH_4) gas at atmospheric pressure, has an internal diameter of 2 cm, and an aluminum window, 6 microns thick. The combined efficiency is the product, expressed as per cent, of the fraction of x-ray intensity transmitted by the window and the fraction absorbed by the P-10, regarded as pure argon. Note the effect to the **K** absorption edge of aluminum near 8 Å. After Henke, Ref. 20*a*, p. 372, Fig. 10.

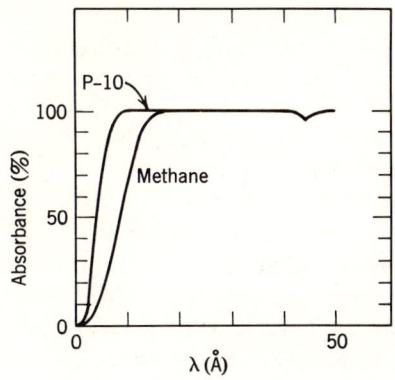

Fig. 2.13–2. Absorption efficiency of two gases in a flow proportional detector that gave a peak counting rate of 6164 cps for carbon $K\alpha$ (wavelength near 45 Å). Detector diameter, 2 cm; gas pressure, 1 atm. The curves show that the filler gas can be used to reject unwanted x-rays by not absorbing them: methane transmits the shorter wavelengths more readily than P-10. The ultrathin window was made by depositing onto a smooth, 200-mesh electroformed grid of nickel (transmittance, 70%) two layers of Parlodion (each about 0.1 μ thick) from a water surface. After Henke, Ref. 20b, p. 474, Fig. 10.

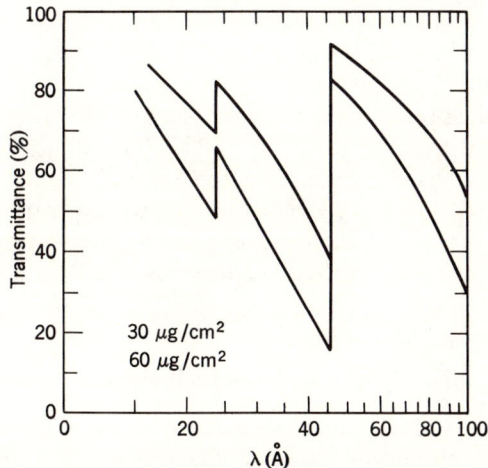

Fig. 2.13–3. Transmittance of Formvar films for flow-proportional-detector windows at two thicknesses (specific weights shown in figure). The heavier film was mounted on a 200-mesh nickel screen of 70% transmittance. As assembled in the vacuum spectrograph, film and screen were pressed against the collimator-blade supporting-system. Perhaps for this reason, an electrically conducting coating on the film was not required. Note the absorption edges of oxygen (near 20 Å) and of carbon (near 45 Å). Composition of Formvar (wt %): O, 32; C, 60; H, 8. Calculations outlined in Chapter 1 will show that this composition gives desirable mass absorption coefficients over a good wavelength range. After Henke, Ref. 20c, p. 280. Fig. 12.

on a metal screen and usually coated (if insulating) with a thin conducting layer (aluminum or carbon) on the side facing the anode, a measure taken to ensure uniformity of the electric field. Ultrathin windows require careful preparation [19]. Whether thin or ultrathin, satisfactory detector operation with such windows requires continuous flow of the filler gas through the detector, the name of which is thus explained. Argon containing 10% methane (the so-called P-10 gas) is a favorite filler. At the flow rates sometimes used (~ 0.5 ft^3/hour), the cost of the gas must be considered.

The art of making thin windows has advanced so rapidly that window thicknesses are now sometimes quoted in angstroms (Å). To illustrate this, we present Figures 2.13–1, –2, and –3 from the notable work of Henke and his colleagues [20a, b, c]. Henke [20d] summarizes the present position by saying that Formvar windows are generally prepared for work in the 10–100-Å region while polypropylene is used when all significant wavelengths exceed the critical absorption wavelength of the carbon K edge. How the location of absorption edges affects the window problem is clear from the figures, where the preference for Formvar, a condensation polymer, is also explained.

In the General Electric emission spectrograph [21a] for ultrasoft x-rays, it has been found that supported Formvar detector windows about 0.1-μ thick could contain gas at pressures up to 15 torr [21b], and that thin (probably near 1-μ thick) polypropylene windows with a transmittance of 77% for C Kα mounted on a 100-mesh nickel grid withstood a pressure differential of one atmosphere and showed no leakage when the spectrograph was evacuated to $5(10^{-6})$ torr [21a].

2.14 Effects of Changes in the Filler Gas

Except when it is used merely as a rapid detector without regard to pulse size—say, under conditions when a Geiger detector might choke—the usefulness of proportional detectors depends on their giving, as nearly as the statistical nature of the multiplication process permits, the same number of electrons in each pulse and with that number proportional to the x-ray-quantum energy. This matter was emphasized in Section 2.12, but without a discussion of the important effects that changes in the filler gas can have on the multiplication process.

Multiplication depends on the interaction of the electrons with the molecules of the filler gas, and on the length of the mean free path for the electrons. It will be remembered that electrons can attach themselves to molecules, and that halogens added to Geiger detectors suppress avalanches by this mechanism. Nothing of this sort is permissible in a proportional detector. Even the interactions that stop short of attachment must be under control. This means that the filler gas must not change composition, either

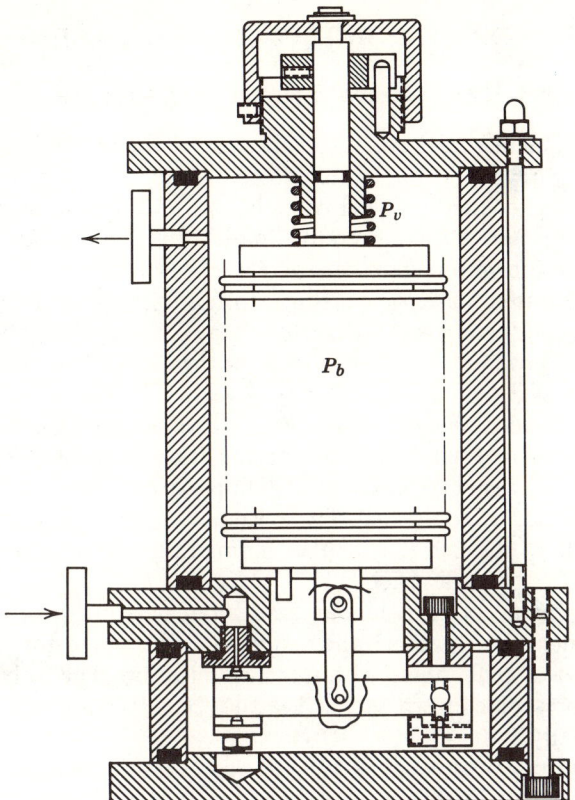

Fig. 2.14–1. Control valve for Philips gas-density stabilizer, PW 1548. The sealed bellows, of practically constant volume, is filled with dry gas to pressure P_b. A lever (bottom of figure) and a conical needle valve (lower left) mounted thereon enable the bellows to maintain $P_v = P_b$ by controlling the flow of incoming gas. A temperature increase calls for greater flow; a pressure increase for less. The inlet of the flow proportional detector is connected to the regulating valve. Gas discharge to the atmosphere is through a fixed orifice. Courtesy Philips Central Service.

by the production of impurities, or by a change in the ratio of its components (e.g., Ar and CH_4) that arises, for example, because the gas storage cylinder is near depletion. Owing to the continuing gas flow, the problem of generated impurities does not exist for a flow proportional detector properly operated.

The mean free path for electrons between collisions with filler molecules limits the velocity of the electrons accelerated by the field, hence determines their energy. This mean free path decreases with increasing gas density. Density in turn increases with pressure and decreases with absolute temperature. A regulating device is needed, such as the gas compensator developed for Philips PW automated spectrographs. See Figure 2.14–1.

LITHIUM-DRIFTED DETECTORS: Si(Li) AND Ge(Li)

2.15 General Information [22–24]

All gas-filled detectors have their drawbacks: in ionization detectors, the pulses are too weak; proportional detectors are too sensitive to impurities and to voltage changes; Geiger detectors are unfit for energy resolution. McKay in 1949 seems to have been the first [25] to think of "the solid state equivalent of an ionization chamber [detector]" as a possibly better device; silicon and germanium are logical solids to use in place of the filler gases.

Solid photoconducting x-ray detectors were known. Suitably activated cadmium sulfide passes very little current in the dark but becomes a good conductor when x-rays strike it. X-ray detectors of this kind are small, simple, instantaneous, rugged, cheap, and serve as warning devices; but they cannot be made reliable enough for analytical chemistry. The reasons are twofold: their crystal lattices are full of imperfections, and their composition cannot be closely controlled. The existence of the huge semiconductor industry attests to the fact that neither germanium nor silicon need suffer seriously from either drawback.

As happens at every turn in our field, nomenclature poses problems. "Gas-filled ionization detector" and "solid ionization detector" are logical names for devices that ionize without multiplication. The only two solid detectors of present concern are Si(Li) and Ge(Li), familiarly known as "Silly" and "Jelly."

To be an improvement over gas-filled detectors, a solid ionization detector would, to begin with, have to:

1. Show adequate x-ray absorption.
2. Produce for each quantum absorbed, a number of ion pairs adequate for satisfactory counting after amplification. There is no internal multiplication in these solid detectors as normally used; but, see Section 2.17.
3. Have this number of ion pairs proportional to the energy of the absorbed x-ray quantum over a wide energy range.
4. Maintain an internal electric field of gradient suitable for charge collection without recombination.

The first requirement demands that mass absorption coefficient be sufficiently high and "sensitive volume" (the volume within which x-ray absorption will give ion pairs) be sufficiently large; a thin film or boundary layer will not do. The second requirement can be met only if the energy needed to produce an ion pair is considerably less than 30 eV. With respect to the third requirement, the solid ionization detector should be competitive with the proportional detector. The fourth requirement bars substances showing metallic conductivity.

LITHIUM-DRIFTED DETECTORS

What of insulators? They conduct very poorly. They have many deep traps for electrical charges; an electric field through them will not have a uniform gradient; and the space charges that build up within them leak away but slowly with time. They must hence be dismissed from consideration. Semiconductors alone remain.

2.16 Silicon and Germanium

Among semiconductors, only silicon and germanium can approach on a practical scale our rigorous requirement of being a nearly perfect single crystal pure enough for the satisfactory control of composition by "doping" as needed.

Both elements have four electrons in the valence band. They differ (Section 4.7) as regards the filled shells between these valence electrons and the atomic nucleus, but this difference is unimportant for present purposes. Electrons in the conduction band, just above the valence band, are free to move; electrons in the valence band, being identified with particular atoms, are more tightly bound. In silicon, the energy difference ("band gap") between conduction and valence bands is 1.1 eV; in germanium, it is only 0.66 eV. This means that, at a given temperature, far fewer electrons will be in the conduction band of silicon owing to thermal excitation than even the small number in the conduction band of germanium. The need to reduce this number by cooling below room temperature (in order to suppress electronic noise) is of course greater for germanium. The band gaps are analogous to the ionization energy in a gas.

Impurities that donate electrons to the *conduction* band can produce an n-type region in a semiconductor; impurities that leave a deficiency of electrons in the *valence* band can produce a p-type region. (Here "n" means negative; "p" means positive.) There is normally a potential difference between an n-type and a neighboring p-type region in the same semiconductor. Antimony (5 valence electrons) in silicon or germanium (4 valence electrons each) will tend to form n-type regions; boron (3 valence electrons) will tend to form p-type. An impurity such as lithium (one valence electron loosely held) will always tend to produce an n-type region: it is not built into the crystal lattice as are boron and antimony.

In semiconductors, the hole-electron ($\oplus - e^-$) pair is the analogue of the ion pair in gases; both can be produced in the same way. But, pair production in silicon needs on the average only 3.5 eV; and even less, or only 2.94 eV, in germanium. Unlike positive ions in gases, the holes have mobilities approaching those of electrons. In germanium at 78°K, the mobility of holes is $1.5(10^4)$ (cm/sec) for a gradient of 1 V/cm; the mobility of electrons is only three times greater.

A difference in the conduction mechanism accounts for the difference just mentioned between gases and semiconductors. In an n-type region, the current is naturally carried almost entirely by electrons (the *majority* carriers) as they are free to move in the conduction band. Any positive holes (the *minority* carriers) exist in the valence band below. In a p-type region, any free electrons are drawn into the valence band by these positive holes. When an electron in such a valence band jumps from one atom to its neighbor, the process is logically regarded as the movement of a positively charged hole in the opposite direction, which makes these holes the *majority*, and the electrons the *minority* carriers of current in p-type regions. This movement of holes is far faster than the movement of positively charged ions in gases at appreciable pressures.

The properties of germanium and silicon just described make them good choices for solid x-ray detectors. In 1969, these materials could not be made into sufficiently pure, "perfect" single crystals large enough for such detectors. Hence it is necessary to examine the "lithium-drifting" process due to Pell [26] by which this difficulty has been overcome.

2.17 Si(Li) and Ge(Li)

Lithium-drifting is done as follows. A slab of p-type Si or Ge, millimeters thick and with resistivity above a few hundred ohm-cm, is coated on one side (the front) with a thin layer of Li, which is allowed to diffuse inward at a temperature high enough to form a junction, an n-p junction, in the semiconductor. A potential difference ("reverse bias") is next applied across the junction to make the lithium ion penetrate further inward; under "forward bias," an n-type semiconductor would send its majority carriers (electrons) inward. As this penetration proceeds, all charge carriers are swept out of the region thus penetrated, and the resistivity rises until it reaches that of the pure semiconductor (attains its "intrinsic value"). What has happened is that the original n^+-p junction ("+" means heavy with charge carriers) has been changed to an n^+-i-p junction ("i" means semiconductor of intrinsic resistivity, just explained). The "i" region is the sensitive region of the detector; see below.

The analytical chemist may prefer to look at the process in an electrochemical way, as in Figure 2.17–1. In process III, we have an electrochemical cell that has created its own electrolyte (the intrinsic zone) and swept all highly mobile charge carriers (mainly ⊕ 's) out of this electrolyte. The driving force for this process is an impressed electromotive force identical in sign with that of the cell; that is to say, the cell is undergoing forced discharge. Figure 2.17–2 shows schematically how the device operates as an x-ray detector. Figures 2.17–1 and 2.17–2 are intended for those unfamiliar with the language of solid state physics, an activity that in principle has much in common with electrochemistry. Figure 2.17–3 shows a con-

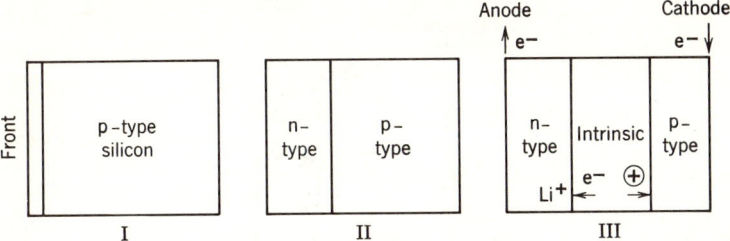

Fig. 2.17–1. Electrochemical visualization of lithium-drifting in silicon. I. Lithium film on front of block of p-type silicon produced under Pell's conditions: 450°C for 45 minutes; painting-on of Li-in-oil suspension [26]. II. Thermal diffusion has created n^+-p junction in block. Done by keeping temperature at 450°C for 5 minutes [26]. III. Applied potential difference with electrodes as shown has created intrinsic zone between n and p regions and made an n^+-i-p diode. Done at 125°C and 24 V [26].

Anode reaction: $\text{Li} = \text{Li}^+ + e^-$ (oxidation of Li).
Cathode reaction: $e^- + \oplus$ = annihilation of hole (reduction of $\oplus$).
Sum $\text{Li} + \oplus = \text{Li}^+$.

Migration of charges: 1. Electrons leave anode, pass through external circuit, meet positive holes at cathode. 2. Positive holes migrate toward cathode. To preserve electrical neutrality in middle zone, each $\oplus$ must be replaced by an Li^+. 3. The forward boundary of the Li^+ migration should be uniform and virtually no $\oplus$'s, which are highly mobile, will exist behind this boundary. The migration will stop when the high (intrinsic) resistance of the middle zone exhausts the driving force (the potential difference). 4. The high mobility of the $\oplus$'s and the requirement that electrical neutrality be maintained throughout the "electrochemical cell" thus ensure that all the $\oplus$'s will be swept out of the middle zone and be replaced by an equal number of Li^+ ions.

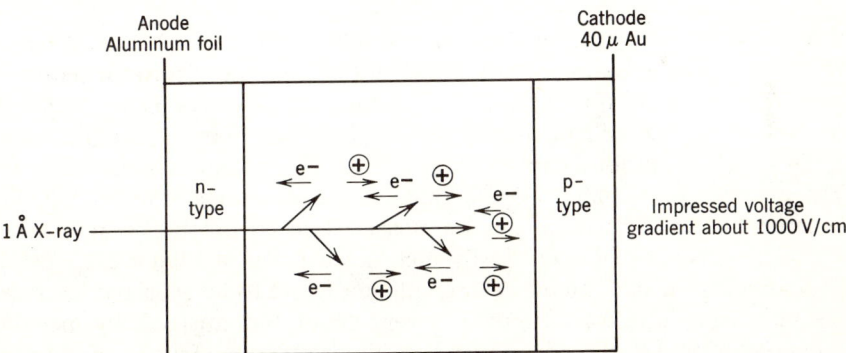

Fig. 2.17–2. Visualized operation of Si(Li) x-ray detector. An x-ray quantum of 1 Å wavelength strikes the sensitive zone (also called here the intrinsic zone and the electrolyte) and creates about $12{,}400/3.5 = 3{,}540$ hole-electron pairs. These highly mobile charge-carriers move rapidly to the electrodes and are collected before recombination can occur. For comparison, note that the same x-ray will produce about 400 ion-pairs in a gas-filled ionization detector. Figures 2.17–1 and 2.17–2 are highly simplified. Note the difference in experimental conditions here and in Figure 2.17–1 (III).

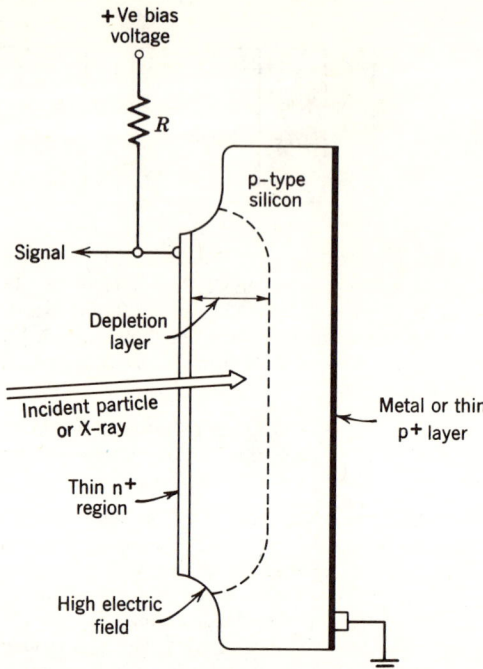

Fig. 2.17–3. Conventional representation of an n^+-p^+ junction solid ionization detector. "Depletion layer" is yet another name for what has been called "sensitive zone," "intrinsic zone," and "electrolyte" herein. After W. M. Gibson, et al., Ref. 23, p. 346.

ventional representation of a detector similar to the one of interest here.

Now, a quick appraisal of Si(Li) and Ge(Li) in terms of the four requirements named in Section 2.15. (1) The absorption of x-rays in unit volume of these materials will obviously be much greater than in any filler gas under usable conditions, and of course greater in germanium than in silicon. The problem of getting a large enough sensitive volume is more difficult because lithium-drifting to depths measured in centimeters is not easy. Progress is being made. (2) Easily met; see caption of Figure 2.17–2. (3) Probably can be met, but electronic difficulties are to be expected because the pulses for long wavelengths are very small. (4) Satisfactorily met by potential difference applied across n^+-p^+ diode separated by semiconductor of intrinsic resistance. Charge carrier lifetime is long enough ($\sim 10^{-3}$ sec) and charge-collection time is short enough ($\sim 10^{-7}$ sec) to preserve individuality of pulses at fairly high counting rates.

The detectors are best operated near 77°K, the temperature of a liquid nitrogen cryostat. A continuous low temperature is advisable to prevent Li redistribution and to reduce conductivity and hence background during

LITHIUM-DRIFTED DETECTORS

operation. The need for cooling is greater for Ge(Li) because of its lower band gap (0.66 eV).

At this writing (1969), Si(Li) seems to be preferred over Ge(Li) as a solid x-ray detector. The chief reason appears to be not the higher band gap of Si(Li) just mentioned, but the facts that the composition of Si(Li) can be better controlled, and that its crystal lattice approaches closer to perfection; it has fewer charge carriers at a given temperature. For very hard x-rays (say, of wavelengths below 0.4 Å), Ge(Li) is almost mandatory because of its higher mass absorption coefficient.

"Avalanche detectors" are not yet satisfactory enough to warrant more than brief mention. They are *solid proportional detectors*—solid-state devices closely related to solid ionization detectors, but capable of electron multiplication that yields internal amplification when the voltage gradient is sufficiently high. At present, the electron multiplication is not uniform across the face of the crystal, and this weakness will have to be corrected before we can expect the detectors to be useful in precise quantitative work. Their outstanding virtues are that they can be very small, and that they are capable of high counting rates with negligible coincidence losses. They may some day replace gas-filled proportional detectors.

2.18 Problems of Solid Ionization Detectors

Consider that a solid ionization detector may have a leakage current of 10^{-9} A at 77°K; this current means a flow of $0.6(10^{10})$ electrons/sec, the electronic charge being $1.6(10^{-19})$ C. Suppose the 3540 electrons produced by the absorption of a 1-Å x-ray (caption, Figure 2.17–2) are collected in 10^{-6} sec: this corresponds to a flow of $0.35(10^{10})$ electrons/sec. Clearly, the signal-to-noise ratio will decrease with increasing x-ray wavelength, for the noise is independent of wavelength. Also, capacitance is needed in detector, preamplifier, and connectors to minimize noise; this condition is not always easy to meet. Detailed treatment of electronics problems is outside the scope of this book.

We must, however, mention the field-effect transistor (FET) [27], without which Si(Li) and Ge(Li) detectors would probably have been doomed for extensive application as x-ray detectors in analytical chemistry. In these transistors, *majority* charge carriers (usually electrons) are swept out of a region in a semiconductor by an electric field; as they are swept out, the resistance of the region increases (see the description of Li-drifting in Section 2.17). Because of this property, a "reverse-biased gate" can control with very little power the flow of current through the device. Amplification can therefore be accomplished by passing this current through an external load resistor. The FETs as first preamplifiers for solid ionization detectors offer the great advantages of minimum noise and small size. Even so, they must be preselected and cooled along with the detectors to which they are joined—a third reason why a cryostat is needed. Because the cryostat must have a window (usually of Be), it is worse than a necessary nuisance; for this window must be stronger, hence of higher absorbance, than that of the flow proportional detector. Consequently, solid ionization detectors lose the advantage of being windowless themselves. At present, it seems that the need for a cryostat window and the electronic noise remaining in even the best preamplifiers will bar solid ionization detectors for quantitative determinations with x-rays longer than 12 Å (energies below 1 keV).

2.19 Early Cases of Unaided Energy Resolution

Solid ionization detector systems joined to a multichannel pulse-height analyzer (see below) promise to become a serious threat to the present x-ray emission spectrographs with wavelength (Bragg) resolution. If the threat is realized, x-ray emission spectrography will undergo a revolution.

The situation is highly fluid. The two principal advantages the new approach has to offer are instantaneous intensity reading of all analytical lines, and a more favorable geometry: the sample to be analyzed and the (highly efficient) detector can be close together. Consequently, the analytical lines need not be excited at high intensity, and a radioactive source may often replace the x-ray tube and the bulky equipment associated with it. Lower capital investment should follow as a third advantage. Counting rates, linear to beyond 10^4 cps, are usually adequate though they may be low for some trace determinations. Stability, given continuous cooling, is good. Other advantages sometimes claimed vary with x-ray wavelength.

Obviously, much depends on the resolution attainable. While resolution

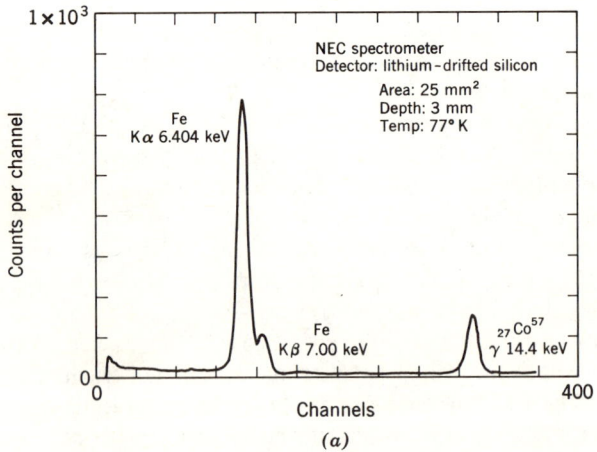

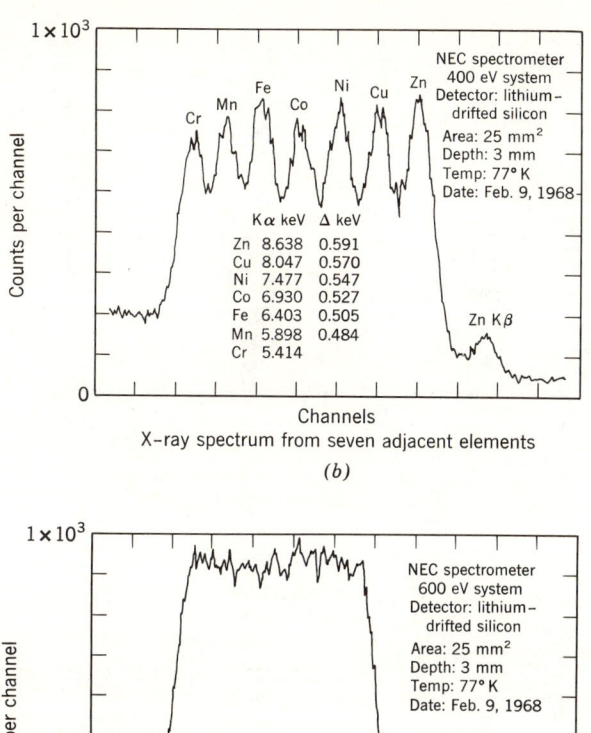
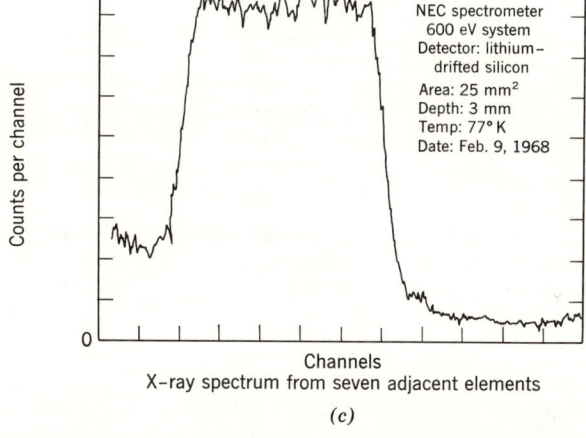

Fig. 2.19–1. Significance of improved resolution achieved by a solid ionization detector. Complete resolution of a characteristic line of iron and of a γ-ray is easily achieved (*a*). The 400 eV system (*b*) shows more promise for analytical chemistry than the 600 eV system (*c*). Further improvement is needed. The resolution achieved with these detectors is affected by detector area (lower for higher area), depth of sensitive region (the deeper, the better), counting rate, and x-ray energy. These four quantities should be specified when detector resolution is quoted. The energy (or wavelength) of the line on which the resolution datum (FWHM; see Sec. 2.21) is based should always be given. The manufacturer of the detectors used for this figure now guarantees a resolution of 280 eV. Courtesy Nuclear Equipment Corporation, San Carlos, California.

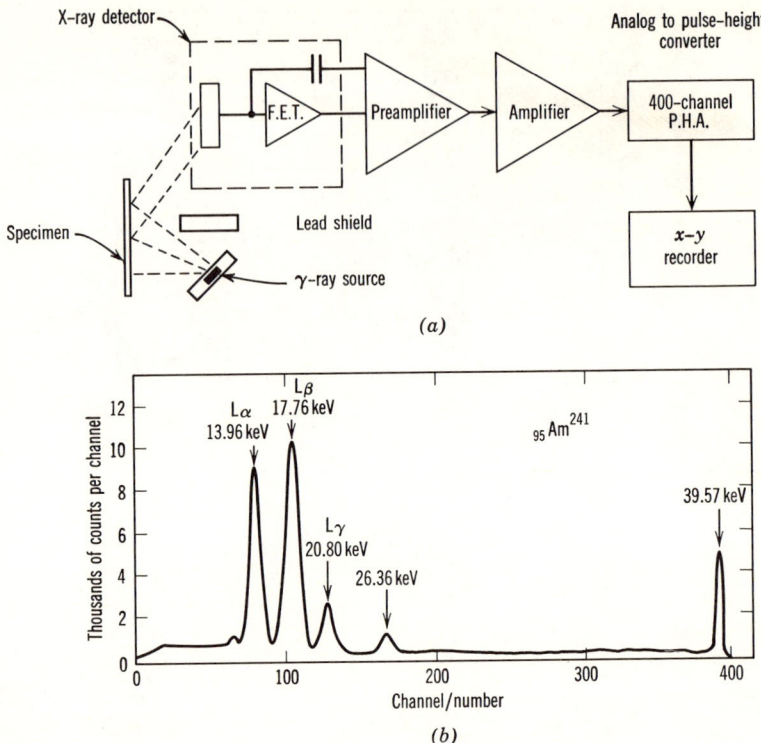

Fig. 2.19–2. (*a*) Diagram of an x-ray emission spectrograph system with unaided energy resolution. Broken lines enclose the low-temperature region. (*b*) Gamma and x-ray spectra of Am^{241} obtained with this system. A 2-mm plastic cover on the radioactive source barred alpha particles from the Si(Li) detector. After Bowman, *et al.*, Ref. 28.

is reserved for later discussion, we mention here that the makers of solid ionization detectors appear to be in a resolution race. This is one way to progress, but it may be salutary to remember that resolution for the purpose of distinguishing between two samples, each of a different element, is one thing; and resolution good enough to establish composition to a precision of 1% in a sample containing other elements, some of which may be unknown, is quite another. Also, the competition is not with Bragg resolution unaided, but with Bragg resolution joined to pulse-height selection (see below). Data are needed from extensive replicate determinations on significant samples, and not only for reasons of resolution, to make valid judgments possible.

Figure 2.19–1 emphasizes the significance of better resolution for analytical chemistry. Figure 2.19–2 is a diagram of what we think is the first

ENERGY RESOLUTION

x-ray emission spectrograph system with unaided energy resolution. The system represents an extension to the x-ray region, by Bowman, Hyde, Thompson, and Jared [28] of technique and experience both highly successful in nuclear physics. Fitzgerald, Keil, and Heinrich [29] have used a solid ionization detector system with an electron microprobe; Giessen and Gordon have applied such a system to x-ray diffraction [30].

Recent progress is described near the close of this chapter.

ENERGY RESOLUTION

As solid ionization detectors, proportional detectors, and scintillation detectors all give pulses of heights proportional to the energies of absorbed x-ray quanta, they offer the possibility of *energy resolution* to replace wavelength (Bragg) resolution. The replacement has much to offer: notably, the simultaneous recording of the intensities of all analytical lines, and the elimination of the serious intensity losses (see Chapter 5) that accompany Bragg reflection. Energy resolution is accomplished by *pulse-height selection* in a *multichannel analyzer*, a technique highly developed in nuclear physics for situations generally simpler than those in analytical chemistry. The *pulse-height distributions* (Figure 2.12-1) for proportional and scintillation detectors are so wide as to make these detectors unpromising for quantitative determinations by *unaided* energy resolution. Fortunately, the solid ionization detector is another story. With the other detectors, energy resolution combined with wavelength resolution is now the accepted way to eliminate interference by lines reflected at orders higher than the first (see Table 2.4-1) and to reduce background.

2.20 Pulse-Height Distributions and the Gaussian

All quantitative observations vary; small variations occur more frequently than large; positive and negative variations tend to balance; huge variations occur seldom, if at all, and are generally unpredictable. This summary of much experience is best represented by the well-known, bell-shaped, Gaussian distribution curve, which is of such great importance to x-ray emission spectrography that it will be discussed more fully in Chapter 8.

For all Gaussian curves, the standard deviation (observed value, s) is the horizontal distance from the mean to the point of inflection, and the *full width* of the curve at *half maximum* (FWHM) is $2.35s$. See Figure 2.20-1. When the Gaussian represents the results of a *single, completely random* process, it has a further characteristic that makes it unique: s, the standard deviation, then equals the *square root of the mean value* of the observations.

When a *monochromatic* x-ray beam is absorbed by a detector whose

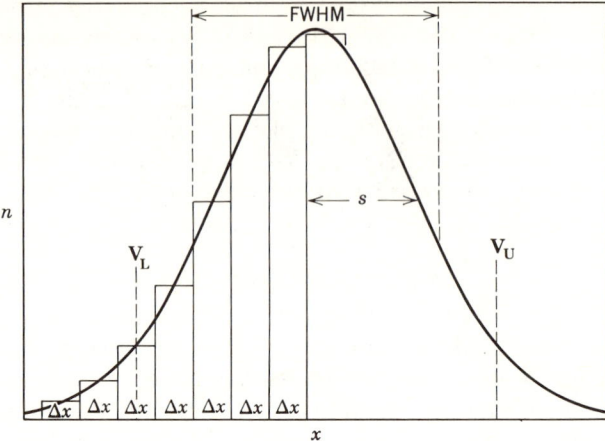

Fig. 2.20-1. An idealized pulse-height distribution superimposed on a Gaussian. The V_U and V_L are the upper and lower voltages defining a *window* in a pulse-height selector.

mean output is proportional to x-ray energy, the pulses in that output vary considerably in height, so that a *pulse-height distribution* results. If the output results from a single random process (ionization and multiplication being regarded as one such process), then the pulse-height distribution should be Gaussian. See Figure 2.20–1.

In the most general sense, y for a given x in Figure 2.20–1 is the *probability* that x lies in the interval dx between two values x_1 and x_2, x_2 and x_3, and so on. The probability is seldom a convenient ordinate. One may use the number (or frequency) of observations for each interval because the probability for any interval is the frequency divided by the total number of observations, which is constant. If intensity is established by counting repeatedly over a fixed time interval, then the frequency with which the counts lie in an interval dx may be plotted; in this case, the ordinate will be counts in time Δt, or cps when Δt is 1 sec. If pulse-height selection is carried out in a multichannel analyzer, and all channels are counted simultaneously for the same *true* counting interval, then one may plot counts per channel against channel number (which is proportional to energy) as abscissa.

Various abscissas are used, the most fundamental being $\bar{n}$, the mean number of electrons in a pulse of energy E. These two quantities are proportional in the linear range of the detector. One may therefore use as abscissa either n or energy; in the latter case, the energy of the monochromatic x-ray is the mean, and the energies of the individual pulses are proportional each to the value of n for the pulse. As the n electrons of a pulse may be used to charge a capacitor to voltage V when the detector operates, V is also a logical abscissa. Thus with a linear detector and linear amplification, n, E, or V deter-

ENERGY RESOLUTION

mines pulse-height. Figure 2.20–1 is therefore a pulse-height distribution for the idealized detection of a monochromatic x-ray, whose energy fixes the mean of the distribution.

Under the simple conditions just outlined,

$$s = \sqrt{E} = s_i \qquad (2.20-1)$$

$$\text{FWHM} = 2.35 s = 2.35 \sqrt{E} \qquad (2.20-2)$$

If other components (e.g., preamplifier, amplifier) of the detector system also introduce random fluctuations, then by the rule for the combination of errors

$$s = \sqrt{s_i^2 + s_p^2 + s_a^2 + \cdots} \qquad (2.20-3)$$

where s, the total standard deviation, is now larger than the intrinsic s_i (2.20–1), because of contributions from the other components designated by subscripts.

For the solid ionization detector, it is experimentally established that

$$s_i < \sqrt{E} \qquad (2.20-4)$$

and this occurs presumably because of fluctuations in the amount of energy degraded to heat in the detector [31, 32]. The theoretical considerations involved, which need not concern us, do not seem obviously applicable to the solid ionization detector. What matters here is the experimental observation that a correction factor—the Fano factor, F—needs to be put under the radical in (2.20–4) to change it into an equality; and that F seems to have a value between 0.10 and 0.15 for solid ionization detectors.

The relationship between pulse-height distribution and cps (or counts for any fixed counting interval) is of first importance. Monochromatic x-ray quanta absorbed in any counting interval will give rise to pulses that may have any height that is included in the pulse-height distribution. Each of these pulses, regardless of its height, *will register as one count* in the rate-meter or in the counting-scaling system. Therefore, the x-ray intensity expressed in cps by the rate meter or as counts per Δt in an accumulative system will be proportional to the *area* of the pulse-height distribution. With x-ray lines of neighboring wavelength, these areas will overlap. Any system of energy resolution will have to reject many pulses in these overlapping areas—a serious disadvantage. The reader is invited to examine Figures 2.20–1, 2.12–3, and 2.19–1 from this point of view. See the next section.

2.21 Sharpness of Energy Resolution

The sharpness of energy resolution is important (1) as a *figure of merit* in rating detectors, and (2) as a criterion of what unaided energy resolution can do in analytical chemistry. Analytical chemistry sometimes makes greater demands than nuclear physics.

The FWHM is satisfactory as a *rough* figure of merit in the rating of detectors, which is done by selecting a reference line [e.g., the γ-ray (0.86 Å; 14.4 keV) of $_{27}Co^{57}$; or the Mn Kα line (2.10 Å; 5.9 keV) from $_{26}Fe^{55}$] and measuring for it the FWHM achieved by a detector. The experimental value thus obtained normally includes any broadening attributable to the electronic circuitry; see (2.20–3). The reference line for which the FWHM was measured should always be given. Obviously the FWHM does not give the *shape* of the distribution curve, and such curves are often far from Gaussian.

The problem in analytical chemistry is much more complex. (1) *Resolution* is a vague concept. Resolution for qualitative identification is more easily attained than resolution for quantitative determination. When the FWHMs for adjacent pulse-height distributions do not overlap, complete resolution can usually be achieved for major constituents from half the maximum height upward, but that is not enough for quantitative work, where complete resolution downward to the abscissa axis is desirable. See the middle of Figure 2.19–1. (2) When unaided energy resolution is used, the presence of other lines belonging to the series of the analytical line will confuse the picture. Notice the Fe **K**β in the top part of Figure 2.19–1. (3) In general, the amounts present of the elements in a sample can vary over a wide range. The concomitant intensity variations complicate the resolution problem; *trace determinations* are unlikely to be done with unaided energy resolution. (4) The presence of background further complicates the problem. (5) The analytical chemist needs to know the resolution achievable over the entire wavelength range in which his analytical lines are found.

In this situation, calculations employing FWHM values are not likely to be of much use; it is usually futile to do more than see how far roughly drawn pulse-height distributions overlap. In terms of Figure 2.20–1, a logical measure of energy resolution is the *relative resolution RR* defined by

$$RR = 2.35 s/\bar{x} = FWHM/\bar{x} \quad \text{(usually expressed as \%)} \quad (2.21-1)$$

where $\bar{x}$ is commonly taken as E, and s as $\sqrt{E}$; see (2.20–2). The *RR* can be used to *estimate* the way in which resolution changes with wavelength or energy, as the following example will show. One manufacturer [33] gives the following specifications (with the observation that these are changing rapidly) for a solid-ionization-detector system: FWHM $\leq$ 600 eV for the 14.4-keV, 0.86-Å γ-ray of $_{27}Co^{57}$ at cps = 20,000 for a detector diameter of 4 mm, which corresponds to

$$RR = 600/14,400 = 0.042 = 4.2\% \quad (2.21-2)$$

ENERGY RESOLUTION

From (2.21–1) and the text that follows, we find for two lines (1 and 2) of different energy or wavelength

$$RR_1/RR_2 = \sqrt{E_2/E_1} = \sqrt{\lambda_1/\lambda_2} \quad [\text{See (1.6-2)}] \quad (2.21-3)$$

The approximate wavelengths of three possible analytical lines are: Na **Kα**, 12Å; As **Kα**, 1.2Å; U **Kα**, 0.12 Å. With RR_2 as 0.042 for 0.86 Å (2.21–2), the approximate RR values for the three analytical lines in the order given above are 0.16 or 16%; 0.05 or 5%; and 0.016 or 1.6%. The high value of RR for Na **Kα** supports the statement at the end of Section 2.18 that Si(Li) detectors are not likely to be useful for x-ray emission spectrography of the light elements.

As the average energy required to produce an ion pair in a gas-filled proportional detector is about ten times that needed to make a hole-electron pair in an Si(Li) detector, the energy resolution attainable within the solid device should be considerably greater, other things equal. The future should see this resolution more closely approached.

The mathematical treatment of *unresolved* pulse-height distributions [34] is an alternative to energy resolution that has proved useful in nuclear physics. The increasing availability of multichannel analyzers and computers makes such treatment more attractive to analytical chemistry. See Chapter 9.

Unaided energy resolution, now marginal with solid ionization detectors, seems scarcely worth considering for the others, as the following RR values show. Proportional detectors: Fe **Kα**, 17.5% [35]; Cu **Kα**, 25% [36]; Mn **Kα**, 60% [37]. Scintillation detectors: Cu **Kα**, 50% [36]; Mn **Kα**, 60% [37]. Though improved RR values may come, these detectors are not likely to rival Si(Li) and Ge(Li) as regards unaided energy resolution.

2.22 The ABCs of Pulse-Height Selection

Pulse-height selectors—old name, "kick sorters," each pulse being a "kick"—were introduced in Section 2.3. Here we try to explain how a selector with one *voltage* window (not to be confused with *detector* windows) operates. Obviously, one window, one channel; that is, we deal here with a single-channel analyzer. The multichannel analyzer operates similarly, but on a grander scale. We assume that pulses of heights proportional to x-ray energy and of proper shape are available for sorting.

The task of these selectors is to make sure that each pulse passes through its proper window. Pulse heights and windows are measured in volts. These are volts of a special kind. They are settings on the instrument, and the number of these volts is set not only by the energy of the x-ray quantum, but also by electronic manipulation. To distinguish these special voltages from all others, they will be written in **bold-face type**. Each window has an upper voltage level (say, $\mathbf{V_U}$ = 100 volts), and a lower voltage level (say,

$V_L = 90$ volts), the difference between which $(V_U - V_L = \Delta V = 10$ volts) characterizes the window: this window should pass all pulses of heights between 90 and 100 volts, and no others. A window was shown in Figure 2.20–1. With multichannel analyzers, V_U and V_L can be tenfold smaller.

The electronic circuits do not concern us. The following simple illustration shows how pulse-height selection might be done. Three x-ray quanta (wavelengths $\lambda_1 < \lambda_2 < \lambda_3$) have generated the three neighboring pulses of heights $V_1 > V_2 > V_3$. [Ideally, according to (1.6–2), λV is a constant.] Let the lower level of the window be maintained at V_L by one pulse-height discriminator, and its upper level be kept at V_U by another. (It will be remembered that such discriminators allow pulses to pass only if their heights exceed the discriminator setting.) Identify discriminators by their settings and pulses by their heights. Let $V_1 < V_L$; $V_L < V_2 < V_U$; and $V_U < V_3$. Then V_1 will not pass V_L; V_2 will pass V_L but not V_U; V_3 will pass both V_L and V_U. Only V_2 is to be counted. The V_1 presents no problem as it is rejected by V_L. How reject V_3? This is done electronically by an *anticoincidence circuit*, which stops all pulses that pass *both* V_L and V_U from being counted. The passing of both discriminators is the *coincidence* to which this last circuit is *anti*.

Perfect windows are generally unattainable. The best window for an analytical line is often difficult to choose [38–40]. A few of the difficulties are as follows. (1) The larger *RR*, the more likely that some pulses in the distribution will go uncounted because it is not practical to make the ΔV of the window large enough to pass them all. See Figure 2.20–1. (2) A pulse can be brought to a given height by changing (a) detector voltage (internal amplification), (b) external amplification, or (c) both. The shape of the pulse-height distribution for a monochromatic beam may depend on how this is done; if the shape changes, so may the fraction of the pulses that pass the window. (3) The background (electronic noise and unwanted x-rays) is often variable. (4) Because the FWHM and $\bar{x}$ (Figure 2.20–1) of a pulse-height distribution change with the wavelength of the analytical line, $\Delta V = V_U - V_L$ must be correspondingly adjusted.

Figure 2.22–1 shows that the pulse heights actually realized may be too low at high counting rates because the pulses come so fast that **V** does not return to its "base-line" value between them. The trouble can be remedied —but not without risk of complication—by use of an electronic "base-line restorer."

The proportional counter is the most easily disturbed of the detectors that qualify for energy resolution. Aside from end effects, pulse-height distributions can be distorted by irregularities at the anode wire. The need for smoothness here was known from the first (Section 2.12). Any particle on the wire will change the gradient of the field in its neighborhood, hence affect the pulse-height distribution. Organic filler gases (such as methane)

ENERGY RESOLUTION

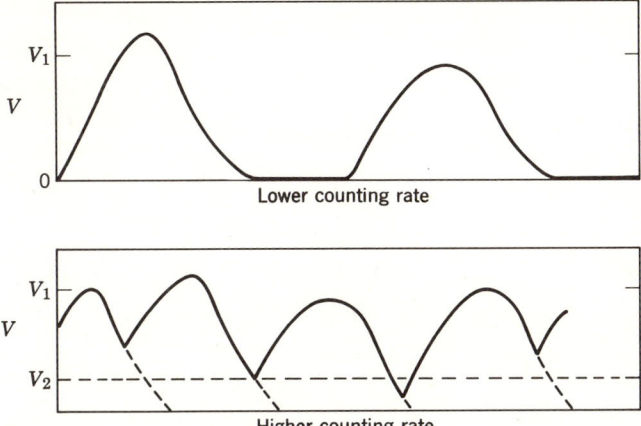

Fig. 2.22–1. Change of effective pulse height with counting rate. In the upper figure the pulses are distinct. In the lower figure, interaction of pulses raises the baseline to V_2 on the average and reduces the average pulse height to $V_1 - V_2$. The pulses are of a variable height as called for by the existence of a pulse-height distribution. After Jenkins and de Vries, Ref. 38, p. 75.

can leave carbonaceous deposits on the anode wire during the life of the detector. Cleaning as needed is in order.

2.23 Energy Resolution with Wavelength Resolution

Energy resolution is now a virtually indispensable aid to wavelength resolution by Bragg reflection. Resolution in the modern x-ray emission spectrograph, though sharper than unaided energy resolution, does not usually approach that obtained photographically by Hevesy (Table 2.4–1). When small amounts must be determined, the presence of a large and variable background precludes obtaining peak intensities that are proportional to these amounts. See Figure 2.23–1. Bragg resolution can use help.

Most problems with Bragg reflection belong in Chapter 5. Here we mention two that can be mitigated by energy resolution; namely, *interference by x-ray lines*, and *presence of background*. Under background, we include both *scattered x-rays* and *electronic noise* (i.e., recorded pulses for which x-ray absorption by the sample is not responsible).

Table 2.4–1 shows that interference by x-ray lines can result from lines reflected in the same or in higher orders; that is, with the same or different values of n in (1.16–2). Hevesy was unable to resolve Pb $L\alpha 1$ and As $K\alpha 1$, for which Siegbahn found a wavelength difference of 0.001 Å. Nor did he resolve Th $L\alpha 1$(II) and Dy $L\alpha 1$, for which Siegbahn's wavelengths were 0.956 and 1.909 Å. (Note (1) that the Roman numerals were used by Hevesy,

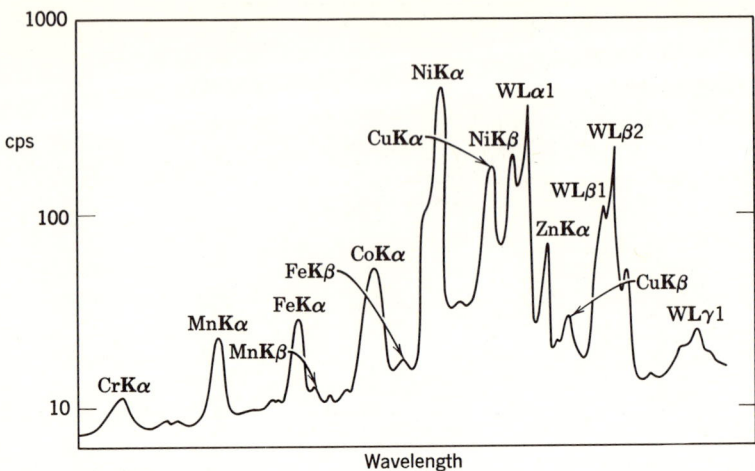

Fig. 2.23-1. The figure is the trace of the x-ray emission spectrum from a paper disk on which had been deposited 2 to 4 μg each of $_{24}$Cr, $_{25}$Mn, $_{26}$Fe, $_{27}$Co, $_{28}$Ni, $_{29}$Cu, and $_{30}$Zn. Experimental conditions did not favor high intensity. Note that characteristic tungsten lines from target of Coolidge tube contribute to spectral interference. Authors' unpublished results.

and will be used herein, to denote $n > 1$; (2) that twice 0.956 is 1.912.) In modern spectrographs with electrical detectors, resolution will not be so sharp as this, which means that interference will be more likely. Energy resolution can help, but only when the interference involves reflections at different values of n—that is to say, lines of different energies.

When an analytical line is excited by a continuous spectrum, scattered x-rays will contribute to the background. This contribution will vary with wavelength as indicated by Figure 4.1-1. To make excitation efficient, the absorption edge that corresponds to the analytical line should be at a wavelength near that of the intensity maximum in the figure. X-rays from the continuous spectrum and of wavelength near that of the analytical line will then be reflected in the first order. Here again, energy resolution cannot help. Energy resolution does eliminate from the background all scattered x-ray lines that have pulse heights smaller than V_L or greater than V_U. Such reduction of background is particularly important for the analytical lines of the light elements.

Energy resolution helps in a special way to reduce background when a scintillation detector is used. The work functions of photocathodes are so low that electron emission is possible without x-ray absorption. Amplification in the photomultiplier will convert these spontaneously emitted electrons into low-voltage pulses that will accompany those generated by the absorption of x-rays of any wavelength whatever. Energy resolution can

ENERGY RESOLUTION

effectively separate the two kinds of pulses if the x-rays absorbed are short enough in wavelength. Fortunately, this situation exists (at least with the K lines) for all but the lightest elements. That is one reason why the scintillation detector must always be supplemented by the flow proportional detector in x-ray emission spectrography intended to cover the largest possible range of atomic numbers.

Energy resolution can clearly help Bragg resolution under many different conditions. Instead of continuing this general discussion, it is more profitable to use the work of Heinrich [40] as an example of a systematic approach to energy resolution, and one that illustrates what energy resolution can accomplish.

Heinrich assumed that it was necessary to determine aluminum (0.1 to 5% by weight) in titanium dioxide. The analytical line Al Kα (8.34 Å) is subject to interference by Ti Kα(III) (2.75 Å; 3 × 2.75 Å = 8.25 Å). Note that no interference in Table 2.4–1 exists at so great a relative wavelength difference.

Figure 2.23–2 is a schematic diagram of Heinrich's equipment, which has as a noteworthy feature that it permits the separate reading and recording of pulses with voltages in the interval $\Delta V = V_U - V_L$ ("within channel"), and those of voltages greater than V_U ("above channel"). In Figure 2.23–2,

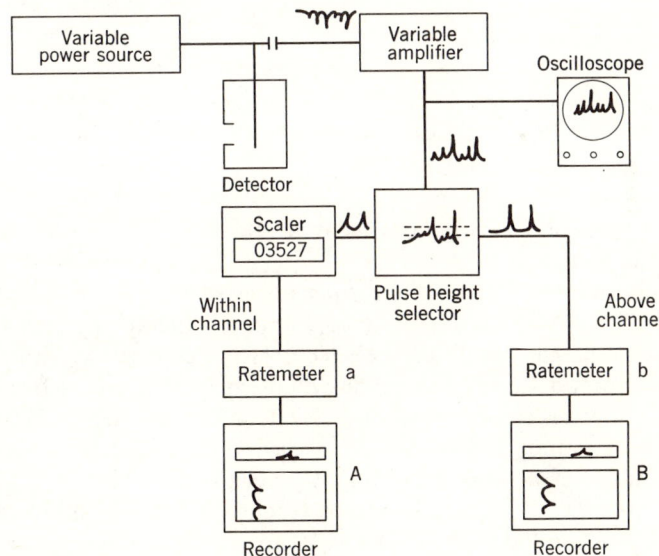

Fig. 2.23–2. Detector system for energy resolution by pulse-height selection as an aid to Bragg resolution. The oscilloscope gives a visual display of the pulses and is useful as a means of control. After Heinrich, Ref. 40, p. 373.

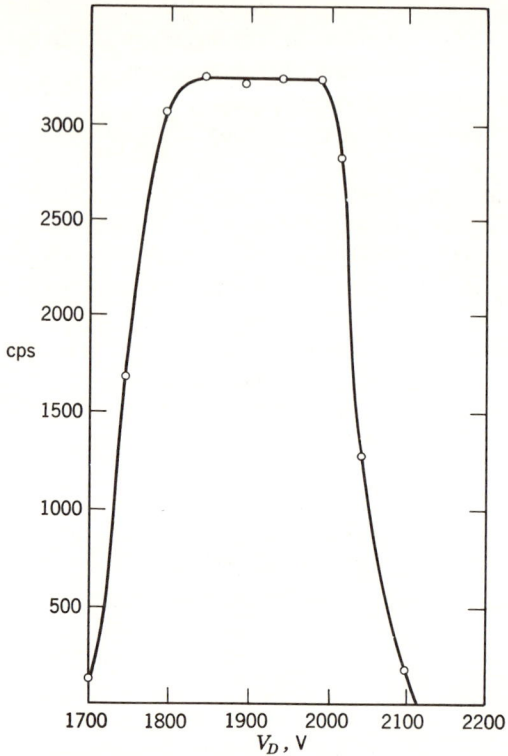

Fig. 2.23–3. Counting rate as a function of detector voltage, V_D. The sample is metallic aluminum and the experiment is carried out with $\Delta V = V_U - V_L$ a maximum. After Heinrich, Ref. 40, p. 374.

the former pulses take path Aa; the latter take path bB. He therefore has a two-channel analyzer.

Heinrich found as follows the proper detector voltage for measuring the intensity of Al $K\alpha$. With a sample of aluminum metal, cps was registered without pulse-height selection. The selector was then inserted with minimum V_L and maximum V_U. The detector voltage was now raised from a value that gave no pulses to one at which pulse height exceeded V_U. The expected behavior was found, as Figure 2.23–3 shows. At a detector voltage $V_D = $ 1700 V, pulses of heights exceeding V_L began to appear; cps for Al $K\alpha$ was near 100. At V_D near 1850 V, cps reached a flat maximum: all pulses now had heights in the interval ΔV. At V_D near 2050 V, some pulses now had heights exceeding V_U and were consequently diverted from path Aa to path bB. Somewhat above $V_D = 2100$ V, all pulses were thus diverted so that cps for the channel became zero. Thus $V_D = 1900$ V was chosen for the

ENERGY RESOLUTION

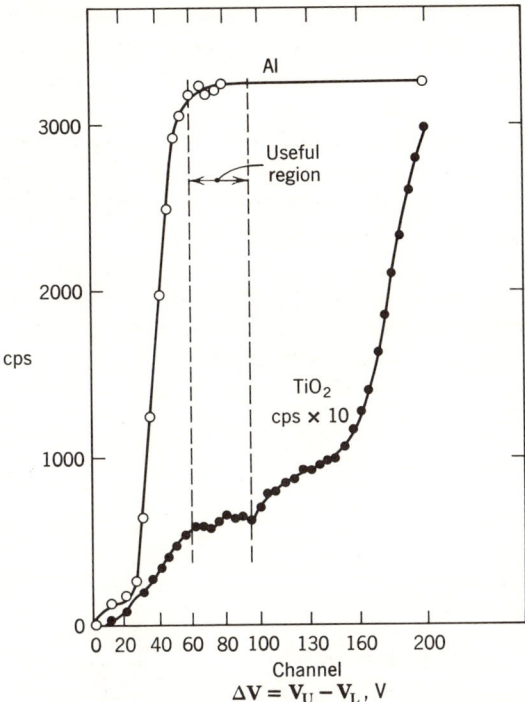

Fig. 2.23–4. Effect of varying ΔV (at minimum V_L) on counting rate. The detector voltage is 1900 V. Note that cps for TiO_2 is one-fortieth that for Al. With modern solid-state circuitry, V_U and V_L in such an experiment might be much reduced. After Heinrich, Ref. 40, p. 375.

determination of aluminum in the presence of titanium, a wise choice as it gives maximum cps at minimum V_D. (The minimum V_D is desirable to reduce the chance that internal amplification in the detector will produce pulses that are not due to Al $K\alpha$ absorption, but are counted nevertheless.)

As a final check, the channel was closed completely ($V_L \geq V_U$), and then gradually opened by increasing V_U. Results for metallic aluminum appear in Figure 2.23–3, which shows that $\Delta V = V_U - V_L$ must be about 60 V or more to give maximum cps for Al $K\alpha$.

With the question of Al $K\alpha$ settled, Heinrich carried out the V_L and V_U experiment just described on TiO_2 as sample. Because the cps values were now very much smaller, the counting interval had to be increased (to 400 sec for every value of V_U). The results are in Figure 2.23–4. The most significant part of this figure is the plateau marked "useful region." At the low-energy end of this plateau ($V_U - V_L = 60$), sensibly all the scattered x-rays that undergo first-order Bragg reflection have entered the channel so that background is virtually constant. As V_U is increased, the cps value increases in two steps as follows: (1) scattered x-rays reflected at higher orders, and (2) Ti Kα (III), which begins to contribute to cps at $V_U - V_L = 160$. Good conditions for the determination of Al in TiO_2 would thus be $V_D = 1900$ V; $V_U - V_L = 80$.

These conditions might be further improved. If the amplifier can give a variable gain over its linear range, it will be possible to achieve the plateau of Figure 2.23–4 by different combinations of amplifier gain and detector voltage: increased gain in the amplifier permits lower voltage across the detector. The combination that gives the lowest background at maximum analytical-line intensity should be chosen: it is to be remembered that amplification within the detector tends to be noise-free.

Heinrich chose to keep V_L at its minimum throughout. In other systems,

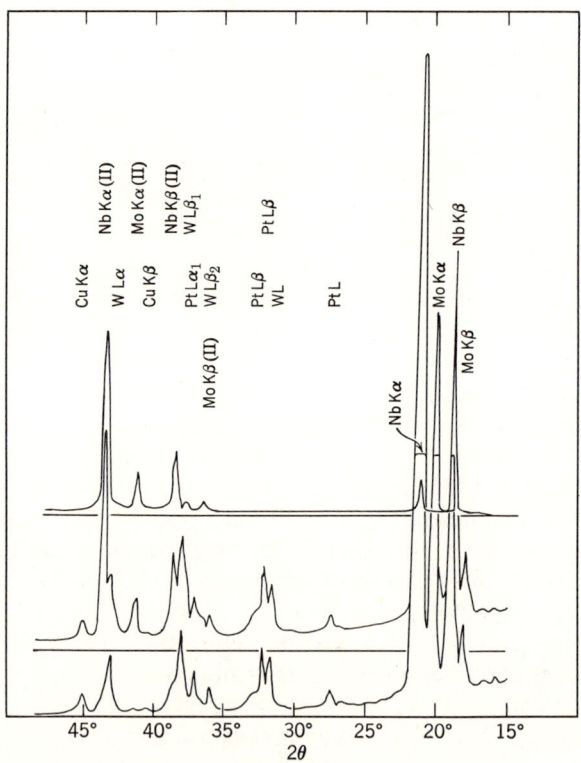

Fig. 2.23–5. Wavelength scan of a sample containing niobium, molybdenum, and tungsten. (1) First-order lines sorted out by energy resolution (lower curve) as an aid to Bragg reflection. (2) Spectrum obtained by unaided Bragg reflection (middle curve). (3) Lines reflected in the second order and rejected by energy resolution (upper curve). These lines followed path bB in Figure 2.23–2. After Heinrich, Ref. 40, p. 376.

ENERGY RESOLUTION

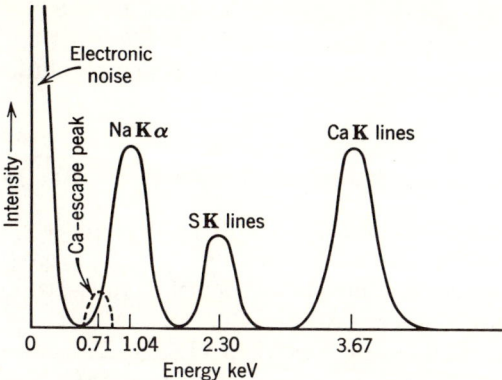

Fig. 2.23–6. Interference with an analytical line (Na **K**α) by characteristic lines (**K** lines of S and Ca) excited in the Bragg crystal by characteristic lines from the sample; the interfering lines appear at all values of 2θ. The lines accomplishing the excitation are not shown, but they must be of wavelengths shorter than the wavelengths of the absorption edges of S and Ca. The S and Ca lines will be counted along with the analytical line when (as will be the usual case) they cannot be rejected by pulse-height selection. At larger energy differences such rejection becomes possible. After Jenkins and de Vries, Ref. 38, p. 84.

changing V_L might be desirable. In general, V_L will have to be changed if cps values are to be obtained for lines varying considerably in energy. In particular, if there is to be a wavelength scan of an x-ray spectrum, both V_L and V_U should change synchronously with the change in θ, the angle of Bragg reflection. This can be done in several ways that are not discussed. Figure 2.23–5 gives the results of scanning experiments by Heinrich, which demonstrate convincingly that energy resolution by pulse-height selection is an invaluable aid to Bragg resolution.

There is another interesting, fortunately uncommon, case in which energy resolution cannot always help Bragg resolution. On occasion, x-rays from the sample excite characteristic lines of elements in the Bragg crystals, and these lines will naturally enter the detector at all values of 2θ. For such lines to have enough intensity to be bothersome, they must usually be excited by characteristic lines of the sample. Figure 2.23–6 presents the problem for a hypothetical example [41]. An actual example is discussed in Chapter 5. In all cases of this sort, it is best to change to another Bragg crystal.

Figure 2.23–6 illustrates another important point. Were 2θ used as abscissa, both the electronic noise and the S and Ca lines would appear in the background at all values of 2θ. This is why 2θ was not discussed as abscissa in Section 2.20 even though 2θ and λ are related by (1.16–2); 2θ is an *operational* variable; λ (or energy) is not. See Chapter 5.

The general information about x-rays already presented in this book shows that interferences in unaided Bragg resolution can sometimes be reduced (but seldom eliminated) in these ways: (1) changing excitation conditions; (2) using filters; and (3) changing the filler gas to reduce its absorption of the unwanted x-rays. These palliatives are usually less satisfactory than energy resolution as aids to Bragg resolution, and they are not discussed.

2.24 Pulses, Bragg Reflections, Escape Peaks, Energy Resolution

Three experiments were done to illustrate some of the discussions in this chapter. In all three, an x-ray emission spectrograph (Chapter 9) was operated in the same way, but the output of its flow-proportional detector system was treated differently.

The sample in the General Electric XRD-6 spectrograph was powdered Cr_2O_3 mixed with enough powdered NaBr to give Cr $K\beta$ (2 Å) and Br $K\alpha$ (1 Å) at comparable intensities. The spectrograph was always operated with the flat LiF crystal set at 62.29° (2θ), the angle for the Bragg reflection of Cr $K\beta$. Under these conditions, Bragg reflection will produce interference of Br $K\alpha$ (II) and Cr $K\beta$. In addition, escape peaks for both Br and Cr will appear in the output as the detector gas contains argon.

Experiment 1. The amplifier output over 0.01 sec was displayed on an oscilloscope. See Figure 2.24–1. In this display, each line is a pulse, for which the height in volts could be measured directly. Such information, it will be recalled, was the basis of Figures 2.12–1 and 2.12–3. *Four* (approximately correct) pulse heights were found. See below.

Experiment 2. As each pulse triggered the oscilloscope, a voltage-time trace was recorded for this pulse over a 2-μ sec interval. The pulses recorded in a total time of 0.04 sec fell into *four* well-defined sets as Figure 2.24–2

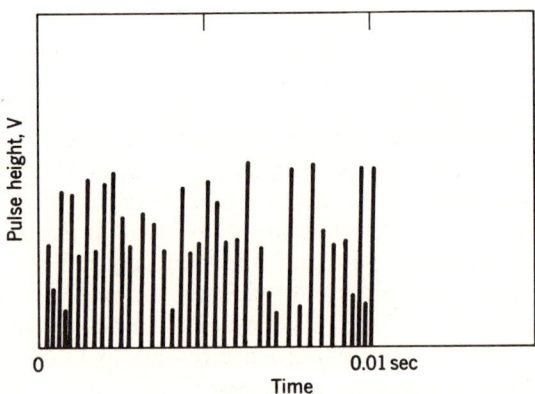

Fig. 2.24–1. Output pulses during 0.01 sec from a General Electric XRD-6 spectrograph as recorded by an oscilloscope. See text. Experiment by P. D. Z.

ENERGY RESOLUTION

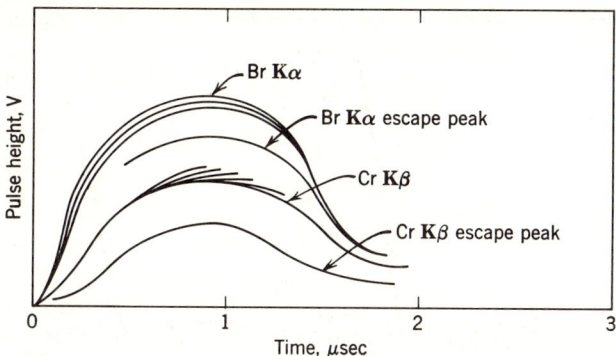

Fig. 2.24–2. Output pulses of 2 μsec duration from a General Electric XRD-6 spectrograph as recorded on one part of an oscilloscope screen. Total time, 0.04 sec. Experiment by P. D. Z.

shows. These are (1) Br Kα (1 Å), (2) its escape peak, (3) Cr Kβ (2 Å), and (4) its escape peak. As the wavelength of argon Kα is 4 Å, the average-pulse-height relationships should be

$$\text{Br K}\alpha \, (1 \text{ Å}): \quad \text{twice the energy, hence twice the height of Cr K}\beta \, (2\text{Å}) \quad (2.24\text{--}1)$$

$$\text{Bromine escape peak: } 3/4 \text{ the height of Br K}\alpha \text{ because}$$
$$1/1 - 1/4 = 3/4 \quad [\text{see } (2.12\text{--}3)] \quad (2.24\text{--}2)$$

$$\text{Chromium escape peak: } 1/2 \text{ the height of Cr K}\beta \text{ because}$$
$$1/2 - 1/4 = 1/4 \quad [\text{See } (2.12\text{--}3)] \quad (2.24\text{--}3)$$

Measurements on the figure confirm these relationships.

Experiment 3. The output of the detector system was fed directly into a 200-channel pulse-height analyzer. The pulses for each channel were recorded for a "live" counting interval of 10 minutes; that is to say, the appropriate correction for coincidence loss was automatically made. Figure 2.24–3 shows that the expected four peaks were found at the correct energies: at 3, 6, 9, and 12 keV; that is, at 4, 2, 1.33, and 1 Å. The incomplete resolution of the bromine escape peak on its high-energy side illustrates the point made earlier about the inadequacy of FWHM as a criterion of resolution. The *RR* value (14%) for FWHM of Br Kα seems acceptable. Because the escape peak is of low intensity, one needs to know not the FWHM but the (much greater) full width at *one-tenth maximum* to judge resolution.

To summarize: The first two experiments show that the output from the spectrograph system contains the four kinds of pulses expected and no others of comparable intensity. The third experiment shows that the

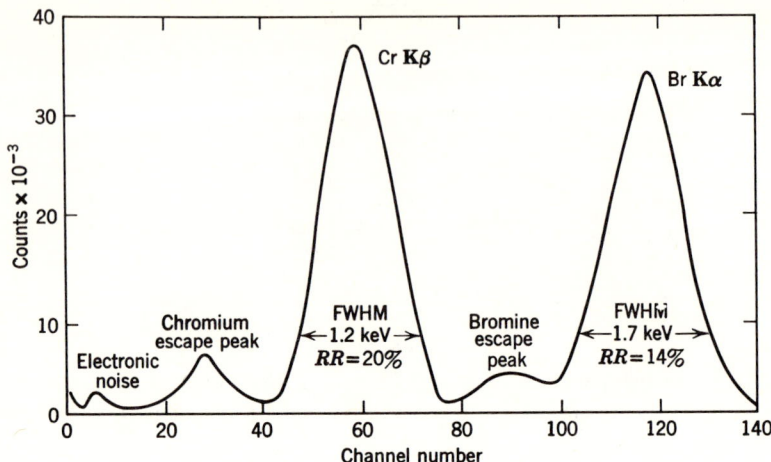

Fig. 2.24–3. Energy resolution by a 200-channel analyzer of the spectrograph output. The mean x-ray energy (keV) is about one-tenth the channel number. For each channel, $\Delta V = 10$ V. Much sharper resolution could be obtained if the flow proportional detector were replaced by a good solid ionization detector. The LiF crystal used in other experiments gave FWHM $\simeq 0.05$ keV for Cr Kβ and 0.200 keV for Br Kα. Experiments by P. D. Z.

energy resolution achieved is inadequate for x-ray emission spectrography. There is no doubt that this resolution could be greatly improved if the proportional detector were replaced by a good solid-state detector with a multichannel analyzer large enough to make the detector resolution-limiting. Fortunately, the price of multichannel analyzers is decreasing, and it should soon be feasible to make available about 10 channels for each analytical line in most samples.

2.25 Detectors in 1969

The year of writing has been given because rapid change is certain if the solid ionization detectors fulfill their promise. Photographic methods for measuring x-ray intensity and wavelength appear in Chapter 6.

A general-purpose x-ray emission spectrograph needs two detectors to cover the maximum number of elements. At present, the most common choices are a flow proportional detector (light elements) and a scintillation detector (all others). The Geiger detector, so prominent up to 1960, is seldom seen in new equipment. Sealed xenon-filled proportional detectors sometimes replace scintillation detectors.

Detectors should give no trouble if properly used with stable electronic circuits. The need for cleaning the anode wire in the flow proportional detector (Section 2.22) can be reduced by operating at the minimum detector

ENERGY RESOLUTION

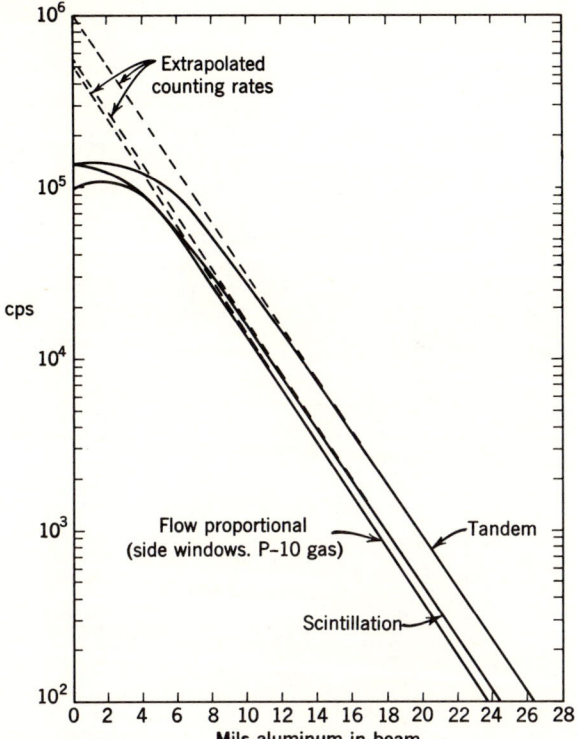

Fig. 2.25–1. Coincidence losses for flow proportional detector and scintillation detector, singly and in tandem. The ordinate is logarithmic. Experimental conditions: Copper as sample in General Electric XRD-6 emission spectrograph; flat LiF crystal; tungsten-target tube; 50 mA, 75 kV with full-wave rectification; pulse-height selection; peak cps values for Cu $K\alpha$ (1.54 Å). Counting rates were increased by removing aluminum foils from the stack initially placed in the x-ray beam. Experiments by P. D. Z.

voltage needed to give the numbers of counts commensurate with the precision sought.

Tandem operation of the two detectors in a spectrograph, with the flow detector naturally in front, is becoming fashionable in spectrographs with *flat* Bragg crystals. In such operation, the counts from the two detectors are registered as a *single* count. Three combinations can usually be obtained by keeping zero voltage on one detector or the other when they are not in tandem. These combinations are tandem arrangement, flow proportional detector alone, and scintillation detector alone. In the third combination, x-ray absorption by the flow proportional detector will reduce output by the scintillation detector below what it would be without the proportional

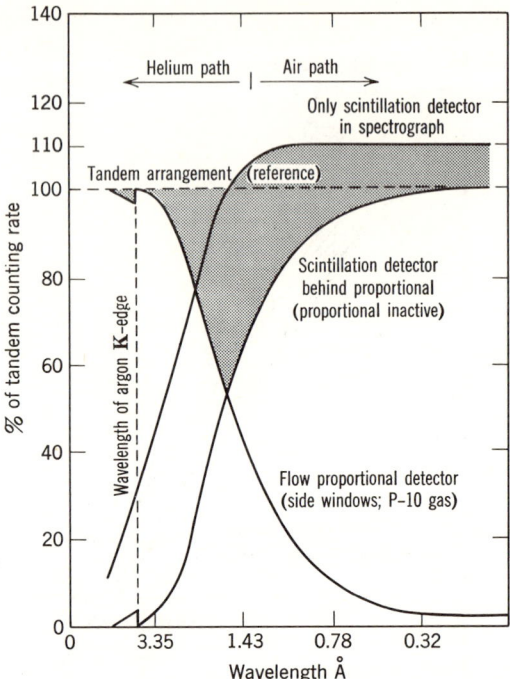

Fig. 2.25–2. Effectiveness of three detector arrangements relative to that of tandem arrangement. The effect of the argon absorption edge appears "inverted" in the curve for the scintillation detector. This detector of course does not show the effect, but the discontinuity results because the ordinate is based on the tandem detector (which does show it) as reference. Shaded area below 100%: tandem arrangement has advantage over either detector alone when both are in position. Shaded area above 100%: scintillation detector has advantage when proportional detector is not in spectrograph. Experiments by P. D. Z.

detector installed. When the proportional detector is present and inactive, counts are therefore lost. See below.

Empirical detector calibration is recommended for precise work because coincidence losses are better determined than calculated (Section 2.12). See Figure 2.25–1. With the tandem detector, these losses at different recorded cps values were 0 at 10^4; 2% at $2(10^4)$; and 10% at $3(10^4)$.

In Figure 2.25–2 are given results based on cps values obtained on a General Electric XRD-6 x-ray emission spectrograph on a series of elements over the range of atomic numbers from about 15 to 80, such values being peak values above background for the tandem detector, the flow proportional detector in tandem position (equivalent to flow proportional detector alone), scintillation detector in tandem position with proportional detector in place, and scintillation detector alone in spectrograph. Figure 2.25–2

ENERGY RESOLUTION

records the *relative* cps value for each of the last three cases with cps for the tandem detector taken as 100%. For **K** spectra, with both detectors installed, the situation is this. Roughly over the range Z = 20 to 60, the tandem arrangement has some advantage (diagonal shading in figure) over the more suitable of the two detectors with the less suitable inactive. At atomic numbers exceeding 30, absorption of x-rays by the proportional detector (inactive) places the tandem arrangement at a disadvantage.

By the time this book appears, solid ionization detectors should have progressed beyond the early applications now to be reported. Even in 1967, Jones and Carpenter [42] had used radioactive iodine and a 400-channel pulse-height analyzer to determine bromine, rubidium, and strontium in aqueous solution. Currently, various organizations are offering or planning to offer complete spectrograph or detector systems. A partial list follows: Kevex Corporation, Burlingame, Calif., 94010; Nuclear Equipment Corporation, San Carlos, Calif., 94070; Ortec Incorporated, Oak Ridge, Tenn., 37830; and Columbia Scientific Industries, Austin, Tex., 78702. The last company was to "release to the market in 1970" a complete x-ray emission spectrograph system at a price about one-half that of conventional systems with similar capabilities [43].

The summary, by Walter and Moore [44], in Figure 2.25–3 gives ex-

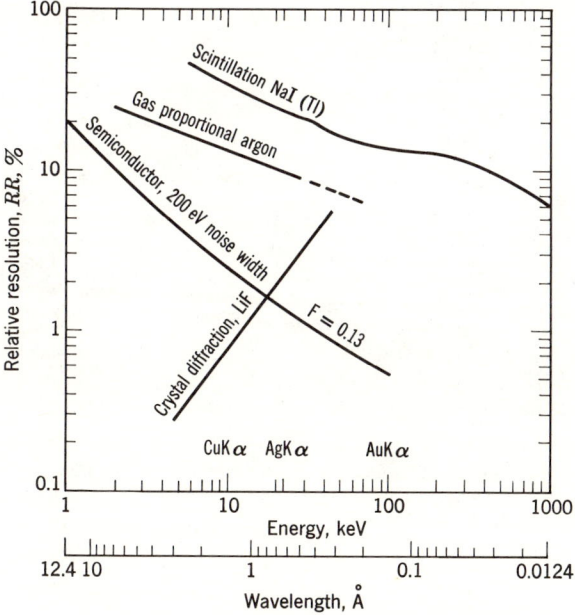

Fig. 2.25–3. RR values according to (2.21–1) for energy resolution by three detectors compared with that for Bragg resolution over a wide range of wavelengths. $F = 0.13$ is the value assumed for the Fano factor; see Section 2.20. After Walter and Moore, Ref. 44, Fig. 12.

Fig. 2.25–4. X-ray emission system with solid ionization detector and radioactive source set up for the non-destructive identification of the pigments in the painting. The bottom of the cryostat is in the upper right-hand corner. Multichannel pulse-height analyzer and other ancillary equipment not shown. Results in Figure 2.25–5. Courtesy of Kevex Corporation.

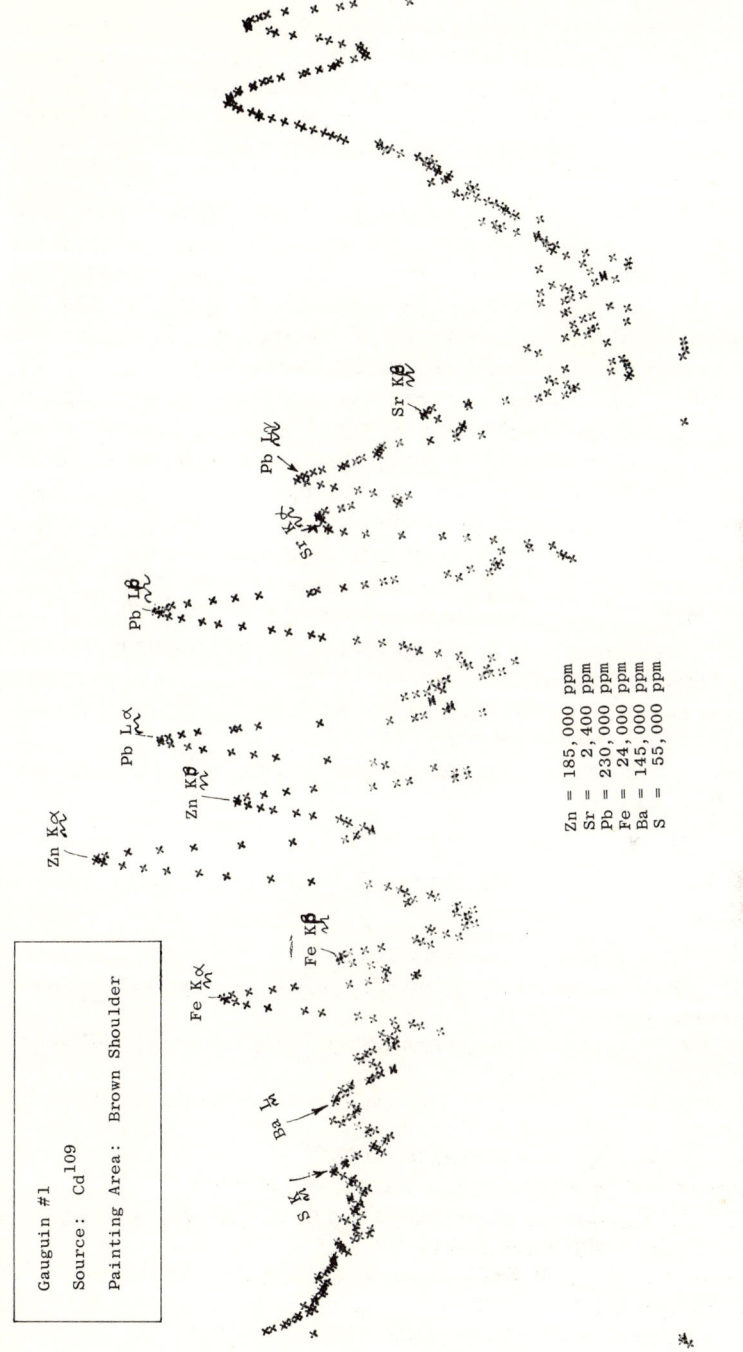

Fig. 2.25–5. Actual readout data from multichannel analyzer in experiment of Figure 2.25–4. The ordinates (peak intensities) are logarithmic. Number of channels (abscissa) not specified. Two unmarked peaks at right presumably from Cd^{109}, the radioactive source. Note increasing importance of background at low atomic numbers. Courtesy of Kevex Corporation.

perimental values of RR, the relative resolution in percentage as defined in (2.21–1), for Bragg resolution with an LiF crystal, and for energy resolution with three detectors. As Chapter 5 shows, the summary depreciates Bragg resolution. This method can give low RR values over a wide wavelength range if properly selected crystals are used and if their surfaces are properly prepared.

Solid ionization detectors not only give good performance today, but they stand ready to serve such laudable causes as truth in art and the interests of the Internal Revenue Service by helping to establish the authenticity of paintings. Of two portraits examined by the Kevex Corporation, that by Velasquez (circa 1658) of Queen Mariana of Austria (age, 23), though the more costly ($1,000,000), yielded less interesting results than the brown shoulder of a South Seas maiden painted by Gauguin. The maiden appears in Figure 2.25–4, and the composition of her shoulder in Figure 2.25–5. Peaks are present for six elements ranging from sulfur to lead, and peak intensities could be translated to semi-quantitative estimates of amounts present. More precise data could probably have been obtained on a better defined sample, and further progress may be expected.

Results like those in Figure 2.25–5 are adequate for much work done on the x-ray emission electron microprobe ("electron microprobe" for short). Because these spectrographs now carry as many as four spectrometers to cover the wavelength range of analytical lines by Bragg reflection, detection systems with solid ionization detectors will no doubt find their most welcome early applications on these microprobes.

For a report of important progress on pure germanium detectors, see Chapter 10, Reference 38.

REFERENCES

1. *Radiography in Modern Industry,* Eastman Kodak Co.
2. B. Lindström, "Roentgen Absorption Spectrophotometry in Quantitative Cytochemistry," *Acta Radiol. Suppl.,* **125** (1955).
3. (a) J. Eggert and W. Noddack, *Z. Physik,* **20**, 308 (1923); (b) *ibid.,* **43**, 222 (1927); (c) *ibid.,* **51**, 796 (1928).
4. (Mrs.) Kathleen Yardley Londsdale, *Crystals and X-Rays,* Van Nostrand, New York, 1949, p. 35.
5. W. Crookes, *Proc. Roy. Soc.,* **71A**, 405 (1903).
6. R. H. Morgan, *Am. J. Roentgenol. Radium Therapy,* **48**, 220 (1942).
7. H. M. Smith, *Gen. Elec. Rev.,* **48**, 13 (March 1945), C.D. Moriarty, *Elec. Eng.,* **64**, 433 (1945); *Gen. Elec. Rev.,* **50**, 39 (February 1947).
8. H. Kallman, in *Electron and Nuclear Counters,* S. A. Korff, Ed., Van Nostrand Co., New York, 1955, Chap. 8.
9. J. B. Birks, *Scintillation Counters,* McGraw-Hill, New York (1953).

REFERENCES

10. J. H. Neiler and P. H. Bell, in *Alpha-, Beta- and Gamma-ray Spectroscopy*, Vol. 1, K. Siegbahn, Ed., North-Holland Publishing Co., Amsterdam, 1966 Chap. 5.
11. S. A. Korff, *Electron and Nuclear Counters*, Van Nostrand, New York 1955, Chap. 8.
12. S. C. Curran and J. D. Craggs, *Counting Tubes*, Academic, New York, 1949.
13. B. B. Rossi and H. H. Staub, *Ionization Chambers and Counters*, McGraw-Hill, New York, 1949.
14. D. H. Wilkinson, *Ionization Chambers and Counters*, Cambridge University Press, London, 1950.
15. S. C. Curran and H. W. Wilson in chapter 6 of the volume that contains Ref. 10 as chapter 5.
16. S. C. Curran, J. Angus, and A. L. Cockroft, *Phil. Mag.*, [7], **40**, 36 (1949).
17. H. Friedman, L. S. Birks, and E. J. Brooks, *Am. Soc. Testing Materials Spec. Tech. Publ.*, **No. 157**, 20 (1954).
18. C. F. Hendee, S. Fine, and W. B. Brown, *Norelco Reptr.*, **3**, 40 (1956); *Rev. Sci. Instr.*, **27**, 531 (1956).
19. E. W. Balis, L. B. Bronk, H. G. Pfeiffer, W. W. Welbon, E. H. Winslow, and P. D. Zemany, *Anal. Chem.*, **34**, 1731 (1962).
20. (a) B. L. Henke, *Advances in X-Ray Analysis*, Vol. 6, Plenum Press, New York, 1963, p.361; (b) Vol. 7, 1964, p. 460; (c) Vol. 8, 1965, p. 269; (d), Vol. 12, 1969, p. 495.
21. (a) R. A. Mattson, *Advances in X-Ray Analysis*, Vol. 8, Plenum Press, New York, 1965, p. 333; (b) R. C. Mattson and R. C. Ehlert, private communication.
22. K. G. McKay, *Physics Today*, **6**, 10 (May 1953).
23. W. M. Gibson, G. L. Miller, and P. F. Donovan in Chapter 6 of the volume that contains Ref. 10 as Chapter 5.
24. ORTEC Catalog Number 1001, ORTEC Inc., Oak Ridge, Tenn., 1968.
25. K. G. McKay, *Phys. Rev.*, **76**, 1537 (1949).
26. E. M. Pell, *J. Appl. Phys.*, **31**, 291 (1960).
27. E. Elad and M. Kakamura, *IEEE NS*, **15**, No. 3, 477 (1968).
28. H. R. Bowman, E. K. Hyde, S. G. Thompson, and R. C. Jared, *Science*, **151**, 362 (1966).
29. R. Fitzgerald, K. Keil, and K. F. J. Heinrich, *Science*, **159**, 528 (1968).
30. B. C. Giessen and G. E. Gordon, *Science*, **159**, 973 (1968).
31. U. Fano, *Phys. Rev.*, **72**, 26 (1947).
32. Ref. 24, p. 49.
33. Ref. 24, p. 69.
34. (a) P. D. Zemany, *Spectrochim. Acta*, **10**, 736 (1960); (b) R. M. Dolby, *Proc. Phys. Soc. (London)*, **73**, 81 (1961); (c) L. S. Birks and A. P. Batt, *Anal. Chem.*, **35**, 778 (1963); (d) L. S. Birks, R. J. Labrie, and J. W. Criss, *Anal. Chem.*, **38**, 701 (1966).
35. H. Friedman, L. S. Birks, and E. J. Brooks, *Am. Soc. Testing Materials Spec. Tech. Publ.*, **No. 157,** 19 (1954).
36. W. Parrish and T. R. Kohler, *Rev. Sci. Instr.*, **27**, 795 (1956).
37. D. M. Miller, *Norelco Reptr.*, **3**, 71 (1956).
38. R. Jenkins and J. L. de Vries, *Practical X-Ray Spectrometry*, Springer-Verlag, New York, 1967, p. 66.
39. D. C. Miller, *Norelco Reptr.*, **4**, 37 (1957).
40. K. F. J. Heinrich, *Advances in X-Ray Analysis*, Vol. 4, Plenum Press, New York, 1961, p. 370.

41. Ref. 38, p. 84.
42. W. B. Jones and R. A. Carpenter, *Advances in X-Ray Analysis*, Vol. 11, Plenum Press, New York, 1968, p. 214.
43. Questionnaire received October 21, 1969, from J. R. Rhodes, Consultant, Columbia Scientific Industries, Austin, Tex., 78702.
44. F. J. Walter and B. J. Moore, paper given at Eastern Analytical Symposium, New York, November 1968.

Chapter 3

Absorptiometry with X-Rays

> I'm full of daze,
> Shock and amaze;
> For nowadays
> I hear they'll gaze
> Thro' cloak and gown—and even stays,
> These naughty, naughty Roentgen Rays.
>
> Wilhelma, in *Photography* [1]

The quotation above testifies to the early impact of *absorptiometry with polychromatic x-ray beams*. Even Roentgen, although he meticulously laid a broad foundation for the technique, must have been surprised at the phenomenal interest it aroused. Because the penetrating power of x-rays made it possible for the first time to learn about the interior of opaque objects without opening or destroying them, applications for the technique soon became legion. On February 19, 1896 it was thought advisable to introduce into the New Jersey legislature a bill "prohibiting the use of x-rays in opera glasses in theaters" [2].

Absorptiometry with polychromatic x-ray beams somewhat resembles colorimetry with white light. When a fluorescent screen serves as detector, the technique is called *fluoroscopy*; when a photographic film on plate is thus used, it becomes *radiography*. By these names, it has become a method for the nondestructive testing or scanning of samples animate or inanimate, and an important method for determining thickness. A polychromatic x-ray beam passes from source through sample to detector: the attenuation of the beam—in general not uniform—gives information about the sample.

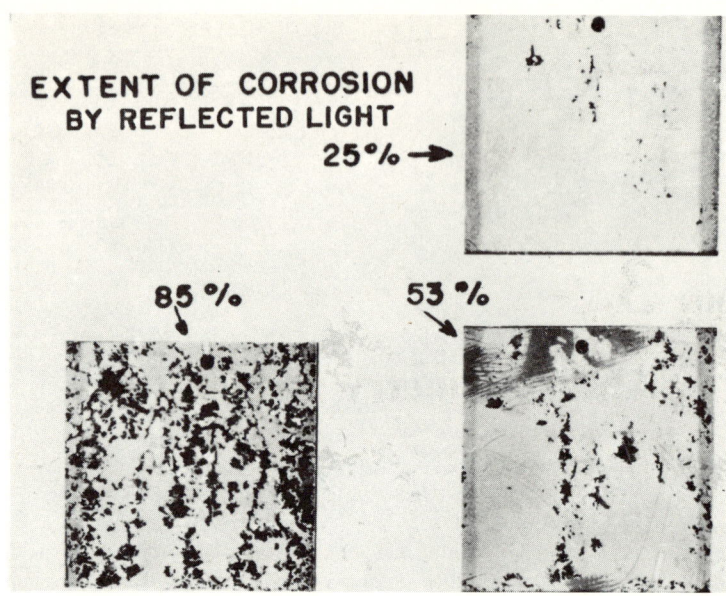

Fig. 3.0–1A

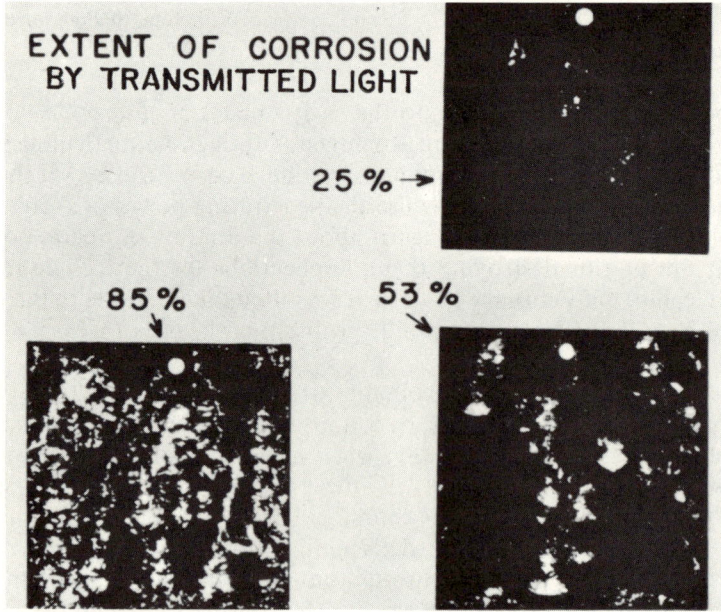

Fig. 3.0–1B

ABSORPTIOMETRY WITH X-RAYS

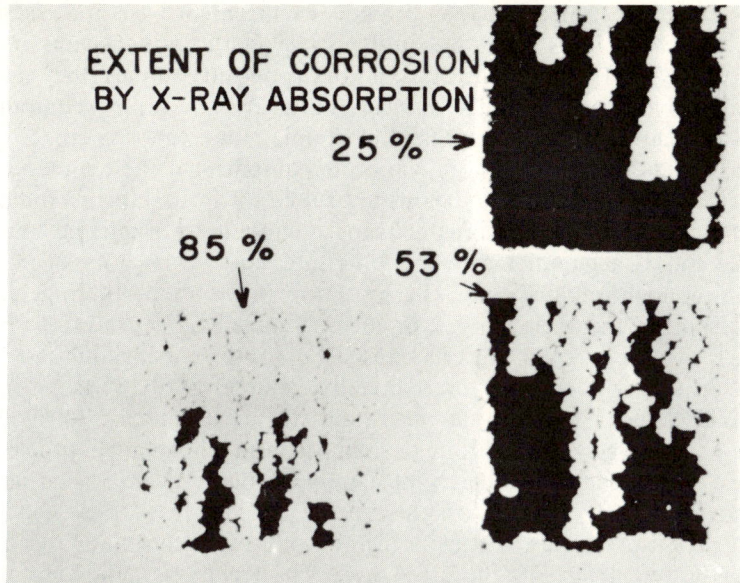

Fig. 3.0–1C

Fig. 3.0–1. X-ray absorption not only reveals extent of corrosion as established by weight loss but shows "pine-tree pits." See H. A. Liebhafsky and A. E. Newkirk, Ref. 3, for experimental details.

As Figure 3.0–1 shows, the method is to be preferred in corrosion experiments over examination by reflected or with transmitted light even when attack has gone far enough to perforate the specimen [3].

X-ray absorptiometry with *polychromatic* beams suffers the great weakness of not being *specific*. It is therefore scarcely surprising that the method has found much wider application in the *control of processes* than in the *characterization of materials*, these being the chief subdivisions of analytical chemistry as here understood. The future promises no change, and two chapters (3 and 5) of the precursor [4] to this book have consequently been combined here into one. The authors believe that x-ray absorptiometry simply carried out has a place in any large analytical laboratory, and they advise the interested reader to consult Ref. 4.

3.1 Monochromatic and Polychromatic Beams

A wholly monochromatic beam of any wavelength does not exist, but some beams are more "monochromatic" than others. For x-ray absorptiometry with such beams, a considerable wavelength range can usually be tolerated.

Two imperfect generalizations provide a convenient basis for assessing the relative usefulness of mono- and polychromatic x-ray beams in absorptiometry. (1) Polychromatic beams are strong and complex; they cannot generally give specific results but can be used with simple equipment. Monochromatic beams are weak and simple; they can sometimes give specific results but require more elaborate equipment if their source is an x-ray tube. (2) Polychromatic beams are suited for instantaneous intensity measurements; for monochromatic beams, quanta must usually be counted and the counts accumulated over a time interval.

With monochromatic beams, there are no worries about filtering, effects of unsuspected absorption edges, or effective wavelengths. And there is far less risk of intensity fluctuations caused by equipment instabilities. This statement is obviously true for radioactive sources of x-rays (see below), but it is also true of conventional x-ray tubes. In an extreme case (low voltage across a Coolidge tube), output currents from a phosphor-photoelectric detector measuring the intensity of a polychromatic beam varied as the *24th* power of the voltage across the tube [5].

For didactic reasons, absorptiometry with monochromatic beams—though last to arrive—is discussed first. The technique subdivides itself logically into (1) simple absorptiometry, not usually specific and governed in straightforward fashion by Beer's law, and (2) differential absorptiometry across an absorption edge (absorption-edge absorptiometry for short), which is specific in principle for all elements with sharp absorption edges [6].

3.2 Monochromatic X-Ray Beams

Four kinds of "monochromatic" beams need to be considered. (The first sentence of Section 3.1 explains the quotation marks.) Each kind is largely taken for granted here and receives further attention in later chapters in connection with topics other than absorptiometry. (See index.) To name the four: (1) beams produced by Bragg reflection; (2) filtered beams; (3) beams in which a characteristic line dominates a background that can be neglected; and (4) beams from radioactive sources. The first three kinds are conventional. The fourth, a by-product of our atomic age, is growing so rapidly in importance that it will soon be conventional too.

In the radioactive x-ray sources of interest here, the following sequence of events may be thought to occur. The nucleus captures an electron from the atom, usually from the **K** shell, and this transmutes the element into the one next nearer hydrogen in the periodic table. The **K** shell vacancy in the new element is now filled by an **L** electron, whereupon a **K** line of the new element is radiated. Although the intensity of this **K** line is usually low, the line is no different for purposes of chemical analysis from the same line generated at the same intensity in any other way, but the use of radioactive x-ray sources can nevertheless bring important advantages.

Many radioisotopes are now known [7], and their number is growing. However, the number of those that meet the requirements of a monochromatic x-ray source satisfactory to the analytical chemist is not large. These requirements are (1) ready availability at reasonable cost, (2) proper half-life (see below), (3) suitable wavelength and adequate spectral purity (emitted electrons, positive or negative, present no problem as they are readily absorbed), and (4) absence of intolerable hazards. The proper half-life in a radioactive x-ray source is inevitably a compromise between a long half-life, which makes for low but (almost) constant intensity; and a short half-life, which makes for an intensity that can be high but decreases rapidly. To qualify as desirable, an x-ray source must be intense enough for the job in hand, and constant enough so as not to require prohibitive recalibration. Hughes and Wilczewski [8] considered all these requirements in choosing $_{26}Fe^{55}$ as an x-ray source for the determination of sulfur.

ABSORPTIOMETRY WITH MONOCHROMATIC BEAMS

3.3 Simple Absorptiometry

The determination of sulfur in hydrocarbons [8–10] is an excellent application to introduce and illustrate simple absorptiometry with monochromatic beams. It is important, has been carefully investigated with a conventional and a radioactive x-ray source, and has been reliably compared with other methods.

The point of departure for the method is (1.10–5), which becomes

$$\mu_U = \mu_H W_H + \mu_C W_C + \mu_S W_S \qquad (3.3-1)$$

where μ_U is the mass absorption coefficient of the unknown, and the other mass absorption coefficients and weight fractions are for the principal elements present: we speak here of sulfur as the element to be determined in a *matrix* containing hydrogen and carbon. For purposes of calibration, the mass absorption coefficient of the standard replaces μ_U in (3.3–1).

The apparatus first used [9] is shown in Figure 3.3–1. It is interesting because it is a modified diffractometer reminiscent of an optical goniometer, the ancestor of the Bragg spectrometer (Section 1.18). The liquid unknown or standard is in the cell on the right, which is supported on the goniometer arm so as to rotate with the Geiger detector against which it abuts. A partially collimated x-ray beam from a molybdenum target enters the figure through a pin-point aperture at the left and passes through the cell and sample to the detector. The time t for molybdenum $K\alpha$ to give 10,000 counts is measured for a sample between measurements of t_0, the time for 10,000 counts with standard of nickel in place of the cell. Because *time to*

Fig. 3.3–1. North American Philips (Philips Electronics) x-ray spectrometer converted to a spectrophotometer for absorption measurements. Courtesy of Hughes and Wilczewski, *Proc. Am. Petroleum Inst.*, **30 M (III)**, 11 (1950).

fixed count and intensity are inversely proportional,

$$\bar{I}_0/I = t/\bar{t}_0 \tag{3.3-2}$$

$\bar{t}_0$ being the mean of the two values taken.

Note that equipment errors tend to be compensated by sandwiching a measurement on the sample between measurements on a standard and using the mean of the latter in the calculations. This makes the method a *comparative method*, of which we shall have many more examples. Also, the detection system here is analogous to a *coulometer* as it *accumulates* counts; not to an *ammeter*, which acts as an *intensity* or *rate* (of electron flow) meter. See the previous discussion in Chapter 2.

To simplify further discussion, we shall assume that (3.3–2) is for one gram of sample. Then, from Beer's law and (3.3–1), we have

$$\log(\bar{I}_0/I) = \log(t/\bar{t}_0) = k'\mu_U \tag{3.3-3}$$

where the empirical constant reflects (among other things) that nickel is being used to determine I_0. A relationship of the middle term of (3.3–3) to

ABSORPTIOMETRY WITH MONOCHROMATIC BEAMS

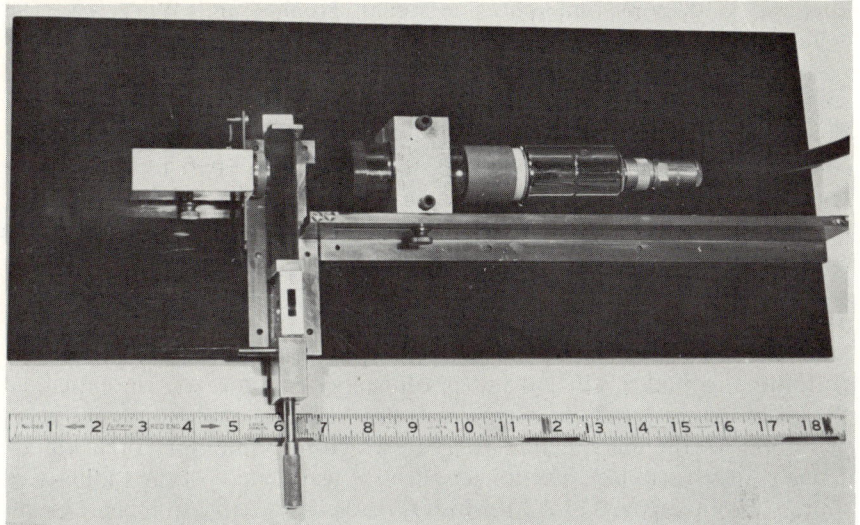

Fig. 3.3–2. An absorptiometer with a radioactive isotope ($_{26}Fe^{55}$) as x-ray source. The button of $_{26}Fe^{55}$ is fused to a thin piece of platinum to protect against oxidation and soldered to a brass rod, which is enclosed in the rectangular brass block on the left of the figure. A spring-loaded shutter closes the exit port in this block except when the cell, below the optical axis in the figure, is in position for measurement. The cell, capacity about 1.5 cc, has 18-mil beryllium windows, and fits snugly into its holder, which slides between the two brass guide rails visible above. The Geiger detector, in the aluminum holder with set screw (right of figure), can be moved closer to the source to maintain constant intensity as the source decays; decay becomes noticeable in about 2 weeks. Background, due mainly to cosmic rays, is appreciable and must be subtracted. See H. K. Hughes and J. W. Wilczewski, Ref. 8.

W_S is needed, and the derivation of this relationship shows how one deals with the matrix in this case. By use of

$$W_H + W_C + W_S = 1 \tag{3.3-4}$$

and

$$W_C/W_H = r_{CH} \tag{3.3-5}$$

(3.3–1) can be transformed into

$$(1 + r_{CH}) \mu_U = \mu_H(1 - W_S) + \mu_C(1 - W_S)r_{CH} + \mu_S W_S(1 + r_{CH}) \tag{3.3-6}$$

which may be written

$$\mu_U = \alpha + \beta W_S \tag{3.3-7}$$

where α and β vary quite slowly with r_{CH}. At constant r_{CH}, (3.3–3) tells us that $\log[t/t_0]$ should give a straight line when plotted against ρW_S as abscissa. With some differences in detail, Hughes and Wilczewski [9] achieved this result and showed the method to be practical—but, see below.

At the wavelength (0.71 Å) of molybdenum Kα, the three mass absorption coefficients of (3.3–6) have the (older) values: μ_H(0.381); μ_C(0.625); μ_S(9.89). The use of molybdenum Kα, desirable to gain high intensity and to decrease counting times, thus brings the minor disadvantages of barring work in the region (near 0.5Å) where variations in r_{CH} could have been neglected because the mass absorption coefficients of carbon and hydrogen are nearly enough identical.

When $_{26}$Fe^{55} became available, Hughes and Wilczewski [8] found it possible to improve the method just described by using this radioactive isotope as an x-ray source. Four millicuries of iron, in the form of a button that initially had 15(10^7) disintegrations per second, was mounted as shown in Figure 3.3–2, the Geiger detector being movable to permit compensation for the unavoidable decay to which this and every other radioactive source is subject. The longer effective wavelength of the radioactive source (2.05 Å versus 0.71 Å for Mo Kα) made it advisable to replace nickel as standard absorber by aluminum. Because μ_S at the longer wavelength is near 206, the cell length could be reduced from 10 to 0.5 cm. Tables 3.3–1 and 3.3–2 show that the $_{26}$Fe^{55} method is so reliable that it seems preferable to three others.

Table 3.3–1. Comparison of $_{26}$Fe^{55}, Bomb-Sulfur, and Lamp-Sulfur Methods for Sulfur in Hydrocarbons

$_{26}$Fe^{55} Method	Bomb or Lamp Method	Difference	$_{26}$Fe^{55} Method	Bomb or Lamp Method	Difference
0.11	0.05	+ 0.06	0.27	0.30	− 0.03
0.27	0.30	− 0.03	0.03	0.05	− 0.02
0.17	0.20	− 0.03	0.62	0.59	+ 0.03
0.16	0.15	+ 0.01	0.73	0.75	− 0.02
1.58	1.58	0.00	1.24	1.22	+ 0.02
0.15	0.07	+ 0.08	0.09	0.06	+ 0.03
0.49	0.48	+ 0.01	0.07	0.07	0.00
0.64	0.60	+ 0.04	0.05	0.05	0.00
0.37	0.34	+ 0.03	0.11	0.13	− 0.02
0.44	0.46	− 0.02	0.02	0.05	− 0.03
2.68	2.76	− 0.08	0.05	0.01	+ 0.04
0.32	0.26	+ 0.06	0.54	0.49	+ 0.05
0.94	0.87	+ 0.07	0.06	0.05	+ 0.01
0.07	0.06	+ 0.01	0.02	0.04	− 0.02
0.50	0.48	+ 0.02	0.07	0.04	+ 0.03

The results by the $_{26}$Fe^{55} method were predictable from (3.3–6) by use of the proper mass absorption coefficients. It is gratifying that empirical factors such as those included in (3.3–7) were not needed. Such empirical factors do not necessarily vitiate a method and need cause no concern when they are understood.

Table 3.3–2. Comparison of Speed, Accuracy, and Cost for Determination of Sulfur in Petroleum Products

Method	Speed			Accuracy Sulfur, %	Relative Cost	
	Elapsed Time, hr	Operator Time, min	Samples per Day		Equipment	Operator Time
Bomb-sulfur	24	36	13	0.03–0.14	5	7.2
Lamp-sulfur	3	29	16	0.03	1	5.8
Old x-ray method [9]	0.12	7	48	0.09	20	1.4
$_{26}$Fe^{55} [8]	0.08	5	60	0.05	6	1.0

Additional data on sulfur determination by simple absorptiometry are given by Hughes and Hochgesang [10] in a paper concerned principally with the successful use of this method for the determination of tetraethyllead in gasoline. In their routine method, molybdenum Kα and a nickel standard were used [9]. They tried both molybdenum Kα and stopped-down polychromatic beams and reached the thoroughly logical conclusion that monochromatic beams are to be preferred under their experimental conditions.

Monochromatic beams should have a clear advantage over polychromatic beams in the absorptiometry of gases whenever it is necessary to use long-wavelength x-rays to ensure appreciable absorption, the reason being that monochromatic rays escape complications due to filtering. A gas analyzer made by Philips [11] uses a monochromatic beam and a Geiger detector.

In general, simple absorptiometry at more than one wavelength is of limited value because, owing to its nonspecific character, the additional information thus obtained is not worth the additional effort. Such measurements at more than one wavelength are occasionally useful as confirmatory evidence or when special circumstances exist as in the following case, which was examined by Coppens [12]. The K edge of $_{40}$Zr is near 0.69 Å; that of $_{39}$Y is near 0.73 Å. Accordingly, xenotime (YPO$_4$) absorbs Mo Kα (0.71 Å) much more strongly than does zircon (ZrO$_2 \cdot$ SiO$_2$), and an absorbance measurement at this wavelength is a sure and easy way of distinguishing between these two minerals, which are otherwise nearly identical in physical and mineralogical properties. Added absorbance measurements at one or more additional wavelengths make it possible to distinguish these two minerals from others, such as sphene ($CaTiSiO_5$), that contain no heavy elements.

3.4 Differential Absorptiometry Across an Edge. The Absorption-Edge Method

An early, interesting experiment, diagrammed in Figure 3.4–1, by de Broglie [13] serves to introduce the method. By continuous rotation of the crystal in a Bragg spectrometer, de Broglie found that he could record on a stationary photographic plate the emission spectrum generated at the tungsten target of a Coolidge tube. The **K** and **L** lines of tungsten all appeared at the proper values of θ (see Figure 3.4–2, in which some of the lines are very faint). There appeared to be "lines" at 0.482 and 0.916 Å, wavelengths somewhat shorter than the **K** lines of shortest wavelengths for Ag and for Br. These "lines," which were on the long wavelength sides of regions of continuous absorption, seemed at first to be associated with x-ray bands. Bragg and Siegbahn [13b] soon interpreted the regions correctly as resulting from the increase of absorption with wavelength for the silver and for the bromine in the photographic film. The spurious "lines" in fact were the absorption edges; they mark the abrupt transition to a region of lower absorbance in the direction of increasing wavelength. The absorption regions under discussion here sometimes complicate the interpretation of photographically recorded spectra or diffraction patterns.

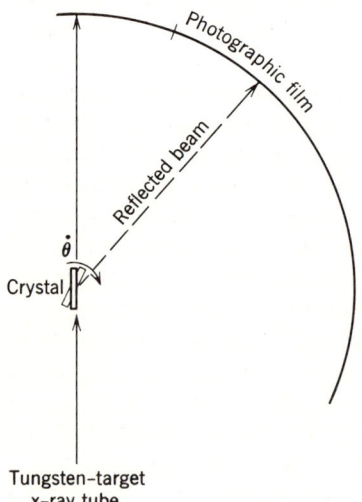

Fig. 3.4–1. Schematic diagram of de Broglie's experiment. With crystal and x-ray beam initially parallel, the beam registered at the left end of the film. As the crystal rotated at angular velocity $\dot{\theta}$, the various characteristic lines of tungsten appeared at the proper values of θ. Registration of the **K** edge of silver and of bromine was an unexpected result. See Figure 3.4–2 and text. The arrangement in the figure somewhat resembles a powder diffraction camera. In the camera, all crystal orientations exist in the sample; above, the orientation of a single crystal is varied by rotation. In both cases, the detector is a stationary photographic film (See M. de Broglie, Ref. 13.)

ABSORPTIOMETRY WITH MONOCHROMATIC BEAMS

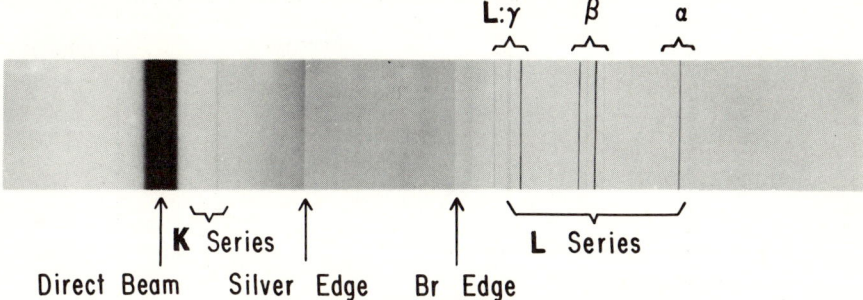

Fig. 3.4–2. The spectrum recorded photographically in the experiment of Figure 3.4–1. Modern values for the wavelengths of the lines and for the critical absorption wavelengths will be found in the Appendices. See M. de Broglie, Ref. 13.

Additional work soon proved that absorption edges of elements other than silver or bromine could be photographed by interposing the element suitably between the x-ray source of a polychromatic beam and the photographic plate. In such records, the blackened portion is, of course, on the *long-wavelength* side of the edge, this being the region in which the sample is more nearly transparent.

In addition to the absence of lines in absorption, a striking feature uncovered by this early work is the sharpness of the absorption edges. This sharpness suggests that differential absorptiometry across the edge might be the means of identifying an element by locating the edge (qualitative) and of determining the amount thereof present by measuring the change in absorbance across the edge (quantitative). Glocker and Frohnmayer [6] appreciated this in 1925 and proved that this analytical method could give excellent results. They used a Siegbahn spectrometer to generate monochromatic x-rays of wavelengths that bracketed absorption edges and used a photographic plate as detector. Representative results are given in Figure 3.4–3, which shows photometric data for a plate that recorded the change in absorbance across the **K** edge of barium in a sample of aqueous barium chloride solution. The reliability of this, the absorption-edge method, can be judged from Table 3.4–1.

Table 3.4–1. Barium in Barium Chloride Solution [6]

Ba present, mg/cm^2	4.5	6.4	9.2	11.2	15.1	20.0	25.2	28.0	36.4	44.8	52.6
Ba found, mg/cm^2	5.8	7.7	9.0	12.4	16.9	19.6	24.3	26.4	36.6	44.0	46.5

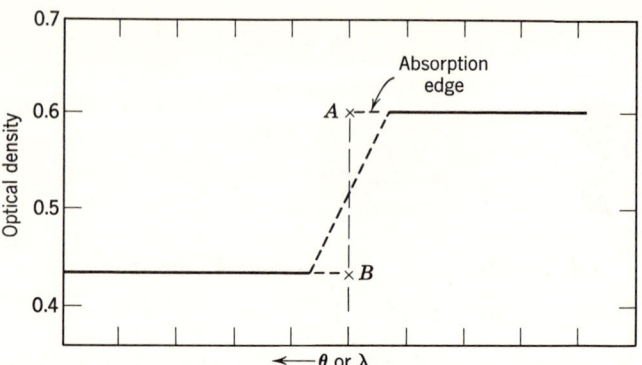

Fig 3.4–3. Analysis by the absorption-edge method. The solid lines represent photometric data from a photograph of the x-ray intensity as a function of angle. The concentration is calculated from the ratio of these lines extrapolated to the absorption edge. Table 3.4–1 gives some typical results.

Table 3.4–1 gives the amounts of barium dissolved per square centimeter of cell cross-section; as the cell length was 1 cm, this is also the amount of dissolved barium in 1 cm³. At this cell length, results for 1.9 and 2.8 mg/cm² were not precise enough to be quoted; in longer cells, greater precision would no doubt have been obtained.

To introduce the absorption-edge method in a modern application, let us assume that total bromine (free or combined), at weight-fraction W_{Br} is to be determined in a sample, the rest of which is considered to be a matrix M. The experimental data will be the results of absorbance measurements at wavelengths slightly greater (λ) and at wavelengths slightly less (λ') than that of the absorption edge of bromine, 0.92 Å. By analogy with the sulfur determination of Section 3.3, we may write for a cell of 1-cm² cross-sectional area containing m grams of sample

$$2.303 \log(I_0/I) = \mu_{Br} W_{Br} m + \mu_M W_M m \quad (\lambda > 0.92 \text{ Å}) \quad (3.4–1)$$

$$2.303 \log(I_0'/I') = \mu_{Br}' W_{Br} m + \mu_M' W_M m \quad (\lambda' < 0.92 \text{ Å}) \quad (3.4–2)$$

where

$$\mu_{Br}' > \mu_{Br} \quad (3.4–3)$$

The more nearly the wavelength interval $\lambda' - \lambda$ approaches zero, the more nearly

$$I_0 = I_0' \quad \text{and} \quad \mu_M = \mu_M' \quad (3.4–4)$$

For such conditions, one finds by substitution and by subsequent subtraction of (3.4–2) from (3.4–1) that

$$2.303 \log(I'/I) = (\mu_{Br} - \mu_{Br}') W_{Br} m = -cW_{Br} m \quad (3.4–5)$$

Under the simple conditions just outlined, only two intensity measurements (I and I') are needed. Under actual conditions, additional measurements may be necessary.

A few general remarks about the absorption-edge method are in order here. For any element, two absorption edges, the **K** and the **L$_{III}$** edges, are most likely to be of practical importance. The **M** and **N** edges are not generally useful; and among **L** edges, **L$_{III}$** has the greatest absorption jump. (See below.) Therefore, the probability of interference due to the presence of neighboring absorption edges is small, although it needs to be considered in special cases: the wavelength of the **K** edge of thallium (0.145 Å) exceeds by less than 0.004 Å that of the **K** edge of lead.

In the literature of physics, an absorption edge is usually characterized by the critical absorption wavelength and by the magnitude of the absorption jump [14b], which in terms of (3.4–5) is the ratio μ'_{Br}/μ_{Br}. The absorption jumps at the **K** and at the **L$_{III}$** edges are generally the only ones large enough to be useful, and their variation with atomic number must be taken into account; the absorption jump at the **K** edge decreases in magnitude with increasing atomic number [14a]. For the present purpose, however, the magnitude of the absorption jump is an incomplete criterion because the difference of absorption coefficients, not their ratio, is important here (3.4–5); as a consequence, the **L$_{III}$** edge of a heavy element may be a better choice than its **K** edge even though the **K** edge shows the greater absorption jump.

Even if c in (3.4–5) is fairly large, an element cannot be precisely determined—or may even escape detection—if it is present in too small amount relative to the matrix. What amount is too small depends not only upon the relative mass (or weight-fraction) of element sought but also upon the mass absorption coefficient of the matrix, as (3.4–1) and (3.4–2) imply.

In the absorption-edge method, two kinds of absorbance change may occur: the jump at the absorption edge; and the change caused by the usual variation of mass absorption coefficient with wavelength, which depends strongly on $\Delta\lambda$, the wavelength interval between absorbance values. When intensity measurements made on each side of an edge are extrapolated to the edge so that the final $\Delta\lambda$ is negligible, then the absorbance change is due to the jump alone, and it measures directly the amount present of the element sought. No correction for the change that involves $\Delta\lambda$ is then necessary. If only one measurement is made on each side of the edge, such a correction is often necessary for good results. Part of the correction is due to the element sought, and this part is provided for if the method is standardized against the element. The part due to the other elements, free or combined, in the sample must be estimated or determined empirically; it is often small. To summarize: the absorption-edge method can give results independent of the matrix associated with the element sought; x-ray emission spectrography often cannot. *Achievable freedom from matrix effects is the most attractive feature of the absorption-edge method.*

Finally, it is obvious that the absorption-edge method will often be at a disadvantage relative to determinations based upon x-ray emission because measuring the intensity of a characteristic line is inherently simpler than making a differential intensity measurement across an absorption edge.

Glocker and Frohnmayer determined the characteristic constant c for nine elements (Ref. 6, Table 4) ranging from $_{42}$Mo to $_{90}$Th. Identical results were obtained with the sample in the polychromatic beam (sample between x-ray source and crystal) or in the monochromatic beam (sample between crystal and detector).

A parenthetical note: The former sample position entails the risk of generating undesired characteristic lines in the sample, results in greater generation of heat in the sample, and involves greater risk of changing it by the action of x-rays. The latter is therefore to be preferred unless the former offers compensating advantages.

Glocker and Frohnmayer applied the absorption-edge method with good results [6] to the determination of barium in glass; of antimony in a silicate; of hafnium in the mineral alvite; and of molybdenum, antimony, barium, and lanthanum in a solution of their salts—for example, 5.45% Ba was found on 90-minute exposure for a glass that gave 5.8% by a conventional wet method. Modern equipment has of course made the x-ray method less cumbersome and much quicker than it was in this pioneering era.

3.5   The Determination of Bromine at Dow

The determination of bromine by the absorption-edge method at the Dow Chemical Company in Midland, Michigan, makes an interesting case history [15–17]. An x-ray spectrophotometer was designed and built for this application by Frevel and North. By 1969, this instrument had been replaced by a Picker Spectrodiffractometer. Determinations were then being done at an annual rate near 300 in a cell of the original design.

In the Frevel and North spectrophotometer, a cone of polychromatic x-rays passes through the sample and strikes a multiple-crystal "lens" comprising four sodium chloride crystals, the monochromatic beams from which are focused on a Geiger detector so that the sum of their intensities can be automatically recorded as an output current. Variation of wavelength is accomplished by having a lathe lead screw move the lens and the detector (at twice the lens speed) along the optical axis.

Voltage is stable to within 0.05% during measurement.... The probable error of the mean of a single intensity measurement is 0.5% for all measurements made for a period of approximately 5 minutes at rates of 10,000 counts per minute and higher. Repeated measurements give agreement within 1%. The time for an I/I_0 measurement is 10 to 15 minutes [16a].

ABSORPTIOMETRY WITH MONOCHROMATIC BEAMS

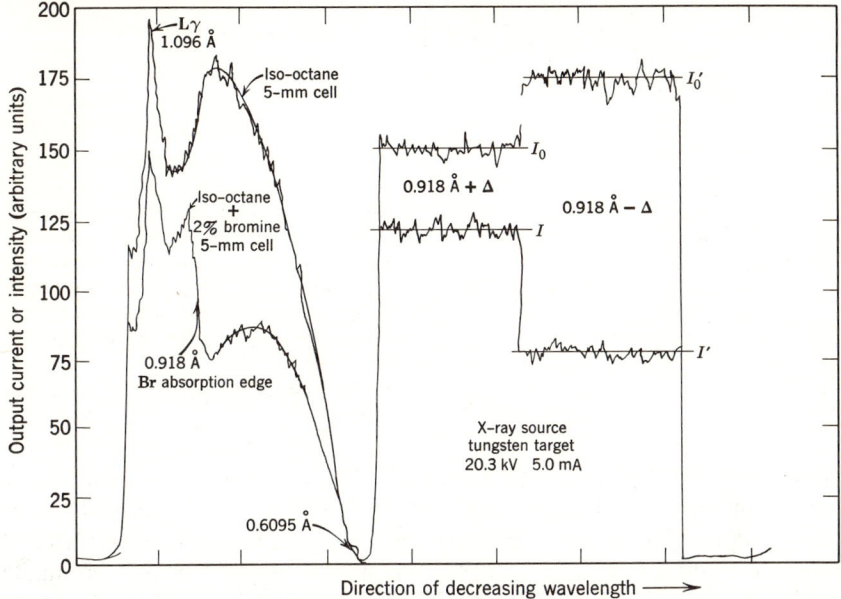

Fig. 3.5-1. Geiger-detector output-currents recorded for the determination of bromine by the absorption-edge method on the Frevel-North automatic x-ray spectrophotometer. Liebhafsky, *Anal. Chem.*, **21**, 17 (1949). Courtesy of Dow Chemical Company.

After the values of log (I/I_0) measured above and below the absorption edge have been extrapolated to the corresponding critical absorption wavelength, the results may be obtained according to (3.4-5).

Typical recorded curves appear in Figure 3.5-1 and show several interesting features. On the left, note the **Lγ** line of tungsten from the target of the x-ray source, and the rapid *apparent* increase in transmittance for the solvent in the region near 0.918 Å, the critical absorption wavelength for the **K** edge of bromine (cf. Figure 3.4-2 and text). This increase is apparent and occurs only because the intensity of the source is increasing rapidly in this wavelength region; it is a tribute to the spectrophotometer that reliable results are obtained with this large variation in intensity. The absorption jump for the bromine sample is clearly evident. On the right of the figure, recorded transmitted intensities at important fixed wavelengths are shown. At each of the two wavelengths, the intensities for the solvent (I_0 and I_0') and for the sample (I and I') are shown. Why $I_0' > I_0$ has just been explained.

3.6 The Absorption-Edge Method at Oak Ridge

The absorption-edge method has had a long history of usefulness at Oak Ridge, where there are analytical problems on materials in the atomic

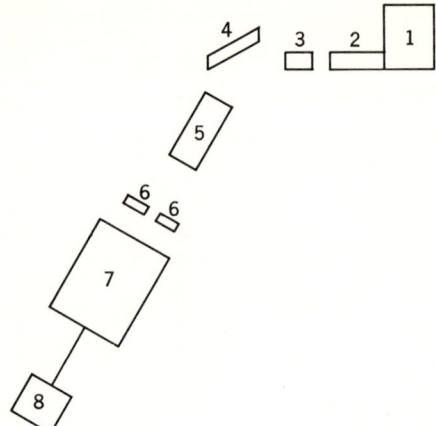

Fig. 3.6–1. Block diagram of General Electric XRD-3 diffractometer as modified by Barringer [19] for absorption-edge determinations [21]. 1. CA-7 x-ray tube; 2. Beam slit; 3. Sample holder; 4. Bragg crystal (NaCl); 5. Soller slit (collimator); 6. Detector slit; 7. Scintillation detector; 8. Preamplifier. See also other references in text.

energy program. By 1952, Peed and Dunn [18] had shown the method to be worthwhile for such problems, and in 1956 Barringer [19] described equipment (Figure 3.6–1) and reported results. A General Electric XRD-3 diffraction unit was used as a spectrometer, and measurements were made across the uranium L_{III} edge with monochromatic beams sorted out by Bragg reflection from the continuous beam obtained at 17 kV from a tungsten target. A scintillation counter served as detector. The 2θ values for the beams were 14.45° and 14.90°, 14.86° being the value corresponding to the edge. The x-ray tube voltage chosen was low enough to eliminate second-order Bragg reflections. The sample was placed between x-ray source and Bragg crystal, not between crystal and detector. The former placing is preferable when the sample is radioactive, or when this procedure reduces the scattered background. Table 3.6–1 indicates the capabilities of the method.

In the determination of molybdenum in uranium alloys containing 2.5% molybdenum [20], the uranium had to be removed because its high absorbance made the method insensitive. After a single solvent extraction, the molybdenum could be determined with a standard deviation of 1.35%, and the x-ray method was judged superior to that using α-benzoinoxime.

Stewart [21] used the same equipment for the determination of thorium in solution, the absorbance measurements being made across the L_{III} edge of the element; critical absorption wavelength, 0.7600 Å. Strontium (K edge, 0.7684 Å) and bismuth (L_I edge, 0.7559 Å) interfere, and the thorium was consequently solvent-extracted from aqueous solution with tri-n-butyl phosphate to eliminate all risk of interference. For a single analysis, the limit of error at the 95% confidence level was found to be ± 2.8% at a concentration of 2-mg thorium/cm³ solution. Duplicate determinations could be completed within 25 minutes.

Table 3.6-1. Absorption-Edge Method for Uranium [19]

Number of Samples	Accepted Value,[a] g U/g sol'n.	Mean of X-ray Analyses,[b] g U/g sol'n.	Bias, %
72	0.053273	0.053238	− 0.06 ± 0.20
32	0.054567	0.05476	0.34 ± 0.42
13	0.054567	0.05450	− 0.12 ± 0.23
64	0.094012	0.094110	0.10 ± 0.14
64	0.017285	0.017304	0.11 ± 0.14
81	2000 ppm	2013.3 ppm	0.66 ± 0.45
64	963	952.8	− 1.07 ± 0.93
64	219	234.1	6.4 ± 2.3

[a] Accepted values known to 0.1%.
[b] Cost of the equipment installed, $32,000. Average cost of a determination, $9.50 (x-ray); $20.00 (wet method).

Finally, the method was extended to accomplish the determination of all three principal constituents in U-Nb-Zr alloys [22]. The following edges were used: U, L_{III}; Nb, **K**; Zr, **K**. The x-ray energies required were below 20 keV. The alloy, in the form of degreased chips, was dissolved; uranium was determined in the strongly acid (HCl, HNO_3, HF) solution, from which the bulk of it was then solvent-extracted so that niobium and zirconium could be determined on the aqueous phase. (The removal of uranium, here as in the molybdenum determination described above, is advisable because of its high absorbance.) The three constituents could be determined on each of 12 samples in 8 hours. Relative standard deviations for alloy-chip samples: U, 0.34%; Nb, 0.90%; Zr, 2.2%.

A similar application has two interesting features that will be met in later work: rotation of the sample to reduce or eliminate effects of nonuniformity and the use of a solid binder as diluent. This work was done by Barieau [23], who adapted a Philips Electronics diffractometer to the absorption-edge method and determined molybdenum and zinc. Barieau mounted solid samples in a rotating holder designed for diffraction work on powders. Among these samples were hydroformer catalysts (8 to 10% MoO_3) and combustion-chamber deposits containing lead and bromine. The deposits, owing to their high absorbance, had to be mixed with a binder (starch or graphite), ground, and pressed into pellets. Analytical results were generally satisfactory, although one set of four molybdenum x-ray values averaged 4 parts per 100 less than those by a wet method.

3.7 Modified Differential Absorptiometry

Various investigators have used differential absorptiometry at two wavelengths that enclose, but do not locate, an absorption edge. This procedure resembles simple absorptiometry at more than one wavelength, and it also resembles the absorption-edge method.

Moxnes [24] used the characteristic lines of one element to bracket the absorption edge of another. He showed, for example, that the addition of 2% ZnO to Al_2O_3 shifted the intensity ratio of tungsten $L\beta 3$ to $L\beta 4$ from 3:2 to about 1:1; the edge is at 1.2833 Å, and the wavelengths of the lines are 1.263 and 1.302 Å.

Engström [25] constructed an ingenious "characteristic-line generator" for measurements of this kind. Collimated x-rays sufficiently high in energy fall upon a sheet of an element mounted (as one of sixteen) upon a disk that can be rotated. The K lines of one element at a time are excited and passed (after filtering, if desired) through the sample to the detector (usually, but not necessarily, a photographic plate). A striking feature of Engström's investigations is the exceedingly small sample size.

Hughes and Hochgesang [26] demonstrated that the industrially important problem of determining tetraethyllead and sulfur simultaneously in a leaded gasoline could be solved satisfactorily in the laboratory by doing differential absorptiometry across the L absorption edges of lead, which are near 0.8 Å. To accomplish this, they used two characteristic lines of thorium, $L\beta 1$ (0.76 Å) and $L\alpha$ (0.96 Å), generated in a gas-filled tube. Two polychromatic beams whose effective wavelengths bracket the L edge of lead can also be used in this way [27], but the method is cumbersome.

ABSORPTIOMETRY WITH POLYCHROMATIC BEAMS

3.8 The Effective Wavelength

The crucial difference between the absorptiometry treated here and that of Section 3.3 is illustrated by the statement that we deal here with relations of the type of (3.3–1), but that we must take into account the variation of mass absorption coefficients with wavelength *for the polychromatic beam*. As this cannot be done exactly and would be highly cumbersome if it could, we resort to the device already introduced in Section 1.6 of the *effective wavelength*.

The effective wavelength λ_e of a polychromatic beam is that of the monochromatic beam exhibiting equivalent behavior in an absorbance measurement. The concept, introduced by Duane [28, 29], has its foundation in the use of absorbance to characterize x-rays (Chapter 1). Although limited in usefulness, the effective wavelength is a valuable guide to the behavior of polychromatic beams. For example, if conditions are simple, μ_{Al} can be determined for such a beam by an absorbance measurement in aluminum. The corresponding λ_e is then obtained from known values of μ_{Al} at different wavelengths. Given this λ_e, it is possible by use of known mass absorption coefficients for the elements in a sample to compute an *effective mass absorption coefficient* by use of the appropriate analogue of (3.3–1).

To illustrate these considerations and to introduce a detector in which measured x-ray intensity is given by an electric current, we use experimental results obtained on the simple laboratory photometer described below. The general approach is broadly applicable in absorptiometry with polychromatic beams.

For a detector that gives current i proportional to x-ray intensity I (i.e., for a *linear* detector),

$$\log(I_1/I_2) = \log(i_1/i_2) = k'(m'_2 - m'_1) \qquad (3.8\text{–}1)$$

where m'_1 and m'_2 refer to the masses (in grams) of two samples of identical composition and not widely different in mass. The proportionality constant, k', is connected with the mass absorption coefficient of the sample by the relation

$$\mu_S = 2.303 \, Ak' \qquad (3.8\text{–}2)$$

where A is the cross-sectional sample area (in square centimeters) perpendicular to the beam. The *filtering* of a polychromatic beam will cause k' and μ_S to decrease.

A brief interpolation: The analytical chemist is accustomed to seeing (3.8–1)—Beer's law—in forms resembling

$$\log(I_0/I) = kc \qquad (3.8\text{–}3)$$

where I is the intensity of longer-wavelength (e.g., visible) radiant energy to which a beam of intensity I_0 is attenuated in passing through a colored solute in a transparent solvent. The obvious difference between the two equations is that one contains mass, the other concentration units; this difference is traceable, of course, to the fact that all the atoms (or ions derived immediately from them) are the absorbing centers for x-rays. However, there is another important difference. With liquids and solids, the intensity ratio, I_0/I, in the case of a polychromatic x-ray beam is usually much higher than in the case of visible light. Values of 1000 or even 10,000 of the intensity ratio I_0/I for x-rays are not unusual. Detectors and the associated electronic equipment are not often linear over these ranges. Nor will k' in (3.8–1) be constant: the effective wavelength of the polychromatic beam certainly changes during attenuation corresponding to these ratios. For these reasons, (3.8–1), which compares two samples with a modest difference in absorption which corresponds to $\Delta m'$, is generally more practical than an x-ray relationship in the form of (3.8–3).

3.9 Simple X-Ray Photometer

Photometers of the type shown in Figure 3.9–1 satisfactorily met the urgent need in World War II for a nondestructive method of testing certain fuzes [30]. Fuzes had to be rejected in which the powder train was so short as to set off prematurely the device they were to activate. This is an example of a *process control* application in which there is no need to establish composition—only mass. Such applications are discussed later; here the purpose is to show that the photometer can quickly give valuable *auxiliary* information about chemical composition [31].

In the phosphor-photoelectric detector of the photometer (A, Figure 3.9–1; see also Chapter 2), the x-ray beam strikes a silver-activated zinc sulfide phosphor to produce blue-violet light that is changed by the photomultiplier tube (Type 931-A) into an electric current that is amplified and read on a suitable micro- or milliammeter. A stable power supply for both x-ray tube and detector circuit are essential, as is clear from the circuit diagrams [30]. That the beam, though broad and divergent, is nevertheless satisfactory for the intended measurements was shown by moving an aluminum foil up and down in the space between x-ray source and detector; this movement did not change the current reading.

Direct absorptiometry on the photometer is carried out as follows. The intensity of the x-ray beam is adjusted to some standard initial value by adjusting the x-ray tube voltage and current until the desired output current is obtained with a standard thickness of aluminum in the beam, the voltage across the detector and the amplifier setting being fixed. The weighed sample contained in the cell is then carefully placed in the beam so as to ensure proper alignment, and an average value of output current then obtained, usually by taking ten readings of the meter with 10-sec intervals between readings. It is often desirable to carry out determinations on at least three weights of sample; this is not prohibitive because each determination requires only a matter of minutes once the sample is in the cell.

Liquids are simplest to handle. With solids, the following precautions must be observed. If the sample is a sheet, its thickness should be uniform to a precision greater than that expected in the analytical result. If the sample is divided, it must be packed into the cell so that the overall density is uniform enough, and the upper surface is level enough, to meet the same precision requirement; this will usually mean that a fine powder should be used.

Gases pose special problems owing to their low densities, and they are discussed later.

No matter what the sample, the data obtained by direct absorptiometry with this photometer are current readings to be evaluated according to (3.8–1). With readings for different weights (or thicknesses) of a sample available, the evaluation is most readily carried out by plotting the readings

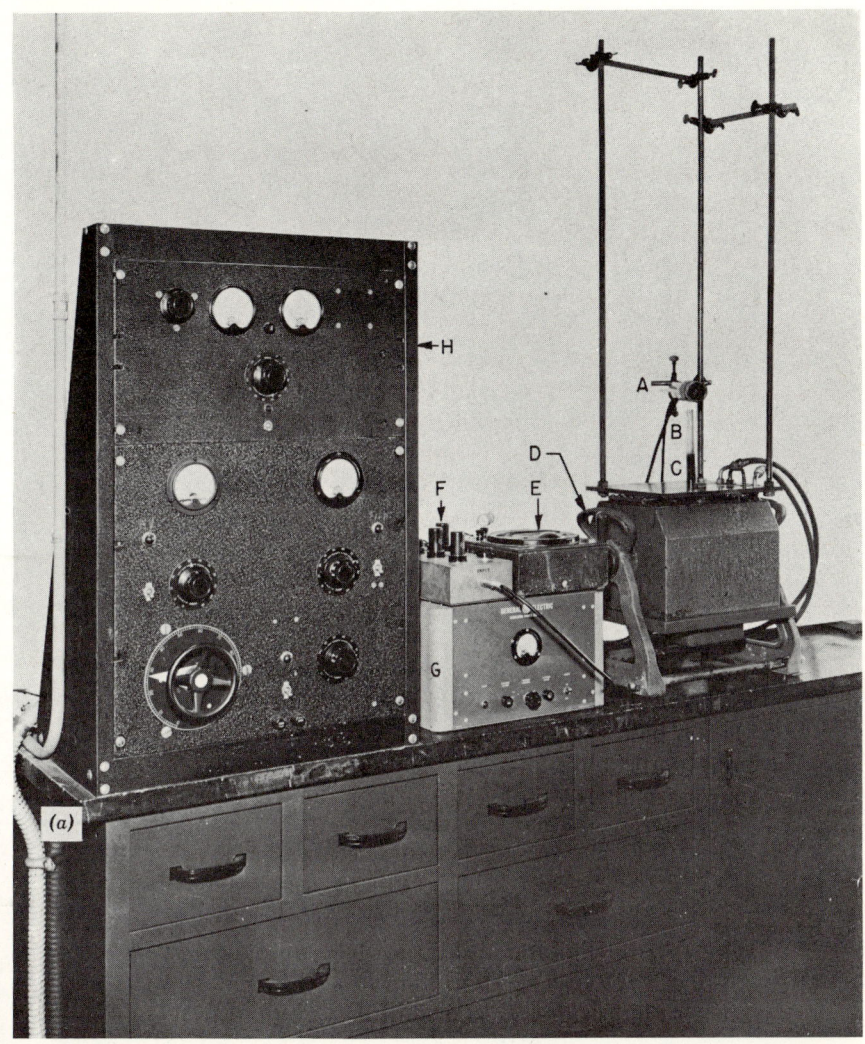

Fig. 3.9–1. Simple laboratory x-ray photometer [30]. (*a*) A, Phosphor-photoelectric detector; B, sample cell; C, sample; D, CA-5 x-ray tube and housing; E, milliammeter; F, amplifier and rectifier vacuum tubes; G, regulated power supply for amplifier tubes and photomultiplier tube; H, control panel.

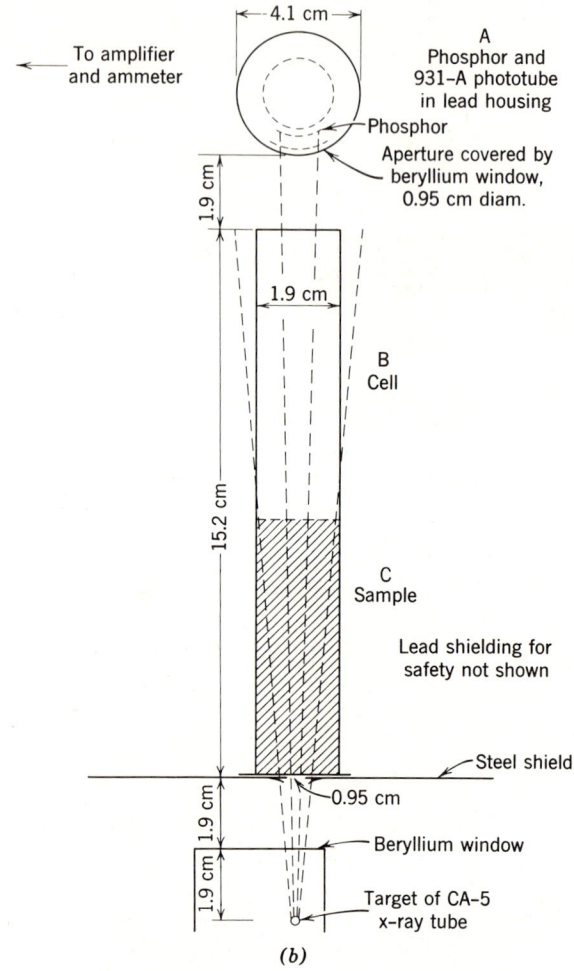

Fig. 3.9–1. Simple laboratory x-ray photometer. (*b*) Schematic diagram. Liebhafsky, Smith, Tanis, and Winslow, *Anal. Chem.*, **19**, 861 (1947).

on a logarithmic ordinate against weight (or thickness) as abscissa. If weights (g) are used, the negative of the slope of the resulting curve is k'; if thickness (cm), the negative of the slope of the resulting curve is $\rho A k'$.

3.10 Determination of Effective Wavelength

Current readings obtained by direct absorptiometry for two series of thicknesses of aluminum are plotted in Figure 3.10–1. For each series, the

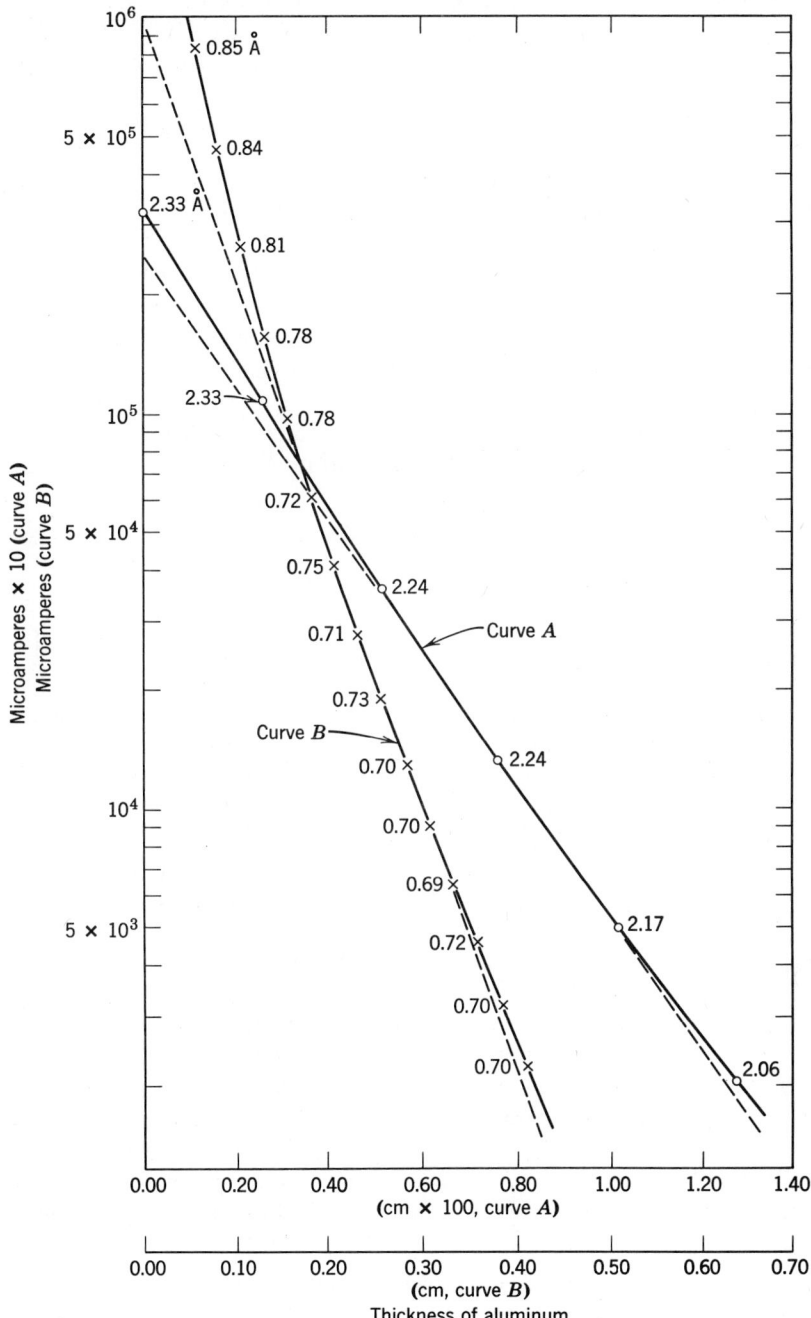

Fig. 3.10–1. Attenuation and filtering of polychromatic x-rays by aluminum. Variation of effective wavelength with thickness. The effective wavelengths shown in the figure correspond to the measured mass absorption coefficients. The change in effective wavelength accounts for the deviations from the (dashed) straight lines. The x-ray intensities used gave 210 μA through 0.0127-cm aluminum (curve A), 3200 μA through 0.381-cm aluminum (curve B). Liebhafsky, Smith, Tanis, and Winslow, *Anal. Chem.*, **19**, 861 (1947).

readings define a curve. To determine whether nonlinearity in the electronic components could have contributed appreciably to this curvature, determinations of effective wavelength were carried out for different thicknesses. It was proved that variations in the effective wavelength accounted for virtually all the curvature [31].

The determination of effective wavelength is of interest here. For each sample of aluminum (of thickness d cm; emergent beam current i_1), an aluminum foil of thickness Δd was placed in the photometer between sample and detector, so that a reduced emergent beam current i_2 could be read. The percentage transmittance of the foil, $100(i_2/i_1)$, was then calculated, and the effective wavelength was read from a plot calculated from known values of μ_{Al} at different wavelengths. This calculation is based upon the relationship

$$2.303 \log(i_1/i_2)_{calc.} = \mu_{Al} \rho \, \Delta d \qquad (d \gg \Delta d) \qquad (3.10\text{-}1)$$

which can be derived from (3.8-1) and (3.8-2) by introducing the density,

$$\rho = \Delta m'/(A \, \Delta d) \, (\Delta m' = m'_2 - m'_1) \qquad (3.10\text{-}2)$$

In the experiments of Figure 3.10–1, Δd was 0.00254 cm for curve A, and 0.0254 cm for curve B. The effective wavelengths found are given in the figure.

Aluminum foil could be used, as previously described, to determine effective wavelengths of x-ray beams emerging from samples other than aluminum.

The effective wavelength provides a useful way of characterizing polychromatic beams that are not appreciably affected by the presence of an absorption edge. One suspects intuitively that the presence of an absorption edge critically located could cause complications in absorptiometry with polychromatic beams. That this does happen has been demonstrated [32], and it limits the usefulness of the effective wavelength. A further limitation arises because the effective wavelength in aluminum may not always be the effective wavelength in other materials—even in the absence of absorption edges. That is to say, the effective mass absorption coefficient (Section 3.8) may be different for different materials. *Comparative absorptiometry* mitigates this difficulty. See Section 3.14.

Figure 3.10–1 shows that the work being discussed here is very closely related to the pioneering investigations of Barkla described in Chapter 1.

3.11 Characterization of Plastics

The experimental data in Figure 3.11–1 can be obtained in a matter of minutes once the materials are at hand, and they illustrate a quick and useful application of direct absorptiometry with simple equipment. Furthermore, they give an insight into the problem of selecting materials to transmit

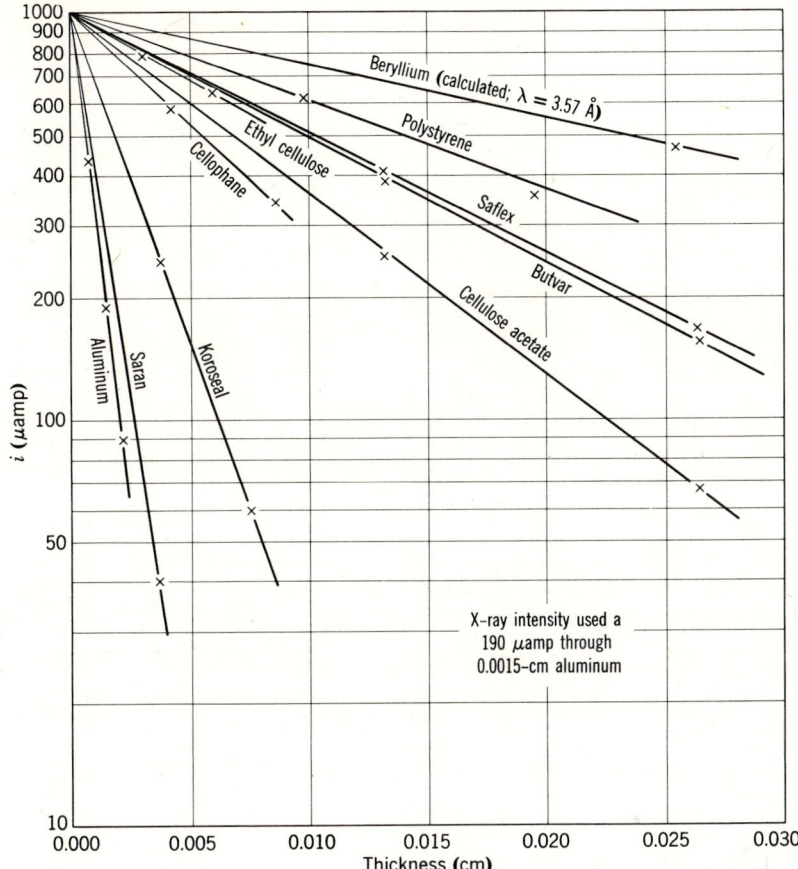

Fig. 3.11–1. Attenuation of a polychromatic beam by various materials. Note superiority of beryllium as a window material. Liebhafsky, Smith, Tanis, and Winslow, *Anal. Chem.*, **19**, 861 (1947).

x-rays (e.g., windows of cells and detectors). No attempt was made to obtain high precision.

Obviously, all the plastics tested are more nearly transparent to 3.5-Å x-rays than aluminum, but more nearly opaque than beryllium. Saran and Koroseal, owing to their chlorine content, are little better than aluminum as regards approach to transparency.

The position of Saran in Figure 3.11–1 relative to that of polystyrene suggests that direct absorptiometry might serve for the determination of chlorine in polymers containing only carbon and hydrogen as other elements.

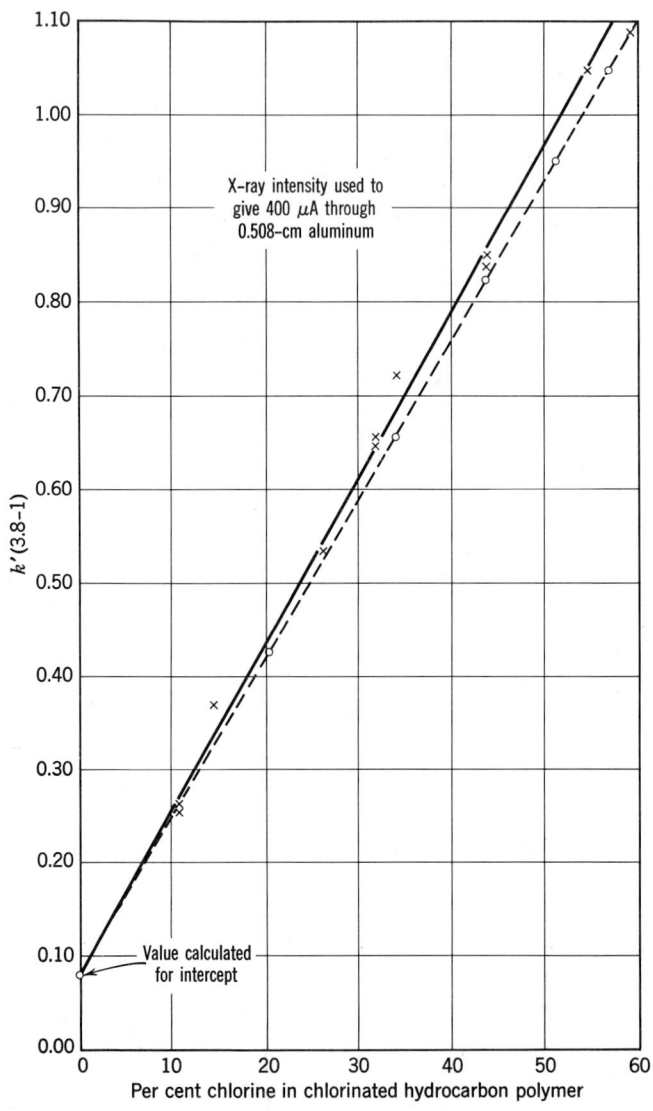

Fig. 3.11–2. Chlorine contents of chlorinated hydrocarbon polymers. Absorptiometric results compared with those from conventional analyses. Crosses = experimental values; dots = calculated values. Liebhafsky, Smith, Tanis, and Winslow, *Anal. Chem.,* **19**, 861 (1947).

ABSORPTIOMETRY WITH POLYCHROMATIC BEAMS

In Figure 3.11–2 are plotted the k' values (ordinates) for various chlorinated hydrocarbon polymers, the chlorine contents as determined analytically after fusion with sodium peroxide being the abscissas. See (3.8–2). The x-ray results were obtained in about *one tenth* the time required for the conventional analyses. The experimental points lie close to the solid line. The deviations from the solid line may be due to any one or any combination of the following causes.

1. Errors in the conventional analyses.
2. Inhomogeneities in the samples.
3. Errors in the x-ray method (including equipment errors).

It is obvious that Figure 3.11–2 serves as a calibration curve with the help of which the chlorine content of unknown, similar materials could be readily determined. The solid line agrees closely with k' values calculated from the mass absorption coefficients of benzene, hydrogen, and chlorine, which shows the usefulness of λ_e in this case. The calculations resemble in principle those based on (3.3–1). See Refs. 4 and 31.

3.12 Gases

Gases lend themselves well to absorptiometry with polychromatic beams [33] provided one recognizes the following differences between such determinations and those made on solids or liquids.

1. Because the mass of the sample is likely to be small, it is advisable to use long-wavelength x-rays to ensure measurable attentuation of the beam.
2. Highly transparent windows are obviously required.
3. With the longer wavelengths, particular attention must be paid to changes in effective wavelength.
4. It is often desirable to calculate the mass of sample from the gas laws. Pressure and temperature must then be known.

For gases, Beer's law is often conveniently used in the form

$$2.303 \log (I_0/I) = 2.303 \log (i_0/i) = \mu \rho d \qquad (3.12-1)$$

where d is the length (in cm) of the sample-containing cell.

X-ray absorption measurements were made on several of the more common gases at atmospheric pressure in a glass cell 60.3-cm long, and 3.8 cm in diameter, with windows of polystyrene 0.010-cm thick. The results for five settings of the x-ray tube voltage are given in Table 3.12–1; only one reading of the output current, i, was taken at each setting.

The second column of Table 3.12–1 contains i_0 values corresponding to the five voltage settings. These i_0 values were calculated from the output

Table 3.12–1. Exploratory Experiments with Various Gases[a]

Setting Number	i_0	H$_2$		CH$_4$		Air		O$_2$		CH$_3$Cl		Cl$_2$	
		i	λ_e	i	λ_e	i	λ_e	i	λ_e	i	λ_e	i	λ_e
1	603,000	600,000		530,000	1.50	320,000	1.59	258,000	1.52	2560	1.25	100	1.17
2	704,500	701,000		620,000	1.50	385,000	1.56	—	—	4100	1.26	195	1.16
3	839,200	835,000		760,000	1.37	480,000	1.52	403,000	1.45	7050	1.22	448	1.13
4	904,400	900,000		805,000	1.45	520,000	1.51	—	—	8600	1.21	598	1.11
5	1,010,000	1,005,000		903,000	1.43	593,000	1.49	500,000	1.43	11,000	1.18	930	1.10

[a] Output current in μA; effective wavelengths in Å.

currents for hydrogen (third column) because a measurement of i_0 on the evacuated cell could not be risked with the plastic windows. In this calculation, which was done by (3.12–1), $\mu_H = 1.00$ was used, hydrogen being almost transparent. Setting 1, which gives the softest beam, is the best for analytical work, since it yields the greatest ratio (6000:1) between the output current for hydrogen and that for chlorine. Effective wavelengths have been included to show the changes that occur in this important variable when one gas is substituted for another. These wavelengths correspond to mass absorption coefficients calculated from (3.12–1) and were obtained by interpolation from values of absorption coefficients for different wavelengths tabulated in the Appendices.

The possibilities of analyzing binary mixtures of known gases may be judged from Table 3.12–1. With mixtures of higher order, absorptiometry is of value in supplying an item of experimental information to be used with other data in arriving at the composition of the gas. It cannot be stressed too strongly that the necessary x-ray data can be obtained with little more effort than is required to measure the pressure or temperature of a gas near standard conditions.

Because gases have a simple equation of state, absorptiometry with polychromatic beams can be used to give information about the state of a gas under conditions (in detonation waves [34], boundary layers [35], or supersonic flow [36]) transient or difficult of access. Temperature measurements [37] have also been made. The technique is a unique method for studying the "fluidization" of a finely divided solid by a gas. Bed density profiles, which reveal the character and effectiveness of fluidization, have been readily determined [38] without disturbing the system as probes would inevitably do.

3.13 Point-to-Point Exploration

X-ray absorptiometry can be used to establish not only the average composition of a material, but its lack of uniformity as well. Owing to the high intensity attainable in polychromatic beams, it is possible to make them small in cross-section without losing rapid detector response. By using a beam of this kind to obtain point-to-point readings, it is possible in favorable cases to establish both composition and approach to uniformity.

The examination of treated carbon brushes used for current collection in aircraft [39, 40] is one important example of point-to-point exploration, which will serve for general illustration. The treatment consists usually of impregnating the brush stock with a solution of a salt (e.g., lead chloride, cadmium chloride, barium fluoride, barium bromide) to an extent that will leave the desired amount, usually a few per cent, of salt in the brush when the water has been removed. The purpose of the treatment is to reduce or eliminate excessive wear of brushes used on motors and generators operating at high altitudes. The speed and reliability with which composition and lack of uniformity could be established were not the only advantages x-ray absorptiometry had to offer in this application. The fact that this method does not sensibly alter the sample—an advantage it has in common with most x-ray methods—was particularly important here because it made possible the subsequent wear-testing of the identical brushes on which composition had been established.

3.14 Comparative Absorptiometry in the Laboratory

Direct absorptiometry is subject to several uncertainties the seriousness of which is greatly mitigated if x-ray absorbance of the unknown (sample) can be properly established relative to that of a standard [5]. The advantages of comparative absorptiometry at longer wavelengths (e.g., infrared) are well known; they are, if anything, greater when x-rays are used.

The uncertainties in direct absorptiometry with polychromatic x-rays arise from unintended fluctuations in the incident beam intensity, from unallowed-for changes in wavelength distribution, and from matrix effects—by which are meant changes in absorbance resulting from the presence in the sample of elements, free or combined, other than the one being determined.

When conclusions about chemical composition are sought by absorptiometry with polychromatic beams, all three of the uncertainties of the preceding paragraph must be dealt with. The simplest way of disposing of them is by making certain that standard and unknown in comparative absorptiometry approach each other as closely as possible in *ultimate* composition, in mass, and in identity of the cells that contain them. When these conditions are met, the absorbances of standard and of unknown for the polychromatic beam will be virtually identical, as will the detector response for each. We may regard this as differential absorptiometry with the difference reduced to zero.

For comparative absorptiometry as outlined above, we obtain by modifying (3.8–1) the fundamental relation

$$\log i^U - \log i^{St} = k^{St} m^{St} - k^U m^U \qquad (3.14-1)$$

where the superscripts St and U refer to standard and unknown. The application of this equation is discussed later.

The addition of the simple stand [5] in Figure 3.14–1, which contains only two aluminum cells as shown, suffices to convert the photometer of Figure 3.9–1 into one for comparative absorptiometry. To permit commuta-

Fig. 3.14–1. Stand for comparative absorptiometry. Comparison is accomplished by manual commutation between standard and unknown. Zemany, Winslow, Poellmitz, and Liebhafsky, *Anal. Chem.*, **21**, 493 (1949).

tion between standard and unknown, each in its cell, the stand was mounted on the steel plate above the housing of the x-ray tube that served as source for the photometer. The stand was pivoted so that a cell in any of the three positions could readily be swung into the x-ray beam and stopped at the proper place when a notch in the base of the stand engaged a spring-arrest on the plate. Reliable output readings could usually be made within 10 sec, which made possible a high rate of commutation between standard and unknown. For detailed treatment of the experimental results, see Refs. 4 and 5.

When the conditions for reliable comparative absorptiometry, which were given in connection with (3.14–1), are satisfied, i^U and i^{St} will be so nearly identical that values of $k^U m^U$ and $k^{St} m^{St}$ for which they are identical can be calculated with high precision. Then (3.14–1) becomes

$$k^{St} m^{St} = k^U m^U \qquad (3.14\text{–}2)$$

and the interpretation of the results no longer involves output currents. As we see later, this laboratory photometer with provision for commutation between standard and unknown is a simple version of a servomechanism system for process control: the difference in output currents between standard and unknown is analogous to an *error signal*. When (3.14–2) holds, the value of the error signal has been *reduced to zero.*

Comparison of unknown with suitable standards is a mainstay of many x-ray methods—especially of x-ray emission spectrography. See Chapter 9.

3.15 Identification of Pure Compounds [5]

Although ultimate analysis by recognized methods is still the court of last resort for establishing the composition of new compounds, it is often possible to obtain sufficient evidence in less tedious ways. When all the elements in a compound are known, an x-ray absorption method can sometimes be used to identify the compound and indicate its purity. Even when an ultimate analysis must eventually be undertaken, x-ray absorption measurements can furnish valuable preliminary information.

Comparative x-ray absorption measurements were used in the identification of various new compounds that could contain at most the following elements: carbon, hydrogen, fluorine, and chlorine. The presumed composition of each compound, known in advance, was duplicated by properly blending carbon tetrachloride, benzotrifluoride, heptane, and benzene; the latter also was used as solvent for the unknown. Under conditions intended to be identical, the amount of unknown was determined by comparative absorptiometry. Table 3.15–1 gives the results.

Table 3.15–1. Identification of Compounds

Presumed Formula	Grams of Unknown Solution Equivalent to 1 g of Standard Solution
$CHCl_3$	1.00_2
$C_2F_2Cl_4$	1.00_2
$C_3F_3Cl_3$	0.99_9
$C_4H_5Cl_3F_2$	1.00_8
$C_5Cl_2F_6$	0.97_4

In every case except the last, the departure from unity shown above could have been due to experimental errors in the x-ray method: these compounds would seem to have had the presumed composition and been of good purity. In the last case, a chlorine determination after peroxide fusion did in fact show chlorine in excess of the theoretical content, thus confirming the indication given by the x-ray data.

3.16 Tetraethyllead Fluid in Gasoline [5]

Because all atoms in a sample absorb x-rays, comparative absorptiometry is not usually suited to a precise determination of constituents present in small amounts. In the case of tetraethyllead fluid in gasoline, however, the absorption coefficient of the fluid sufficiently exceeds that of the base stock (the unleaded gasoline) to make the determination of this fluid a highly promising application. This fact was recognized long ago by Calingaert, and by Aborn and Brown [41].

In the work described here, tetraethyllead fluid in four gasolines supplied by the Ethyl Corporation was determined by the comparative method on weighed samples, aluminum serving as standard. These tetraethyllead fluids could have contained dibromo- or dichloroethane, or both, in addition to the lead compound.

When the x-ray absorption measurements were made, nothing was known about the samples except that AOT-1 and B62M-1 were the only two base stocks involved. The sixteen values of equivalent thickness obtained were interpreted by those who supplied the samples.

The results in Table 3.16–1 subdivide themselves naturally into four groups, each group containing the same type of tetraethyllead fluid. Accordingly, the experimental data in column 2 establish four equations of the slope-intercept type, which are given in the table. Column 3 contains values of the equivalent thickness calculated from these equations, and the excellent agreement of calculated with observed values shows at once that high precision was attained in the work. Each value of x in column 5 was calculated from the corresponding observed equivalent thickness and the proper slope-intercept equation.

The values of x in column 4 were obtained by the Ethyl Corporation by a chemical method, for which the estimated precision is $\pm$ 0.02 ml of tetraethyllead fluid per gallon. Comparison of columns 4 and 5 shows agreement within these limits for all samples except B62M-3; the reason for the considerably greater discrepancy here is unknown. The precision of the x-ray work is better than was expected. (The precision is sufficiently great to warrant consideration of the difference in the x-ray absorption of the base stocks, samples AOT-1 and B62M-1.) Furthermore, the accuracy of the x-ray method must be at least comparable to that of the chemical method.

Table 3.16–1. Determination of Tetraethyllead Fluid in Leaded Gasoline

Sample	Equivalent Thickness y, cm Al		TEL Fluid, x, ml/gal	
	Obsd.	Calcd.	Chemical	X-Ray
AOT-1	0.381_5	0.381_5	0	0.0_0
AOT-2	0.415_5	0.415_5	0.9_2	0.9_2
AOT-3	0.445_0	0.445_1	1.7_2	1.7_2
AOT-4	0.486_7	0.486_6	2.8_4	2.8_4
Experimental equation: $y = 0.3815 + 0.0370x$				
AIT-1	0.425_7	0.425_4	0.8_1	0.8_2
AIT-2	0.479_6	0.479_6	1.8_1	1.8_1
AIT-3	0.534_2	0.534_3	2.8_2	2.8_2
Experimental equation: $y = 0.3815 + 0.0542x$				
A62M-1	0.419_6	0.420_0	0.8_0	0.7_9
A62M-2	0.465_8	0.465_2	1.7_4	1.7_5
A62M-3	0.512_1	0.512_3	2.7_2	2.7_2
Experimental equation: $y = 0.3815 + 0.0481x$				
B62M-1	0.383_0	0.383_0	0	0.0_0
B62M-2	0.429_0	0.429_7	1.0_1	1.0_0
B62M-3	0.470_7	0.475_9	2.0_1	1.9_0
B62M-4	0.522_2	0.521_6	3.0_0	3.0_1
Experimental equation: $y = 0.3830 + 0.0462x$				

INDUSTRIAL APPLICATIONS. PROCESS CONTROL

3.17 General Remarks. Servomechanisms. Automated Control.

In industry, analytical chemistry is seldom done for its own sake. Usually it is done to furnish information needed as a basis for decision, and often the decision depends upon how much and in what direction a material differs from what is expected or desired: for example, how a compound deviates from purity or how the thickness of a sheet of metal deviates from specification. In such cases, the information sought via analytical chemistry may be regarded as an *error signal*—as an estimate of the difference between an unknown and a standard.

Servomechanism systems correct errors in processes on the basis of error signals they generate. Such systems may be animate, animate-inanimate

INDUSTRIAL APPLICATIONS. PROCESS CONTROL

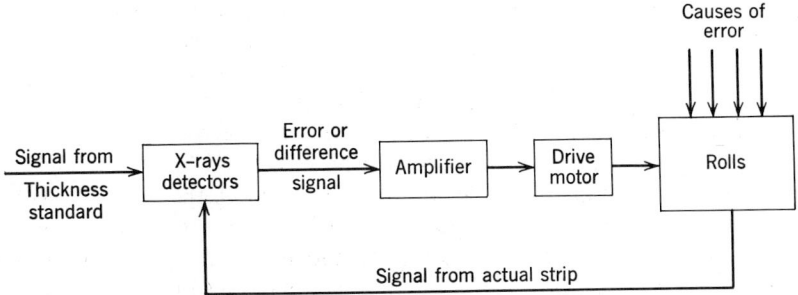

Fig. 3.17–1. Schematic diagram to show the flow of *information* in a servomechanism system that is a simplified version of one in which an x-ray thickness gauge automatically controls by absorptiometry of a polychromatic beam the setting and operation of a rolling mill. Energy sources (such as that for the motor) are not shown.

(mixed), or inanimate (automated): *animate*, as is the system by which an animal controls its body temperature; *animate-inanimate*, as when an operator reads the gauge on an x-ray photometer to establish whether a solution has the desired composition and initiates remedial action when it does not; and *inanimate*, as is the system in which the thickness of sheet metal is continuously monitored by x-ray absorptiometry, and the error signal amplified so that it operates a motor that accomplishes corrective action.

From this broad point of view, many analysts may be regarded as animate links in mixed servomechanism systems. This statement is not intended to be unflattering—merely realistic. The development of instrumental methods started a movement toward inanimate systems, which is being greatly accelerated by the computer with its memory for instructions and its tireless ability to make instantaneous comparisons. X-ray methods are admirably suited to inanimate servomechanism systems. When the intensity is high, an *analog* (continuous) error signal results, and this must be converted to *digital* form for the computer. When the intensity is low enough so that x-ray quanta are counted, the information is already in digital form (though not in the *binary* system the computer uses).

Servomechanism systems owe much to military necessity—witness, for example, the use of radar to detect unseen aircraft and activate a completely inanimate servomechanism system that launches a missile along a trajectory automatically calculated by a computer. There is neither need nor intent to discuss the many complexities of these systems here. Our much more modest aim is to leave the reader with a qualitative understanding of the simplest possible servomechanism system. To do this is the function of Figure 3.17–1, which diagrams the flow of *information* (not of *energy*) in a system drawn to introduce the steel thickness gauge described later. The figure shows that the system functions by comparing the thickness of strip from the rolling mill (the *unknown*) with the desired thickness of a *standard* to generate an error signal on the basis of which the computer commands the motor to adjust the rolling mill so as to reduce the error to zero; the entire process is of course continuous so that changing causes of error are automatically compensated. A similar procedure can be envisaged for the comparative absorptiometry described in Section 3.14. In general, the operator will find a slight difference between i^U and i^{St}, which may be regarded as an error signal. The operator can then compute and make the change required to reduce the error signal to zero and give validity to (3.14–2). Operator and photometer then form an animate-inanimate servomechanism system.

We find that other x-ray instruments, notably x-ray emission spectrographs, are being used increasingly as components of inanimate servomechanism systems for automated process control. To amplify a point made earlier: this move in the direction of automated analytical chemistry will continue. Analytical chemistry as practiced in the laboratory will not suddenly disappear, but automated analytical chemistry will do more and more of the *routine analyses* required in *large numbers* for process control [42]. Not all of these analyses are as simple as thickness gauging.

3.18 Commercial X-Ray Photometers

The photometer of Section 3.9 equipped for comparative work is far from being a commercial instrument. It requires a voltage source too nearly constant, its rate of commutation is too small, it would be cumbersome to automate, and so on. Instruments were built that could operate satisfactorily on the ordinary 110-V supply of alternating current, and in which the comparison between standard and unknown was either continuous (see below) or carried out at, say, 30 times a second by use of a "chopper" either in the polychromatic x-ray beam [43] or between two phosphor screens (one for standard; one for unknown) and a single photomultiplier tube [44]. Such instruments could conceivably function as part of inanimate servomechanism systems for process control: for example, in the preparation of leaded gasoline, base stock might flow through one cell and the leaded product through the other.

Figure 3.18–1 is an annotated block diagram of a photometer that found extensive use. A noteworthy feature is the use of ionization detectors that respond to electric current. The substitution of these for the phosphor-photoelectric detectors, such as that used in the photometer of Section 3.9, increased simplicity and improved operation. Improved external amplifiers made the substitution worthwhile.

A survey by the authors indicates that the use of commercial x-ray photometers has not grown and may have shrunk. The interested reader is con-

INDUSTRIAL APPLICATIONS. PROCESS CONTROL

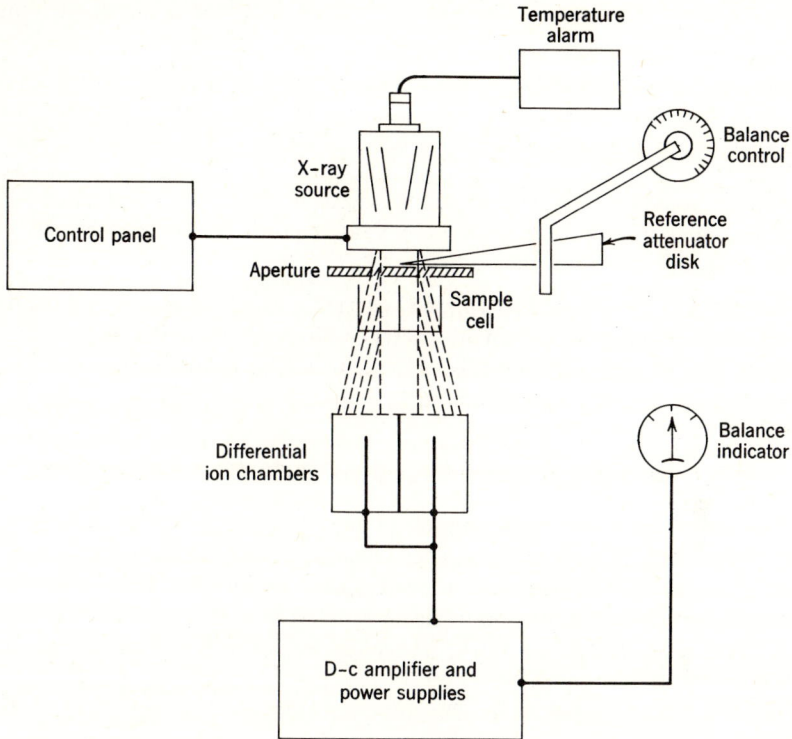

Fig. 3.18–1. Block diagram of later General Electric X-ray Photometer. Courtesy of General Electric Company.

sequently asked to consult Ref. 4 for a fuller discussion to supplement what follows.

In the petroleum industry, the analytical problems solved by this x-ray method have one thing in common: the petroleum product itself, being a hydrocarbon, shows relatively low x-ray absorbance. Note, in this connection, that μ_H and μ_C do not differ greatly in the region near 0.5 Å, which therefore is often the wavelength region chosen for such work. Consequently, variations in absorbance resulting from changes in the hydrocarbon content in a given volume of petroleum product are more likely to be due to differences in density than in composition—particularly at an effective wavelength near 0.5 Å. Among the determinations carried out were: sulfur in hydrocarbon mixtures, tetraethyllead fluid in gasoline, additives (such as metal soaps) in lubricating oils, and the metal content of metallo-organic derivatives [45–50].

The following quotation [51] summarizes the early usefulness of comparative absorptiometry in the nuclear energy program at Hanford, Washington.

Without disclosing security information we can state that at present we are making 10,000 heavy metal ion determinations annually by means of the x-ray photometer and that we expect this number to increase as we find new applications among essential materials, process reagents, and others. We estimate that use of the instrument saves approximately 3000 man-hours annually. This figure is estimated on the basis that analysis of a single solution by x-ray requires 20 minutes, whereas other methods require twice this time.

Bartlett [52] and Lambert [53, 54] carried out extensive investigations of the usefulness of x-ray absorptiometry with polychromatic beams in the nuclear energy program, details of which will be found in reports of the U.S. Atomic Energy Commission.

3.19 The General Electric Raymike® 2000 Thickness Gauge [55]

The automated thickness gauging of steel strip shows x-ray absorptiometry at its best. During the thickness measurement, the steel strip may be hot (1500° to 1750°F), moving (about 2000 feet/minute horizontally with possible vertical vibrations up to several inches in amplitude), and subjected to a spray of cooling water. In 1955, this application was made fully automatic; that is, the error signal was used to readjust tandem cold reduction mills of the U.S. Steel Corporation. Automatic control proved significantly more effective than manual control. Under the drastic operating conditions, the guaranteed accuracy of the present gauge over its measurement range is: within $\pm$ 2 mils between 140 and 200 mils; within 1 % of thickness between 200 and 400 mils; and within $\pm$ 4 mils from 400 mils to the upper limit of the range; that is, 1999 mils. Not bad for the conditions!

Gauge proper and carriage appear in Figure 3.19–1. Figure 3.19–2 is a schematic diagram to show how the gauge works. When it is part of an automated system that includes computer and rolling mill, we then have a system that operates in principle like that of Figure 3.17–1.

Points of interest in Figure 3.19–2 are as follows.

1. With two ionization chambers as detectors, the gauge continuously compares unknown (the steel strip) and standard (the wedge—No. 1 in the figure). The wedge is part of its own servomechanism system to ensure precise setting and easy, rapid calibration. The wedge is set so that an error signal of zero results when the strip has the desired thickness.

2. Within limits, differences in chemical composition (that is, of mass absorption coefficient) can be allowed for in presetting the wedge. Known standard samples stored in the gauge make possible calibration at different compositions.

Fig. 3.19-1. The General Electric Raymike®2000 Thickness Gauge. Steel strip intercepts a polychromatic x-ray beam (see Figure 3.19-2) as it passes between the jaws of the "C," which are widely separated (84 in.) to keep the gauge from being damaged owing to the severe operating conditions. A motorized carriage permits positioning of the beam and scanning of the strip. Compare with Section 3.13. Courtesy of General Electric Company.

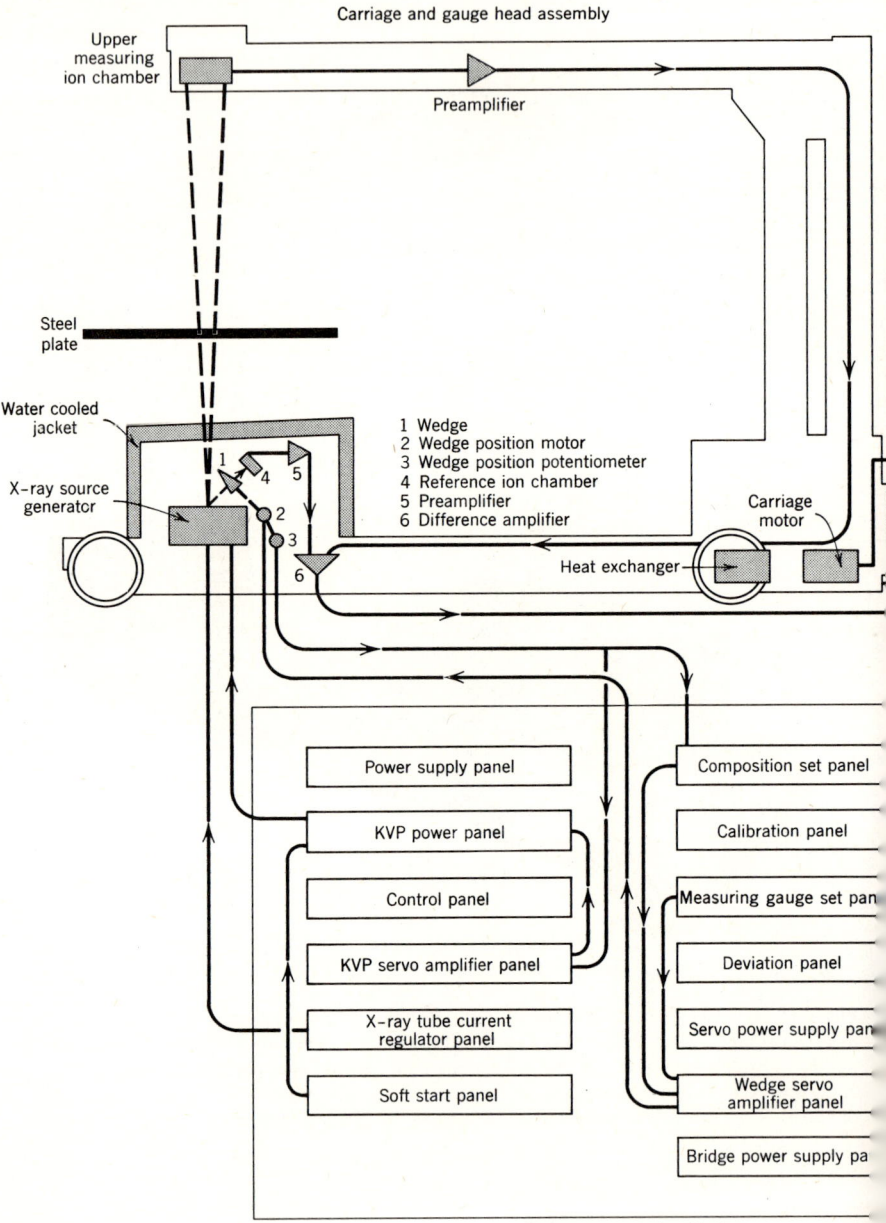

INDUSTRIAL APPLICATIONS. PROCESS CONTROL

3. The ionization chambers give a-c output signals that are preamplified and fed through a difference amplifier (No. 6 in the figure), the resultant signal from which is converted to an analog output proportional in magnitude to the difference between the original signals. There results a d-c error signal that appears on the deviation meter of the operator's control panel.

4. When the gauge is joined to ("interfaced with") computer and rolling mill, the error signal just mentioned is used to accomplish automated control of the mill.

5. Thanks to solid-state circuitry, gauge response is rapid and background ("noise") is low. The gauge is operational after a 3-minute warm-up from a

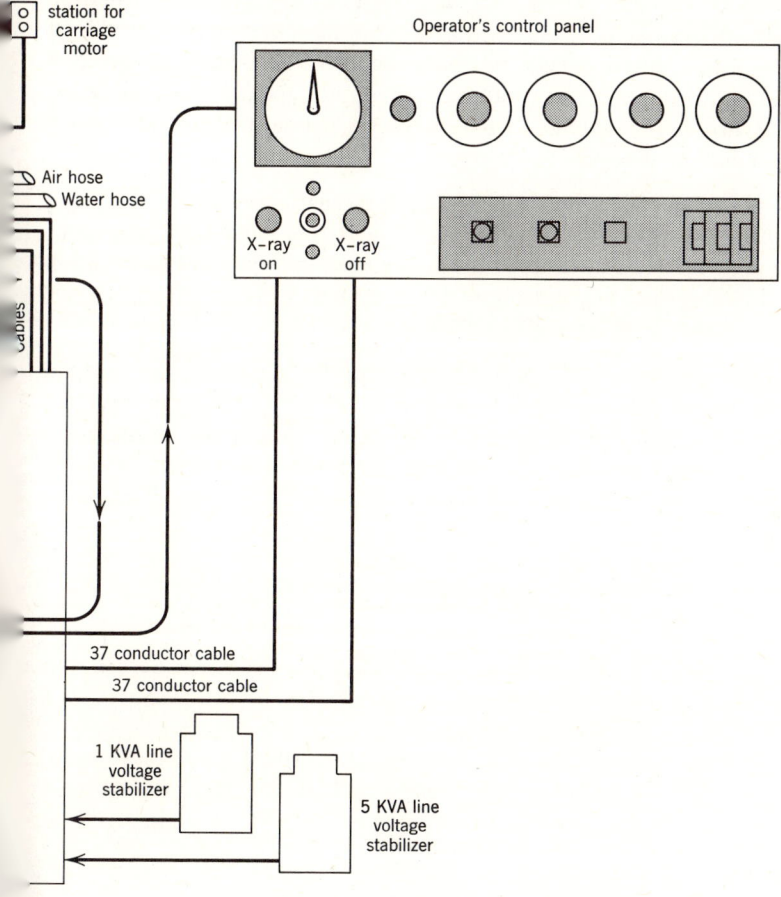

Fig. 3.19–2. Schematic diagram of the gauge system for attended (not automated) operation of a rolling mill. Courtesy of General Electric Company.

cold start, and 5 sec after the x-ray-tube voltage has been turned on. It responds to almost two-thirds of a 5% thickness change within 0.05 sec.

6. The Measuring Gauge Head and Carriage (Figure 3.19–1) weighs about 7000 lb; the equipment control cabinet, about 1500 lb; and the operator's control panel, about 80 lb.

CONCLUSION

A final word is necessary. The spectacular growth of x-ray emission spectrography has made x-ray absorptiometry in the laboratory into a poor relation—perhaps to a greater extent than justified. The absorption-edge method tends to be cumbersome because it requires more than one intensity reading, yet it is free of the matrix effects that often bedevil x-ray emission spectrography. X-ray absorptiometry with polychromatic beams has the great weakness of not being specific: what Hevesy [56] said about it, though perhaps too harsh, is worth quoting:

It would even fall within the range of possibility to analyze a chemical sample by determining the absorption coefficient of its characteristic radiation in aluminum or other suitable matter. Such a primitive X-ray analysis would, however, fail to detect any but the chief constituents of the substance investigated, and would involve a very clumsy and unhandy procedure.

Yet it can offer the virtues of high intensity, simple equipment, and rapid (if not instantaneous) readout. Large laboratories might profit from having a simple laboratory x-ray photometer for the quick comparison of materials —after all, doctors and dentists still rely on x-ray absorptiometry with polychromatic beams in the kind of analytical chemistry they practice on us!

REFERENCES

1. O. Glasser, *Wilhelm Conrad Röntgen*, John Bale, Sons and Danielsson, London, 1933, p. 44.
2. *Elec. Eng.*, **21**, 216 (February 16, 1896).
3. H. A. Liebhafsky and A. E. Newkirk, *Corrosion*, **12**, 92t (1956).
4. H. A. Liebhafsky, H. G. Pfeiffer, P. D. Zemany, and E. H. Winslow, *X-Ray Absorption and Emission in Analytical Chemistry*, Wiley, New York, 1960.
5. P. D. Zemany, E. H. Winslow, G. S. Poellmitz, and H. A. Liebhafsky, *Anal. Chem.*, **21**, 493 (1949).
6. R. Glocker and W. Frohnmayer, *Ann. Physik*, **76**, 369 (1925).
7. J. M. Hollander, I. Perlman, and G. T. Seaborg, *Rev. Modern Phys.*, **25**, 469 (1953).
8. H. K. Hughes and J. W. Wilczewski, *Anal. Chem.*, **26**, 1889 (1954).
9. H. K. Hughes and J. W. Wilczewski, *Proc. Am. Petroleum Inst.*, **30 M (III)**, 11 (1950)

REFERENCES

10. H. K. Hughes and F. P. Hochgesang, *Anal. Chem.*, **22**, 1248 (1950).
11. *Rev. Sci. Instr.* **26**, 630 (1955).
12. R. Coppens, *Compt. rend.*, **232**, 1681 (1951).
13. (a) M. de Broglie, *Compt. rend.*, **157**, 924 (1913); (b) *ibid.*, **158**, 1493 (1914).
14. (a) F. K. Richtmyer, *Phys. Rev.*, **27**, 1 (1926); (b) *ibid.*, **30**, 755 (1927).
15. *Chem. Eng. News*, **26**, 993 (1948).
16. L. K. Frevel, Dow Chemical Co., Midland, Mich.; letters to H. A. Liebhafsky, (a) October 16, 1948, (b) November 5, 1948, (c) October 9, 1968.
17. L. T. Hallett, *Anal. Chem.*, **20**, 391 (1948).
18. W. F. Peed and H. W. Dunn, *U.S. Atomic Energy Commission, Rept. ORNL-1265*, 1952.
19. R. E. Barringer, *U.S. Atomic Energy Commission, Rept. TID-7516*, 1956.
20. W. C. Dietrich and R. E. Barringer, *U.S. Atomic Energy Commission, Rept. Y-1153*, 1957.
21. J. H. Stewart, Jr., *Anal. Chem.*, **32**, 1090 (1960).
22. J. H. Stewart, Jr., T. H. Barton, Jr., and M. R. Ferguson, *Anal. Chem.*, **40**, 27 (1968).
23. R. E. Barieau, *Anal. Chem.*, **29**, 348 (1957).
24. R. Glocker and W. Frohnmayer, *Ann. Physik*, **76**, 369 (1925); N. H. Moxnes, *Z. physikal. Chem.*, **A144**, 134 (1929); **A152**, 380 (1931).
25. A. Engström, *Nature*, **158**, 664 (1946); *Rev. Sci. Instr.*, **18**, 681 (1947); *Acta Radiol.*, **31**, 503 (1949).
26. H. K. Hughes and F. P. Hochgesang, *Anal. Chem.*, **22**, 1248 (1950).
27. H. A. Liebhafsky and E. H. Winslow, *Am. Soc. Testing Materials Bull.*, No. 167, 67 (1950).
28. W. Duane, *Proc. Natl. Acad. Sci. U.S.*, **13**, 668 (1927).
29. W. Duane, J. C. Hudson, and H. N. Sterling, *Am. J. Roentgenol. Radium Therapy*, **20**, 241 (1928).
30. H. M. Smith, *Gen. Elec. Rev.*, **48**, 13 (March 1945).
31. H. A. Liebhafsky, H. M. Smith, H. E. Tanis, and E. H. Winslow, *Anal. Chem.*, **19**, 861 (1947).
32. P. D. Zemany, E. H. Winslow, G. S. Poellmitz, and H. A. Liebhafsky, *Anal. Chem.*, **21**, 493 (1949).
33. E. H. Winslow, H. M. Smith, H. E. Tanis, and H. A. Liebhafsky, *Anal. Chem.*, **19**, 866 (1947).
34. G. B. Kistiakowsky, *J. Chem. Phys.*, **19**, 1611 (1951).
35. R. N. Weltmann and P. W. Kuhns, *Natl. Advisory Comm. Aeronaut., Tech. Notes*, No. 3098, 1954.
36. E. M. Winkler, *J. Appl. Phys.*, **22**, 201 (1951).
37. G. J. Mullaney, "Temperature Determination in Flames by X-Ray Absorption Using a Radioactive Source," *Rev. Sci. Instr.*, **29**, 87 (1958). The x-ray source was $_{26}Fe^{55}$.
38. E. W. Grohse, *AIChE. J.* **1**, 358 (1955).
39. A. C. Titus, *Power App. Sys.*, No. 14, 1160 (October 1954).
40. A. C. Titus and E. H. Winslow, *Tech. Information Series R53MG301*, General Electric Co., Schenectady, N.Y., 1953.
41. R. H. Aborn and R. H. Brown, *Ind. Eng. Chem., Anal. Ed.*, **1**, 26 (1929).
42. H. A. Liebhafsky, *Anal. Chem.*, **34**, 23A (1962).
43. (a) T. C. Michel and T. A. Rich, *Gen. Elec. Rev.*, **50**, 45 (February 1947); (b) H. A. Liebhafsky, *Anal. Chem.*, **21**, 17 (1949).
44. R. W. Cranston, F. W. H. Matthews, and N. Evans, *J. Inst. Petroleum*, **40**, 55 (1954).

45. R. C. Vollmar, E. E. Petterson, and P. A. Petruzzelli, *Anal. Chem.,* **21**, 1491 (1949).
46. W. L. Kehl and J. C. Hart, *Proc. Am. Petroleum Inst.,* **28 (III),** 9 (1948).
47. G. Calingaert, F. W. Lamb, H. L. Miller, and G. E. Noakes, *Anal. Chem.,* **22**, 1238 (1950).
48. A. Y. Mottlau and C. E. Driesens, Jr., *Anal. Chem.,* **24**, 1852 (1952).
49. R. W. Cranston, F. W. H. Matthews, and N. Evans, *J. Inst. Petroleum,* **40**, 55 (1954).
50. H. D. Terrell and J. C. Davidson, Pittsburgh Conference on Analytical Chemistry and Applied Spectroscopy, Pittsburgh, Pa., March 1955.
51. A. H. Bushey, General Electric Co., Richland, Wash., letter to H. A. Liebhafsky, July 25, 1952.
52. T. W. Bartlett, *U.S. Atomic Energy Commission, AECD Repts. 2765, 2766* (1949).
53. M. C. Lambert, *U.S. Atomic Energy Commission, Rept. HW-30634* (1954).
54. M. B. Leboeuf, D. G. Miller, and R. E. Connally, *Nucleonics,* **12**, 18 (August 1954).
55. *Product Data Sheet A6350,* General Electric X-Ray Department, Milwaukee, Wis., 53201, June 1968.
56. G. (von) Hevesy, *Chemical Analysis by X-Rays and Its Applications,* McGraw-Hill, New York, 1932, p. 26.

Chapter 4

X-Ray Spectra

> The results already obtained show that such data [frequencies of x-ray lines] have an important bearing on the question of the internal structure of the atom, and strongly support the views of Rutherford and Bohr. [Words in brackets supplied.]
>
> H. G. J. Moseley [36a, Chap. 1]

The introductory treatment of x-ray spectra in Chapter 1 is expanded here to deepen understanding of their excitation by electrons and by x-rays, to show the great contribution their study has made to our knowledge of atomic structure, and to lay the foundation for a discussion of x-ray optics.

4.1 The Continuous Spectrum

The following discussion extends Sections 1.5 and 1.6.

The simplest continuous spectra are generated by bombarding thin metallic targets with monoenergetic electrons. Ideally, such a target should be so thin that an incident electron collides with not more than one atom (ion). For an ideal thin target, Webster [1] has used measured results on a massive target to deduce that the distribution of intensity with wavelength should be governed by

$$I_\lambda = c_1/\lambda^2 \qquad \lambda \geq \lambda_0 \qquad (4.1-1)$$

$$I_\lambda = 0 \qquad \lambda < \lambda_0 \qquad (4.1-2)$$

λ_0 being the short-wavelength limit (1.5). The sharpest curve in Figure 4.1–1

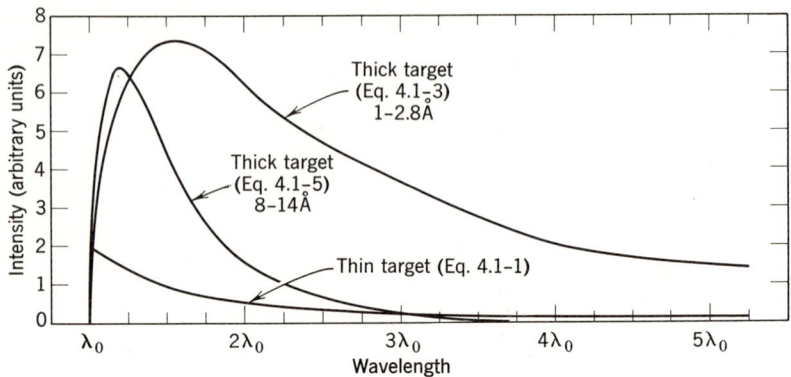

Fig. 4.1–1. Calculated intensity-wavelength distributions for continuous spectra produced by electron bombardment of various targets. The ordinate scale is arbitrary.

represents this distribution, which has been verified by direct measurement [2].

The massive target may be looked upon as a stack of thin targets. Most monoenergetic electrons incident upon the uppermost target in this imagined stack will lose energy owing to scattering processes. As the electrons penetrate, each imagined thin target will therefore receive electrons whose average energy has been reduced to an extent increasing with the distance of the target below the surface. The average short-wavelength limit for each successive target will therefore be at a successively longer wavelength.

As it penetrates, the electron loses energy in many small steps that lead to electron displacement in, or excitation or ionization of, the atoms with which it collides. An energy loss leading to an emitted x-ray quantum is much less probable than are these small steps. For this reason, the efficiency of x-ray production by electron bombardment (1.6–3) is low.

Figure 4.1–2 shows graphically that the continuous spectrum from a massive target can be approximated by superposing the individual spectra of thin targets in a stack. The shift of the short-wavelength limit is striking.

Kulenkampff [3] gives

$$I_\lambda = \frac{c_2 Z}{\lambda^2}\left[\frac{1}{\lambda_0} - \frac{1}{\lambda}\right] + \frac{BZ^2}{\lambda^2} \qquad (4.1\text{--}3)$$

for the intensity-wavelength distribution at wavelengths between 1 and 2.8 Å for massive targets of eight representative metals (aluminum to platinum). The first term on the right of (4.1–3) was deduced theoretically by Kramers [4] for simplified conditions. Furthermore, differentiation with

X-RAY SPECTRA

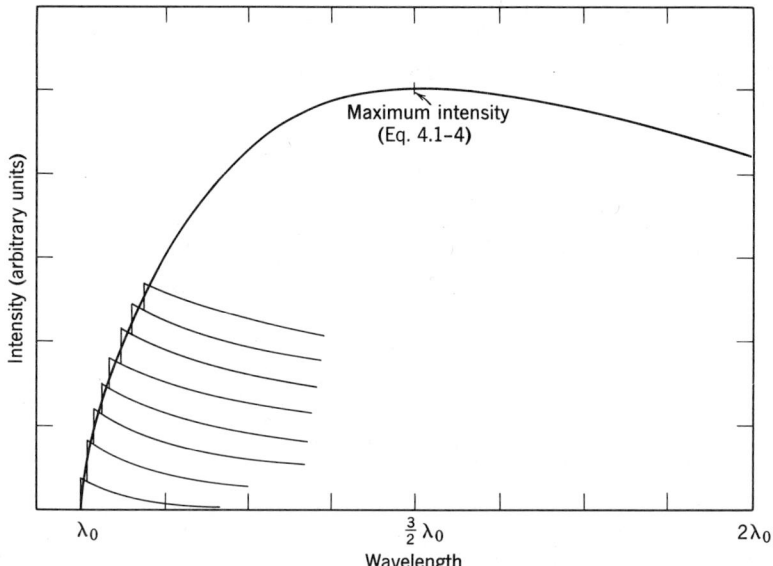

Fig. 4.1–2. Continuous spectrum from a massive target obtained by superposition of spectra from thin targets. After Compton and Allison, *X-Rays in Theory and Experiment*, Van Nostrand, New York.

respect to wavelength of this term gives for the wavelength of maximum intensity

$$\lambda_{I_{max}} = 3\lambda_0/2 \qquad (4.1\text{–}4)$$

under conditions where the contribution from the second term is negligible. Equation 4.1–4 will be remembered as a rough rule introduced just after (1.6–2); Figure 4.1–2 shows the intensity maximum to be broad indeed.

Between 8 and 14 Å, Stephenson and Mason [5] found

$$I_\lambda = \frac{c_3 Z}{\lambda^5} \left(\frac{1}{\lambda_0} - \frac{1}{\lambda} \right) \qquad (4.1\text{–}5)$$

as an analogue of (4.1–3). A distribution curve calculated from each equation is plotted in Figure 4.1–1. The curve for the longer wavelengths more nearly resembles that for the thin target, in part because the mass absorption coefficient for these wavelengths is much higher—the more strongly the *generated* x-rays are absorbed, the smaller the mean depth at which the *emitted* x-rays are generated. We see later that *absorption effects* such as this, though very small here because this mean depth is so small, are important in thickness gauging and in x-ray emission spectrography.

4.2 The Characteristic Spectrum from a Target

For the generation of $K\alpha$ lines by electron bombardment, Jönsson [6] found the proportionality

$$I_K \propto (V - V_{\min})^2 \qquad (4.2\text{–}1)$$

where $V_{\min}$ is the minimum voltage required to excite these lines. When V increases to above about six times this minimum, the expected intensity is no longer realized: the efficiency of excitation by electrons becomes subject to the law of diminishing returns, probably because the energy lost in scattering processes becomes disproportionately high as the electrons (because of their higher energies) penetrate further and further into the target.

The following ratios [7] are a rough guide to the *relative* line intensities within certain series:

For all elements: $K\alpha 1 : K\alpha 2 \approx 100 : 50$
For tungsten: $K\alpha 1 : K\alpha 2 : K\beta 1 \approx 100 : 50 : 35$
For copper: $K\alpha 1 : K\alpha 2 : K\beta \approx 100 : 50 : 25$
For tungsten: $L\alpha 1 : L\alpha 2 : L\beta 1 : L\beta 2 \approx 100 : 11 : 52 : 20$

Intensities in *different* series are not to be compared: for example, $L\alpha 1$ does *not* necessarily have the same intensity as any $K\alpha 1$ though both are given the value 100 above.

4.3 Conditions Relating to X-Ray Excitation

See Section 1.20, in which x-ray and electron excitation were compared.

Usually, high intensity is desirable in a continuous spectrum to be used for x-ray excitation. The efficiency of x-ray production,

$$\text{Efficiency} = 1.4(10^{-9})ZV \qquad (4.3\text{–}1)$$

can be reextracted from (1.6–3) by dividing tube power (iV) into beam power, or integrated intensity. At 50 kV, the efficiency is 0.52 % for a tungsten target ($Z = 74$). Tungsten is a preferred target material because it is the element of highest atomic number that meets the rigorous requirements (ready availability at high purity, high melting point, suitable mechanical properties) of target service.

Because the efficiency is low even with tungsten, it is necessary to look to increased tube voltages for the high intensities often required in x-ray excitation. It is usually desirable that the tube voltage be at least so high as to place the intensity maximum in the continuous spectrum at a wavelength short enough to excite the desired line. This means according to (4.1–4) that

$$3\lambda_0/2 = \lambda_{I_{\max}} < \lambda_c \qquad (4.3\text{–}2)$$

X-RAY SPECTRA

where λ_c is the critical absorption wavelength in question. This tube voltage can be calculated from (1.6–2).

Characteristic lines from a target sometimes cause trouble in x-ray excitation. They are welcome when their wavelengths are shorter, but not much shorter, than the critical absorption wavelength for the line to be excited. The characteristic target lines are then very effective because they are highly absorbed from a beam in which they can be quite intense. Trouble comes when the target line, either unmodified or modified by Compton scattering, interferes with an *analytical line* from the sample. Such interference occurs when the target line is scattered by the sample, and it occurs not only when the two lines are of about the same wavelength, but also when a higher-order reflection of the target line takes place at or near the Bragg angle for the sample line. The interfering lines may belong to different series, as when an L line from a tungsten target interferes with a K line of germanium excited in a sample.

Such interferences are likely enough to warrant having available, for x-ray excitation, beams from at least two kinds of targets. Tungsten ($_{74}$W), the preferred target material in a Coolidge tube, is well supplemented by $_{42}$Mo. Table 4.3–1 shows the more common characteristic lines with which the lines from these two targets will interfere; in most cases, lines from both targets do not interfere with the same line. Interference problems are discussed in later chapters.

Table 4.3–1 Common Interferences by Tungsten and Molybdenum Lines[a]

Target Line	First-Order Interferences	Second-Order Interferences	Third-Order Interferences
W Lα (1.48)	Ni Kβ (1.50)		Cl Kβ1 (4.40)
W Lβ1 (1.28)	Pt Lα (1.31)		
	Au Lα (1.28)		
W Lβ2 (1.24)	Ge Kα (1.26)	V Kα (2.50), Pr Lα (2.46)	K Kα (3.74)
	Hg Lα (1.24)		
W Lγ1 (1.10)	Se Kα (1.11)		Te Lα (3.29)
Mo Kα (0.71)		Zn Kα (1.44)	
Mo Kβ (0.63)		Hg Lα (1.24), Ge Kα (1.26)	
		Au Lα (1.28)	

[a] Numbers in parentheses are wavelengths in angstroms.

The spectrum from a Coolidge tube often contains lines traceable to impurities, those of copper, nickel, and iron being the most common. The impurities may be present in the target of the new tube, but they are more likely to be deposited on the target during operation. It is consequently

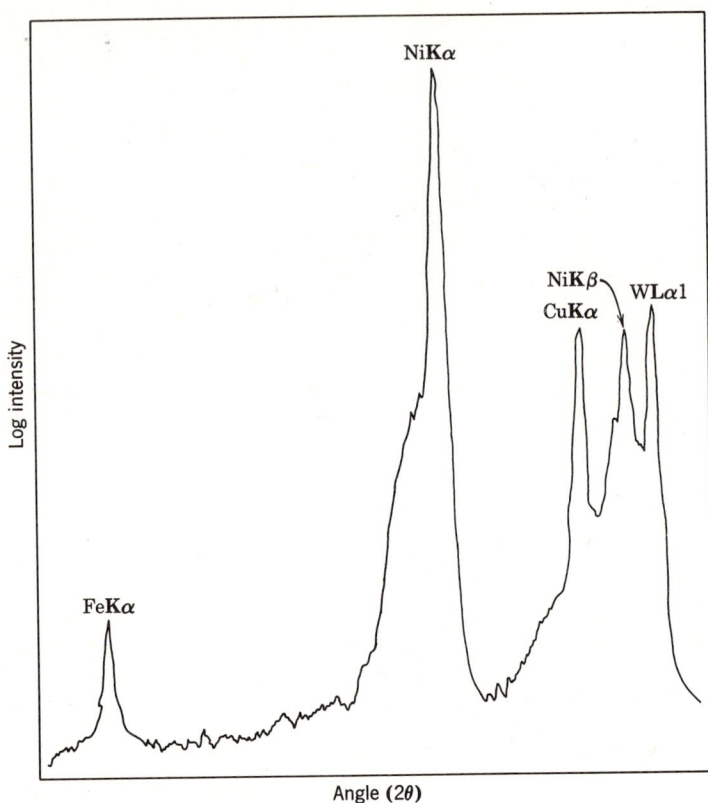

Fig. 4.3–1. Lines from copper, nickel, and iron impurities which appeared in the spectrum of an x-ray tube after the tube had been operated for several hundred hours. X-rays from the tube were scattered by filter paper in the sample holder. The spectrum was recorded over the 2θ range from 41 to 59°, which corresponds to a wavelength range from 1.45 to 2.0 Å. Author's unpublished results.

desirable that the analytical chemist be always familiar with the spectrum of his x-ray source.

Figure 4.3–1, which records in part the spectrum of an old OEG-50 tungsten-target tube, makes this point effectively. Lines of copper and nickel are present at significant intensities.

4.4 Efficiency of X-Ray Excitation

This section contains an illustrative, highly approximate calculation of the intensity of cobalt $K\alpha$ lines produced by x-ray excitation. Attempts at exact calculation are not likely to give highly reliable results.

X-RAY SPECTRA

This intensity can be obtained as the product of three quantities as follows:

Characteristic-line intensity = photons absorbed per second × fraction of absorbed photons available for excitation × characteristic-photon yield   (4.4–1)

1. For this calculation, let us assume a simple case in which a suitable monochromatic beam of intensity I_1 falls perpendicularly upon unit area (1 cm^2) of a monolayer of cobalt atoms resting upon a transparent substrate. The first term in (4.4–1), ΔI, is given by

$$\Delta I = I_1 \mu \rho \Delta d \quad \text{(density, } \rho\text{; thickness, } \Delta d\text{)} \quad (4.4\text{–}2)$$

This equation is obtained from (1.10–1) by substituting

$$\Delta I/I_1 \text{ for } \ln(I_1/I_2) \quad \text{(permissible for } \Delta I \ll I_1 \text{ or } I_2\text{)}$$

2. Clearly, not all absorbed photons are available for excitation even though they are short enough in wavelength (Figure 1.13–1). The fraction thus available is conveniently expressed in terms of r, the absorption-jump ratio, which (along with the critical absorption wavelength) characterizes an absorption edge. We may regard the absorption jump as abrupt and discontinuous, and this makes it possible to define r as the ratio of a maximum to a minimum coefficient:

$$r = \mu_{max}/\mu_{min} \quad (4.4\text{–}3)$$

the maximum being for the short-wavelength side of the edge. See Figure 4.4–1. Each of these coefficients contains a scattering and a photoelectric component (1.13–1), the former being identical for both coefficients at the edge. The photoelectric components for higher-wavelength absorption edges are also identical on both sides of the absorption edge under consideration. Under these conditions, the fraction of absorbed photons available for the excitation of all **K** lines is

$$(\mu_{max} - \mu_{min})/\mu_{max} = (1 - 1/r) \quad (4.4\text{–}4)$$

provided the incident monochromatic beam is of wavelength just shorter than the critical absorption wavelength. (This phrase defines "suitable" as used in (1) above.)

3. The characteristic-photon yield (1.23–1) is the fraction of photons absorbed according to (4.4–4) that will actually appear as **K**α lines. This characteristic-photon yield is the product of $w_{K(Co)}$ and 0.9 because only 0.9 of all **K** photons are **K**α photons, the other 0.1 being **K**β.

Now (4.4–1) has become

$$I_{K\alpha} = (I_1 \mu_{max} \rho \, \Delta d)(1 - 1/r) w_{K\alpha} \quad \text{(per cm}^2\text{)} \quad (4.4\text{–}5)$$

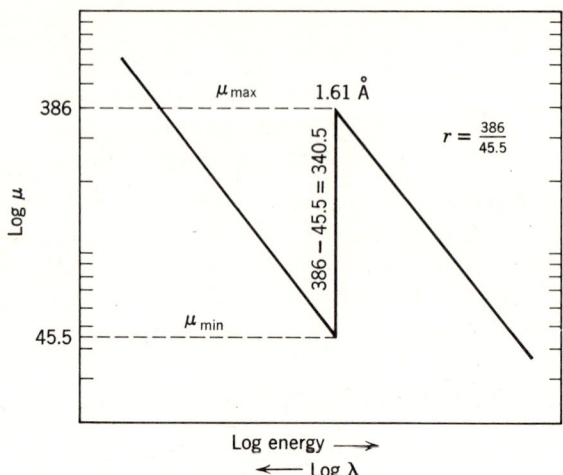

Fig. 4.4–1. Representation of absorption jump for the problem in text. See (4.4–3) and (4.4–4). Note that on the long-wavelength side of an absorption edge the absorbance is *less* than on the short-wavelength side until the wavelength has increased sufficiently to compensate for the decreased absorbance measured by the absorption jump. There is of course no conflict with (1.13–2).

The three coefficients relating to x-rays in (4.4–5) may conveniently be combined into a "mass emission coefficient," $\varepsilon_{K\alpha}$. If this is done, the equation becomes

$$I_{K\alpha} = I_1 \varepsilon_{K\alpha} \rho \, \Delta d \qquad \text{(per cm}^2\text{)} \qquad (4.4-6)$$

The data required for our illustrative calculation are summarized below.

Sample: monolayer, 1 cm^2 of cobalt atoms, 0.2×10^{-6} g,
$\Delta d = 2.3$ Å, $\rho = 8.9$ g/cm^3
Incident wavelength: 1.60 Å, photon energy, $1.24(10^{-15})$ W-sec
Excited wavelength: cobalt Kα, 1.79 Å
Critical absorption wavelength: cobalt K edge, 1.61 Å
X-ray source: tungsten-target tube at 50 kV, 50 mA
X-ray coefficients: $\mu_{max} = 386$, $r = 8.5$, $w_{K\alpha} = 0.36$, $\mu_{min} = 45.5$

The assumption implicit above, that the x-ray source delivers a monochromatic beam of wavelength 1.60 Å, is unrealistic. It is discussed later. The size of the quantum for wavelength 1.60 Å is the product of Planck's constant and the corresponding frequency.

According to (1.6–3), the x-ray power produced is 13 W. The corresponding number of photons per second is $13/[1.24(10^{-15})]$, or $10.5(10^{15})$. In an actual spectrograph, which might for example have a 1-cm^2 sample at 3 cm from the target of the x-ray tube, the geometrical relationship between source and sample will reduce the incident intensity to perhaps 1% of that at the source. Therefore,

$$I_1 = 10.5(10^{13}) \text{ photons/sec} \tag{4.4-7}$$

According to (4.4-5) and (4.4-7), with numerical values substituted, the intensity we seek is

$$I_{K\alpha} = 10.5(10^{13}) \times 386 \times 8.9 \times 2.3(10^{-8}) \times (1 - 1/8.5) \times 0.36$$
$$= 2.64(10^9) \text{ photons/sec} \tag{4.4-8}$$

If the experiment to which the calculation refers were actually attempted, several differences would appear. Most important, the x-ray power would be expended over a wide spectrum. The intensity in (4.4-2) would be the integrated intensity from the short-wavelength limit to the critical absorption wavelength. Also, μ_{max} would need to be adjusted to the wavelength range of the integrated intensity. The net effect of these differences would be to reduce $I_{K\alpha}$ below the value of (4.4-8), perhaps by as much as tenfold.

For the purposes of the calculation, a single value of the exciting wavelength was chosen; in actual practice, the exciting x-rays will include a large range of wavelengths which excite the characteristic x-rays with varying efficiency. At wavelengths short compared to the absorption edge, two effects ordinarily operate; both reduce the efficiency of excitation. The first is the deeper penetration of the short-wavelength x-rays, which decreases the fraction of the excited x-rays that actually escape the sample. The second effect is the relatively larger part that scattering plays at shorter wavelengths, which reduces the role of photoelectric absorption. With a monolayer as sample, only the second effect needs to be considered.

The photoelectric absorption coefficient increases as the cube of the wavelength in regions remote from the absorption edge while the scattering coefficient increases much more slowly (say, as $\lambda^{1/2}$). Consequently, the shorter wavelengths are very much less efficient in exciting characteristic photons. Also, the increased background resulting from *relatively* high scattering coefficients reduces precision.

4.5 The Origin of Characteristic Lines

A review of Sections 1.4, 1.12, 1.20, 1.21, and 1.22 is advisable at this point.

Figure 4.5-1 is a convenient schematic diagram that illustrates the generation of characteristic lines by using a simple model of the Bohr atom. The results obtained on the magnetic photoelectron spectrograph make it easier to understand the diagram.

In Figure 4.5-1, a nucleus carrying Z positive charges is surrounded by

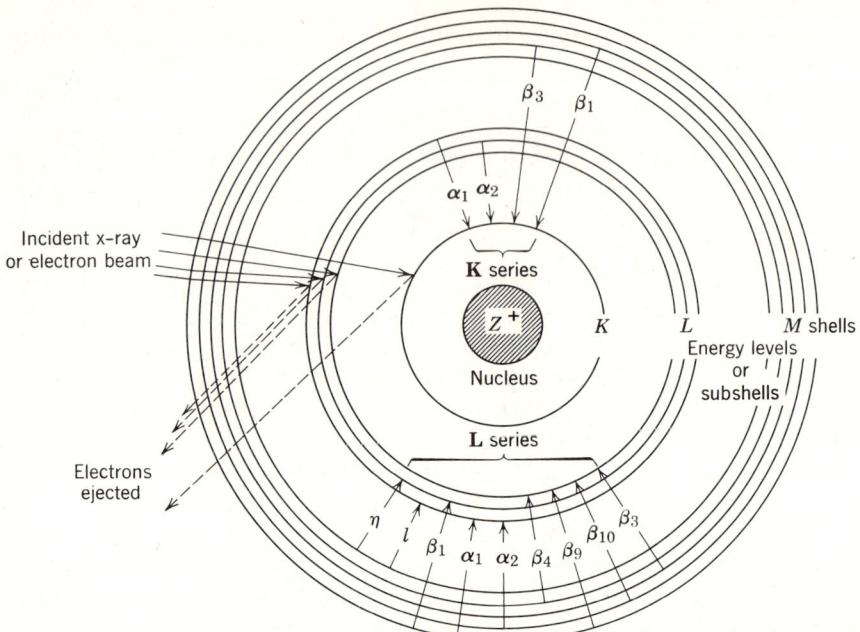

Fig. 4.5–1. Schematic diagram showing electron or x-ray excitation of characteristic lines in an element of atomic number 29 (copper) or greater. The **K** series, for example, results when an electron from an outer shell falls into a hole made in the **K** shell when a **K** electron is ejected from the atom, either by electron bombardment or x-ray absorption. The diagram shows why all **K** lines will be excited simultaneously, and why this is not true for lines in the **L** series. See text. **K**, **L**, **M** shells are characterized by total quantum numbers 1, 2, and 3, respectively. The two outer shells are subdivided into energy levels or subshells. The structure shown above has been confirmed by use of the magnetic photoelectron spectrograph (Section 1.12). The designation of the emitted x-ray lines is at best confusing.

circles, each representing an *energy level*, in which electrons are to be found. A family of energy levels (one **K**, three **L**, and five **M**) is an *electron shell*. The directional lines labeled with Greek letters are the *characteristic lines*; each family of lines is a *series* (**K** and **L**). The diagram has general applicability for the generation of *analytical lines*: it is drawn to represent $_{29}$Cu, the first element in which the **M** shell is full, but it also represents so far as it goes all the heavier elements. To make it represent lighter elements, the nuclear charge must be reduced and nonexistent outer energy levels must be deleted. Copper has a single (valence) electron in the **N** shell, not shown in the figure. This point is of course important in atomic structure as we see later. The energy levels shown contain 28 electrons.

Let **K***, **L***, and **M*** represent the energies required to remove to infinity an electron from one of these shells. Any such energy can be determined

X-RAY SPECTRA

absorptiometrically as it is identified with a critical absorption wavelength λ_C and an absorption edge. Complete identification of **L*** and **M*** requires a subscript (see below). With the energy of the normal atom arbitrarily equated to zero, **K***, **L***, and **M*** are then also the energies of the atom in **K**, **L**, and **M** states; to repeat, **K***, for example, represents both the energy of the **K** edge and that of the atom in the **K** state.

M. de Broglie [8] brought clarity into the early work with the magnetic photoelectron spectrograph by applying the Einstein equivalence law, given in Chapter 1 as

$$h\nu = \tfrac{1}{2}m_e v_e^2 + W_e \qquad (1.4\text{--}3)$$

He realized that x-ray excitation at a frequency ν high enough to eject **K** (and hence **L** and **M** ...) electrons from an atom should give an electron spectrum for which

$$\tfrac{1}{2}mv_1^2 = h\nu - \mathbf{K}^* \quad \text{(K electron ejected at velocity } v_1) \qquad (4.5\text{--}1)$$

$$\tfrac{1}{2}mv_2^2 = h\nu - \mathbf{L}^* \quad \text{(L electron ejected at velocity } v_2) \qquad (4.5\text{--}2)$$

$$\tfrac{1}{2}mv_3^2 = h\nu - \mathbf{M}^* \quad \text{(M electron ejected at velocity } v_3) \qquad (4.5\text{--}3)$$

(similarly for **N, O, P, Q** electrons)

and

$$\tfrac{1}{2}mv_{\max}^2 = h\nu \quad \text{(``Surface'' electron ejected)} \qquad (4.5\text{--}4)$$

Comparison with (1.4–3) shows that W_e is zero in (4.5–4); and that it equals, respectively, **K***, **L***, or **M*** in the other equations if (as is assumed) the electrons lose no velocity on their way out of the target. As ν is known with high precision from x-ray spectrometry and the v's can be precisely measured on a magnetic photoelectron spectrograph (Figure 1.12–2), **K***, **L***, and so on can be established. Values obtained in this way agree excellently with the (less precise) absorptiometric results. The highest precision is attainable by Robinson's "difference method" [9, 10], in which experiments are done with different monochromatic x-rays (e.g., of frequencies ν and ν') so that

$$(\tfrac{1}{2}mv^2) - (\tfrac{1}{2}mv_e^2)' = h\nu - h\nu' \qquad (4.5\text{--}5)$$

The difference method was used to determine h, the value of which was then uncertain to several tenths per cent. Three means of 10 or 11 observations each were internally consistent to about 0.01% and yielded $h = 6.62_9\,(10^{-27})$ erg-sec, in excellent agreement with today's accepted value of 6.6256 ± 0.0005 for the preexponential term. There is no better tribute to the value of the magnetic photoelectron spectrograph and to the consistency of x-ray data.

The evidence supporting Figure 4.5–1 is completed by the proof that the energy of a characteristic line is the energy *difference* of two levels. That this is true was mentioned in Section 1.22; (1.22–1) in general form with the symbols introduced above would read

$$\mathbf{K}^* - \mathbf{L}^*_{III} = \mathbf{K}^*\alpha 1 \qquad \text{(further explanation follows)} \qquad (4.5\text{–}6)$$

with the third asterisk used to denote the energy of the emitted line. As the figure shows, the excitation of this line requires the ejection of an electron from the **K** level followed by the transition of an electron from the **L**$_{III}$ level (or subshell); the transition is indicated in the figure by the directional line labeled α1. The **K** series is thus a series containing lines that originate from a *single* initial state (the **K** state), and the wavelengths of the lines differ because the emission of each leaves the atom in a *different* final state (**L**, **M**, or whatever).

The role that the **K**, **L**, **M** electrons play in generating the corresponding series is not obviously predictable from a knowledge of visible or ultraviolet spectra. Strictly speaking, neither hydrogen nor helium has a **K** series, although each has **K** electrons. Why? Because, again strictly speaking, the **K** series is generated only when the **K** shell contains a hole that is filled by an electron that leaves one of the outer (**L**, **M**, ...) shells; thus the generation of the **K** series requires (1) the *absence* of a **K** electron, and (2) the *presence* of an outer-shell electron whose transition to the **K** shell is permitted by the selection rules. This picture explains why—no matter what the method of excitation—all **K** lines have the *same excitation threshold* so that all **K** lines appear together if they appear at all. This statement is not true for all **L**, **M**, ... lines. (Explanation later.)

An atom with a **K** electron removed from the **K** shell is usually a singly charged positive ion, but an ion different (for the elements beyond helium) from the kind formed when one valence electron is lost. In rare cases, the **K** electron in question will lodge in an unfilled external shell, and the atom will remain uncharged. If a **K** electron is ejected from a conventional ion (e.g., Na$^+$), then the usual charge will increase by one. We shall say that an atom or ion with a single hole in the **K** shell is in the **K** state no matter what its charge. The **L**, **M**, ... states are similarly defined.

The generation of the **K** lines is often represented as follows.

$$\mathbf{K} \text{ state} \rightarrow \mathbf{L} \text{ state} + \mathbf{K}\alpha \qquad (4.5\text{–}7)$$

$$\mathbf{K} \text{ state} \rightarrow \mathbf{M} \text{ state} + \mathbf{K}\beta \qquad (4.5\text{–}8)$$

Because the energy of any **M** state is less than that of any **L** state, it follows that **K**β will have a shorter wavelength than **K**α. In this symbolic representation, which is readily extended to cover the other series, the letter identifying the initial state is identical with that describing the emitted line; and the letter for the final state identifies the shell that supplied the electron whose transition generated the line. For example,

$$\text{L state} \to \text{M state} + \text{L line} \qquad (4.5\text{--}9)$$

As will be explained, three *groups* of lines originate from L states, each group corresponding to a different L energy level; taken together, these three groups make up the L series. There are five groups of lines corresponding to the five M states. Because the K state has only one energy level, the K series contains only one group of lines. We may accordingly say that there are one K state, three L states, and five M states. Owing to the relationships indicated in the equations, L and M lines always accompany K lines.

The existence of related lines within a series is taken into account by introducing Roman-numeral subscripts to identify the states; thus

$$\text{L}_\text{I} \text{ state} \to \text{M}_\text{III} \text{ state} + \text{L}\beta 3 \text{ line} \qquad (4.5\text{--}10)$$

$$\text{L}_\text{III} \text{ state} - \text{O}_\text{I} \text{ state} + \text{L}\beta 7 \text{ line} \qquad (4.5\text{--}11)$$

Needless to say, an equation analogous to (4.5–6) applies for each equation in the group (4.5–7) to (4.5–11), and for all similar equations; the equation

$$\text{K state} \to \text{L}_\text{III} \text{ state} + \text{K}\alpha 1 \qquad (4.5\text{--}6a)$$

completes this group.

Even though the maximum number of emitted lines corresponding to an energy-level diagram as outlined above is not very large, the number of interest to the analytical chemist is reduced considerably below this modest maximum. This fortunate situation exists for two reasons. First, the operation of the selection rules excludes many of the formally possible electron transitions. Second, many of the transitions permitted by the selection rules give rise to lines so weak as to make them practically unimportant. However, in special cases, such as the determination of traces, weak lines of an element present in large amount may have to be considered because they interfere with the line of the element being determined, or because they enhance the background.

An examination of the energy-level diagram of uranium (Figure 4.5–2) will drive home the preceding discussion. With obvious modifications, this diagram can be used for other elements, a fact which again emphasizes the fundamental simplicity of x-ray emission spectra. The diagram does not contain lines, all weak, that result from energy transfers more complex than those described above.

Figure 4.5–2 amplifies the previous discussion of the simplicity of absorption spectra. An inner electron is forced to leave the atom in a single jump because it cannot usually find a hole in an outer level within the atom. The hole thus formed normally initiates a series of steps: for example, (1) an

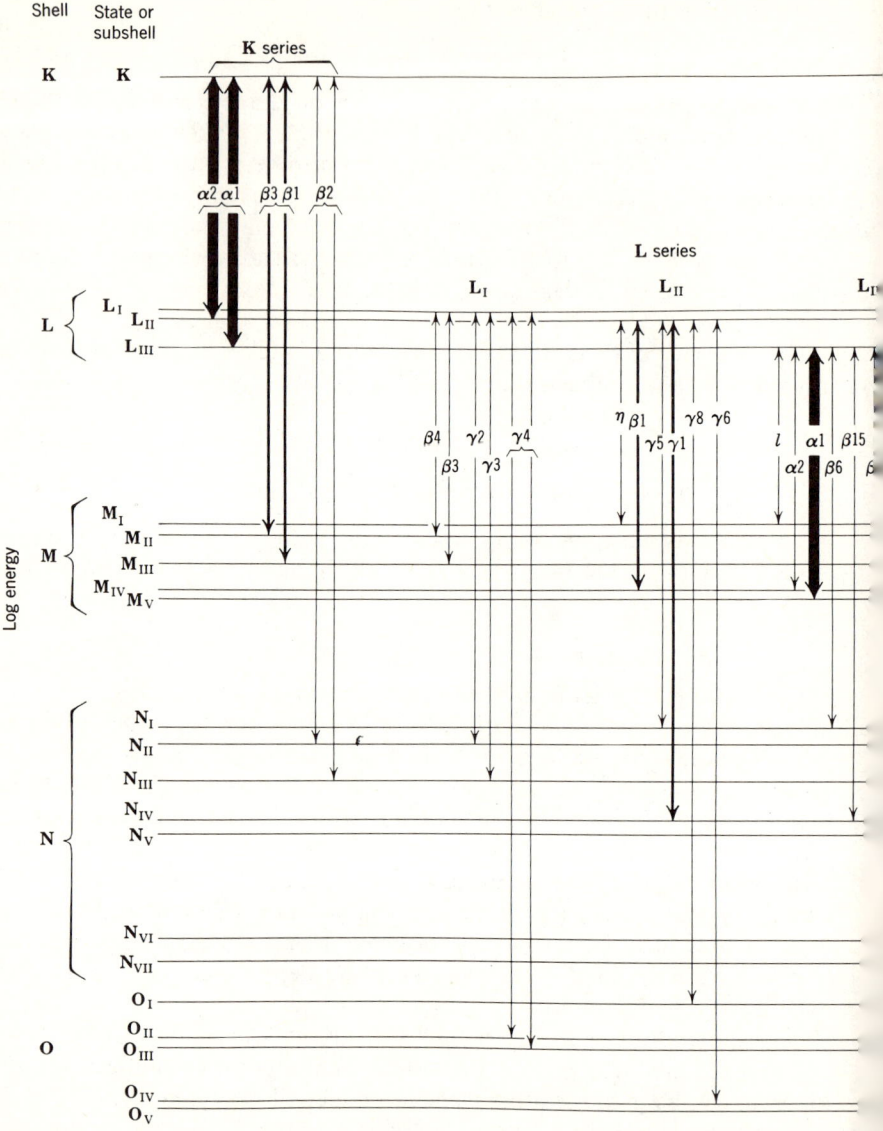

Fig. 4.5–2. The **K**, **L**, and **M** x-ray energy-level diagram for a heavy element (uranium). The heaviest lines are those of major analytical interest. Lines of occasional analytical interest are of medium weight. The energy of a state is that which an atom has when an electron is

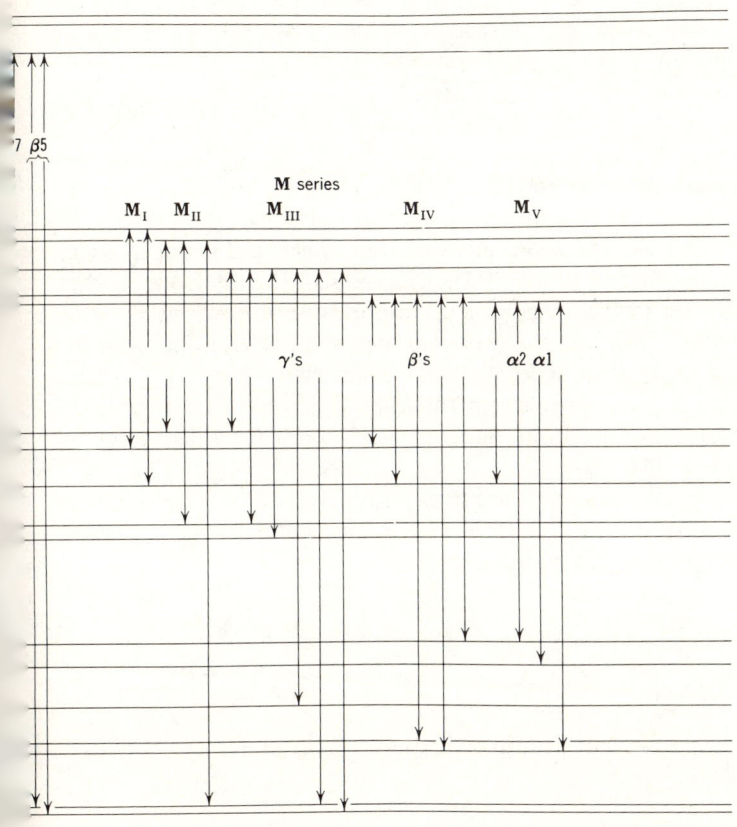

missing from the level corresponding to that state. All lines with intensity greater than 1% of the strongest line in the spectral series are shown for the **K** and **L** states. As the line is emitted, the electron moves up the arrow; the hole in the atom and the state of the atom move down.

L electron falls into the hole that exists in the **K** shell of an atom in the **K** state; (2) another moves from the **M** to the **L** shell to fill the hole formed in (1); (3) and so on. This succession of processes can also be described by saying that the hole leaves the atom in a succession of steps.

4.6 X-Rays and the Bohr Atom

The great similarity of (1.21–1) to (1.21–2) is historically important for the powerful support it gave the Bohr theory of atomic structure. Further discussion of Moseley's work is needed to show the nature of this support.

The interpretation of all *atomic* spectra, x-ray or optical, makes use of the Rydberg constant

$$R = \frac{2\pi^2 m e^4}{ch^3(1 + m/M)} \quad (4.6\text{–}1)$$

in which m and e are the mass and charge of the electron, and M is the mass of the atom. (The small mass ratio m/M may often be disregarded.) Use of R in the formulas for spectral series gives *wave numbers*, and it was in terms of these quantities that Rydberg (1890) expressed the hydrogen series discovered by Balmer in 1885. The general importance of R was established by Ritz in 1908 when he found the wave number of a spectral line to be the difference of two terms (e.g., $R/2^2 - R/3^2$), each term fitting into a large number of combinations. This empirical knowledge received a brilliant theoretical interpretation by Bohr, and Moseley's work provided powerful support for this interpretation.

Moseley quite sensibly chose to use frequencies instead of wave numbers and did this by substituting the "Rydberg fundamental frequency" $v_0 = cR$ for R. Thus expressed, his results are

$$v = \left(\frac{1}{1^2} - \frac{1}{2^2}\right) v_0 (Z - 1)^2 \quad \text{for} \quad K\alpha \text{ lines} \quad (4.6\text{–}2)$$

and

$$v = \left(\frac{1}{2^2} - \frac{1}{3^2}\right) v_0 (Z - 7.4)^2 \quad \text{(approximately)} \quad \text{for} \quad L\alpha \text{ lines} \quad (4.6\text{–}3)$$

We show how Moseley arrived at these equations by presenting his results for the $K\alpha$ lines. He computed values of

$$Q = [v/(3/4 v_0)]^{1/2} \quad (4.6\text{–}4)$$

for each element and listed the results (Ref. 36a, Chap. 1, p. 1028) as in the first two lines below:

X-RAY SPECTRA

	Ca	Sc	Ti	V	Cr	Mn	Fe	Co	Ni	Cu	Zn
Q	19.00	—	20.99	21.96	22.98	23.99	24.99	26.00	27.04	28.01	29.01
Z	20	21	22	23	24	25	26	27	28	29	30

Evidently, $Q = Z - 1$. These data show the fundamental character of Z even more impressively than Figure 1.21–2.

What of the "screening constants" subtracted from Z in (4.6–2) and (4.6–3)? Moseley had this to say (Ref. 36b, Chap. 1, pp. 712–713): "The fact that the numbers and arrangements of the lines in the **K** and **L** spectra are quite different, strongly suggests that they come from different vibrating systems, while the fact that [the screening constant] is much larger for the **L** lines than for the **K** lines suggests that the **L** system is situated further from the nucleus." We can add little to this explanation except to point out how reasonable it is in light of present knowledge. In the **K** state, an atom has only one electron in the **K** shell to "screen" the total nuclear charge, which is Z^+; a net charge of $(Z - 1)^+$ remains. In the **L** state, the atom has 9 electrons to screen the nucleus: 7.4 is a good approximation to 9. These x-ray data make logical the Bohr assumption of Coulombic attraction between nucleus and the electrons in the shells: for example, an **L** electron that fills a hole in the **K** shell sees a net nuclear charge of $Z - 1$ before it fills the hole; an **M** electron that falls into an **L** energy level sees a net nuclear charge of $Z - 7.4$.

Coulombic attraction of electrons and nucleus had been assumed by Bohr, as had the validity of Newton's laws of motion for the electrons in orbits. In addition, he assumed that the electrons normally moved only in certain permitted orbits, that they radiated no energy under these conditions, and that energy was radiated or absorbed in whole quanta of size $h\nu$ when an electron left such an orbit. Assume the electron moves from an orbit of higher to an orbit of lower energy. The frequency of the spectral line then emitted is

$$\nu = Z^2 \left(\frac{2\pi^2 m e^4}{h^3} \right) \left(\frac{1}{n_2^2} - \frac{1}{n_1^2} \right) \qquad (4.6\text{–}5)$$

in which the n's are the *total quantum numbers* of the orbits, with n_1 the larger. Bohr's great triumph was that the second term on the right of (4.6–5) agreed exactly (except for the factor m/M, which is negligible here) with the product cR, R being given its empirically determined value. When numerical values replace the symbols in this term, (4.6–5) becomes (1.21–2). The similarity of (4.6–2) and (4.6–3) to (4.6–5) appears when one remembers that $\nu_0 = cR$;

the similarity is so close as to require no comment once the screening constants have been explained. Furthermore, (4.6–2) shows that the *total quantum number* of the **K** shell is 1; (4.6–3) shows that of the **L** shell is 2; the other shells follow as expected.

If $1/n_1^2$ is made zero ($n_1 = \infty$) in (4.6–5), and the remainder of the equation is multiplied by h, we shall have a value of the energy for the shell corresponding to n_2. It is easy to see that this leads to **K*** = 4**L*** = 9**M***— in other words, that it predicts a rapid decrease in the energy of an electron shell as one moves outward from the atomic nucleus. For molybdenum, the critical absorption wavelengths show that the energy ratio **K*/L*** is nearer 7 than 4. However, it is important that the simplest possible application of the Bohr theory calls for energies not too far from those actually observed in the electron shells around the nucleus. We shall see that the difference between **K*** and **L***, for example, is much greater than the energy differences (such as $L_I^* - L_{III}^*$) between levels within the **L** shell; this statement holds for the other shells also.

4.7 X-Rays and the Periodic Table

Figure 4.5–1 suggests that x-ray processes, because they involve changes of inner electrons, should help in understanding the periodic table. Kossel [11] deserves credit for envisioning the **K, L, M**, and so on *shells* on the basis of the theoretical work of Bohr and the experimental work of Moseley. He recognized that Moseley had found the Ritz combination principle applicable to emitted x-rays; for example, that in wave numbers or in frequencies (which are used here):

$$\nu K\beta - \nu K\alpha = \nu L\alpha \qquad (4.7\text{–}1)$$

Recognition of the general validity of relations analogous to (4.7–1) led him to postulate that the nucleus of an atom was surrounded by electron *shells*.

The Pauli exclusion principle [12] made it possible to continue toward understanding the architecture of the periodic table, and the experimental evidence for the *energy-level* structure (shell *substructure*) in Figure 4.5–1 gave this principle powerful support. In Pauli's words, the principle reads:

Es kann niemals zwei oder mehrere äquivalente Elektronen im Atom geben, für welche in starken Feldern die Werte aller Quantenzahlen n, l, m_j (oder, was dasselbe ist, n, l, m_1, m_s) übereinstimmen. Ist ein Elektron im Atom vorhanden, für das diese Quantenzahlen (im äusseren Felde) bestimmte Werte haben, so ist dieser Zustand 'besetzt.' [Certain symbols changed.]

Before we use this principle to build part of the periodic table, let us examine the supporting x-ray evidence in Table 4.7–1.

X-RAY SPECTRA

Table 4.7–1. Absorption Edges for Selected Elements[a]
(Critical Absorption Wavelengths in X-Units)[b]

Element	Absorption Edge or Energy Level[c]								
	K	L_I	L_{II}	L_{III}	M_I	M_{II}	M_{III}	M_{IV}	M_V
$_{14}$Si	6731.0								
$_{24}$Cr	2065.9								
$_{34}$Se	977.73								
$_{44}$Ru	558.4	—	4169.3	4360.4					
$_{54}$Xe	—	2272.4	2425.3	2587.5					
$_{64}$Gd	246.2	—	1558.7	1706.2					
$_{74}$W	178.22	1020.5	1071.3	1211.6	4365	4800	5416	6475	6708
$_{79}$Au	153.20	860.9	899.6	1038.6	3742	4085	4677	5306	5711
$_{83}$Bi	136.78	755.9	787.8	922.1	—	—	3893	4568	4762
$_{92}$U	106.58	568.0	591.3	720.8	2228	2385	2877	3327	3491

[a] Values from Ref. 8, p. 222 of Chap. 1 this book.
[b] In this table, take an X-unit to be 10^{-3} Å.
[c] The **K** edge of course corresponds to the **K** shell; the other edges, to levels.

The problem is to allocate electrons according to the Pauli exclusion principle to the **K** shell and the **L** and **M** energy levels known to exist for the various elements such as those shown in Table 4.7–1. *Four* quantum numbers are needed to describe each electron.

As the Pauli quotation shows, at least two choices are open for this set of four quantum numbers; we shall take the first. The total quantum number n has already been introduced and related to the energy of an atomic state (e.g., **K***); this is tantamount to relating it to the binding energy of the (ejected) electron. The other three quantum numbers deal with momenta that can be combined vectorially; the vector representing this momentum will appear in bold type that corresponds to the quantum number in *italics*; thus l specifies *orbital* angular momentum, the vector for which is **l**. Electrons have *spin* angular momentum that is conveniently combined with **l** to produce **j**, the vector for *total* angular momentum specified by j; see Figure 4.7–1. In addition, there is needed a quantum number m_j that characterizes the behavior of each electron in a magnetic field so strong that each electron acts independently of the others. The corresponding vector $\mathbf{m}_j$ is obtained by a projection of j in the direction of the magnetic field. See Figure 4.7–2.

The quantum numbers can have only the values given by the following rules:

n: any integer (not zero)
l: 0, 1, 2, 3, ... $(n - 1)$
j: $(l + \frac{1}{2})$ and $(l - \frac{1}{2})$; but $j = +\frac{1}{2}$ when $l = 0$
m_j: all half-integer values between and including $-j$ and $+j$, which makes a total of $(2j + 1)$ values.

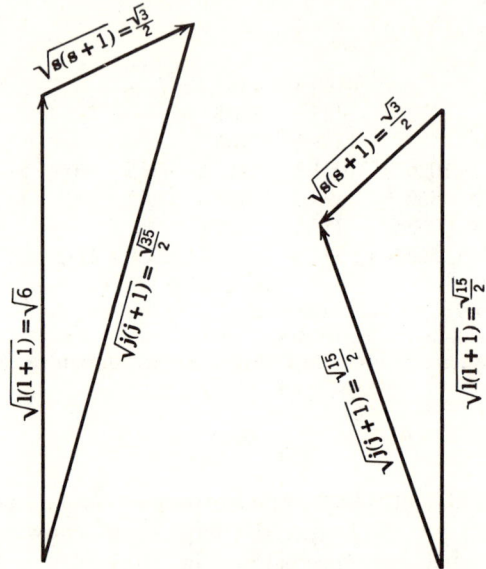

Fig. 4.7–1. Vectorial combination of orbital (inner) angular momentum (**l**) with angular momentum (**s**) due to electron spin (additive or subtractive) to form the vector for total angular momentum (**j**). This combination is governed by the wave mechanics, and the numbers are for $l = 2$ and spin quantum number $s = +\frac{1}{2}$ and $s = -\frac{1}{2}$; s can have no other values.

We next assign quantum numbers by these rules to the energy levels that correspond to the x-ray absorption edges of Table 4.7–1 and then apply the Pauli exclusion principle to give the maximum number of electrons permitted in each energy level in Table 4.7–2.

Table 4.7–2 shows that the Pauli exclusion principle allocates to the shells of Table 4.7–1 the correct maximum number of electrons (**K**, 2; **L**, 8; **M**, 18) required by the periodic table of the elements. Note, however, that there is no inkling of *chemical periodicity* in Table 4.7–2: the x-ray data are primarily for the *inner* electrons. Further light on this matter comes from Table 4.7–3.

In Table 4.7–3, the first valence electron to impart metallic character (hydrogen is considered metallic) is shown in italics. For the first eighteen elements, the electrons go in to fill shells and energy levels so as to leave no gaps.

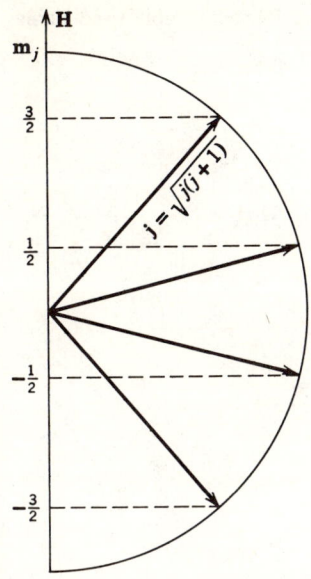

Fig. 4.7–2. Possible values of m_j for $j = 3/2$. The projection of j in the direction of magnetic field H is made according to the wave mechanics. For a single electron (the case shown), j and m_j are required to be odd half-integers.

Table 4.7–2. Quantum Numbers and Electrons for Various Energy Levels (See Table 4.7–1)

Shell	Energy Level	Values of Quantum Numbers				Electrons (Max. No.)[b]	Optical[a] Designation
		n	l	j	m_j		
K		1	0	$\frac{1}{2}$	$+\frac{1}{2}, -\frac{1}{2}$	2	$1s$
L	L_I	2	0	$\frac{1}{2}$	$+\frac{1}{2}, -\frac{1}{2}$	2	$2s$
	L_{II}	2	1	$\frac{1}{2}$	$+\frac{1}{2}, -\frac{1}{2}$	2	$2p$
	L_{III}	2	1	$\frac{3}{2}$	$+\frac{3}{2}, +\frac{1}{2}, -\frac{1}{2}, -\frac{3}{2}$	4	$2p$
M	M_I	3	0	$\frac{1}{2}$	$+\frac{1}{2}, -\frac{1}{2}$	2	$3s$
	M_{II}	3	1	$\frac{1}{2}$	$+\frac{1}{2}, -\frac{1}{2}$	2	$3p$
	M_{III}	3	1	$\frac{3}{2}$	$+\frac{3}{2}, +\frac{1}{2}, -\frac{1}{2}, -\frac{3}{2}$	4	$3p$
	M_{IV}	3	2	$\frac{3}{2}$	$+\frac{3}{2}, +\frac{1}{2}, -\frac{1}{2}, -\frac{3}{2}$	4	$3d$
	M_V	3	2	$\frac{5}{2}$	$+\frac{5}{2}, +\frac{3}{2}, +\frac{1}{2}, -\frac{1}{2}, -\frac{3}{2}, -\frac{5}{2}$	6	$3d$

[a] This is the optical designation of the individual electrons.
[b] The maximum number of electrons for any one value of l is $2(2l + 1)$. This simple rule follows because l can have $(2l + 1)$ values and the two values of s ($+\frac{1}{2}, -\frac{1}{2}$) are possible for each l. Note, however, that more than one energy level can have the same value of l.

Table 4.7–3. Electrons, X-Ray Energy Levels, and the Periodic Table (See Table 4.7–2)

Period	Element	K	L		M			N[a]
					Energy Level (or Subshell)			
		L_I	$(L_{II} + L_{III})$[b]	M_I	$(M_{II} + M_{III})$	$(M_{IV} + M_V)$		N_I
1	$_1$H	*1*[c]						
	$_2$He	2						
2	$_3$Li	2	*1*					
	$_4$Be	2	2					
	$_5$B	2	2	*1*				
	$_6$C	2	2	2				
	$_7$N	2	2	3				
	$_8$O	2	2	4				
	$_9$F	2	2	5				
	$_{10}$Ne	2	2	6				
3	$_{11}$Na	2	2	6	*1*			
	$_{12}$Mg	2	2	6	2			
	$_{13}$Al	2	2	6	2	*1*		
	$_{14}$Si	2	2	6	2	2		
	$_{15}$P	2	2	6	2	3		
	$_{16}$S	2	2	6	2	4		
	$_{17}$Cl	2	2	6	2	5		
	$_{18}$Ar	2	2	6	2	6		
4	$_{19}$K	2	2	6	2	6		*1*

[a] Only first N electron is shown.
[b] As is often done when the energy differences become small, certain of the energy levels are paired. (The smaller such energy differences, the less certain it is to which level an added electron goes.) They are kept separate in Tables 4.7–1 and 4.7–2.
[c] The first valence electron to impart metallic character is shown in italics. See text.

If this were to continue with potassium, however, its valence electron would be an **M** electron, for which $n = 3$. Because of the close chemical relationship of potassium to sodium, it is clearly necessary that the potassium electron should be an **N** electron ($n = 4$). Where valence electrons are concerned, chemical considerations override. As there is no need to describe the periodic table to chemists, we shall not discuss the way in which electrons place themselves in the higher shells. The object here is merely to show that x-ray evidence has played a vital role in settling atomic structure. X-ray absorption

X-RAY SPECTRA

data show how the apartment house is built, but they cannot in all cases settle the order in which apartments are rented.

Two other matters relating to absorption remain to be pointed out. First, the changing energy differences in the energy-level structure (Section 4.6). Consider the data for $_{74}$W in Table 4.7–1. Energy is of course proportional to reciprocal wavelength. Clearly, the energy difference for $M_{IV} - M_V$ is almost negligible as compared with that for $K - L_{III}$; this justifies considering these M levels together as in Table 4.7–3. Second, the matter of all L (or M or N...) lines not necessarily being excited at once (Section 4.5). Continuing with the data for $_{74}$W as example, we see from Table 4.7–1 that wavelengths greater than 1.2116 Å will excite no L lines; that wavelengths greater than 1.0713 Å but not greater than 1.2611 Å will excite those L lines associated with the L_{III} edge, but no others; that wavelengths greater than 1.0205 Å but not greater than 1.0713 Å will excite the lines associated with the L_{II} and the L_{III} edges; and that, finally, absorption of a wavelength of 1.0205 Å or less will excite the lines associated with all three L levels. These expectations are borne out, as is the expectation that all K lines appear if any does. The correctness of the views about x-rays and atomic structure are thereby additionally confirmed.

4.8 Selection Rules and X-Ray Emission Spectra

Experience shows that fewer emitted lines of significant intensity are observed than are permitted by energy considerations and the number of electrons in atoms (Table 4.7–2). As in the case of optical spectra, it turns out that most of the lines actually emitted are those for which the *selection rules*

$$\Delta l = \pm 1 \tag{4.8-1}$$

$$\Delta j = 0, \pm 1 \tag{4.8-2}$$

are *both* satisfied: that is, an x-ray line is emitted if the rules give the difference in the initial and final states for the electron transition that transfers the hole and leads to the emission of the characteristic quantum. To illustrate (see Figure 4.5–1): transitions from the L_{II} and L_{III} levels are permitted when the atom is in the K state, but the transition from the L_I level is not permitted because for it Δl is zero. How good are the selection rules? They are reliable for all but weak lines. Figure 4.5–1 shows two lines β_9 and β_{10} not permitted by the selection rules. Figure 4.5–2 shows none.

The usefulness of j in (4.8–2) explains why the set of quantum numbers including it was chosen for the application of the Pauli exclusion principle in Section 4.7.

4.9 X-Ray and Optical Spectra. General

The comparison to be made is between the optical spectra of atoms (electronic energy levels) and x-ray spectra. The vibrational and rotational spectra of molecules have no x-ray analogues.

The main concern up to now has been x-ray spectra of atoms with atomic numbers so high that the energies of transition for valence electrons were negligible in comparison with the energies of transition for inner electrons. Such elements make one limiting case. Hydrogen, which has no inner electrons so that all its spectra must result from valence-electron transitions, stands at the other limit.

For the elements of high atomic number, x-ray and optical spectra are distinct. As regards absorption, we repeat that x-ray spectra show no absorption lines. Such lines, common in optical spectra, are not found in the x-ray region because an electron ejected to infinity is not likely to return to its former home. In the optical region, where a valence electron is raised to a permitted outer energy level when absorption occurs, this electron can

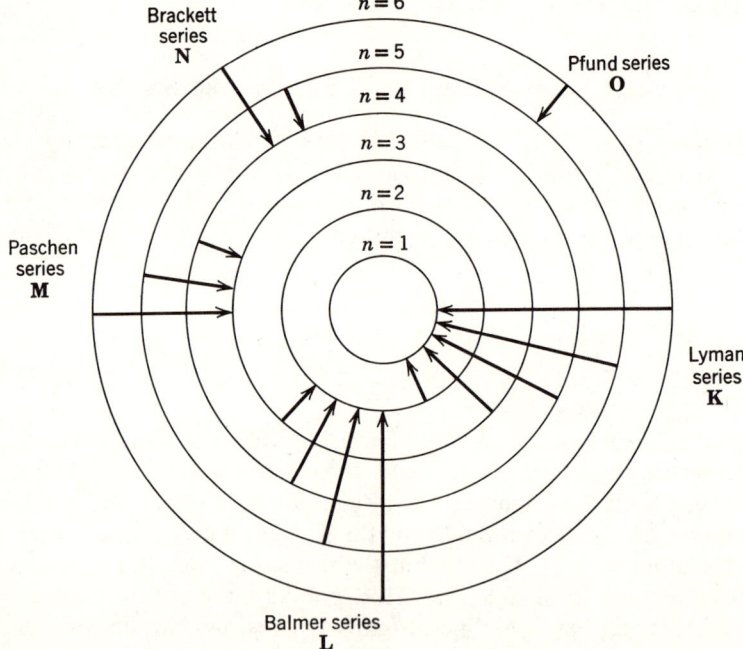

Fig. 4.9–1. Analogy of optical emission spectra of hydrogen with the x-ray emission spectra of the heavier elements. The spectral series for hydrogen are named. The letters identify the analogous x-ray emission series; the analogy rests on the identity of total quantum numbers (n's) for the two cases.

X-RAY SPECTRA

always fall to a lower level with the emission of a spectral line. X-ray absorption and emission must involve at least *two* different electrons; the optical analogue may be regarded as involving the same electron and often the same energy levels.

Hydrogen at the other limit is an interesting case. The minimum voltage required to remove the electron from a hydrogen atom is 13.58 V, which corresponds to a quantum of wavelength 913 Å (1.6–2). This wavelength is in the ultraviolet. This alone makes it awkward to regard it as an x-ray (a **K** line of hydrogen) even though Moseley's relation (1.21–1) in the absence of a screening constant gives

$$\lambda = c/v \approx 3(10^{10})/(0.25)(10^{16})(1^2) = 12(10^{-6}) \text{ cm} \qquad (4.9\text{–}1)$$

or 1200 Å. Note that (1.21–1) becomes (1.21–2) when the screening constant disappears.

Figure 4.9–1 shows the analogy existing between the optical emission spectra of hydrogen and the x-ray emission spectra of the heavier elements.

4.10 The Fine Structure of Spectra

According to Figure 4.7–1 and the selection rules, (4.8–1) and (4.8–2), the existence of electron spin increases the number of permitted spectral lines. Furthermore, it leads to the existence of pairs of energy levels that differ only in spin quantum number; one having $s = +1/2$ and the other, $s = -1/2$. The energy difference between two such levels is relatively very small. This leads to spectral lines that also differ very little in energy.

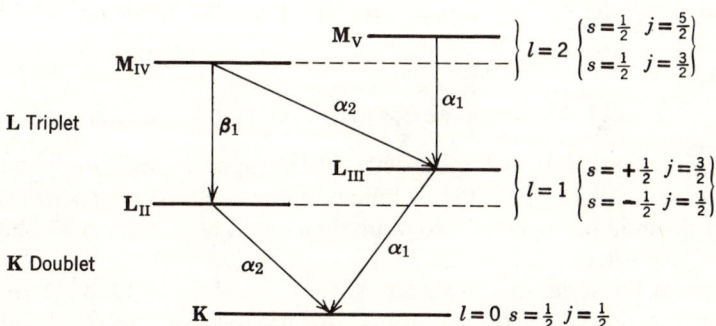

Fig. 4.10–1. Diagram to show fine structure of x-ray lines according to the selection rules. Compare with Figures 4.5–1 and 4.5–2. The diagram above is for a "one-hole" x-ray spectrum, which is analogous to a one-electron optical spectrum. The Kα1,α2 doublet and an **L** triplet are shown. In the triplet, Lα2 and β1 may be regarded as a doublet. Note that the energy difference $[L_{II} - L_{III}]$ for the **K** doublet is identical with that for this **L** doublet. Such regularities help in the identification of spectral transitions.

The situation is described by saying that atomic spectra have a "fine structure"; or by saying that doublets, triplets, and higher multiplets exist in such spectra.

Optical spectra of the alkali metals show the simplest fine structure. They are generated by transitions of a single valence electron outside a filled electron shell: they are "one-electron" spectra. In these, higher multiplets are not permitted by the selection rules. The doublets and triplets that are permitted are actually found. Perhaps the best known doublet contains the yellow lines of sodium with wavelengths of 5895.93 Å (D_1) and 5889.96 Å (D_2).

The most common x-ray spectra are "one-hole" spectra. Pauli pointed out that the spectral terms in this case are what they would be if the missing electron alone occupied the subshell (in the **K** state, the shell) from which it was ejected. This means that the one-hole x-ray spectra should be identical in fine structure (though not in quantum numbers; see next paragraph) with the one-electron optical spectra. And so it is. This fine structure for the x-ray case is illustrated by Figure 4.10–1. The Kα1,α2 doublets are the best known, and their wavelength-differences (generally about 0.001 to 0.005 Å) appear in the Appendices.

That x-ray doublets do exist was implicitly assumed in Section 4.5 and is clear from Figure 4.10–1. Similar figures for optical spectra are inverted because of the man-made difference—mentioned earlier—between zero energies for the two cases. Here common sense seems to favor the x-ray position, which is that an atom gains energy when it absorbs a quantum so that, for example, the **K** state is of higher energy than the normal. In the optical case, the potential energy of the hydrogen atom is customarily taken as zero when proton and electron are an infinite distance apart—that is, when the atom no longer exists. The position is logical because the Coulombic energy $-Ze^2r$ is zero at infinity, which makes for *negative* energy at finite r.

4.11 X-Ray-Like Spectra in the Optical Region

The hot spark (ultraviolet) spectra of "stripped atoms" were found by Bowen and Millikan [13–15] to follow the laws for x-ray spectra as completely as could be expected. No more than an inkling of this beautiful work can be given here.

Consider the series of species Na, Mg^{+1}, Al^{+2}, Si^{+3}, P^{+4}, S^{+5}; these (except for the Na) are "stripped atoms." All have similar electron configurations and the same total number of electrons (11) outside the nucleus; the nuclear charge and the atomic number increase as one goes from 11(Na) to 16(S^{+5}). This amounts to a man-made variation of nuclear charge for constant inner-electron structure. The question is whether the ultraviolet emission spectra under these conditions obey the laws for x-ray spectra, which are the law of regular doublets (hinted at in this chapter), the law of irregular doublets (not elsewhere mentioned), and Moseley's law for the variation of frequency with atomic number (discussed in Chapters 1 and 4). The answer is that the first and third laws are satisfactorily obeyed, and that nothing can be said about the second because the regular doublet separations are too small to be observed. Particularly impressive are Moseley plots ($\sqrt{v/R}$ against Z) for the different spectral lines (7 lines are given in Ref. 15, Figure 1) emitted by the series of "stripped atoms" named above.

REFERENCES

1. D. L. Webster, *Phys. Rev.*, **9**, 220 (1917).
2. W. W. Nicholas, *J. Res. Natl. Bur. Stand.*, **2**, 837 (1929).
3. H. Kulenkampff, *Ann. Physik*, **69**, 548 (1922).
4. H. A. Kramers, *Phil. Mag.* [6], **46**, 836 (1923).
5. S. T. Stephenson and F. D. Mason, *Phys. Rev.*, **75**, 1711 (1949).
6. A. Jönsson, *Z. Phys.*, **43**, 845 (1927).
7. "Röntgenstrahlen," in *Handbuch der Physik*, Vol. XXX, Julius Springer, Berlin, 1957, p. 11.
8. M. de Broglie, *J. Phys.*, **6**, No. 2, 265 (1921).
9. H. R. Robinson, *Phil. Mag.* [7], **18**, 1086 (1924).
10. C. J. B. Clews and H. R. Robinson, *Proc. Roy. Soc.*, **176A**, 28 (1940).
11. W. Kossel, *Verh. Deutsch. Phys. Ges.*, **16**, 899 and 953 (1914); *ibid.*, 339 and 396 (1916). Later references also.
12. W. Pauli, *Z. Phys.*, **31**, 765 (1925).
13. I. S. Bowen and R. A. Millikan, *Phys. Rev.*, **24**, 209 (1924).
14. R. A. Millikan and I. S. Bowen, *Phys. Rev.*, **24**, 223 (1924).
15. I. S. Bowen and R. A. Millikan, *Phys. Rev.*, **25**, 295 (1925).

Chapter 5

The Selection of X-Ray Wavelengths

5.1 Introduction

Information obtained by working with x-rays—whether it relates for example to composition, structure, or thickness—is usually more certain the more nearly monochromatic the rays: wavelength selection is thus generally desirable and often mandatory.

There are no monochromatic x-ray beams. Such a beam would have zero width on a wavelength scale, and therefore zero *integrated intensity* at finite height. This limiting condition is approached the more closely, the more nearly monochromatic the x-ray—that is, the more rigorous the selection process. In other words, *reduced intensity* is the price that must always be paid in the selection process. Sometimes this price is so great that intensity can no longer be read instantaneously, but must be established accumulatively, as by counting for a fixed time. As always, the price paid should be the minimum needed to get the work done.

This chapter is concerned mainly with selection by filtering and by wavelength resolution. Another way of producing a "monochromatic" beam—that is, by use of a *characteristic-line generator* (e.g., irradiating a piece of zinc foil to excite Zn $K\alpha$ for subsequent use) is not really a selection process and has so much in common with the generation of the characteristic line of a target (Chapter 1) and of a sample (Chapters 7 and 9) that no special description is needed.

5.2 Selection by Filtering

Filtering is the simplest way of increasing the relative intensity of desired wavelengths. This method is strikingly successful in simple cases such as the following. When a copper-target Coolidge tube is used in x-ray diffraction work, the **K**β lines often interfere, only the **K**α lines being desired. At 1.54 Å (copper **K**α), μ for nickel is only 48, but it rises to 286 at 1.39 Å (copper **K**β) because the **K** absorption jump of nickel is at 1.49 Å. Thus a thin nickel filter will suppress the **K**β line without prohibitively reducing the intensity of copper **K**α [1].

The sequential use of two filters, as in the balanced-filter method of Ross [2], brings greatly improved wavelength selection over Figure 5.2–1*a* at intensities far higher than could be achieved by Bragg reflection. Two filters, preferably elements of atomic numbers Z and $Z-1$, are used *sequentially* to isolate wavelengths in the shaded area *BCED* of Figure 5.2–1*b*, called the *pass band* by Kirkpatrick [3a]: Zn **K**α ($\sim$ 1.44 Å) could be isolated with the filters of the figure. The pass band is the difference between *ABCEF*, which represents the transmitted intensity with the copper filter in place, and *ADEF*, the corresponding intensity for the nickel filter. Ideal balancing is attained if the thicknesses of the two metals can be so adjusted that each attenuates identically x-rays of all wavelengths outside the pass band. *Simultaneous* measurements on a split beam with one detector for each filter make it possible to obtain the intensity of the beam transmitted by the pass band as a *difference signal*.

With only two filters, the balancing is not usually as good as in Figure 5.2–1*b*; possible causes of mismatch mentioned by Kirkpatrick [3a, b] are differing wavelength dependence of photoelectric absorption by the filters; differing wavelength dependence of scattering; appearance of unwanted (and unneeded) absorption edges; and different absorption jumps for the filters. The last seems to be the most important. Kirkpatrick showed that imperfect balancing could be much improved by the addition of a third filter (e.g., Al) that has no absorption edges in the wavelength region of interest. Young's admirable extension [3c] of Kirkpatrick's work should be consulted.

Further description of filtering is unnecessary because what it can do is predictable from the absorption equations in Chapter 1 by use of data in the Appendices. It comes to this: a filter is useful if it increases the ratio I_{Des}/I_{Ob} where I_{Des} is for a desired wavelength (such as that of an analytical line in x-ray emission spectrography) and I_{Ob} is for an objectionable line (interfering line or background)—always subject to the proviso that I_{Des} after filtering must still be high enough for precise, convenient measurement. Suppose I_{Ob} originates in the x-ray source; the logical place for the appropriate

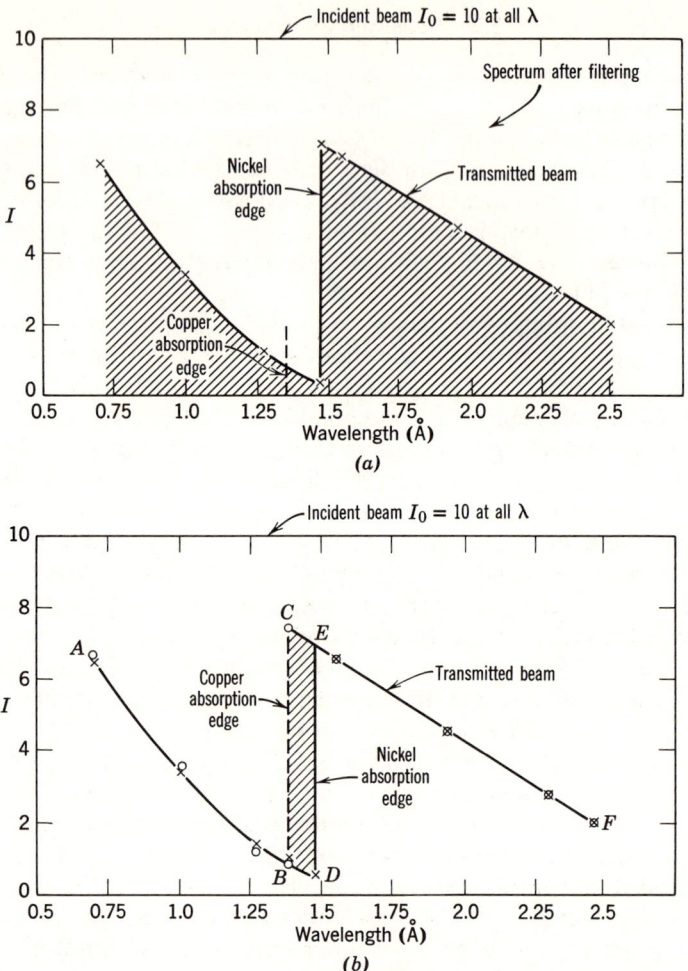

Fig. 5.2–1 Shaded areas show integrated intensities from ∼ 0.75 to 2.5 Å of filtered beams produced in two following ways from incident beams for which $I_0 = 10$ (arbitrary units) at all wavelengths: (a) Single filter, 0.01-mm Ni (b) Ross balanced filters: (1) 0.0092-mm Cu; (2) 0.01-mm Ni. See text.

filter then is between source and sample. If it originates in the sample, the filter belongs between sample and detector, usually next the detector. Salmon [4] and Gunn [5] have demonstrated the usefulness of filters in interesting examples of x-ray emission spectrography.

The best and simplest filters are metal foils. Elements not available as foils may be used as compounds and/or placed on substrates, or be absorbed in dissolved form by filter paper. See the discussion of supported samples in Chapters 7 and 9.

Filtering can thus select (and therefore produce) "monochromatic" x-rays, but it is limited because it cannot be used conveniently at all wavelengths and because it cannot achieve high spectral purity at any wavelength. Simplicity, its greatest virtue, makes filtering useful as an adjunct to more sophisticated methods, or as a substitute for these when they are not available, as in a portable x-ray emission spectrograph for use in the field.

BRAGG REFLECTION BY A FLAT CRYSTAL

5.3 Collimation in the X-Ray Spectrograph

Collimation selects direction, not wavelength, but it is a necessary adjunct to Bragg reflection from a flat crystal because satisfactory wavelength resolution cannot be achieved in this way unless neither incident nor reflected beam is appreciably divergent. According to Bragg's law, the wavelength singled out for first-order reflection by a crystal of spacing d depends only on θ, the angle of incidence. One may turn this around to say that a crystal fixed in position with respect to the optical system will reflect different λ's for different values of θ. As Figure 5.3–1 shows, a range of θ values from θ_1 to θ_3 is possible with a divergent, uncollimated beam when the crystal

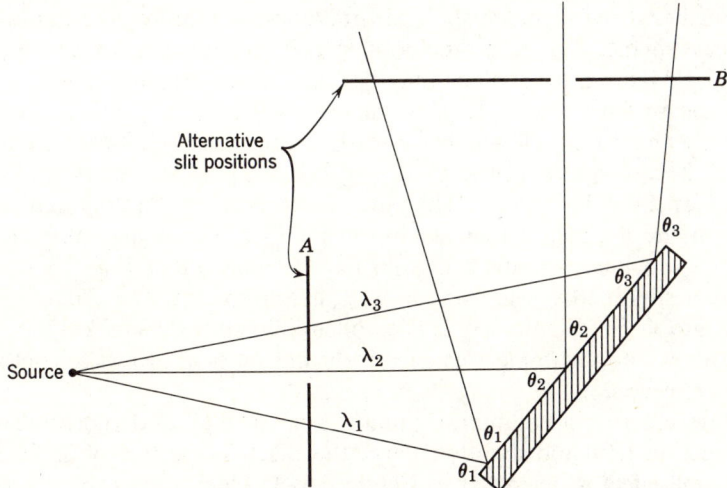

Fig. 5.3–1 Diffraction of a beam from a point source by a large crystal. The crystal is positioned for the Bragg reflection of wavelength λ_2 at angle θ_2. Without a slit, Bragg reflection of all wavelengths between λ_1 and λ_3 will occur because the crystal receives x-rays at all angles between θ_1 and θ_3. A slit at A or B will collimate the beam and remove the unwanted wavelengths.

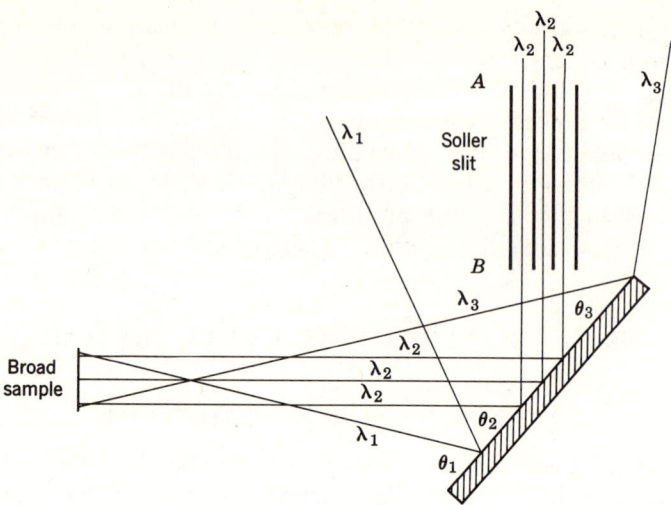

Fig. 5.3–2. Diffraction of a divergent beam from a broad sample by a large crystal. Collimation of this beam requires the Soller slit system shown. This system is equivalent to simple slits at A and B with separators provided to reduce divergence in θ-plane (plane of paper). Divergence in channels not shown. See Figure 5.3–3. Sample, crystal, and slits all show extension in plane perpendicular to paper.

is set to reflect λ_2 at angle θ_2. Some of these unwanted x-rays would enter the detector (not shown) if collimation were not provided as in the figure.

In further discussion, we shall call the plane containing the goniometer arc the θ-plane. The goniometer axis is perpendicular to the θ-plane. In a spectrograph, the θ-plane may be vertical or horizontal.

The beam from a sample of usual area will require a collimator more elaborate than a simple slit because the sample is both broad and long, and the x-rays from it will show divergence both in the θ-plane and in the plane perpendicular thereto. The latter divergence can be tolerated. Divergence in the θ-plane is reduced by means of a Soller [6] collimator, or slit system, which is a stack of thin parallel plates that forms a series of extended slits in the beam direction, thus intersecting the θ-plane of the spectrograph. See Figure 5.3–2. The Soller collimator thus makes it possible to use an extended sample area effectively and yet establish wavelength with adequate precision.

X-rays entering a collimator channel near 0° may undergo total reflection (Section 1.19) and slightly enlarge the effective aperture of the channel. This small effect is neglected in Figure 5.3–3, which shows the divergence that remains in the θ-plane in a collimator of the dimensions there given. This divergence in angular measure is $2(\Delta\theta) = 0.3°$, where

$$\Delta\theta = \text{arc tan}\,(0.01/4) = 9' = 0.15° \qquad (5.3\text{–}1)$$

BRAGG REFLECTION BY A FLAT CRYSTAL

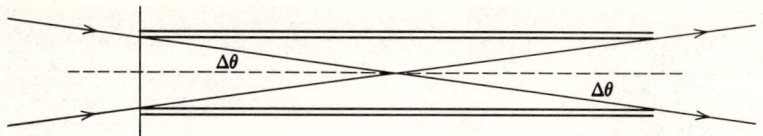

Fig. 5.3–3. Angular divergence in a single channel of a Soller collimator assumed to be 4 in. long, made of metal sheets 1 mil thick and spaced to give a 10-mil (0.01 in.) opening. Only rays of maximum divergence are shown.

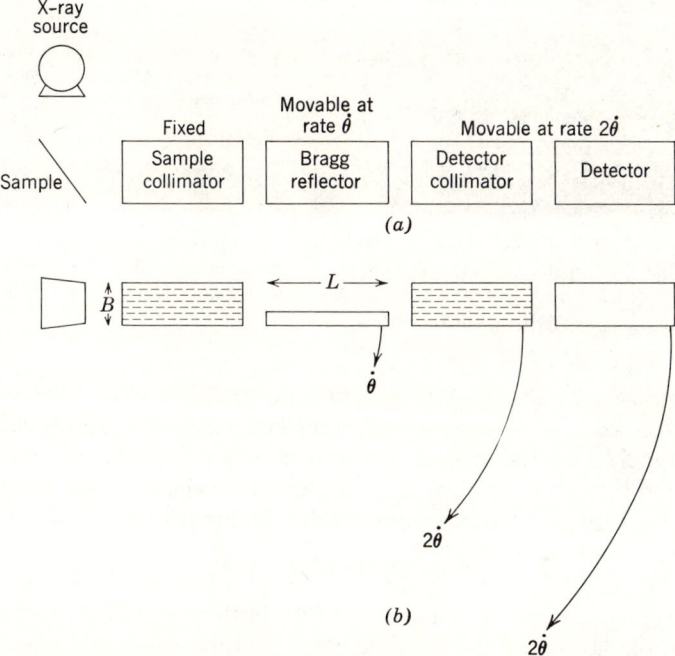

Fig. 5.3–4. Schematic views of spectrograph components. (*a*) Side view. θ-plane $\perp$ to page. Motion in θ-plane at angular velocities shown. (*b*). View of same parts from above. θ-plane coincides with page. Arrows show motion and angular velocity during scanning from $2\theta = 0°$ (initial value). X-ray source not shown in (*b*).

The beam is uniform over the exit of the collimator and absent behind the plates at that point, but beams from adjacent channels begin to overlap about 0.2 in. further on. Other complications exist. There results a beam that has an intensity distribution (I vs θ) resembling a truncated triangle— in other words, an approximately Gaussian distribution to which the divergence 2 ($\Delta\theta$) is a guide. See below.

As Figure 5.3–4 shows, a fixed collimator, called the *sample collimator*, is often used along with a *detector collimator*, which must move with the detec-

Fig. 5.3–5. Actual transmittance of a Soller collimator consisting of iron plates 1-mil thick with nominal 9-mil openings. Transmittance for x-rays recorded on film. Ideal pattern would have been lines 1-mil thick separated by white spaces 9 mils wide. Note how far the collimator departs from ideality. Experiments by P.D.Z.

tor to conserve the 2θ-relationship. The figure also shows that the useful 2θ-range is limited near $0°$ because part of the beam from the sample collimator will be blocked by the side of the crystal, and not all of the rest will be intercepted by the crystal face. The smallest θ at which a crystal of length L will intercept all of the beam from a Soller collimator of total breadth B is

$$\theta = \text{arc sin}\,(B/L) \sim 8° \qquad (5.3\text{–}2)$$

The θ given is for $B = 0.5$ in. and $L = 3.5$ in., both reasonable values in which considerable latitude is permissible. As 2θ approaches $180°$, the detector and its collimator are obviously on a collision course with other equipment.

The foregoing discussion is idealized. Collimators are far from ideal as Figure 5.3–5 shows. The plates warp; dirt settles on them; they scatter x-rays and may even show diffraction peaks and emit characteristic lines. The edges of the plates cause complications. The finer the collimator, the more trouble it will give. A 50% loss in intensity in a single collimator is not unusual.

Also, the intensity of the beam leaving the collimator cannot be uniform because the sample-source distance varies over the sample. When an x-ray beam strikes the sample at a nominal angle of $60°$, the actual angles of incidence for the entire sample may cover a $15°$ range, which corresponds to a variation in intensity of some 10%. With variation in angle comes a variation in critical depth: see Chapter 7. The total intensity from a small sample will therefore vary with its position, and the x-rays from such a sample may not pass through all the channels of a Soller collimator: see Figure 7.12–2.

The collimator should be selected on the general principle that a spectrograph should operate at the maximum intensity and minimum resolution required to get the work done. In practice, one will have to compromise. Adler [7] has found a satisfactory compromise to be sample collimator, 4×0.005 in.; detector collimator, 1.5×0.023 in.

5.4 Intrinsic Line Width

When an x-ray line undergoes Bragg reflection, it can only be broadened, and the broadening (increase of FWHM) can have various causes. To assess them, we need a starting point, which must be the intrinsic ("natural" or minimum) line width. Zero width is a physical impossibility (Section 5.1). The intrinsic line width can be calculated classically [8] or according to the quantum mechanics [9]. Only the results are needed here.

The classical treatment assumes the line to be emitted by a damped electronic oscillator and predicts a line shape (relation between intensity I and oscillator frequency v) given by

$$I_\omega / I_{max} = \frac{1}{1 + [(\omega - \omega_0)/b]^2} \qquad (5.4\text{–}1)$$

where

$$v = \omega / 2\pi \qquad (5.4\text{–}2)$$

and b is the damping factor. At the FWHM, the intensity ratio is $\frac{1}{2}$, and (5.4–1) yields (in wavelength units)

$$\text{FWHM} = 4\pi e^2 / 3mc^2 = 1.18(10^{-4})\,\text{Å} \qquad (5.4\text{–}3)$$

in terms of the electronic charge and mass, and of the velocity of light. For Mo Kα, the measured value is

$$\text{FWHM} = 2.94(10^{-4})\,\text{Å} \qquad (5.4\text{–}4)$$

The numerical discrepancy and the experimental observation that the intrinsic width of x-ray lines depends on the properties of the initial and final transition states both indicate the need for a quantum-mechanical treatment.

The quantum-mechanical treatment yields a similar line shape corresponding to an equation resembling (5.4–1) except that it contains, in lieu of the damping constant, a transition probability from initial to final state. This transition probability is taken to equal the FWHM and is subject to the Heisenberg uncertainty principle. The quantum-mechanical treat-

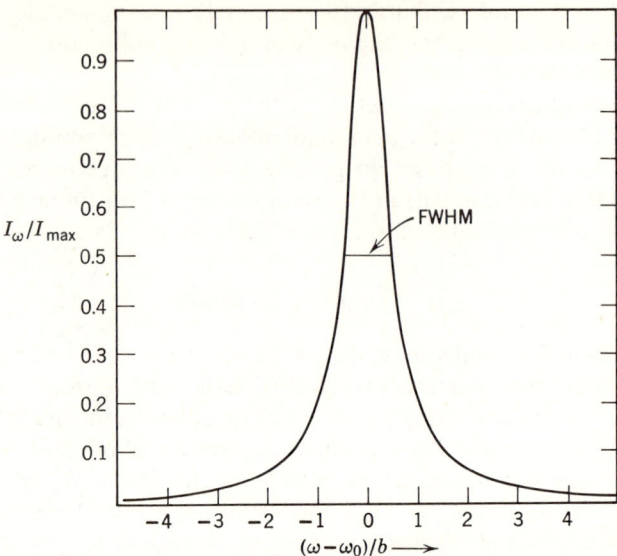

Fig. 5.4–1. The shape of a spectrum line emitted by a classical electronic simple harmonic oscillator with electromagnetic damping. After Compton and Allison. Ref. 8, p. 273, Figure IV–1.

ment is thus an improvement on the classical and of special interest because it foreshadows a possible chemical effect on x-ray spectra in that both final and initial states influence the transition probability, and consequently the line profile; the influence is marked for the light elements.

Let us calculate the angular divergence that corresponds to the measured FWHM for Mo Kα. Differentiation of Bragg's law gives

$$n d\lambda = 2d \cos \theta \, d\theta \qquad (5.4\text{–}5)$$

which, when divided into Bragg's law, yields

$\lambda/d\lambda = \tan \theta/d\theta \qquad$ or $\qquad d\theta/d\lambda = \tan \theta/\lambda \qquad$ (in radians per angstrom)
$$(5.4\text{–}6)$$

For $\lambda(\text{Mo K}\alpha) = 0.71\text{Å}$, $n = 1$, $d = 1$ Å, (5.4–6) gives for $d\lambda = 3(10^{-4})$ Å

$$d\theta = [3(10^{-4}) \times 0.3792]/0.71\text{Å} = 1.6(10^{-4}) \text{ radians} = 9.1(10^{-3})°$$

Figure 5.4–1 is calculated from (5.4–1) and may be taken as the starting point for the discussion of FWHM increases caused by Bragg reflection. Two such increases are unimportant in x-ray emission spectrography. They are (1) the increase associated with the inexactness of Bragg's law in its simple form (see Section 1.19), and (2) the increase associated with ab-

BRAGG REFLECTION BY A FLAT CRYSTAL

sorption of incident and reflected ray. See discussion of film thickness determinations in Chapter 7. Under favorable conditions, the intensity loss on Bragg reflection will be as low as 1%.

5.5 The Rocking Curve

Imperfections in the large crystals used in x-ray emission spectrographs are the principal reason why Bragg reflection of a monochromatic line in these instruments gives an $I - \theta$ distribution curve with FWHM considerably greater than that in Figure 5.4–1. These imperfections are described by saying that the crystal has a "mosaic structure," the surface showing many *mosaic blocks* of about 500 atoms on one side; or by saying that the crystal shows a "high lattice dislocation density." No matter what words are used, the gist of the matter is that Bragg reflection from an imperfect crystal will at one λ occur over a range of θ's in accord with the first part of Section 5.3. The result is a *broader* beam with an increased *integrated reflected intensity* (advantage) but more difficult to resolve (disadvantage). Imperfections can be deliberately produced and their surface density to some extent controlled (see below). Figure 5.5–1 illustrates the situation.

The curves in Figure 5.5–1 are called "rocking curves" of the crystals, a name selected when front porches were more common than they are now. Measurement of such a curve requires two Bragg crystals in series, with usually the second movable [8, p. 709] so that it can be "rocked" through the angular range over which reflected intensity exceeds background.

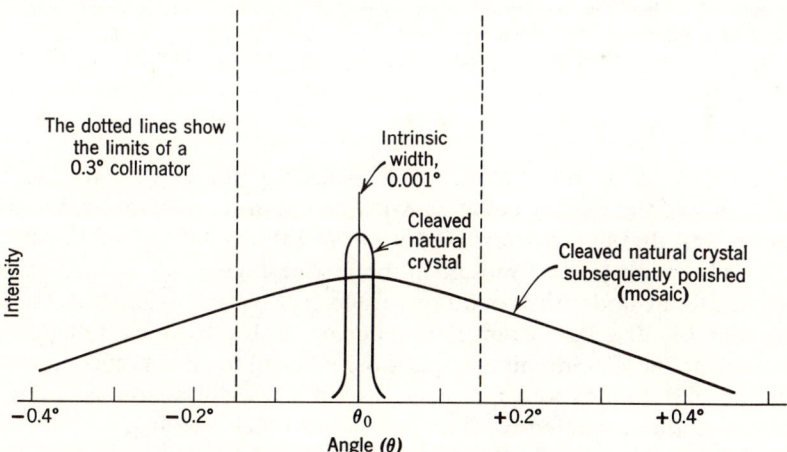

Fig. 5.5–1. This diagram shows the intensity variation with angle for a rock salt crystal in the region near the Bragg angle, θ_0, for an incident monochromatic beam. The area under the mosaic crystal curve could be thirty times greater than the ideal. The intrinsic width is an estimate; see (5.4–6). After Renninger [10].

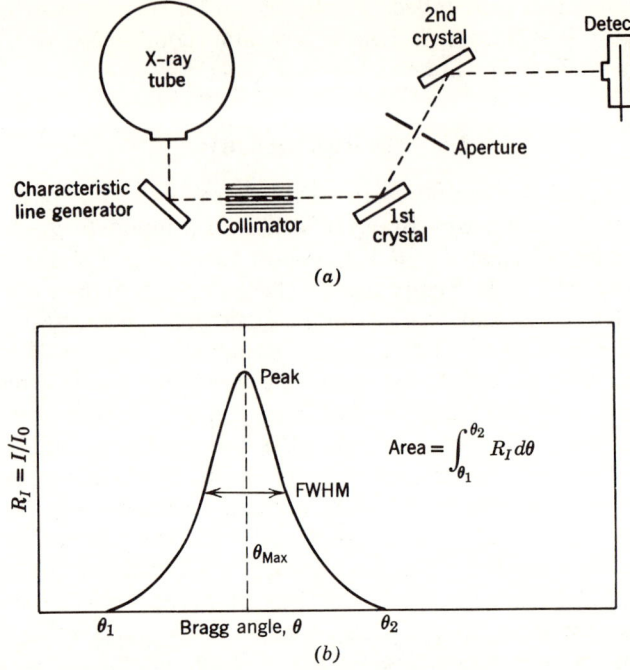

Fig. 5.5–2. (*a*) Double-crystal spectrometer with replaceable characteristic-line generator to give highly monochromatic x-ray beam for measurement of rocking curves. Characteristic-line generators give a series of nearly monochromatic **K** lines in the range 0.7–2.1 Å; the aperture intercepts the **Kβ** line. The first crystal, as nearly perfect as possible, acts as monochromator; the second is the crystal for which the curve is measured. (*b*) Idealized rocking curve. The ordinate is the ratio of reflected to incident intensity. After Vierling, Gilfrich, and Birks, Ref. 11, p. 342, Figure 1.

The detector must not broaden the measured curve. Recognizing the importance of the rocking curve for x-ray emission spectrography, Vierling, Gilfrich, and Birks [11] determined it for an LiF crystal treated in various ways and for hot-pressed mosaic graphite. See Figure 5.5–2. LiF crystals were measured under the conditions listed in Figure 5.5–3. Abrasion was done with 15–20 μ wet carborundum; flexing, with subsequent reflattening, was done at 300°C. Graphite [12] is more efficient in Bragg reflection than LiF, and it is coming into increasing use in x-ray diffraction as a way of eliminating the unwanted **Kβ** lines (see Section 5.2). Vierling, Gilfrich, and Birks [11] point out that the graphite is not a single crystal; that it is so efficient because orientation of its crystallites is highly preferred in the (002) direction; and that efficient Bragg reflection can be obtained either by treating a crystal (LiF) initially too nearly perfect so as to produce dis-

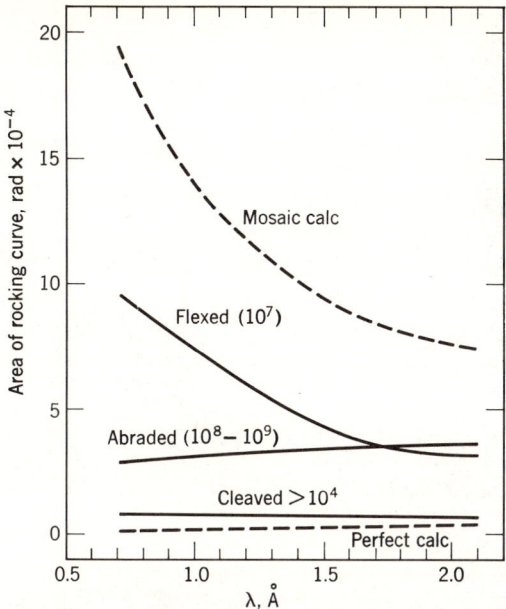

Fig. 5.5-3. Measured and calculated values of rocking-curve areas ($\int R_I d\theta$, see Figure 5.5-2) as functions of wavelength for LiF crystals in the conditions shown. The numbers in the body of the figure are estimated dislocations per square centimeter of surface. Dislocations found at greater depth in flexed than in abraded crystals. After Vierling, Gilfrich, and Birks, Ref. 11, p. 343, Figure 4; the reference explains how the calculations were made.

locations, or by treating crystallites so as to make a hot-pressed mass approach perfection in one direction (graphite).

5.6 Resolution (Dispersion) in the Spectrograph

Dispersion originally was a qualitative term used to describe, for example, the action of a prism on sunlight. Then came linear dispersion, $dx/d\lambda$, x being distance along a spectrum thus broken up; and angular dispersion, $d\theta/d\lambda$, to measure angular separation. The concept has been further generalized, and it is now being used in connection with Bragg's law: for example, (5.4-6). But, "dispersion theory" means, among other things, attempts to explain the variation of refractive index with wavelength, x-ray wavelengths included; this has nothing to do with (5.4-6). Finally, "dispersion" is not really needed to describe wavelength resolution by Bragg reflection. We shall not use it in this sense.

Resolution at its simplest, when quantitatively applied to a spectrum, is the ratio of the average wavelength of two spectral lines that can just be

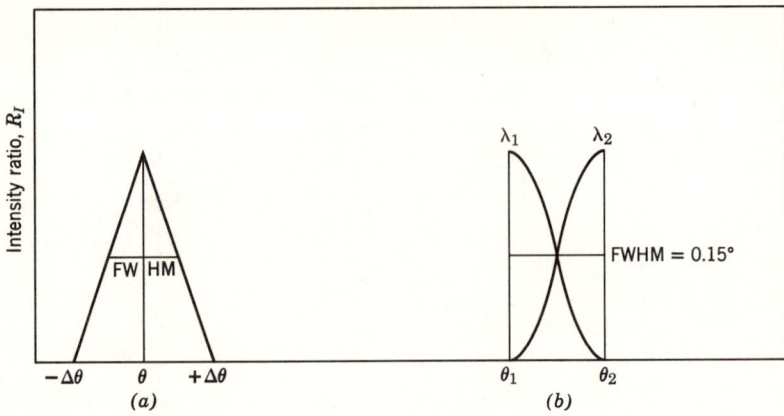

Fig. 5.6–1. Estimate of attainable resolution. (a) Idealized $R_I - \theta$ relationship at the exit of a single slit in a Soller collimator. FWHM $= \Delta\theta$. (b) Diagram to illustrate calculation (5.6–2). Gaussian curves assumed. Note that the peaks of the curves are "resolved," but that the tails are not.

differentiated as a doublet, to their wavelength difference: $R = \lambda_{av}/\Delta\lambda$. The simplicity disappears when lines become broadened, and when pulse-height distributions (Chapter 2) and energy resolution enter the picture. Analytical chemistry adds further complications (see Section 2.21): for example, traces must sometimes be determined in the presence of interfering major constituents.

With all these complexities, refined resolution calculations are pointless. In what follows, we shall estimate (1) the FWHM of a monochromatic beam *as it leaves the Bragg crystal*, and (2) the corresponding wavelength difference for two *monochromatic beams* on the basis of Bragg's law. The estimates will be in degrees θ.

The FWHM in question results mathematically from intrinsic-line width (Figure 5.4–1); from angular divergence in the collimator (Figure 5.3–3); and from the rocking curve (Figure 5.5–2b). The individual FWHMs for the first and last cases are shown in the figures.

The FWHM for collimation is estimated as follows. The angular divergence of the collimator is an *intensive* property; it is that of a single slit and independent of the number of (assumed identical) slits. The maximum intensity at the exit of a slit will be at its center. The $R_I - \theta$ relationship will therefore approach the form of an isosceles triangle (Figure 5.6–1a). For a family of *similar* triangles, base and altitude remain proportional. Hence the FWHM in the figure equals $\Delta\theta$, or 0.15° for the collimator of Figure 5.3–3.

As in Section 2.20, we consider the three curves to which the three FWHMs

BRAGG REFLECTION BY A FLAT CRYSTAL

relate as Gaussian, so that the relationship between FWHM and standard deviation s is given by (2.20–2). We then apply the rule for the combination of errors in the form

$$\text{FWHM} = \sqrt{(\text{FWHM}_i)^2 + (\text{FWHM}_c)^2 + (\text{FWHM}_r)^2} \quad (5.6\text{--}1)$$

where the subscripts mean intrinsic width, collimator, and rocking curve; (5.6–1) is merely (2.20–3) with the proportionality constant 2.35 introduced as in (2.20–2).

With $\text{FWHM}_r = 0.03°$, which is close to measured values [13], and with other values as given above, we have

$$\text{FWHM} = \sqrt{(0.01)^2 + (0.15)^2 + (0.03)^2} = 0.15° \quad \text{or} \quad 0.0026 \text{ radians} \quad (5.6\text{--}1a)$$

Here, better resolution demands a finer collimator. See Campbell, Leon, and Thatcher [14] for experimental demonstrations of what can be done to improve resolution with various collimator arrangements, especially for lines of different intensities. Fine collimation of course always means

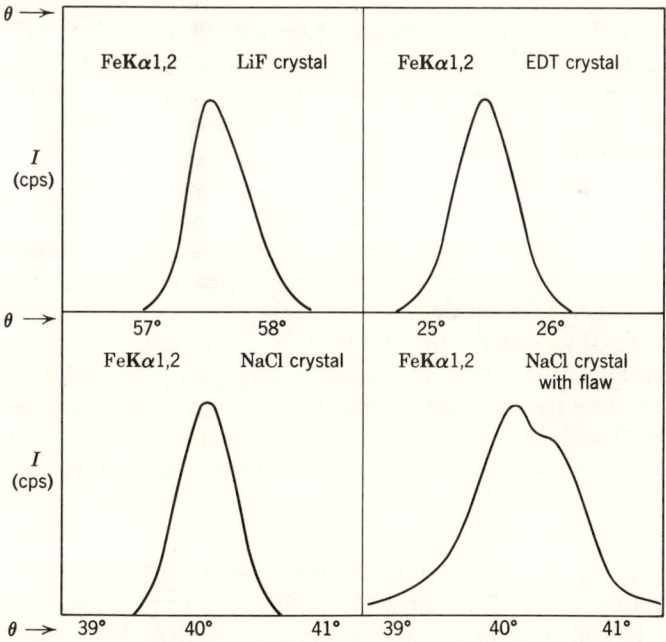

Fig. 5.6–2. Chart recordings to show FWHM values for various crystals. For collimator alone, $\text{FWHM}_c \sim 0.2°$. Sealed proportional detector. Authors' unpublished results, 1958.

reduced analytical-line intensity and may mean greater difficulty in adjusting the spectrograph.

Bragg's law has four variables. Let us take $n = 1$, $2d = 4$ Å, and calculate $\Delta\lambda = \lambda_1 - \lambda_2$ for $\Delta\theta = 0.15°$ when $\lambda_1 = 2$ Å. This could be done by calculating $\Delta\theta$ as the difference $\theta_1 - \theta_2$ from Bragg's law. Because such a calculation involves $\sin \theta$ values more precise than those ordinarily given, it is more convenient to transform (5.4–6) and obtain

$$\Delta\lambda = (\Delta\theta \times \lambda)/\tan \theta_1 = (0.0026 \times 2)/0.5774 = 0.009 \text{ Å} \quad (5.6\text{–}2)$$

or about half the wavelength difference between $K\beta 1$ of $_{24}$Cr and $K\alpha 1$ of $_{25}$Mn. Whether these two lines would be "resolved" under these conditions depends upon how "resolution" is defined. See Figure 5.6–1b.

These calculations, modeled upon those of Birks [13], give useful insights into the resolution problem, but they are incomplete in that they do not include possible broadening of the curves by the detector. This broadening could be allowed for by including a fourth term under the radical in (5.6–1). As the data needed are lacking, the section is closed with Figure 5.6–2, which shows measured curves that include any effect of the detector.

5.7 The Choice of Bragg Reflector

Reflector will often be used henceforth because devices other than crystals can accomplish the kind of reflection discovered by Bragg: for example, gratings, layers of metals, soap films—the last called *Langmuir-Blodgett gratings*. Only the last are important in analytical chemistry. From now on, $2d$ and 2θ will be used increasingly.

There is no lack of crystals that could serve as Bragg reflectors for intermediate wavelengths; only those of greatest interest in analytical chemistry will be considered. The performance of even one kind of crystal varies with its rocking curve, which is in turn a function of wavelength. The choice of crystal proper for a spectrograph depends upon the other components as (5.6–1) indicates; this does not tell the whole story. Fortunately for the analytical chemist, manufacturers now commonly incorporate into the spectrograph the kind and number of Bragg reflectors called for by the wavelength range of the instrument.

Let us examine resolution on the basis of (5.4–6) in the form

$$\Delta\theta/\tan \theta = \Delta\lambda/\lambda = 1/R \quad (5.7\text{–}1)$$

which has the advantage that resolution is expressed as a *fraction*. Discussions based on $\Delta\lambda$ encounter the difficulty that a $\Delta\lambda$ meaningful for Na $K\alpha$ could well be comparable with the *wavelength* of W $K\alpha$. See (2.21–1).

1. The significance of changing θ (n, d constant) is brought home by

BRAGG REFLECTION BY A FLAT CRYSTAL

citing two absurd limiting cases:

$$\text{At } \theta = 0°, \quad \tan\theta = 0, \quad \text{and} \quad \Delta\theta/\tan\theta = (\Delta\lambda/\lambda) \to \infty \quad (5.7\text{--}2)$$
$$\text{At } \theta = 90°, \quad \tan\theta = \infty, \quad \text{and} \quad \Delta\theta/\tan\theta = (\Delta\lambda/\lambda) \to 0$$

We progress virtually from no resolution whatever to infinitely fine resolution as θ increases from 0 to 90°. Clearly, (5.4–5) and (5.4–6) get into difficulty when a change of $\Delta\theta$ produces an appreciable relative change in $\cos\theta$ or $\tan\theta$.

2. The significance of changing from d_1 to d_2 ($n = 1$). Use (5.4–5) in the form

$$\Delta\theta/\Delta\lambda = 1/(2d_1 \cos\theta_1) \quad \text{(similarly for } d_2, \theta_2\text{)} \quad (5.7\text{--}4)$$

Resolution is better at the smaller d: compare topaz and lead stearate in Table 5.7–1. In fact, for low θ, resolution varies inversely as d to a usually satisfactory degree of approximation. As θ approaches 90°, (5.7–4) becomes progressively less useful because of the difficulty mentioned above. For an assumed $\Delta\theta$, $\Delta\lambda/\lambda$ can be calculated from (5.7–4) and Bragg's law. To summarize: The calculations just described are valuable mainly as qualitative guides for *one component* in the wavelength selection and detection processes.

Over how wide a 2θ-range is a Bragg reflector useful? A practical answer to this complex question is: "Over the widest range that geometry permits." At low 2θ, a reflector will not intercept all the beam until some angle near 20° is reached (5.3–2). At 2θ somewhere below 180°, the detector will collide with other components, and reflected intensity often decreases. In Table 5.7–1, the bars have been drawn to show geometrical limitation from 8 to 16°, and absence thereof from 16 to 145°. If a better answer is needed and more than one crystal is available, the performance of each crystal should be tested on a standard sample.

Here, as elsewhere in x-ray emission spectrography, the light elements cause trouble. Bragg's law shows that a Bragg reflector cannot deal with wavelengths exceeding $2d$, for $\sin\theta$ cannot exceed unity. Reflectors with $2d$ values in the range $\sim$ 10 to 100 Å are therefore needed. Furthermore, the analytical lines that lie in this wavelength range will have high mass absorption coefficients so that the reflector must have a smooth surface if the negative absorption effect (see Chapters 5, 7, and 9) is not to be prohibitive. But smooth surfaces are poor Bragg reflectors. To illustrate: Campbell, Leon and Thatcher [14] found the intensity ratio $I_{etched}/I_{polished}$ for quartz (101) to be 2.3 at 0.49 Å and only 0.40 at 4.7 Å, which is the wavelength of Cl Kα, and Cl is not yet a light element. The multilayered soap films that we call Langmuir-Blodgett gratings are the best Bragg reflectors for the light elements (more precisely, for wavelengths in the 10–100 Å range). See Figure 5.7–1. In Chapter 7, we shall encounter a *dual curved* Bragg reflector (Henke-type) in which a lead-stearate overlay reflects the long wavelengths, and a mica substrate, the short. See Figures 7.15–1 and 7.15–2.

Table 5.7-1. Flat Bragg Reflectors for X-Ray Emission Spectrography
A. Spacings and Useful Ranges

Bragg reflector	$2d$ (Å)
PbSt	100
PbMy	80.5
OHM	63.5
KAP	26.63
Gypsum	15.19
ADP	10.65
EDDT	8.808
PET	8.742
Ge	6.532
Si	6.276
NaCl	5.614
LiF (200)	4.028
LiF (220)	2.848
Topaz	2.712

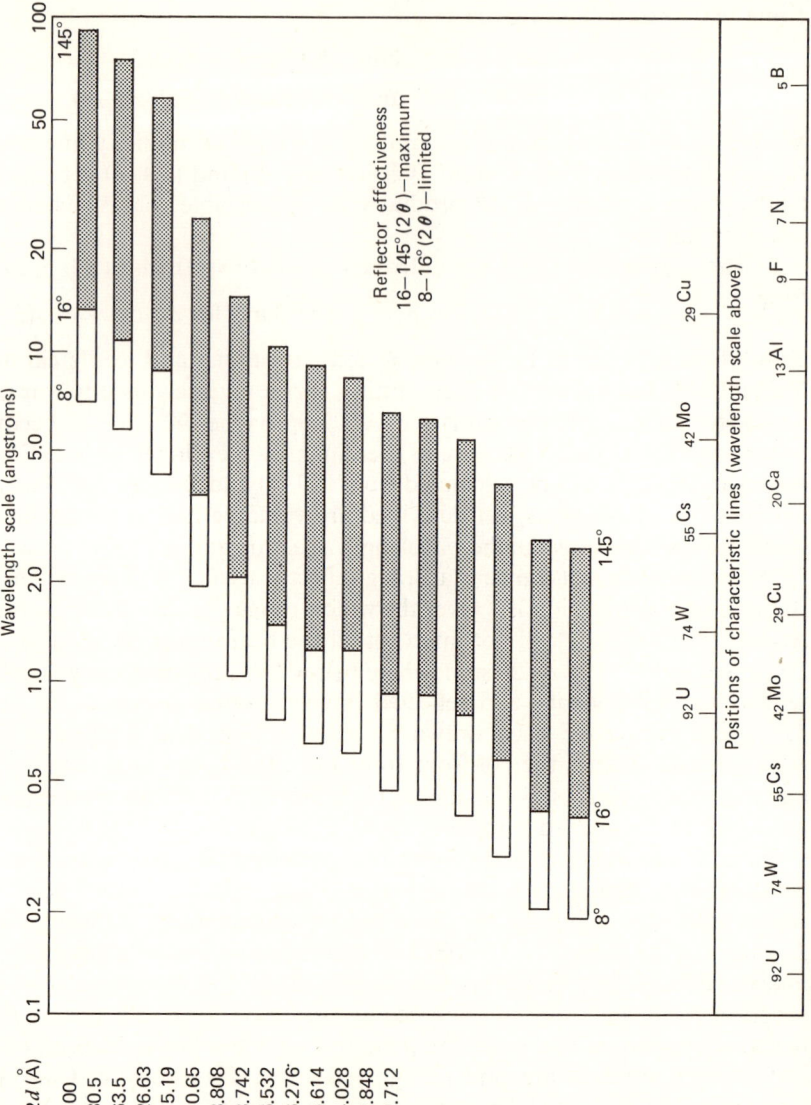

B. Further Information

Bragg Reflector	Name	Remarks
PbSt	Lead stearate	Langmuir-Blodgett grating
PbMy	Lead myristate	All following reflectors are crystals
OHM	Octadecyl hydrogen maleate	Good for determining carbon
KAP	Potassium acid (hydrogen) phosphate	Preferred for determining F or Na
Gypsum	Calcium sulfate dihydrate	May effloresce. Store properly
ADP	Ammonium dihydrogen phosphate	Low intensity
EDDT	Ethylenediamine ditartrate	Low intensity
PET	Pentaerythritol	Higher intensity than the preceding two
Ge	Germanium	Both elements: even orders missing
Si	Silicon	in reflection. Intensity comparable
NaCl	Sodium chloride	Useful but hygroscopic. Store properly
LiF (200)	Lithium fluoride. Note Miller indices	Best general-purpose crystal
LiF (220)	Lithium fluoride. Note Miller indices	Good intensity. High resolution
Topaz	Aluminum fluosilicate (hydrated)	Best resolution

216 THE SELECTION OF X-RAY WAVELENGTHS

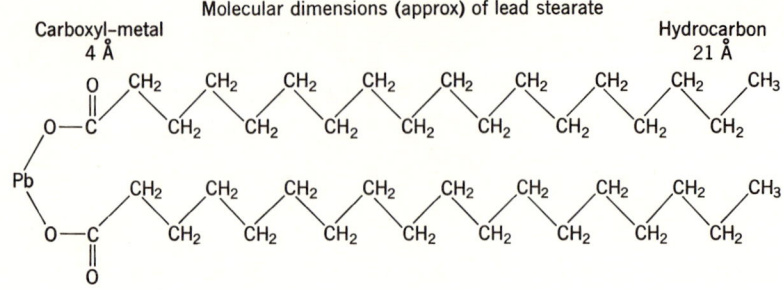

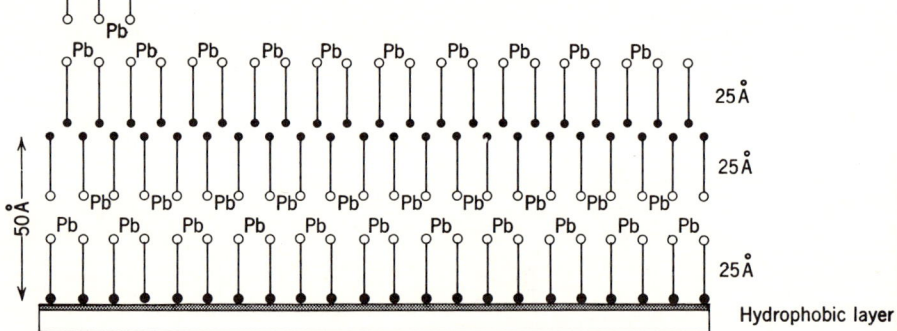

Fig. 5.7–1 A typical (lead stearate) Langmuir-Blodgett grating. Note rod-like character of molecule with one end (O) hydrophilic, the other (●) hydrophobic. Prepared by successively dipping into, and withdrawing the substrate plate from, water that carries a monomolecular film of the soap so that the molecules are vertical with the (●) end exposed. Provided the substrate surface is *initially hydrophobic*, the film on the substrate will have the structure shown above with *precise* spacings ($d = 50$ Å) as indicated: between these spacings, the film resembles a liquid crystal. Each immersion-withdrawal cycle thus thickens the film by 50 Å. Approximately 100 cycles make a good Langmuir-Blodgett grating.

5.8 Digest of Operating Details for Flat Bragg Reflectors

GENERAL. Any reflector should be mechanically sound (no undue strains) and reflect x-rays uniformly (no twinning) over its operating surface. Its dimensions should be suited to the spectrograph; a 1-in. width is customary, and the length (say, 3 in.) is usually chosen with (5.3–2) in mind. The reflecting surface must be protected always; note that many organic materials (PET, soap films) are soft. Composition changes are to be avoided or minimized; special measures are needed if the reflector shows hygroscopicity, efflorescent character (see Table 5.7–1), or tendency toward being damaged by x-rays (organic materials especially, undue exposure of which should be avoided so as to lengthen their useful life). Good Bragg reflectors are costly

BRAGG REFLECTION BY A FLAT CRYSTAL

and deserve the best of care. The performance of a Bragg reflector under standard conditions must be measured often enough to ensure that it has not changed in a way that will reduce the reliability of the determinations made; some such assurance can be provided by the *comparative method* (Section 9.18).

TEMPERATURE. An *increase* of temperature will increase the interplanar distances in a Bragg reflector, causing θ to *decrease*, and it will broaden the reflected line owing to increased thermal vibration in the reflector. Only the first effect is important. A temperature change in the Bragg reflector usually results from a change in room temperature, but it could be caused by heat flux from the x-ray tube.

With x-ray emission spectrography in mind, Lee and Campbell [15] investigated the effect on θ of changing the temperature T of several common Bragg reflectors. Differentiation of Bragg's law ($n\lambda$ constant) gives

$$0 = 2d[\partial(\sin \theta)/\partial T] + 2[\partial(d)/\partial T] \sin \theta \qquad (5.8-1)$$

or, in practical form

$$0 = d(\cos \theta) \Delta\theta/\Delta T + (\Delta d/\partial T) \sin \theta \qquad (5.8-2)$$

which transforms into

$$-\Delta\theta = [1/d \, \Delta d/\Delta T](\tan \theta) \Delta T = (\alpha \tan \theta) \Delta T \qquad (5.8-3)$$

The last expression in brackets is the coefficient of thermal expansion α (in the direction perpendicular to the reflecting plane in question). Lee and Campbell measured $\Delta\theta$ as a shift in the diffracted peak intensity of known characteristic lines for known ΔT. Equation 5.8–3 gave values of α for LiF (200) and for ADP (101) that agreed with those accepted. EDDT gave unexplained high results, and this led Lee and Campbell to suggest that organic Bragg crystals be systematically investigated. Computer calculations were made for a number of Bragg crystals.

TYPICAL RESULTS. For LiF (200) an *increase* of 5° in T produces the following *percentage decreases* in measured I for the FWHM collimator values and Bragg angles (both expressed as degrees 2θ) given: *0.34* for 0.15° and 30°; *4.58* for 0.15° and 90°; *0.86* for 0.35° and 90°. Four conclusions emerge: (1) the effect of temperature change on θ can easily become comparable with chemical effects (e.g., $\Delta\theta$ difference in peak intensity for Fe and Fe^{+++}). (2) Air conditioning to maintain laboratory temperature within $\pm$ 3°C is generally advisable and sometimes mandatory. (3) The finer the collimation, the more precisely ought the temperature of the Bragg reflector to be controlled. (4) No simple rule will cover all situations; [15] and [16] are useful guides; as above, comparison of standard and unknown could be useful.

UNWANTED WAVELENGTHS. 1. See Figure 2.23–6 and related text. The Bragg reflector can act as a generator of unwanted characteristic lines. These may be attenuated (e.g., by air path or filter) enough to make them harmless before they reach the detector. If not, there are two possibilities: (a) *No pulse-height selection.* The unwanted lines (and their escape peaks) will appear as background over the entire 2θ-range; it will sometimes be possible to cope with this background. See Chapters 8 and 9. (b) *Pulse-height selection employed.* The unwanted lines (and their escape peaks) will contribute to background when the pulses they generate pass voltage windows along with pulses from the analytical line (Section 2.23). This is the price that must be paid for eliminating the unwanted lines at other settings of the voltage window. Peaks from unwanted lines will begin to distort those of analytical lines as their energies approach each other. Suggested procedure: On basis of fundamental information available (Chapters 1, 2, and Appendices) decide whether the unwanted lines could be harmful. If so, see whether they could be eliminated by attenuation or by adjusting voltage window. If all else fails, get another Bragg reflector.

2. Abnormal Bragg reflections give rise to unwanted peaks that are of low intensity and usually broader than those from analytical lines. Such reflections are uncommon, seldom if ever found with Bragg crystals of high symmetry (e.g., LiF as opposed to topaz or EDDT), and may not cause trouble when they do appear. They occur because the Bragg crystal has planes along its length or width, the normals of which are slightly tilted with respect to the normal of the plane that reflects the analytical line. See Jenkins and de Vries [17], who also discuss the possibility of Laue diffraction from samples that are *single crystals*.

3. In some Bragg crystals (e.g., Si, Ge), the lattice has atoms so placed as to remove certain higher-order (here, the even orders) reflections by destructive interference. This property can sometimes *remove* unwanted spectral lines. Thus Rose, Adler, and Flanagan [18] used a *curved* Ge crystal in an Applied Research Laboratories nine-channel x-ray spectrograph to eliminate the interference of Ca $K\beta1$(II) with P $K\alpha$, the analytical line, in an important investigation of methods for determining light elements in rocks and minerals. Pulse-height selection might also have mitigated this interference.

BRAGG REFLECTION FROM CURVED CRYSTALS

5.9 Introduction

As we have seen, a flat Bragg reflector needs a parallel x-ray beam. Although such a reflector can intercept analytical lines from the *entire area*

BRAGG REFLECTION FROM CURVED CRYSTALS

of a sample that is both broad and long, the need for collimation greatly reduces intensity. Curved Bragg reflectors can give excellent resolution with a good image, but they require beams that are divergent or convergent: *no collimator is needed.* A curved reflector must view x-rays that originate from a *line* or a *point*, not an *area*. If the sample is both broad and long, a *slit* must be used; this of course reduces intensity, for the analytical lines generated over much of the sample area will not reach the reflector. On the other hand, if the analytical lines are generated at a point, as with a small sample, or in the electron microprobe, there need be little loss of analytical-line intensity. The curved reflectors then appear at their best.

The advantages of curved Bragg reflectors were appreciated early [19], but they could not be realized in x-ray spectrographs because the needed reflectors were not available. The situation has now changed, in part because of the increasing popularity of the electron microprobe. The work on cylindrical lenses (curved and ground crystals) at the Applied Research Laboratories [20] has been of particular importance to analytical chemistry. Curved Bragg reflectors to meet most requirements are now available.

Compton and Allison [21] summarize the early work; that of DuMond and Kirkpatrick [22] is particularly important. They realized that the principle of the Rowland grating [23] could be made applicable to x-rays even though such application demanded that two conditions at first sight mutually exclusive be satisfied; for a cylindrical reflector of radius R and with slit S parallel to the cylinder, the conditions are:

1. *Rowland condition.* Slit S and its real image must lie on the circumference of a circle of radius $R/2$, and the ordinary laws for the reflection of light must be obeyed. This circle is the *focal circle*. To meet this requirement, the reflector *must have the radius R*, which defines the *crystal circle*.

2. *Bragg condition.* For Bragg's law to be obeyed and the Rowland condition to be met, the reflector must everywhere be tangent to the focal circle. For a reflector of large aperture, this means that the reflector *must have the radius $R/2$*. One reflector must have two radii!

With Rowland gratings, which are ruled on the surface, these conditions cannot both be met. DuMond and Kirkpatrick realized that with x-rays, one can have his cake and eat it too because x-ray reflection involves a crystal lattice of exceedingly small spacing. It is therefore possible to bend the crystal to radius R and grind the concave face to radius $R/2$ (or vice versa). They discussed the schemes in both Figures 5.10–1 and 5.11–2, but they had to use fifty small crystals to simulate Figure 5.10–1 because satisfactory large ones were unavailable.

We shall consider only three ways in which curved crystals can be used. The figures will represent right-sections of cylindrical crystals, and the slit through which the x-rays reach the crystals will appear as points.

5.10 The Cauchois Arrangement. Transmission

Mlle. Cauchois [24] used in transmission with a convergent beam a thin crystal bent cylindrically with the Bragg planes about as shown in Figure 5.10–1. Mica, gypsum, and quartz proved suitable because they could be made thin and *bent*.

Consider in Figure 5.10–1 the convergent x-rays of wavelength λ_1 for which the Bragg angle is θ_1. The three reflected rays (one from the center of the crystal, and one from each edge) come to imperfect focus near F_1 as shown. Actually, rays reflected from neighboring planes intersect, and the intersections lie on the "caustic circle." The useful crystal aperture is limited by the imperfection in focusing that can be tolerated.

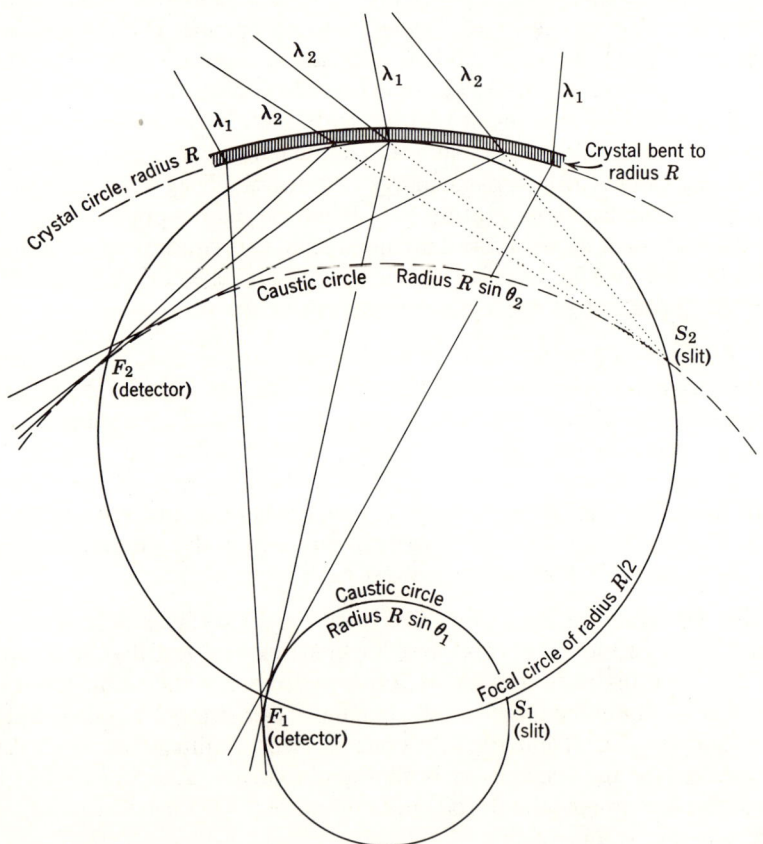

Fig. 5.10–1. The Cauchois bent crystal spectrograph. Reflecting planes normal to surface. Note that the focusing imperfection for λ_2 is about equal to that of λ_1 even though a smaller area of the crystal is used for λ_2.

Repeat the foregoing construction with x-rays of wavelength λ_2 for which the Bragg angle θ_2 is twice θ_1. The imperfect focus is now near F_2, and there will be a caustic circle with the same center as the other but with a radius $R \sin \theta_2$.

An important point is that F_1 and F_2 for the beams reflected from the center of the crystal lie on the focal circle. Foci for other reflected wavelengths will lie on the same circle, as can be shown graphically. Another important point is that the virtual sources of λ_1, λ_2 and of other reflected wavelengths (the S's in the figure) also lie on the same circle. It is consequently possible to reverse the scheme by placing a suitable source of x-rays (from a slit, pinhole, or properly shaped sample) on the focal circle and using an extended detector beyond the crystal.

5.11 The Johann and the Johansson Arrangements. Reflection

The Johann arrangement [25], shown in Figure 5.11–1, is the x-ray analogue of the Rowland diffraction grating [23]. The imperfect focusing in the x-ray case is analogous to spherical aberration, which is at a minimum when $u = v$ (see figure). In both cases, good focusing is possible only over the region where reflector and focal circle are tangent, or nearly so. The

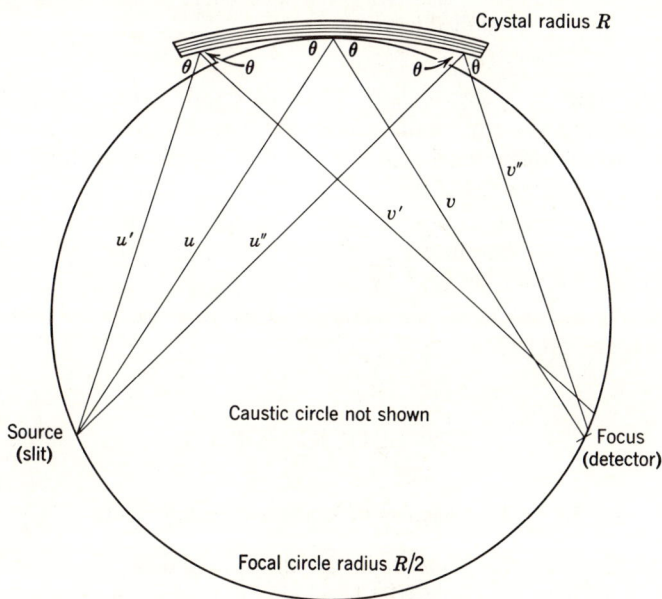

Fig. 5.11–1. The Johann arrangement. Slit-crystal distances, u, u', u''. Crystal-detector distances v, v', v''.

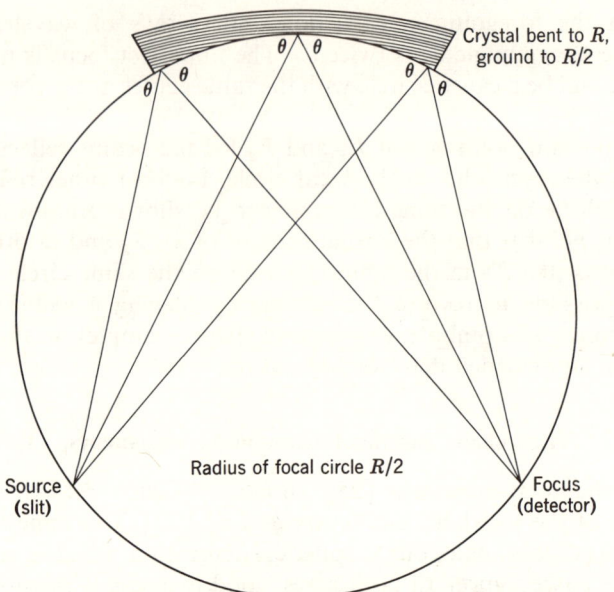

Fig. 5.11–2. The Johansson arrangement. The x-rays make the same angles with the reflecting crystal planes an in Figure 5.11–1, but the crystal now lies on the focal circle. Note the absence of a caustic circle and the improved focusing.

short wavelength of x-rays means that this condition is met only if the reflector aperture is very small. Consequently, the Johann arrangement should ordinarily not be used except with reflectors for which the Johansson arrangement is not possible.

Johansson [26] put into practice the ideas of Du Mond and Kirkpatrick by grinding and then bending his crystals. A cylindrical crystal of this kind can, under the best conditions, focus almost perfectly an x-ray beam that diverges from an extended slit only in directions perpendicular to the slit. See Figure 5.11–2.

CONCLUDING TOPICS

5.12 Comparison of Various Arrangements

The flat-reflector arrangement has the simplest geometry—a "one-circle" geometry in which this circle (the focal circle) is tangent to the Bragg reflector. The kind of reflector is the chief design variable.

By contrast, Figure 5.10–1 shows a "three-circle" geometry, the radius of

CONCLUDING TOPICS

the focal circle and that of the caustic circle depending in different ways on that of the crystal circle; this is true also of Figure 5.11-1. The radius of the caustic circle varies with θ; hence, for a given crystal, with λ. Additional design variables are three: R, u, and v. Adjustment of these three variables to provide exact focusing at all wavelengths is impracticable. The R can be changed either by bending a single crystal or by using several crystals. Ingenious compromises must be made. The Johansson arrangement has no caustic circle, which makes it a two-circle arrangement and simplifies the focusing problem, but leaves it still complex. In the Applied Research Laboratories design, which uses this arrangement, the detector path is one loop of a curve known as a "four-leaf" rose.

In spite of these complexities, curved Bragg reflectors have gained in importance relative to flat Bragg reflectors in recent years. Table 5.12-1 is a thumbnail summary.

Table 5.12-1. Short Comparison of Spectrometers for X-Ray Spectrographs

Arrangement	Optics	Advantages	Disadvantages
Flat reflector	Reflector forms tangent to focusing circle.	Simple geometry; line broadening not too dependent upon geometry.	Collimation necessary. Resolution limited to divergence allowed by collimator.
Johann	Crystal circle has radius R. Focal circle has radius $R/2$. Caustic circle has radius $R \sin \theta$.	X-rays diffracted to line focus, therefore only a simple slit is required to intercept scattered x-rays. Fairly large acceptance angle.	Serious defocusing of analytical line.
Johansson	Reflector has external radius R and internal radius $R/2$. Focal circle has radius $R/2$. No caustic circle.	Similar to Johann but with defocusing eliminated.	Some crystals excluded for mechanical reasons.
Cauchois	Radii as for Johann (above). Used in transmission. Reflecting planes normal to surface.	Large solid angle of acceptance for *either* detector or sample.	Serious attenuation of x-ray beam as it passes through the crystal.

5.13 Comparison and Nomenclature of X-Ray Instruments

The current nomenclature of x-ray instruments and methods seems taken from *Alice in Wonderland*. To forestall confusion, and to show the relation-

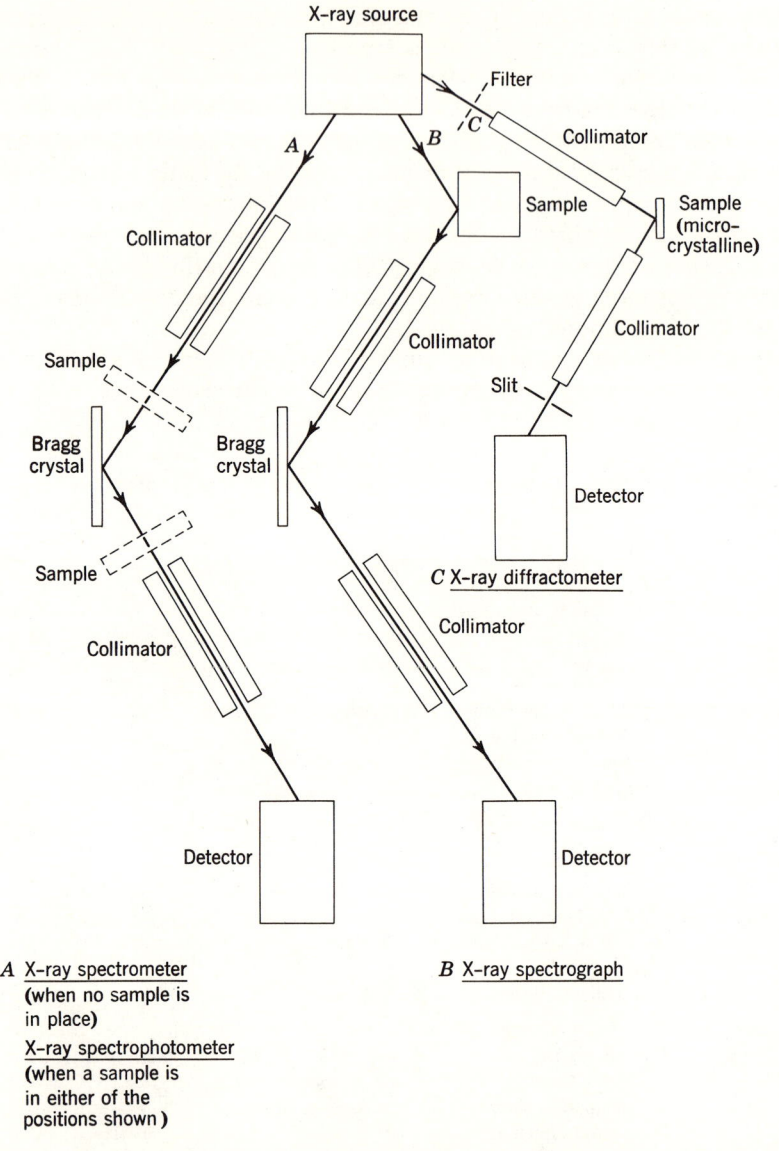

Fig. 5.13–1. Nomenclature and relationships of various x-ray instruments. Only flat Bragg reflectors are shown, and these are assumed to be crystals. Note how the spectrometer is related to the other instruments, with which it ought not to be confused.

CONCLUDING TOPICS

ships among the various instruments, these are shown schematically in Figure 5.13–1.

The simplest instrument, an *x-ray photometer*, which need comprise only *x-ray source, sample*, and *detector* is not shown. By adding to it as in the figure, one obtains the other instruments. Of course, what happens to the sample depends upon whether an absorption, emission, or diffraction measurement is being made. As the figure is self-explanatory and gives the names of the instruments, only three comments will be made: (1) The *spectrometer* (no sample) is a *component* of the *spectrograph* and of the *spectrophotometer*. (2) The Bragg crystal is often called a *monochromator* when it selects a wavelength from an x-ray source, as it does when it is placed between source and sample in the spectrophotometer. (3) Monochromatizing at the source minimizes heating and x-ray damage of the sample.

5.14 Losses in an X-Ray Spectrograph

The losses in an x-ray spectrograph limit its sensitivity—a matter particularly important for trace determinations. Estimates of these losses contribute to an understanding of the instrument. For these reasons, oversimplified estimates of *all* significant losses will be made for comparison with the experimental result for a sample (a monolayer of cobalt) on which other illustrative calculations have been made. Better than order-of-magnitude agreement between calculated and experimental results is meaningless.

The calculations are made in the steps described below and summarized in Table 5.14–1. Figure 5.14–1 is a guide to the steps. They will be for unit area of sample under conditions where neither the total area of sample nor the detector aperture limits the calculated result.

TUBE POWER AND CURRENT (STEP I). Calculation of power for root-

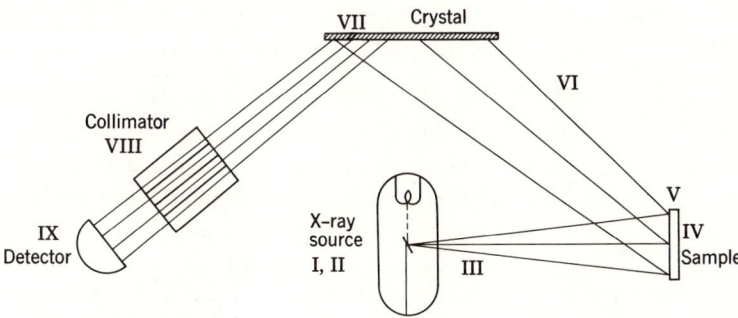

Fig. 5.14–1. A schematic view of the losses in an x-ray spectrograph. The Roman numerals are keyed to the text.

mean-square voltage, which is $\sqrt{2}/2$ times peak voltage (50 kV). The number of electrons per second in the tube current (50 mA) is calculated for an *order-of-magnitude* comparison with the residual x-ray quanta per second shown in Table 5.14–1.

Table 5.14–1. Loss Calculations for Spectrograph used in Cobalt Determination

Step	I	II	III	IV	V
Reduction factor	—	—[b]	0.93	0.009	$4.5(10^{-6})$
Power	1765^a	$3.3(10^{15})$	$3.1(10^{15})$	$2.8(10^{13})$	$1.3(10^{8})$

Step	VI	VII	VIII	IX
Reduction factor	0.9	0.1	$2.9(10^{-6})$	0.8
Power	$1.2(10^{8})$	$1.2(10^{7})$	35	28

[a] Electrical power in watts. All other power values expressed as quanta per second. Tube current of 50 mA is a flux of $3.1(10^{17})$ electrons/sec.
[b] See text. Fraction of electrical power that appears as 1-Å x-rays is 0.0037.

X-RAY POWER (STEP II). Calculated from (4.3–1) for 0.7071×50 kV and from (1.6 – 1a) for a beam of effective wavelength 1 Å.

ATTENUATION (STEP III). Calculation of fraction transmitted as for Table 2.2–1.

GEOMETRIC FACTOR FOR UNIT AREA OF SAMPLE (STEP IV). Calculated for target as point source 3 cm distant from sample considered as lying on surface of sphere.

FRACTION OF 1-Å X-RAYS ABSORBED BY SAMPLE AND REEMITTED AS Co Kα (STEP V). See Section 4.4. Value of 0.2 taken for $w_K(1 - 1/r)$.

ATTENUATION BY AIR PATH, SAMPLE TO DETECTOR (STEP VI). Take fraction transmitted to be (approximately) 0.9. See Table 2.2–1.

BRAGG CRYSTAL REFLECTION EFFICIENCY (STEP VII). Assume 0.1.

GEOMETRIC AND LOSS FACTORS FOR COLLIMATOR (STEP VIII). Assume one collimator slit (rectangle, 0.009×0.750 in.) lies on surface of sphere with radius 13 in., the sample-collimator distance, this being the distance traveled by a light beam from the sample to the collimator exit, and have a mirror replacing the Bragg crystal. This geometric factor is an *intensity factor*; hence the calculation is for one slit; its value is $3.2(10^{-6})$. This value must be multiplied by the *collimator efficiency*, assumed as 0.9 (loss factor, 0.1).

DETECTOR EFFICIENCY (STEP IX). Assume 0.8. The attenuation by the detector window is included.

CONCLUDING TOPICS

The calculated value of 28 quanta per sec is to be compared with 15 quanta per sec, which was measured under conditions approaching those assumed above. The significance of the agreement was assessed at the beginning of this section.

5.15 X-Ray Diffraction. Parafocusing

X-ray diffraction in chemical analysis is concerned so largely with microcrystalline (powdered) samples that we shall not consider others at this point. The relationship of this to other x-ray methods has been indicated in Figure 5.13–1. What follows should be read in conjunction with Chapter 6 (particularly Sections 6.5 and 6.19).

Collimation and focusing are different matters in powder x-ray diffraction than in x-ray emission spectrography. The differences arise because x-ray diffraction uses monochromatic x-rays (λ not a variable), and because the randomly oriented microcrystals in the sample reflect in all directions: there is no one θ-plane as there is when a single crystal selects a wavelength according to Bragg's law in x-ray emission spectrography.

Focusing with microcrystalline samples differs from that with single crystals and has been given the distinguishing name *parafocusing* by Brentano in a classic paper [27]. Unfortunately, *para* as prefix has many meanings, more than one of which (e.g., faulty, irregular, disordered, alongside of) could apply here. But, Brentano says "it may be helpful ... to set out the distinction between real focusing and the ray-collective properties of powder arrangements which more appropriately can be described as parafocusing." Accordingly, as concerns Bragg reflection, we place parafocusing *alongside of* true focusing and identify the former with microcrystalline samples, and the latter with single crystals or large crystal faces.

The quality of parafocusing will vary from one arrangement to another, the highest quality requiring that the microcrystalline sample be distributed on a segment of toroidal surface. Such a surface, having double curvature, is impractical. Several satisfactory compromises are available in which the sample is distributed over a short region on the inner cylindrical surface of a camera; this surface in right section through the sample is the *parafocusing circle*. The only parafocusing camera important in Chapter 6 is of the type due to Guinier, in which a bent crystal is used as monochromator for a convergent x-ray beam that passes through the sample. See Figure 6.19–1. The Debye-Scherrer-type camera (Figure 6.5–1) is not parafocusing. For many x-ray diffraction determinations in analytical chemistry, the Debye-Scherrer camera suffices. However, Chapter 6 shows that parafocusing of high quality, which presupposes highly monochromatic x-rays, becomes more desirable as the number of possible components in a sample increases and as the library of x-ray standards becomes more extensive.

Because a diffractometer must scan over a 2θ-range, satisfactory collimation and parafocusing are more difficult to achieve than in a camera. See

228   THE SELECTION OF X-RAY WAVELENGTHS

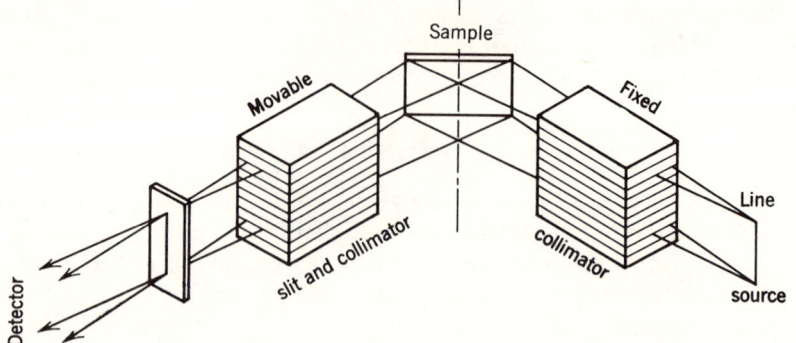

Fig. 5.15–1. Collimation scheme for diffractometer. Filter or other means of selecting wavelength from source (fine-focus x-ray tube) not shown. Rotation of sample and detector described in next figure. Goniometer plane horizontal. Measured θ-plane horizontal.

Figure 5.3–4. In the diffractometer, the detector and Bragg reflector still move at the angular velocities $2\dot{\theta}$ and $\dot{\theta}$, but the Bragg reflector is now the microcrystalline sample, which reflects in all directions. The collimation problem is consequently changed. In Figure 5.3–4, the need was to sort out the proper θ (that corresponding to the desired wavelength) in a single θ-plane; here the need is to sort out a single θ-plane from a multitude in

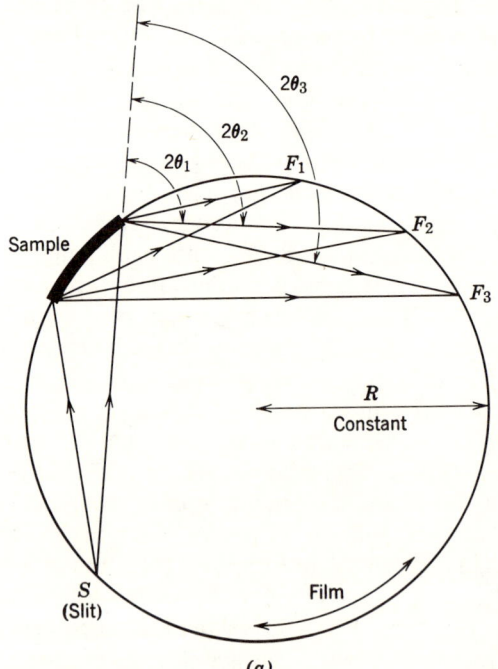

Fig. 5.15–2.   (a)

CONCLUDING TOPICS

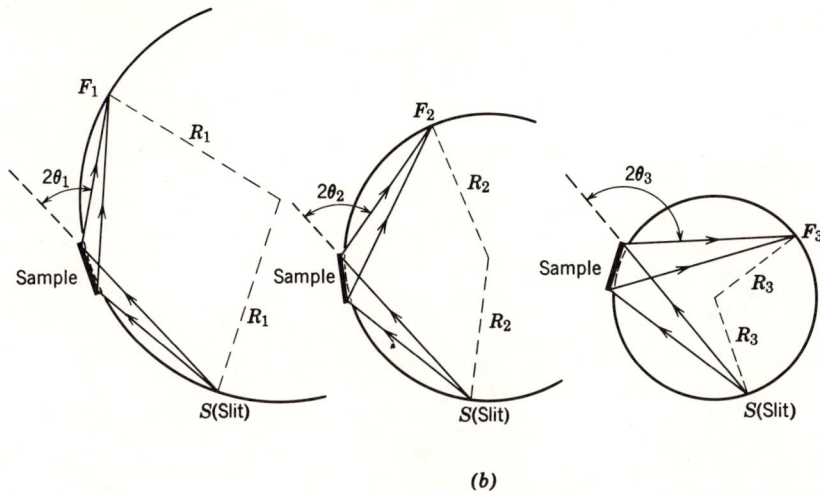

(b)

Fig. 5.15–2. Two-dimensional sketches to explain parafocusing powder diffractometer (*b*) on basis of Seeman-Bohlin parafocusing camera (*a*). In the camera, the parafocusing circle has a fixed radius R. Parafocused lines corresponding to interplanar distances d_1, d_2, and d_3 appear at F_1, F_2, and F_3 on the circle at positions corresponding to $2\theta_1$, $2\theta_2$, and $2\theta_3$. In the diffractometer, the sample rotates at angular velocity $\dot{\theta}$ around the center of a parafocusing circle of *variable radius*. A *flat* sample is used so that it can remain tangent to this circle. As θ increases, this radius decreases from R_1 to R_2 to R_3, and the detector moves at angular velocity $2\dot{\theta}$ from F_1 to F_2 to F_3. The sample-detector distance remains constant throughout. Relationship of d's to 2θ's to F's as in Figure 5.15–2a. Parafocusing achieved in the diffractometer is a satisfactory compromise. After Klug and Alexander, Ref. 28, p. 237.

space, all of the same θ. The θ-plane thus sorted out must obviously parallel that of the goniometer. The scheme in Figure 5.15–1 provides satisfactory collimation for a diffractometer. *Horizontal* divergence is minimized by using a line source of x-rays generated by a fine-focus tube. *Vertical* divergence is minimized by the use of Soller-slit collimators in which the slits naturally parallel the measured θ-plane. (Contrast with Figure 5.3–4!) One or more vertical slits intercept scattered x-rays to reduce background. Figure 5.15–2 shows schematically the satisfactory compromise (flat sample tangent to parafocusing circle) made to achieve parafocusing.

For further information, see Klug and Alexander [28].

5.16 The Birks Edge-Crystal Spectrograph [29]

The description of the simple, small, and ingenious Birks edge-crystal spectrograph is a good link between this chapter and the next, for this spectrograph (Figure 5.16–1) has much in common with a diffraction camera,

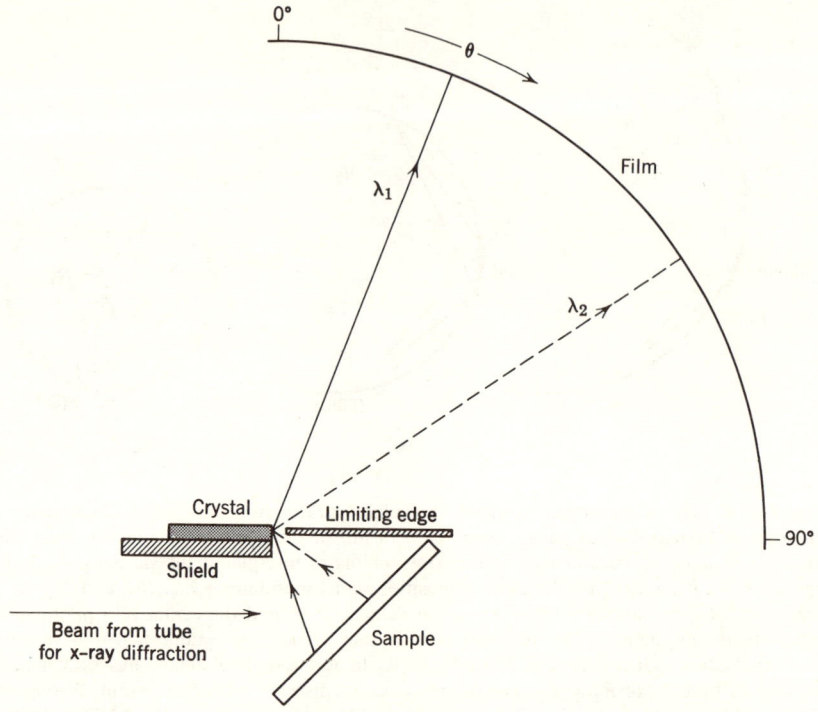

Fig. 5.16–1. The Birks edge-crystal spectrograph. After Birks, Ref. 29c, p. 32, Fig. 4–6.

and it is useful in providing quick, semiquantitative determinations of elements as an aid to diffraction determinations on powdered samples.

As Figure 5.16–1 indicates, Bragg reflection of the analytical lines occurs from planes parallel to the edge of a *thin* crystal (e.g., 5-mil LiF), and the line width on the film is the projected width of the edge. With excitation by the beam from an x-ray tube for diffraction use, the exposure time needed is 10–30 minutes. Each analytical line recorded on the film originated in a different part of the sample. The more nearly homogeneous the sample, the more nearly quantitative the results.

REFERENCES

1. H. P. Klug and L. E. Alexander, *X-Ray Diffraction Procedures,* Wiley, New York, 1954, p. 100.
2. P. A. Ross, *J. Opt. Soc. Am. Rev. Sci. Instr.,* **16**, 433 (1928).
3. (a) P. Kirkpatrick, *Rev. Sci. Instr.,* **10**, 186 (1939); (b) *ibid.,* **15**, 223 (1944); (c) R. A. Young,

REFERENCES

Z. *Kristallogr. Miner.*, **118**, 233 (1963); (d) U. W. Arndt and B. T. M. Willis, *Single Crystal Diffractometry*, Cambridge University Press, New York, 1966, pp. 182–187.
4. M. L. Salmon, *Advances in X-Ray Analysis*, Vol. 6, Plenum Press, New York, 1963, p. 301.
5. E. L. Gunn, *Anal. Chem.*, **36**, 2086 (1964).
6. W. Soller, *Phys. Rev.*, **24**, 158 (1924).
7. I. Adler, *X-Ray Emission Spectrography in Geology*, Elsevier Publishing Co., Amsterdam, 1966, p. 92.
8. A. H. Compton and S. K. Allison, *X-Rays in Theory and Experiment*, Van Nostrand, New York, 1935, p. 268.
9. W. Heitler, *The Quantum Theory of Radiation*, Oxford University Press, New York.
10. M. Renninger, *Z. Krist.*, **89**, 344 (1934).
11. J. Vierling, J. V. Gilfrich, and L. S. Birks, *Appl. Spectr.*, **23**, 342 (1969).
12. R. W. Gould, S. R. Bates, and C. J. Sparks, *Appl. Spectr.*, **22**, 549 (1968).
13. L. S. Birks, *X-Ray Spectrochemical Analysis*, 2nd ed., Interscience, New York, 1969, p. 38, Table 4–2.
14. W. J. Campbell, M. Leon, and J. W. Thatcher, *Advances in X-Ray Analysis*, Vol. 1, Plenum Press, New York, 1957, p. 193.
15. F. S. Lee and W. J. Campbell, *Advances in X-Ray Analysis*, Vol. 8, Plenum Press, New York, 1965, p. 431.
16. T. A. Davies, *J. Sci. Instr.*, **35**, 407 (1958).
17. R. Jenkins and J. L. de Vries, *Practical X-Ray Spectrometry*, Springer-Verlag, New York, 1967, pp. 39–42.
18. H. J. Rose, I. Adler, and F. J. Flanagan, *Appl. Spectroscopy*, **17**, 81 (1963).
19. M. de Broglie and F. A. Lindemann, *Compt. rend.*, **158**, 944 (1914).
20. J. W. Kemp and G. Andermann, Fifth Annual Conference on Industrial Applications of X-Ray Analysis, Denver Research Institute, University of Denver, Denver, Colo., August 1956.
21. A. H. Compton and S. K. Allison, *X-Rays in Theory and Experiment*, Van Nostrand, New York, 1935, pp. 750–756.
22. J. W. M. Du Mond and H. A. Kirkpatrick, *Rev. Sci. Instr.*, **1**, 88 (1930).
23. R. Glazebrook, *A Dictionary of Applied Physics*, Vol. IV, Macmillan and Co., London, 1923, p. 48.
24. Y. Cauchois, *J. Phys. Rad.* [7], **3**, 320 (1932).
25. H. H. Johann, *Z. Phys.*, **69**, 185 (1931).
26. T. Johansson, *Z. Phys.*, **82**, 507 (1933).
27. J. C. M. Brentano, *J. Appl. Phys.*, **17**, 420 (1946).
28. H. P. Klug and L. E. Alexander, *X-Ray Diffraction Procedures*, Wiley, New York, 1954, pp. 211–218 and 236–239.
29. (a) L. S. Birks, U. S. Pat. 2,842,670 (1958); (b) L. S. Birks and E. J. Brooks, *Anal. Chem.*, **27**, 1147 (1958); (c) Ref. 13, pp. 31–32.

Chapter 6

X-Ray Diffraction in Chemical Analysis

> The method consists in reducing to powder form the substance to be examined, placing it in a small glass tube, sending a beam of monochromatic X-rays through it, and photographing the diffraction pattern produced. The only apparatus required is a source of voltage, an X-ray tube, and a photographic plate or film. The amount of material necessary for a determination is one cubic millimeter. The method is applicable to all chemical elements and compounds which are crystalline in structure.   A. W. Hull [1]

> The identification of a completely unknown material is therefore, in general, not possible. Except in special cases it is necessary to have some additional information such as an approximate chemical composition. Difficulties in identification are encountered when more than one crystalline compound is present.   B. E. Warren [2]

> The powder diffraction pattern of a sample does not in any sense indicate either the chemical elements or the chemical compounds in the sample; *it is characteristic solely of the crystalline phases present.*   B. E. Warren [2]

6.1 Introduction

A review of Sections 1.16–1.19 is advisable at this point. The quotations given above from the early literature summarize the advantages [1] and the limitations of x-ray diffraction in analytical chemistry. Photographic film then served as detector. Now, electronic detectors are used in the

X-RAY DIFFRACTION IN CHEMICAL ANALYSIS

diffractometer, which closely resembles the *x-ray emission spectrograph*. Because the latter receives much attention later in the book, and because the two instruments are often associated, we give more space here to photographic than to diffractometric methods.

The reader will remember from Section 1.16 that Bragg "reflection" is diffraction that results from the coherent, unmodified scattering of x-rays by the electrons in scattering centers; that intensity measurements in Bragg reflection can establish electron distribution; and that the most spectacular achievements of x-ray diffraction—such as the establishment of the structure of insulin with 777 atoms in its molecule by Dr. Dorothy Hodgkin [3], already a Nobel Laureate for earlier structure work—is outside the realm of analytical chemistry, where the major uses for x-ray diffraction are for qualitative and quantitative determinations based on measurements of interplanar spacings and x-ray intensity. In addition, we concern ourselves with certain information obtainable by measuring the sharpness of the lines in the diffraction patterns of polymers and of crystalline powders.

X-ray diffraction is closely related to other x-ray methods as the following indicates: (1) because diffraction can at best establish crystal structure, it must in most cases be supplemented by a method that identifies at least some of the elements in the sample. For this, x-ray emission spectrography is almost ideal for the heavier elements and gaining in value for the light elements as analytical lines of longer wavelength become easier to use. (2) In some cases analytical lines are generated when diffraction measurements are made, but they now disappear into the background: simultaneous recording of diffracted and emitted lines for analytical purposes may some day be possible. (3) Monochromatic $K\alpha$ lines for diffraction work are obtained by filtering out the shorter-wavelength $K\beta$; this relates diffractometry and absorptiometry. (4) In diffractometry, there is absorption by the sample of incident and diffracted beams; or, there is an *absorption effect* similar to those in x-ray emission spectrography. In both methods, *internal standards* are used to cope with these effects.

Because x-ray diffraction cannot command more than a single chapter in this book, interested readers are urged to dig deeper by reading in Klug and Alexander's *X-Ray Diffraction Procedures* [4] or in Azároff and Buerger's *The Powder Method in X-Ray Crystallography* [5].

6.2 Crystallography and Bragg's Law

Nearly all solids are crystalline and, if pure, show the sharp fusion or sublimation temperatures that one would expect for a highly ordered structure. They are characterized by simple *external* and *internal* geometrical relationships. The former, not always obvious, served as basis for the discovery of the latter: that is to say, Bragg (Section 1.19) based the first calculation of an x-ray wavelength upon a value of the *interplanar spacing* or *repeat distance* that was deduced from crystallographic knowledge of the *external* relationships existing among the faces of a sodium chloride crystal, and from a knowledge of Avogadro's number. Thereafter, x-ray diffraction became an indispensable way of revealing the *internal* relationships, which crystallographic methods could not reach. The external relationships are complicated by the fact that crystals come in all sizes, are seldom simple, and seldom have all their faces fully developed. X-ray diffraction usually shrugs off these difficulties.

The first law of crystallography says that the angle between corresponding crystal faces, no matter how well or how poorly developed these faces, is invariable. This law was established by using a reflecting goniometer, an instrument that Bragg modified (Section 1.17) to make his first x-ray spectrometer.

The second law of crystallography—the law of rational indices—says that for every crystal there is a set of axes for which the indices of all crystal faces are small integers or zero. What does all this mean, and how is the law established?

1. It means first that to describe crystal symmetry in three dimensions one must have three coordinate reference axes with known angles between each pair. These axes (X, Y, Z) are to be the most convenient possible with angles between them as follows: α between Y and Z; β between X and Z; and γ between X and Y.

2. It means further that a crystal face must be selected to serve as unit plane. This plane must intersect all three axes and should again be the most convenient among the possible choices.

3. To establish the law, the intercepts of the unit plane on the axes are measured with a goniometer. Their values will be relative only because they depend upon the size of the crystal. Denote them for the three axes as follows: $a(X)$; $b(Y)$; $c(Z)$. Now measure the intercepts of the other faces. They will be ma, nb, and pc, where m, n, and p are either small integers or ∞, the infinite value belonging to a face that is parallel to one axis. As infinity is awkward to deal with, the Miller indices (h, k, l) were invented and defined as proportional to the intercept reciprocals; that is, h to $1/m$; k to $1/n$; and l to $1/p$.

Among the thousands upon thousands of crystals investigated, none has ever been found with h, k, and l not either small integers, or zero. The law of rational indices and the symmetry of crystals are both firmly established.

Figure 6.2–1 and Table 6.2–1, both from Ref. 4, p. 16, show lucidly the relation between face intercepts and Miller indices.

As $X, Y,$ and Z extend also in the negative directions, there must be Miller

X-RAY DIFFRACTION IN CHEMICAL ANALYSIS

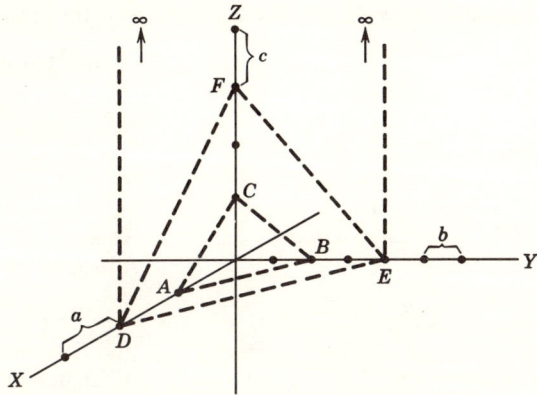

Fig. 6.2–1. Three faces (*ABC*, *DEF*, and *DE*∞) that intersect a set of orthorhombic axes (*X, Y, Z*). The corresponding Miller intercepts appear in Table 6.2–1. After Klug and Alexander, Ref. 4, p. 16.

Table 6.2–1. Derivation of Miller Indices for the Face Intercepts in Figure 6.2–1

Face	Intercepts	Reciprocals of Intercept Multiples	Cleared of Fractions	Miller Indices; *hkl*
ABC	$1a:2b:1c$	$\frac{1}{1}\ \frac{1}{2}\ \frac{1}{1}$	2 1 2	(212)
DEF	$2a:4b:3c$	$\frac{1}{2}\ \frac{1}{4}\ \frac{1}{3}$	6 3 4	(634)
DE	$2a:4b:\infty c$	$\frac{1}{2}\ \frac{1}{4}\ \frac{1}{\infty}$	2 1 0	(210)

indices with negative values. They are distinguished from the positive by use of a bar: thus $\bar{1}$ (negative) in contrast with 1 (positive). The result is an increase in the number of possible crystal faces.*

Only the six (principal) symmetry systems in Table 6.2–2 are needed to describe all known crystals.

Each of the systems in Table 6.2–2 has *crystallographic interplanar spacings*, denoted by d_{hkl}. We confine ourselves to the simplest system, the cubic, for which

$$d_{hkl} = a_0/\sqrt{h^2 + k^2 + l^2} \qquad (6.2\text{–}1)$$

where a_0 (sometimes designated a) must now have the value for the *unit cell*, the three-dimensional repeating unit basic for the determination of crystal structure. We repeat that the unit cell is not essential to analytical chemistry, which can usually make do with interplanar spacings instead.

---

*There are eight faces, namely (*hkl*), ($\overline{hk}l$), ($\bar{h}kl$); ($h\overline{kl}$), ($h\bar{k}l$); ($\bar{h}k\bar{l}$); ($hk\bar{l}$), ($\overline{hk}l$); of the general form (*hkl*): parallel faces have complementary signs and are enclosed by semicolons.

Table 6.2–2. Descriptive Data for the Six Crystal Symmetry Systems

System	Axial Ratios	Angles Between Crystal Axes
Triclinic (anorthic)	$a:b:c$	α, β, γ (generally all = 90°)
Monoclinic (oblique, monosymmetric)	$a:b:c$	β ($\alpha = \gamma = 90°$)
Orthorhombic (rhombic, prismatic)	$a:b:c$	All angles 90°
Tetragonal	$a:c$ ($b=a$)	All angles 90°
Hexagonal		
(A) Hexagonal division		
(1) Hexagonal axes	$a:c$ ($b=a$)	$\gamma = 120°$ ($\alpha = \beta = 90°$)
(2) Orthohexagonal axes	$a:b:c$ ($b=a\sqrt{3}$)	All angles 90°
(B) Rhombohedral division (trigonal)	$a = b = c$	$\alpha = \beta = \gamma \neq 90°$
Cubic (regular, isometric)	$a = b = c$	$\alpha = \beta = \gamma = 90°$

The enormous body of work on crystal structure by x-ray methods shows that the crystallographic interplanar spacings and the corresponding Bragg values from (1.16–2) are in fact identical; that is

$$d_{hkl} = d \text{ (Bragg)} \qquad (6.2-2)$$

The work also has shown that the prominently developed crystal faces are thickly populated with scattering centers; have low values of h, k, l; and that they give predominant lines in diffraction patterns. See Figure 6.2–2. The same considerations apply to *cleavage faces*, which are formed when natural fracture occurs easily because the interatomic or intermolecular forces are at a minimum in the direction perpendicular to the plane of the cleavage face. These points are illustrated in Figure 6.2–2.

The *number* of possible interplanar spacings can be calculated from (6.2–1) by substituting all possible values of h, k, l in (6.2–1). The key word is *possible*. In the cubic system, the experimentally established structures are of three types: the simple cube (P), the face-centered cube (F), the body-centered cube (I). The diamond structure (D) is a special F type. Table 6.2–3 shows the possible reflections for each, *all* reflections being possible for the simple cubic lattice (P's omitted).

The Bragg equation

$$n\lambda = 2d \sin \theta \qquad (1.16-2)$$

has four variables with all of which the reader is familiar. (Note, however, that each crystalline species has a *set* of d values, which means that many interplanar spacings may be found for a sample that is a mixture of such

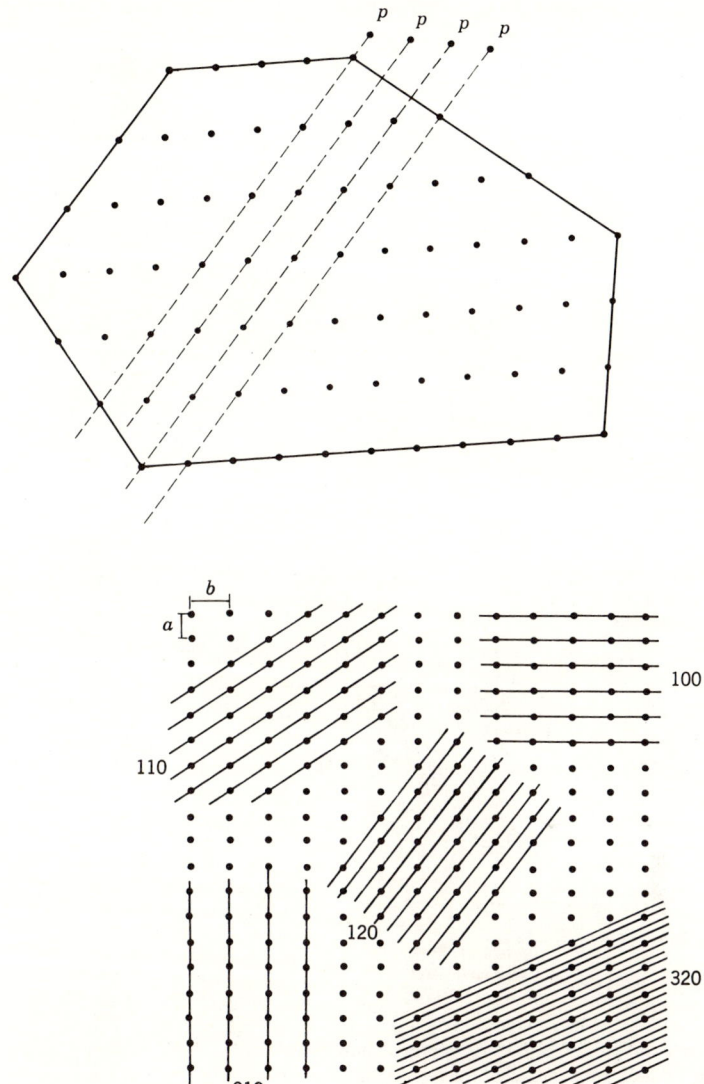

Fig. 6.2–2. Parallelism of external faces, either natural or cleavage, to important internal planes (above). Various sets of planes, with their Miller indices, in a crystal. After Lonsdale, Ref. 6, p. 52 and p. 55.

Table 6.2–3. Possible Bragg Reflections from Crystals with Cubic Symmetry[a]

$h^2 + k^2 + l^2$	Lattice	hkl
1		100
2	I[c]	110
3	F[d]D	111
4	IF	200
5		210
6	I	211
7		[b]
8	IFD	220
9		300, 211
10	I	310
11	FD	311
12	IF	222
13		320
14	I	321
15		[b]
16	IFD	400
17		410, 322
18	I	410, 320
19	FD	331
20	IF	420
21		421
22	I	332
23		[b]
24	IFD	422
25		500, 430
26	I	510, 431
27	FD	511, 333
28		[b]
29		520, 432

[a] Only for values of $a^2/d_{hkl}^2 = h^2 + k^2 + l^2$ of 29 and below. Higher values in Ref. 4, p. 679, the source of the values given above.

[b] Mathematically impossible. No number of the form $4^x(8y + 7)$ can be expressed as the sum of the squares of three integers or zero so long as x and y are themselves integers or zero. Such numbers (7, 15, 23, 28, etc.) therefore cannot meet the requirement that the Miller indices be small integers or zero.

[c] For body-centered cube (I), lattice extinctions (destructive interferences) occur that impose the requirement h, k, and l have an *even sum*.

[d] For face-centered cube (F), considerations mentioned above impose the requirement that h, k, and l have integer values either *all odd* or *all even* (zero excepted).

species.) To these variables, the intensity I of the reflected lines must be added as a fifth variable to be considered in diffraction work. We shall return to the intensity in Section 6.7, but several points of particular interest will be noted here. The *multiplicity* of crystal faces helps determine intensity. Thus the symbol $\{101\}$, which includes all faces related in this way, represents *twelve* faces as follows: (011) (0$\bar{1}$1) (0$\bar{1}\bar{1}$) (01$\bar{1}$) (101) ($\bar{1}$01) (10$\bar{1}$) ($\bar{1}$0$\bar{1}$) (110) ($\bar{1}$10) (1$\bar{1}$0) ($\bar{1}\bar{1}$0). The intensity depends of course on the numbers and types of scattering centers in the different planes—more about this when the analysis of mixtures is discussed. Scattering centers half-way between planes interfere destructively; silicon and germanium in this way reduce interference resulting from Bragg reflections of *even* order when these are used as diffracting crystals in x-ray emission spectrography. Finally, the greater the thermal vibrations of the scattering centers, the lower the intensity, a conclusion first reached by von Laue when he observed that diamond, the atoms of which are hard to make vibrate, gave back-reflected x-rays of relatively high intensity.

In the main, the analytical chemist needs to know that a set of d's can *under favorable circumstances* tell him what crystalline phases are present in a sample, and that amounts of these are related to the relevant line intensities.

Almost all the analytical samples dealt with by x-ray diffraction methods are either powders or reduced to powders. Diffraction determinations to establish the *composition* (as well as structure) of single crystals could be done by modern methods, but such determinations are rarely needed and are not discussed.

A final note. On the basis of measured ratios of interplanar spacings, x-ray diffraction methods on powders can now establish axial ratios (a/b, etc.) for crystals with greater reliability than is possible with the two-circle goniometer or by x-ray single-crystal diffraction as usually practiced [7].

X-RAY DETERMINATIONS ON POWDERED SAMPLES

6.3 Introductory Remarks

The computer revolution now under way makes it difficult to write about x-ray methods—most particularly about x-ray diffraction, as this is the method for which the amount of stored information that must be scanned is greatest, and the changes to be expected from full use of the computer consequently the most profound. In support of this statement, we point to the *Powder Diffraction File* of the Joint Committee on Powder Diffraction Standards [members: the American Society for Testing Materials, the American Crystallographic Association, the Institute of Physics (British), and the National Association of Corrosion Engineers], which, begun in 1938, contained 13,000 standard patterns (10^7 characters) in 1967 and was growing at the rate of 2000 standard patterns per year [8]. Also, the reader may wish to examine the three large volumes of the *International Tables of X-Ray Crystallography* [9].

Of the various ways in which diffraction determinations can be done on

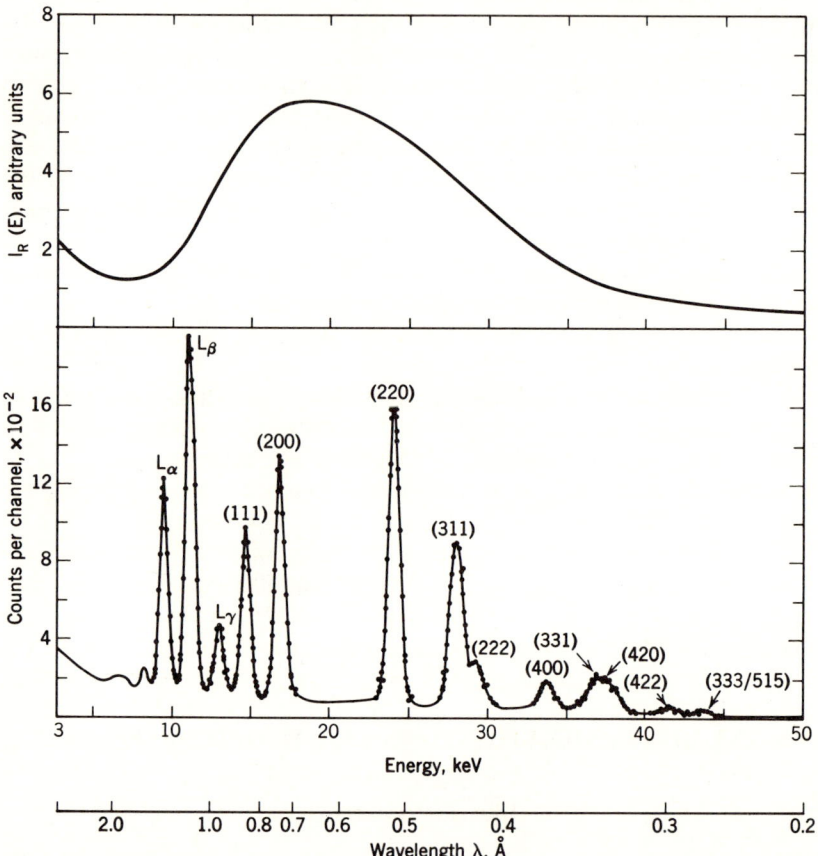

Fig. 6.3–1. Combined x-ray diffraction and emission spectrography on a sample of platinum sheet by unaided energy resolution. (See that topic in Chapter 2.) The experiment was done in a Siemens diffractometer at the setting for $\sin \theta = 0.1861$ with an iron anode (8 mA at 45 kV, the short-wavelength limit). The Si(Li) detector was connected to a 4096-channel pulse-height analyzer. The counting interval was 1.5 hours. The upper curve shows the response of the system to the undiffracted polychromatic beam and may be compared with Figure 4.1–1. The lower pattern shows the L lines of platinum and the various diffraction peaks identified by their Miller indices. The peaks for the L lines have FWHM values of about 0.7 keV. After Giessen and Gordon, Ref. 10, p. 974.

powders, only the Hull-Debye-Scherrer method (usually called by the last two names) and the diffractometric method are described. The important difference between the two stems from the nature of the detectors: photographic film as opposed to Geiger or proportional detector.

The use, by Giessen and Gordon [10], of an Si(Li) detector to record *simultaneously* the L lines of platinum and the Bragg peaks from that metal is important even though the precision with which interplanar spacings were determined lags far behind that presently realized by use of the methods now standard. The investigators aptly name their technique *powder-diffraction spectrography*. Its combination with x-ray emission spectrography would be most welcome, but the chances that it will displace the standard methods soon are so small that extended discussion seems unwarranted. The great speed of the method (a time of 0.3 sec for a recognizable diffraction pattern is expected) and the advantage (which it shares with the photographic plate) that the method gives all peaks at once may lead to its early application on samples at extreme pressures and temperatures, with which standard methods cannot cope. Figure 6.3–1 records the results on platinum.

Hull [1] gave what seemed then a sound basis for the powder diffraction method: to wit, "that the same substance always gives the same diffraction pattern; and that in a mixture of substances, each produces its pattern independently of the others, so that the photograph obtained with a mixture is the superimposed sum of photographs that would be obtained by exposing each of the components separately for the same length of time." Though this basis is still sometimes acted upon, matters actually are more complex, as we shall see later.

6.4 The Hull-Debye-Scherrer Method Outlined

Credit for the method belongs to Hull [11], and to Debye and Scherrer [12]; it seems to have been put to use in analytical chemistry first by Hull [1]. It consists of recording on a cylindrical strip of film the diffraction pattern produced when a collimated beam of monochromatic x-rays strikes a small finely powdered cylindrical sample, often *rotated* about its axis. The two cylindrical axes coincide. The diffraction pattern thus obtained is compared with that of a standard, or the interplanar spacings and intensities derived from the pattern are compared with information stored for known substances: the method was accordingly called the "fingerprint method" in the older literature. As regards Bragg's law (1.6–2), the method uses a known λ and a measured θ to obtain *simultaneous* values of all ratios d_{hkl}/n for the sample. This is possible because the finely divided sample, *if* it has *random crystal orientation*, properly presents *all* crystal faces to the x-ray beam. The slow rotation of the sample (about 1 rpm) about its axis virtually ensures that this condition is met.

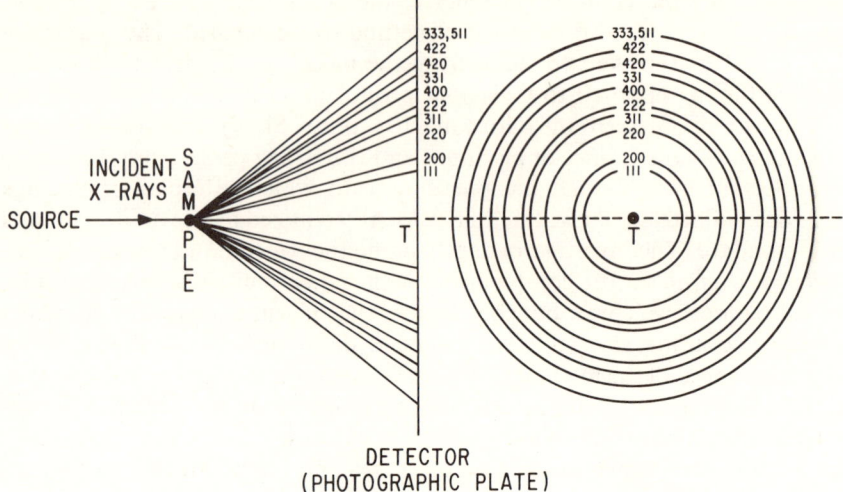

Fig. 6.4–1. Photographic plate as detector to illustrate the Hull-Debye-Scherrer method. The lettering is keyed to Figure 6.5–1 to facilitate comparison. The Miller indices for the ten rings in the figure are identical with those in Table 6.2–3 for F, the face-centered cubic lattice. The agreement shows that the sample can contain only a single crystalline phase, and that the phase has this structure. One arrives at the Miller indices through (1.6–2) and (6.2–1), measured values of θ being the starting point. After Lonsdale, Ref. 6, p. 77, Fig. 58.

With a flat photographic detector, the method is almost self-explanatory; see Figure 6.4–1. The flat plate is impractical, however, because it cannot efficiently record patterns at high values of 2θ; a cylindrical film with the sample at its axis not only does this, but gives patterns in back as well as in forward reflection. On such a film, the circles of Figure 6.4–1 become ellipses; the diffraction pattern on a strip of such a film is shown in Figure 6.4–2.

The geometry underlying the measurement of θ on a film such as that of Figure 6.4–2 is simple provided one remembers that θ appears in Bragg's law while 2θ is directly recorded on the film. In Figure 6.5–1, take $r = SL$ as the radius of the camera. Then

$$\text{arc } LD = 2\pi r \times 2\theta/2\pi = 2r\theta \tag{6.4-1}$$

Suppose, for convenience, that the camera diameter $2r$ is to be chosen so that 1 mm on the scale of Figure 6.4–2 represents 2 degrees in 2θ or *one degree in* θ. Then the (circumferential) strip will have to be 180 mm long, for 2θ has a maximum value of 360°, and the camera diameter will be $180/\pi$ or 5.73 cm.

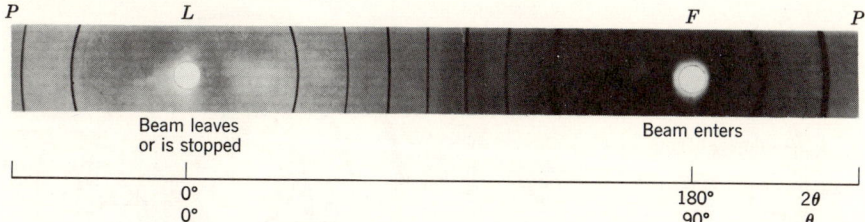

Fig. 6.4–2. Diffraction pattern of tungsten powder taken with Cu Kα. The lettering is keyed to Figure 6.5–1. Length of film is assumed equal to circumference of camera. The back-reflection pattern has sharper lines and lower background. The factors governing the appearance of powder photographs are many and complex; see Ref. 4 and Ref. 5, Chapter 16. Courtesy of L. M. Osika, General Electric Company.

6.5 Selected Experimental Details of the Method

Examination of a sample by the Hull-Debye-Scherrer method consists of three steps: (1) sample preparation; (2) generation and interpretation of the diffraction pattern; and (3) comparison of the pattern with that of known substances.

Samples, as received, often are unsatisfactory, and the analyst must exercise considerable care to make them suitable. If the sample is not a mass with crystals of the right size range, it must be reduced to a satisfactory size (325 mesh) without introducing foreign substances, segregating the components, or producing strains and distortions in the crystal lattice. The most common foreign substance to be guarded against is material from the mortar in which samples are ground.

Powders much coarser than 325 mesh produce individual diffraction spots in the lines and contribute graininess to the pattern. Most substances can be ground quite satisfactorily, but some that are too soft or too malleable require special treatment (Ref. 4, p. 194). Care also must be exercised that the softer components of a powder are not preferentially selected.

The powder, having been reduced to the proper fineness, must be mounted in the x-ray beam without introducing a support that contributes a pattern of its own, or that scatters x-rays too strongly. A massive polycrystalline sample can be mounted directly in the x-ray beam, but a fine powder needs some support. Where possible, the sample may be mixed with a binder and molded into small cylinders. Other approaches are to coat a small amorphous rod or hair with the powder, or to pack it into a glass or polymer tube. The sample diameter must, in all cases, be kept as small as possible to reduce absorption and scattering effects. In fact, for heavy elements it is sometimes necessary to introduce a diluent. The absorption effects reduce the intensity of the low-angle lines relative to those in the back-reflection

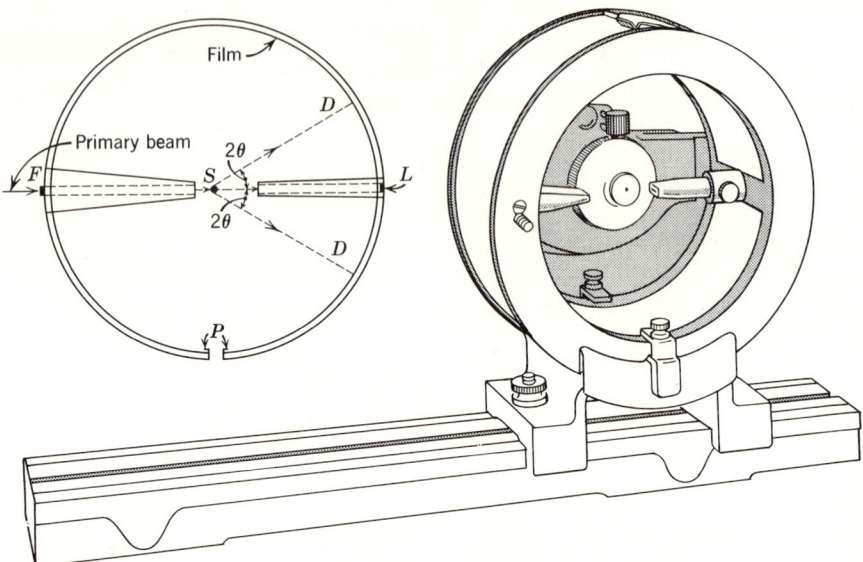

Fig. 6.5–1. A drawing and schematic diagram (insert) of a powder-diffraction camera. At F is a filter for the x-ray beam; S is the sample; θ is the Bragg angle; DD is the cone of diffraction that intercepts the cylindrical film. L is an absorber for the direct beam; and the film stops are at P. Better patterns are often obtained if L is eliminated and the x-ray beam allowed to escape.

region ($\theta > 45°$), and scattering raises the general background level, reducing the sensitivity of the method. Other methods of mounting are discussed in more detail by Klug and Alexander [4].

A typical Debye-Scherrer-type powder camera is shown in Figure 6.5–1. The shape is cylindrical, with provision for mounting the sample on the axis of the cylinder. The extended x-ray entrance and exit ports are designed to hold to a minimum the scattered x-rays that reach the film from the primary beam.

Commercial cameras include a means for aligning the sample in the center of the pinhole system. Good technique requires that the entrance pinhole be filled uniformly with the x-ray beam. With constructions now popular, this limits the entrance pinhole to about 1.2 mm in diameter. The sample should be completely in the direct beam (Figure 6.5–1) and should therefore be slightly smaller in diameter than the pinhole. Provision for the rotation of the sample is usually present and helps to decrease the necessary fineness of grinding. The finer details of camera design are beyond the scope of this chapter; for further discussion and references, see Klug and Alexander [4].

X-RAY DETERMINATIONS ON POWDERED SAMPLES

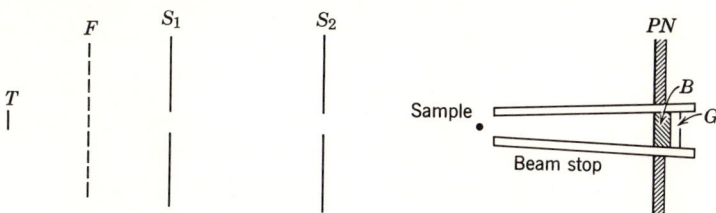

Fig. 6.5–2. Schematic diagram showing the slit system and sample position for powder diffraction. At T is the target of the x-ray source; F is a filter; and S_1 and S_2 are the collimating slits. Note that the sample is completely bathed in the direct beam. P is black paper; N is a photographic film; B is a fluorescent screen; and G is a lead glass. The fluorescent screen is useful for adjusting the camera.

The camera first must be located so that the optical system views the target of the x-ray source at a small angle (usually about 6°) so that the target appears almost square. The camera mounting also must be adjustable vertically, laterally, and about the sample axis. The adjustment is facilitated by the fluorescent screen of Figure 6.5–2. The camera support must be designed so that the camera can be removed and returned to precisely the same position.

The sample, mounted in a cylinder smaller in diameter than the pinhole system, must be carefully centered in the powder camera. This may be accomplished by observing the sample through the entrance slits and rotating it, with adjustment of the centering screws on the sample holder, until the sample shows no lateral motion. The camera may then be loaded with film and placed in position.

The film is held against the inside wall of the camera by spring clips, film stops, beveled rings, or by other means. Holes are made in the film where the x-ray beam enters and leaves the camera, to reduce local blackening and general scattering. The ends of the film are placed symmetrically about the entrance slit—or asymmetrically (Straumanis) so that the complete angular range is included without a break in the film (Figure 6.5–1). With camera diameters of 5.73 (or 11.46 cm), 1 mm along the film is equivalent to 1 (or $\frac{1}{2}$) degree in the Bragg angle, θ.

The choice of x-ray beam for the powder-diffraction pattern depends on the nature of the substance to be examined, the complexity of the pattern, and the interplanar spacings. For general use, Cu Kα (1.542 Å) probably is the best. For organic compounds with low absorption coefficients and large unit cells, Cr Kα (2.291 Å) is useful; and for inorganic compounds with small unit cells and high absorption coefficients, Mo Kα (0.711 Å). Tubes with sufficiently pure targets of these materials are commercially available. It is advisable to check the spectral purity of a diffraction tube periodically by taking the pattern of a known substance, such as sodium chloride. The most common target impurities are iron, tungsten, nickel, and copper.

The filter shown on the entrance side of the camera optical system is there to remove the $K\beta$ lines from the beam. As generated in the target, the $K\beta/K\alpha$ intensity ratio is just below 0.2 for molybdenum. A filter with its K absorption edge between the two lines can be used to reduce this ratio by another factor of 10, without a catastrophic increase in exposure time. The filter thickness, th, in centimeters, is given for this case by

$$th = 2.303/(\mu_2 - \mu_1)\rho \qquad (2.303 = \ln 10) \qquad (6.5\text{-}1)$$

where μ_1 and μ_2 are mass absorption coefficients for Mo $K\alpha$ and Mo $K\beta$ for a filter with density ρ. See (1.10–2). Achieving the exact calculated thickness is not important.

The proper filter materials for molybdenum, copper, and chromium are zirconium, nickel, and vanadium, respectively. The most convenient filter is in the form of a foil, but a layer plated on a light substrate (Al or Mylar film), or a powder dispersed in a plastic or wax, can be used to give the right concentration of the element per square centimeter. The ratio of exposure times with and without filter is given by

$$t_1/t_0 = \exp \mu_1 \rho th \qquad (t_0, \text{ without filter}) \qquad (6.5\text{-}2)$$

When enough is known about a sample so that allowance can be made for the effect of the $K\beta$ lines on the diffraction pattern, the unfiltered x-ray beam will give satisfactory results, and the exposure times will be reduced. Typical exposure times (from Ref. 4, p. 210) are given in Table 6.5–1 for

Table 6.5–1. Typical Exposure Times for the Hull-Debye-Scherrer Method

Camera Diameter, mm	Pinhole Diameter, mm	K Filter	Metal Filings	Simple Salts and Minerals	Organic Compounds, Complex Inorganic Compounds, and Minerals
57.3	1.0	No	5 min	10 min	20 min
57.3	1.0	Yes	10 min	20 min	40 min
57.3	0.5	No	15 min	30 min	1 hr
114.6	1.0				
57.3	0.5	Yes	30 min	1 hr	2 hr
114.6	1.0				
114.6	0.5	No	1 hr	2 hr	4 hr
114.6	0.5	Yes	2 hr	4 hr	8 hr

X-RAY DETERMINATIONS ON POWDERED SAMPLES

targets of molybdenum, silver, copper, iron, or cobalt. For chromium targets, exposure times up to twice as long are needed.

6.6 Diffractometry

The close relationship of an x-ray diffractometer to an x-ray emission spectrograph was indicated in Chapter 5: the early modern spectrographs were in fact converted diffractometers. The first modern diffractometer with a Geiger detector was developed by Friedman [13], and the modern spectrograph followed. As we saw in Chapter 2, the proportional and the scintillation detectors are now replacing the Geiger. The diffractometer normally has parafocusing geometry in order to maximize intensity (Chapter 5 and Figure 6.6–1), and it uses a flat powdered sample that is rotated to cover the required range of θ values. It thus produces (See Figure 6.6–2) a recording of intensity (cps) as a function of θ. Though identical *in principle* with the x-ray spectrometer (Section 1.17), it differs in that it uses a *known wavelength* to establish *unknown interplanar spacings*; in that it often deals with pow-

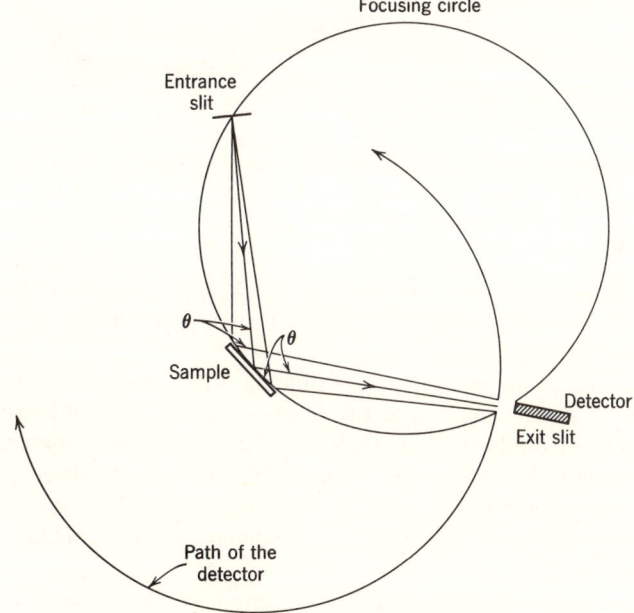

Fig. 6.6–1. Schematic diagram of a parafocusing diffractometer. As the sample rotates through an angle θ, the detector moves along the path shown at twice the angular velocity of the sample. The sample-to-detector distance remains constant. The entrance slit determines the horizontal divergence of the x-ray beam; the exit-slit angular aperture determines the resolution.

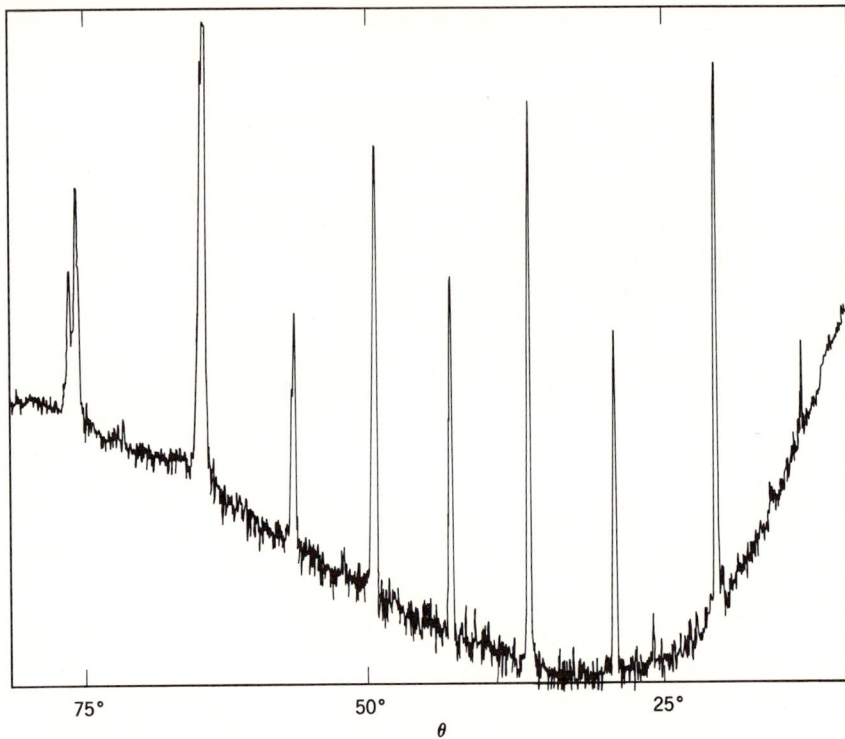

Fig. 6.6–2. Diffractometric trace for a sample of powdered tungsten. Courtesy of L. Bronk, General Electric Company.

dered samples, sometimes amorphous; and that it never uses an ionization detector. These differences are profound enough to justify *diffractometer* as a distinguishing name.

With parafocusing geometry and a flat sample, absorption effects are independent of Bragg angle, and this makes for increased intensity relative to photographic methods. The effective sample area also is increased by the use of a flat sample. At very low values of θ, the sample subtends too small a part of the primary beam for full intensity in the parafocusing arrangement. Detailed alignment procedures are given by the manufacturers of the spectrometers.

The powder should be ground to a considerably smaller size than the 325 mesh satisfactory for photographic work. The spectrometer does not usually allow the sample to be rotated, so that graininess or preferred orientation cannot be averaged out. Large intensity deviations are possible with the larger crystals. The mounting of the sample also is critical because

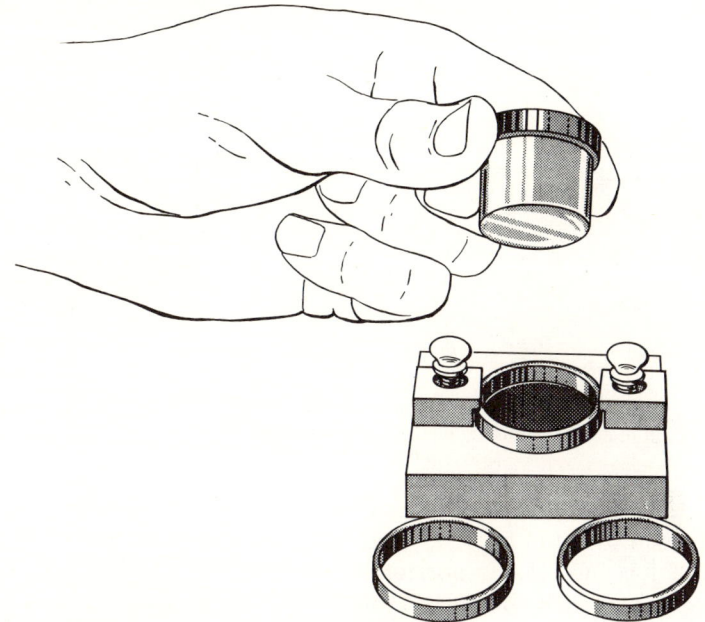

Fig. 6.6-3. Typical sample holder for a parafocusing diffractometer. Courtesy of the General Electric Company.

the path of the x-ray beam in the sample decreases, by a factor proportional to $1/\sin \theta$, as the Bragg angle increases. If the sample is thick enough, so that essentially all of the primary beam is absorbed at high values of θ, trouble from this source is eliminated. Further discussion appears in a later section.

A sample consisting mainly of carbon, oxygen, and hydrogen will have the minimum thicknesses for maximum intensity in the ratio of 1:5:15 for Cr $K\alpha$: Cu $K\alpha$: Mo $K\alpha$—another good reason why the longer wavelengths are generally better for organic samples.

Sample holders usually are pieces of aluminum, plastic, or glass, with a circular depression sufficiently deep to provide the minimum thickness. The sample is loaded into the depression and is smoothed with spatula or other flat edge. If difficulty is experienced in holding the powder in the cavity, a small amount of binder should be added. At small Bragg angles, where the beam strikes at a glancing angle, the cavity must be long (Figure 6.6-3).

The line resolution and the area of the sample irradiated both depend on the divergence of the x-ray beam. With a flat specimen, the vertical divergence must be limited, both between target and sample and between

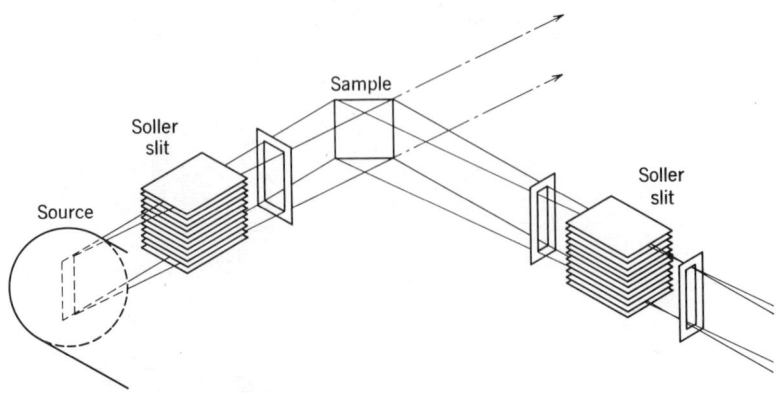

Fig. 6.6–4. Schematic diagram showing how two Soller slits limit the vertical divergence of the x-ray beam in an x-ray diffractometer. After Klug and Alexander, Ref. 4, p. 241.

sample and detector. This is done (Figure 6.6–4) with two Soller slits, in effect the equivalent of a large number of slits, arranged so as to add their transmitted beams. In the horizontal plane in the figure, the x-ray tube focal spot is viewed at an angle so small that it appears as a line source. The horizontal divergence is usually kept below 2°.

Most commercial instruments have coupled drives for the sample and detector, with the detector output recorded on a strip chart. Considerable care must be taken in choosing the optimum combination of drive rates and of the integrating-time constant in the detector.

If the integrating-time constant of the detector is too long, the recording of the diffraction line is shifted in the direction in which the scan is going. If the time constant is too short, the statistical fluctuations make the data too erratic. A time constant of less than half the time it takes the receiving slit to travel its own width should normally be used to avoid a large shift in the line peak and the concomitant lack of resolution.

Table 6.6–1. Suitable Values of the Time Constant, RC, for Various Scanning Speeds and Receiving-Slit Widths [4]

Scanning Speed, degrees/minute	Receiving-Slit Width, degrees	Time Width of Slit, W_t, sec	Maximum Recommended RC, sec
2.0	0.20	6	3
1.0	0.13	6	3
0.5	0.35	6	3
0.2	0.025	7.5	3.8

The relation between scanning speed and slit width for different types of determinations is given in Table 6.6–1 [4, p. 310].

6.7 The Intensity of Bragg Reflection for Powdered Samples

When visual estimates of intensity (either direct or by comparison with a standard) do not suffice, methods using photographic film are usually at a disadvantage *vis-à-vis* diffractometry. The advantage becomes progressively less serious, and may sometimes disappear as regards analytical chemistry, when intensity measurements made by microphotometering a film are automated (see below), and when many intensities must be measured for a powdered sample. Microphotometering in principle resembles x-ray absorptiometry (Chapter 3); visible light replaces x-rays and the absorbing substance is the silver in the finished film. For a discussion of microphotometering, see Ref. 4, pp. 368–376.

It is now necessary to extend the statements about intensity made in Section 1.5. To repeat, the analytical chemist is concerned with *relative* not *absolute* intensities. Strictly speaking, there are no monochromatic lines, for a monochromatic line would have an FWHM of zero: it is of course true that some lines are more nearly monochromatic than others: a $K\alpha 1$ line is more nearly monochromatic than the $K\alpha$ doublet. In Section 1.15, the integrated intensity mentioned was that obtained by integrating over the wavelength range of a polychromatic beam; no such integration is needed in the case of Bragg reflection. However, as was pointed out in Chapter 5, the FWHM of a line is increased by Bragg reflection to an extent that increases with the imperfection of the crystal. Especially in the diffractometer, this increase is augmented by the action of other factors. Therefore, we have always a *peak* and an *integrated* intensity to consider; the integration here is not over a wavelength interval, but over an interval in θ for the diffractometer, and over an interval of distance on the photographic film. The intervals for these integrations are broader than the corresponding wavelength intervals, but the integration over wavelength is included in them.

Not only in x-ray diffraction, but also in x-ray emission spectrography, we almost always deal with peak intensities. It is easy to see why. Consider the diffraction pattern of tungsten in Figure 6.6–2. The integrated intensity

$$I_{\text{int}} = \int_{\theta_1}^{\theta_2} I_\theta \, d\theta = N_T \text{ counts} \qquad (6.7\text{–}1)$$

is obtained by accumulating the counts for the interval $\theta_2 - \theta_1$ that includes virtually all the quanta of the diffracted line. The first difficulty is the proper choice of θ_1 and θ_2. The second difficulty is more serious. The N_T includes

N_B, the counts registered by the background. A glance at Figure 6.6–2 will show that N_B over the interval $\theta_2 - \theta_1$ is difficult to establish; in addition, the fluctuations in N_B will introduce uncertainties. (These statistical matters are discussed in a later chapter; they are implicit in Figure 2.20–1).

The peak intensities are the differences $N_T - N_B$ for a given counting interval at the value of θ for the peak; when divided by Δt, these differences become counting rates. The peak intensities are thus more quickly and more easily obtained than the integrated intensities, and this explains why peak intensities are generally preferred for determinations both by diffraction and by emission.

The preceding discussion is not a blanket endorsement of peak intensity readings even when, as is usual in analytical chemistry, the determination is based on a comparison of standard and unknown. It is true that such a comparison increases reliability, and that this procedure is usually adequate in x-ray emission spectrography even though the peak intensity is more sensitive than the integrated to variations in equipment. In x-ray diffraction, however, the nature of the diffracting surface is an additional variable. The peak intensity and the FWHM of a diffracted line can be affected by, for example, the mosaic character of the diffracting surface (see Chapter 5), which can vary from one sample to another. It comes to this: the analytical chemist must not rely blindly on peak intensities and must be prepared to check dubious results by using integrated intensities as well. *Only the integrated intensity can contain all the quanta.* When a photographic film is properly microphotometered, these questions do not arise because integrated intensities are then obtained.

Several features of the background in powder diffractometry are of general interest. Residual **Kβ** lines will contribute to the background if filtration is insufficient. The absorption edge of silver and that of bromine can appear as in de Broglie's experiment (Figure 3.4–2); those of elements in the sample may be recorded on film also. Compton scattering (Section 1.15) contributes to the background. Characteristic lines excited in the diffracting crystals will appear as general background as they do in x-ray emission spectrography without pulse-height selection (Section 2.23). For extended path lengths within the sample (small θ), and particularly for long wavelengths, and for samples containing elements of high atomic number—see (1.13–2)—a Hull-Debye-Scherrer line may actually be split (to form a pseudodoublet) owing to complete absorption of its center portion. Contamination of the target of a Coolidge tube by tungsten from the filament (Section 1.3) can lead to diffraction of the tungsten **L** lines; such diffraction can be misleading. Scattering of x-ray by air in the camera and scattering caused by lattice imperfections also contribute to background. All in all, expert knowledge is needed for the correct interpretation of all but the simplest powder diffraction patterns. See Ref. 4.

THE INTERPRETATION OF THE DIFFRACTION PATTERN

The factors that govern intensity [4, pp. 132–161, 364–379] are likewise of general interest though not of immediate concern to the analytical chemist. They are listed with brief comment below.

1. *The polarization factor.* Gives the total energy as a function of θ for Thomson scattering. See Section 1.15. Given in Ref. 4, Appendix VIII, Table 1.

2. *The Lorentz and "velocity factors."* Not discussed in this book. Measure the opportunity of a plane to accomplish Bragg reflection as governed by its orientation and by the length of time (for a moving sample) it is in position to reflect. Combined with polarization factor in Ref. 4, Appendix VIII, Tables 2 and 3.

3. *The temperature factor.* Mentioned in Section 6.2. Consequence of quantum theory as applied to heat capacities of solids by Debye. Tabulated in Ref. 4, Appendix IX, Table 1.

4. *The atomic scattering factor.* Classical case of Thomson scattering in Section 1.15. Rigorous treatment much more complex. Tabulated in Ref. 4, Appendix VII, Tables 1 and 2.

5. *The structure factor.* Takes into account the effects on the diffracted beam (in both phase and amplitude) of having different kinds of scattering centers at different positions in the crystal lattice. These effects change intensities and can lead to the complete extinction of certain classes of reflections. For example, a face-centered cubic lattice will never show Bragg reflections for which the Miller indices are mixed; the indices must be either all odd or all even. See Table 6.2–3. The structure factor is too complex for this book. See Ref. 4, pp. 141–150 and 157–161.

6. *The multiplicity factor.* Simple example in Section 6.2, where {101} represents 12 faces.

7. *The absorption factor.* Measures the absorption effect. Extreme example is the splitting of a reflected line, which was described above. In the Hull-Debye-Scherrer method, the absorption factor is high at low θ, as this example shows. In general, the factor is a complicated function of θ and of the dimensions of the sample. In this respect, the diffractometric method has the advantage. For it, the absorption factor is simple (later discussion) and independent of θ.

THE INTERPRETATION OF THE DIFFRACTION PATTERN

6.8 General

In the powder methods considered here, the record of the diffraction pattern is either a film (Figure 6.4–2) or a chart (Figure 6.6–2). This record in each case must be converted into useful form for interpretation. For routine work, the most rapid interpretation can be made by direct comparison with a stan-

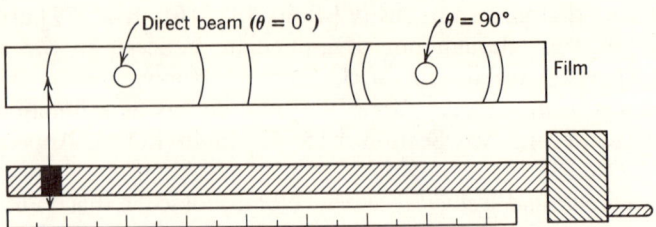

Fig. 6.8–1. Diagram of film-measuring mechanism. The hand wheel includes a vernier scale for accurate setting.

dard film or chart. If no deviations are found, the identification is complete.

If no reference pattern is available, the 2θ values from the film or chart must be converted to d values for identification. A special film-measuring device (Figure 6.8–1), which can be prepared readily or purchased, is essential. Most commonly, this consists of glass plates uniformly illuminated from below, with movable cross hairs or pointers on a film holder. The indicators are attached to a scale and can be driven by a screw thread. When a comparator of this type is used, care must be taken to avoid backlash on the screw thread by approaching each measurement from the same direction.

The 0 and 90° positions in θ can be determined accurately from the arcs that are centered about each of these positions. It is then a simple task to get the other angles by direct proportion. If the camera diameter is 5.73 or 11.46 cm, the 90° in θ corresponds to 9.0 or 18.0 cm, and the measured spacings can be converted to angles directly. For accurate work, the variable film shrinkage introduces a slight correction. If a spectrometer is used, the θ values are given by the chart directly. The angles are then converted to spacing (d) values by the Bragg equation. Comparison of these values with the spacing values of known structures is essential.

Before the comparison can be made, the intensities of the lines must be estimated and tabulated with the d spacings. This intensity estimate can be made by eye or with a photometer. One of the most satisfactory methods is to prepare a film with a series of lines of graded densities for direct comparison. This procedure will enable the operator to keep the exposures in the linear region of the film and put all of the intensity measurements on the same basis. The peak heights from the diffractometer chart recordings usually are a sufficiently good estimate of the line intensities, and can be used directly.

6.9 The Hanawalt-Rinn-Frevel Identification System [14]

The publication in 1938 by these authors of tested diffraction data for 1000 compounds is a landmark in analytical chemistry. It provided a useful

THE INTERPRETATION OF THE DIFFRACTION PATTERN

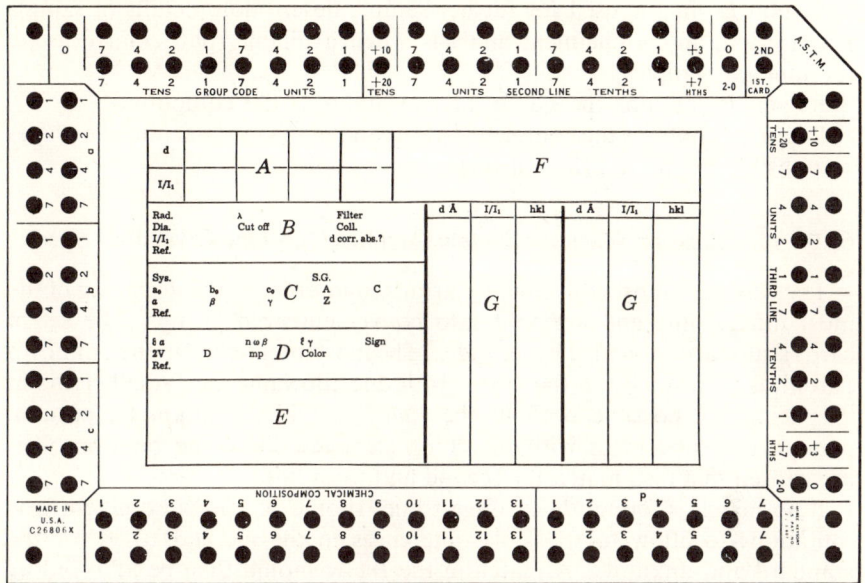

Fig. 6.9-1. Card used in the Hanawalt-Rinn-Frevel classification system.

library of reference patterns, gave the basis for a useful classification scheme, and was the genesis of the modern *Powder Diffraction File* mentioned in Section 6.3. Though the computer will in the future relieve most analytical chemists of the chore of comparing their diffraction patterns with the information stored in the *File*, they will still need to understand the rationale of the comparison.

The classification system rests on a punch card with seven subdivisions, each one for a different kind of information needed in the comparison. See Figure 6.9–1.

Region A of the card contains the d spacings and relative intensities of the most intense reflections (with the strongest given an intensity value of 100). Region B contains the essential features of the pattern recording. The radiation used, filter conditions, and type of instrument can all affect the intensity estimates and the appearance of spurious lines. It is essential to remember that low-wavelength, high-angle, and all parafocusing spectrometer patterns are less affected by absorption than low-angle, long-wavelength patterns. With spectrometers, precautions should also be taken against line shifts attributable to the time constants of the integrating circuits.

Regions C and D contain the essential crystallographic and physical data. The data in this part of the card can be used in connection with a polarizing microscope and hot stage to prove a doubtful identification.

Region *E* often is used for further sample description, details of sample preparation, and a chemical analysis when available (particularly useful for mineral samples).

Region *F* contains the names and formulas of the compound.

Region *G* contains the complete *d*-value and intensity data, plus the planes responsible for the diffraction lines.

6.10 Identification When the Sample Contains Only One Crystalline Species

The cards are most conveniently arranged in order of the *d* spacing of the most intense line and are split into convenient-sized groups. The set of cards is also accompanied by an *Index*. The most intense line in the unknown pattern is then sought in the *Index*, with due allowance that small errors in *d* spacing are possible both in the *Index* and in the unknown pattern. Among the substances with matching strongest lines, the one, or ones, are chosen that also match the second and third lines.

If no match is found, the strongest lines should be tried in different permutations to allow for possible differences in the methods used for the standard and unknown. A match in the *Index* should then be followed by comparison with the individual cards.

If more than one card is found to match the strongest lines, a complete comparison of the patterns is necessary (region *G*, Figure 6.9–1).

6.11 Identification of Crystalline Species in Mixtures

If no match is found by the aforementioned methods, a careful inspection of the powder under a microscope is desirable. If the powder seems to be composed of a mixture of substances, the identification can be attempted on this basis. At this point a semiquantitative elemental analysis can be of great help. An x-ray-emission spectrograph can provide a rapid analysis for the elements above sodium. Optical-emission spectrography also can be used to advantage. If some guess can then be made as to the compounds present, the proper cards can be compared.

With all of the other available information assembled, it is possible to proceed as before and match the strongest lines. In the case of a mixture, not all the lines will be found to match a single diffraction pattern. One component may, nevertheless, be identified if a reasonable number of the lines of the unknown pattern are matched to a known pattern in respect to spacing and intensity, and none of the reasonably intense lines of the known are missing. The remaining lines are then treated as a new pattern and the usual identification procedure is followed. It must be kept in mind that any lines of the second substance that coincided with those of the first will not be included in the residual list, and one should check back for lines apparently missing. The identification, which may be tentative, should

THE INTERPRETATION OF THE DIFFRACTION PATTERN

check with the analytical and crystallographic data. More complex mixtures often cannot be identified without additional information. Further discussion is found later.

Common difficulties that arise with complex mixtures are sample segregation and reactions between the components during sample preparation. In the final diffraction pattern, components present in small amounts may cause confusion because they contribute only a line or two.

In cases of such complexity, the time often is best spent on a mechanical separation—laborious as it may be—and then on the individual analysis of the separated fractions by the usual method.

6.12 Straumanis' Determination of the Lattice Constant (a_0) for Aluminum

The reader will have noticed in the previous section some hedging regarding the capabilities of diffraction methods. That this is necessary soon appears. Meanwhile, it seems only fair to present an example that shows the precision and certainty possible with such methods when the sample is simple and pure, and when the work is done with great care by a master in the field.

The work was done by Straumanis [4, 15] on a cylindrical sample of aluminum, 0.18 mm in diameter and 99.998% pure, exposed for 8 hours in a camera 57.7 mm in diameter at 23.10° C. The film was in the (Straumanis) asymmetric position. See Figure 6.12–1, which shows the diffraction pattern

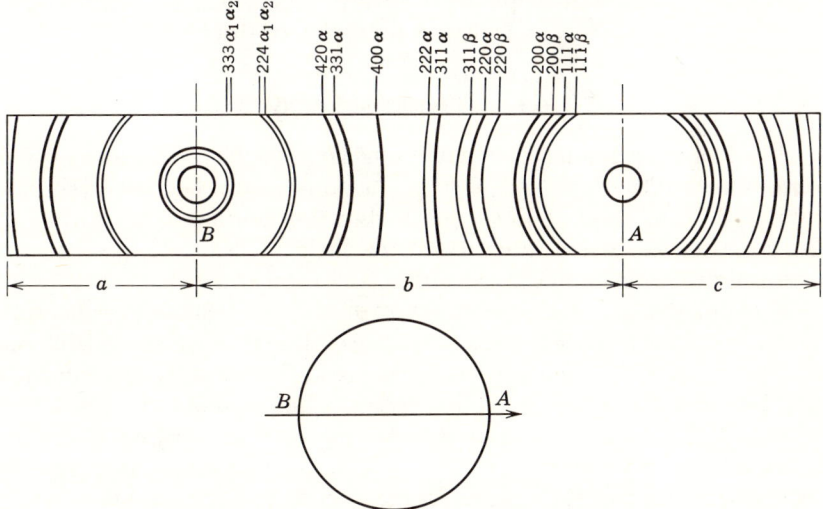

Fig. 6.12–1. Diffraction pattern of aluminum by Straumanis. As the most precise results were expected in the back-reflection region (about B), unfiltered Cu Kα was chosen as the incident beam, and the patterns from the *resolved doublet* (Cu Kα1 and Cu Kα2; wavelength difference, about 0.003 Å) could be used for the calculation of a_0 in (6.2–1). After Ref. 4, p. 459.

and the corresponding Miller indices. Comparison of these indices with Table 6.2–3 will show that aluminum has beyond doubt a face-centered cubic lattice; notice particularly that the *absence* of $hkl = 100$ eliminates the simple cubic lattice as a possibility. The final result, $a_0 = 4.04945$ Å, was based only on the 333 lines; it is the mean of 4.04944 (Cu **K**α1) and 4.04946 (Cu **K**α2).

ABSORPTION EFFECTS. INTERNAL STANDARDS

The importance for the measurement of film thickness (Chapter 7), for x-ray emission spectrography, and for other x-ray methods of the following material warrants the individual center title. In x-ray diffraction, we have a relatively simple situation because no *emission* of characteristic x-rays by the sample need be considered, and because the x-rays in question are virtually monochromatic; we do have the complexities inherent in dealing with powders. The use of internal standards is in many cases the best way of carrying out determinations by the *comparative method* (cf. certain absorptiometric methods in Chapter 3, in which the comparison was between a standard and a separate unknown.) For analytical chemistry, the most important paper is that of Alexander and Klug [16], which is followed closely in the discussion given below. The samples are assumed flat and finely enough divided so that the absorption of x-rays can be calculated by using macroscopic mass absorption coefficients: "microabsorption" [17] is negligible, and there is no preferred orientation.

6.13 Absorption Effects

When an x-ray beam is incident at angle θ upon a flat, powdered, crystalline sample, diffraction and absorption both occur. The intensity of the diffracted beam at any point x in the direction normal to the surface will be proportional to the intensity of the incident beam at that point. As absorption attenuates the incident beam, the locally diffracted beam becomes weaker and weaker. As the diffracted beam is also attenuated by absorption on its way out of the sample, the local contribution to the measured diffracted beam soon becomes negligible as x increases. The value of x for which this first becomes true is the *critical thickness*, th_c; for present purposes, samples of critical thickness or greater may be regarded as *infinitely thick*. Such samples give maximum measured diffracted intensity. (The reader will have noticed that th, not d, is used as the symbol for thickness in this chapter so that there will be no confusing of thickness with interplanar spacing.)

The value of th_c depends on experimental conditions and especially on the sensitivity of the detection system. An accepted choice was made on the

ABSORPTION EFFECTS. INTERNAL STANDARDS

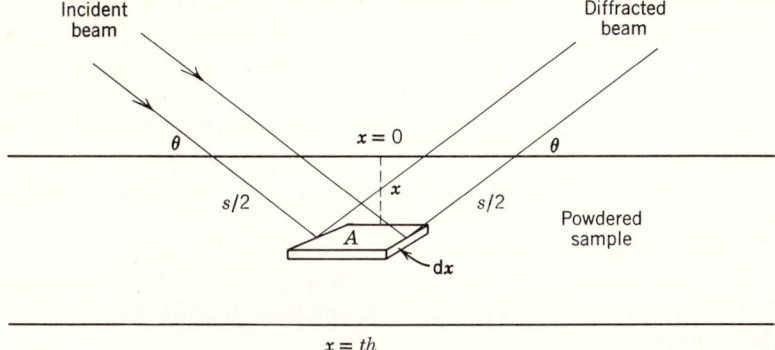

Fig. 6.13–1. Schematic diagram of powdered sample in diffractometer. The incident beam has cross-sectional area A_B, and it irradiates sample area A_{irr} at depth x. The emergent diffracted beam is the local contribution to the measured beam from the portion of sample for which $dV = A_{irr}\, dx$. Also, $dx = (ds \sin \theta)/2$.

basis that the local contribution I_x is negligible when it is less than $1/600$ of the measured diffracted intensity [18]. The resulting criterion is

$$\mu_l s = \mu \rho s \geq 6.4 \quad \text{see (1.10–3)} \tag{6.13–1}$$

where s is the maximum path length of x-rays within the sample and the absorption coefficients are averages for the sample, interstices included.

Figure 6.13–1 shows schematically the geometrical conditions for the sample in the diffractometer. For a sample of thickness th_c,

$$s_c/2 = th_c/\sin \theta \tag{6.13–2}$$

Also, if x-ray absorption in the interstices is negligible, and μ has its value for the solid, then (6.13–1) becomes

$$th_c \geq 3.2/\mu\rho \times \rho/\rho' \times \sin \theta = 3.2 \sin \theta/\mu\rho' \tag{6.13–3}$$

This criterion is met by all samples under discussion here. Beyond this point, the interstices in the sample need no longer be considered. Also, the density of the sample may be regarded as that corresponding to zero volume change on mixing the constituents.

From Figure 6.13–1

$$dV = A_{irr}\, dx = A_B\, ds/2 \tag{6.13–4}$$

For a sample that is a mixture, what is the local contribution dI_i to the measured beam of component i in volume dV at depth x? To arrive at this we let f_i be the volume fraction (assumed constant for the sample) of this component, and (for convenience) denote by $(I_0)_i$ the contribution that unit volume of pure component i would have made to the measured diffracted beam had there been no absorption in the sample. The answer to our question is

$$dI_i = (I_0)_i f_i A_B e^{-\mu_l s} \, ds \tag{6.13-5}$$

in terms of the *linear* absorption coefficient. Integration from $s = 0$ to $s = \infty$ now gives

$$I_i = (I_0)_i f_i A_B / 2\mu_l = K_i f_i / \mu_l = K_i f_i / \mu\rho \tag{6.13-6}$$

for the *total* contribution made to the measured diffracted beam by component i. The constants for given experimental conditions have been lumped in K_i for convenience. The absorption coefficients and the density are those of the sample.

As the contribution I_i is independent of θ, the absorption factor—which helps determine this contribution—must also be independent of θ. This is the basis for a statement in Section 6.7.

By definition,

$$f_i = (w_i/\rho_i)/(w/\rho) = (w_i/w)(\rho/\rho_i) = W_i(\rho/\rho_i) \tag{6.13-7}$$

where W_i is the weight fraction of i and the quantities without subscripts are for the sample of weight w.

By substituting from (6.13–7) into (6.13–6) one obtains

$$I_i = K_i W_i / \mu \rho_i \tag{6.13-8}$$

Now, divide the sample into constituent i and matrix M. Then, according to (1.10–5)

$$\mu = W_i \mu_i + (1 - W_i) \mu_M$$

and, finally

$$I_i = K_i W_i / \rho_i [W_i \mu_i + (1 - W_i) \mu_M] \tag{6.13-9}$$

Let us apply (6.13–9) to the following five important special cases:

CASE I. Pure substance. Example: quartz. $W_i = 1$; $i = 1$.

$$(I_i)_{pure} = K_i / \rho_i \mu_i \quad [\text{Cf., (6.13-8)}] \tag{6.13-10}$$

CASE II. Two crystalline phases of identical composition (polymorphs). Example: quartz and cristobalite. W_i variable; $\mu_i = \mu_M$.

$$I_i = K_i W_i / \rho_i \mu_i = W_i (I_i)_{pure} \tag{6.13-11}$$

CASE III. Pure substance with transparent diluent. Example: quartz and starch as (almost) ideal diluent or binder. Starch not counted in sample weight. $W_i = 1$; $\mu_M = 0$ (approximately).

$$I_i = K_i / \rho_i \mu_i = (I_i)_{pure} \quad \text{(approximately)} \tag{6.13-12}$$

ABSORPTION EFFECTS. INTERNAL STANDARDS

CASE IV. Crystalline phase i and matrix M with $\mu_i > \mu_M$. Example: quartz mixed with beryllia. W_i variable.

$$I_i > K_i W_i / \rho_i \mu_i \quad \text{(except at } W_i = 1 \text{ or } W_i = 0\text{)} \quad (6.13\text{–}13)$$

Case of a *positive absorption effect*.

CASE V. Like case IV except $\mu_i < \mu_M$. Example: quartz mixed with potassium chloride. W_i variable.

$$I_i < K_i W_i / \rho_i \mu_i \quad \text{(except at } W_i = 1 \text{ or } W_i = 0\text{)} \quad (6.13\text{–}14)$$

Case of a *negative absorption effect*.

6.14 Experimental Tests

Case III will be taken for granted. Cases II, IV, and V were tested by Alexander and Klug [16] on the basis of relative intensities with case I as standard. Calculated and measured relative peak intensities were compared.

CASE II. Three synthetic mixtures were prepared of micronized quartzite and Johns Manville Celite, a commercial thermal insulating powder that is mainly finely divided, slightly impure, α-cristobalite. The impurities— no quartz among them—had virtually the mass absorption coefficient of silica, which is 34.9 for Cu Kα. Determinations were made in triplicate on a Norelco diffractometer by measuring with manual scanning the peak intensity for quartz at 3.34 Å, sufficient counts (N_T) being taken to give a low counting error and correction being made for the coincidence error of the Geiger detector. In Figure 6.14–1, the three measured results lie virtually on the straight line for which the ordinate is the calculated quotient of (6.13–10) *into* (6.13–11). In this case, the ordinate is simply W_i, and the line must have a 45° slope, as it does. The slope is independent of the mass absorption coefficient.

CASE IV. The samples were properly prepared mixtures of quartz and beryllia. Intensity measurements were made as just described. In this case, μ_i is 34.9 and μ_M (beryllia) is 8.9 (again, Cu Kα). With these data, the relative intensities were calculated as the quotient of (6.13–10) *into* (6.13–9). Again, as Figure 6.14–1 shows, the three measured points are in satisfactory agreement with the calculated curve. The reader will note that the mixture for which $W_i = 0.5$ has a relative intensity of 80%, of which 30% is a *positive absorption effect*.

CASE V. The procedure for case IV was repeated with potassium chloride ($\mu = 124$ for Cu Kα) in place of beryllia. And yet again, the results lie very near the calculated curve in Figure 6.14–1. The mixture for which $W_i = 0.5$

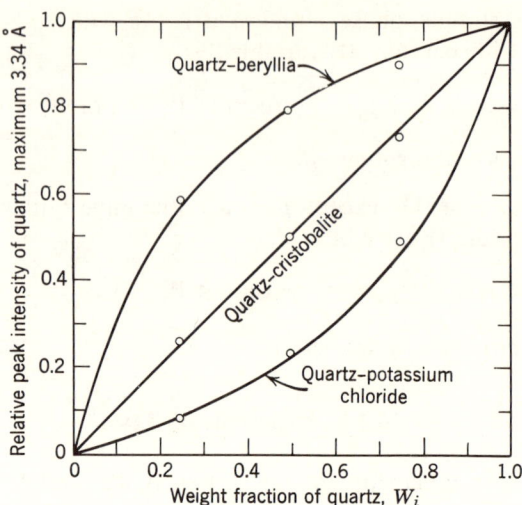

Fig. 6.14–1. Measured (circles) and calculated relative intensities of three binary mixtures containing quartz. Note the positive absorption effect (upper curve) and the negative absorption effect (lower curve). After Alexander and Klug, Ref. 16, p. 888.

has a relative intensity of only 22% owing to the presence of a strong *negative absorption effect*.

Figure 6.14–1 is an important contribution to the understanding of x-ray methods in chemical analysis.

6.15 Internal Standards

Alexander and Klug [16] point out that using an internal standard to carry out determinations by the comparative method, even in x-ray diffraction, did not begin with them. Of particular interest in this connection is the work of Clark and his colleagues [18, 19]. But the first complete treatment of such a standard for x-ray diffraction is that by Alexander and Klug [16], which follows slightly modified. The primed quantities are for a sample to which has been added the internal standard in known, constant proportion by weight and in constant amount; the small w's are actual weights; and the matrix M includes neither i nor internal standard St.

By definition,

$$W_i = \frac{w_i}{w_i + w_M} \qquad (6.15\text{–}1)$$

$$W_i' = \frac{w_i}{w_i + w_M + w_{St}} \qquad (6.15\text{–}2)$$

ABSORPTION EFFECTS. INTERNAL STANDARDS

$$W'_{St} = \frac{w_{St}}{w_i + w_M + w_{St}} = \text{constant} \qquad (6.15\text{--}3)$$

the last condition being fulfilled when w_{St} is constant, and the sum $w_i + w_M$ is also constant. Then

$$W_i/W'_i = \text{constant}. \qquad (6.15\text{--}4)$$

From (6.13–8),

$$I'_i = K_i W'_i/\mu' \rho_i \qquad (6.15\text{--}5)$$

and

$$I'_{St} = K_{St} W'_{St}/\mu' \rho_{St} \qquad (6.15\text{--}6)$$

Whence,

$$I'_i/I'_{St} = \text{const}' \times W'_i \qquad (6.15\text{--}7)$$

$$= \text{const} \times W_i \quad [\text{by } (6.15\text{--}4)] \qquad (6.15\text{--}8)$$

To test (6.15–8), Alexander and Klug [16] mixed quartz and calcite to give samples containing 30, 60, and 100% by weight of quartz. Calcium fluoride as internal standard was then admixed so that $W'_{St} = 0.20$. Peak intensities for quartz at 3.34 Å and fluorite at 3.16 Å were obtained as for Figure 6.14–1. Values of the three ratios I'_i/I'_{St}, each an average of ten determinations, were plotted as in Figure 6.15–1 to give complete verification of (6.15–8).

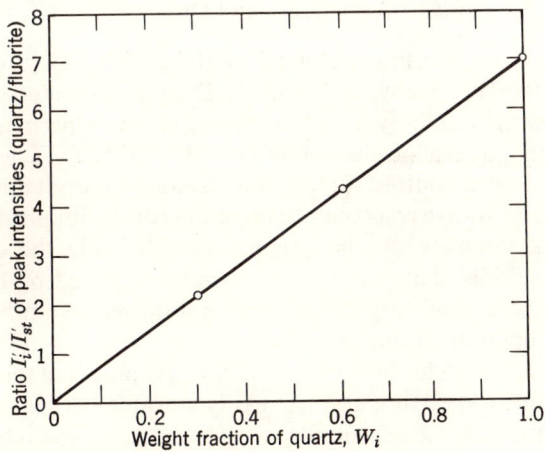

Fig. 6.15–1. A calibration curve for the determination of quartz (matrix, calcite) with fluorite as internal standard. After Alexander and Klug, Ref. 16, p. 888.

AN ASSESSMENT OF X-RAY DIFFRACTION METHODS

The reader is asked to review the quotation from Hull (1) at the head of this chapter; those from Warren (2), which immediately follow (1); and a second from Hull (3), which appears in Section 6.3. Note especially the italicized portion of (2), which warns us that the determination of interplanar spacings and of diffracted-line intensities is all we can expect from the diffraction methods so far discussed. In the years since (3), many complications have appeared that Hull could not have foreseen, and Frevel [20, 21] has been a leader in pointing these out. Some of these complications affect the background on a photographic film and the intensities (Section 6.7). Sometimes these complications yield to a change in experimental method (for example, the use of a Guinier instead of a Debye-Scherrer camera—see below). But others cited by Frevel [20, 21]—isomorphism, isotypism, isotopic substitution, and the formation of solid solutions are inherent in the samples themselves. For such samples the uniqueness of diffraction patterns tacitly assumed by Hull (3) often does not exist. Frevel [21] mentions unknown metallic beads that gave a diffraction pattern satisfactorily matching that of platinum. But there are *four* solid solutions with interplanar spacings virtually identical to those of platinum. In Frevel's case, x-ray emission spectrography showed that the sample contained 83.5 w/o Pd and 16.5 w/o Ag, being one of the solid solutions and not platinum at all. In spite of such limitations, diffraction methods are useful, but they often need help.

6.16 Advantages and Disadvantages

As always in comparing analytical methods, it is risky to generalize in dealing with the Hull-Debye-Scherrer (H-D-S) method *vis-à-vis* diffractometry. Is the sample unknown and complex, and is a permanent record of the diffraction pattern needed? Use the H-D-S method with its photographic film. Is it a routine sample that contains a crystalline constituent for which a quantitative result can be unambiguously obtained by measuring a single peak intensity on the original sample? Use the diffractometer. Samples are seldom that simple. Some are best treated by both methods: for example, an exploratory film exposure followed by an intensity determination on the diffractometer.

The examples cited do show the principal advantages of the photographic method (complete spectrum at once; easily storable permanent record) and of the diffractometric (speed; no film to develop; easy and reliable measurement of peak intensity.) Speed is not a simple matter: the time for sample preparation, and for calculation and interpretation of results must be

AN ASSESSMENT OF X-RAY DIFFRACTION METHODS

included: working time is one thing; elapsed time, another. Also, a single x-ray generator can accommodate more than one camera; and such a generator need not be nearly as stable as one that serves a diffractometer. For complex unknown samples, the complete diffraction pattern must usually be interpreted; even if the sample contains only a single crystalline phase, identification is most directly accomplished by matching its powder diffraction photograph with that obtained on a standard under similar conditions. The discussion could easily be prolonged. It seems to come to this. All analytical laboratories of any size should have both diffraction methods available. The diffractometer and the x-ray emission spectrograph go well together. A laboratory that has neither may prefer to begin with the H-D-S method, for which the capital investment is the lowest. If this route is followed, the equipment (especially the x-ray source) should be bought with later addition of diffractometer and x-ray emission spectrograph in mind.

6.17 Illustrative Applications

There is not room here for a comprehensive list of the many different applications that have been made of diffraction methods. The reader can find such information in Refs. 4, 5, and 19 (particularly comprehensive). For recent work, see the well known *Advances in X-Ray Analysis*, Plenum Press, and especially the excellent reviews in *Analytical Chemistry* by Merritt and Streib [22]. The few applications discussed below have been selected to show how diffraction methods fit into the scheme of analytical chemistry.

We shall classify the possible applications according to whether diffraction methods alone are *necessary and sufficient; sufficient and convenient,* or *insufficient.* For applications of the first class, diffraction methods are indispensable; for the second kind, they must compete with others; in applications of the third kind, they must be helped by others.

NECESSARY AND SUFFICIENT. Here determinations of interplanar spacings and intensities give the only satisfactory answer. The determination of graphite in carbon, and the proof that diamonds were man-made in the early experiments at the General Electric Company are simple examples. An important application, useful here as an illustration because it is related to Sections 6.14 and 6.15, is the determination of silica in siliceous dusts.

Silicosis as an industrial hazard is of course the reason why this determination has become important. See Ref. 19, pp. 199, 435, 442, and 536; and Ref. 4, pp. 430–433. The health problem itself is highly complex; the words "necessary and sufficient" are used here only in connection with the determination, there being no intent to imply that a satisfactory determination is more than a necessary first step in solving the health problem. This much is certain: as quartz, the most abundant crystalline form of silica in the dusts, is the main culprit, the chemical methods of determining silica are useless. Furthermore, diffraction methods are indispensable in establishing whether the other crystalline forms are entirely free from blame.

To illustrate this diffraction method, we use an example given by Klug and Alexander [4]. The first step was the making of the diffractometric trace in Figure 6.17–1, which established (1) the presence of quartz in a concentration near 21 w/o, (2) the presence of kaolinite and illite as major constituents, and (3) the absence of lines near the fluorite peak at 3.16 Å, which suggests the use of that substance as internal standard. Accordingly, the proper amount of fluorite was added to the sample, and the final quartz determination made by manual counting in the manner of Section 3.15. The reported result was 19 w/o quartz, suspected of being slightly high owing to partial superposition of the 10.2-Å (III) illite reflection on the 3.34-Å quartz peak. (Notice that these are *d spacings*, not *wavelengths*, hence 10.2/3 should equal 3.34/1; λ is fixed.)

SUFFICIENT AND CONVENIENT. Here we mention diffraction determinations on soils; in such determinations, the complications (isomorphism, etc.), cited by Frevel (see above) seem not to be generally troublesome.

The nature of a soil is largely fixed by its minerals, especially by those in the clay fraction, in which primary and secondary silicate minerals are of first importance. After pointing this out, Whittig [23] goes on to say, "x-ray diffraction has contributed more to mineralogical characterization of clay fractions of soils than has any other single method of analysis." The great care needed in pretreating soil samples shows the complexity of this kind of geochemical analysis. Kunze [24] states that removal during pretreatment of free iron oxide, of soluble salts, of carbonates, and of organic matter from a soil is either desirable or necessary for each of the constituents named if good x-ray diffraction patterns are expected. A monograph [25] is a guide to earlier work in this field.

A more specific example is the identification by the "fingerprint method" of various organic reaction products by Meyers, Warwas, and Hancock [26] in a study of the relative reactivities of various solid benzoic acids. Powder patterns were taken on a Debye-Scherrer camera (diameter 57.3 mm; Mo Kα; Zr filter) for each of the individual acids and salts; for the mixtures of reactants; and for the expected products. By matching of the proper known and unknown patterns, the substances that produced the latter could in all cases be identified, which made it possible to establish what (if any) reaction had occurred when various solid benzoic acids and various solid potassium benzoates were mixed, one with another.

INSUFFICIENT. An example listed as Case 2 by Frevel [21] is an excellent illustration to show that a knowledge of composition, obtained here by x-ray

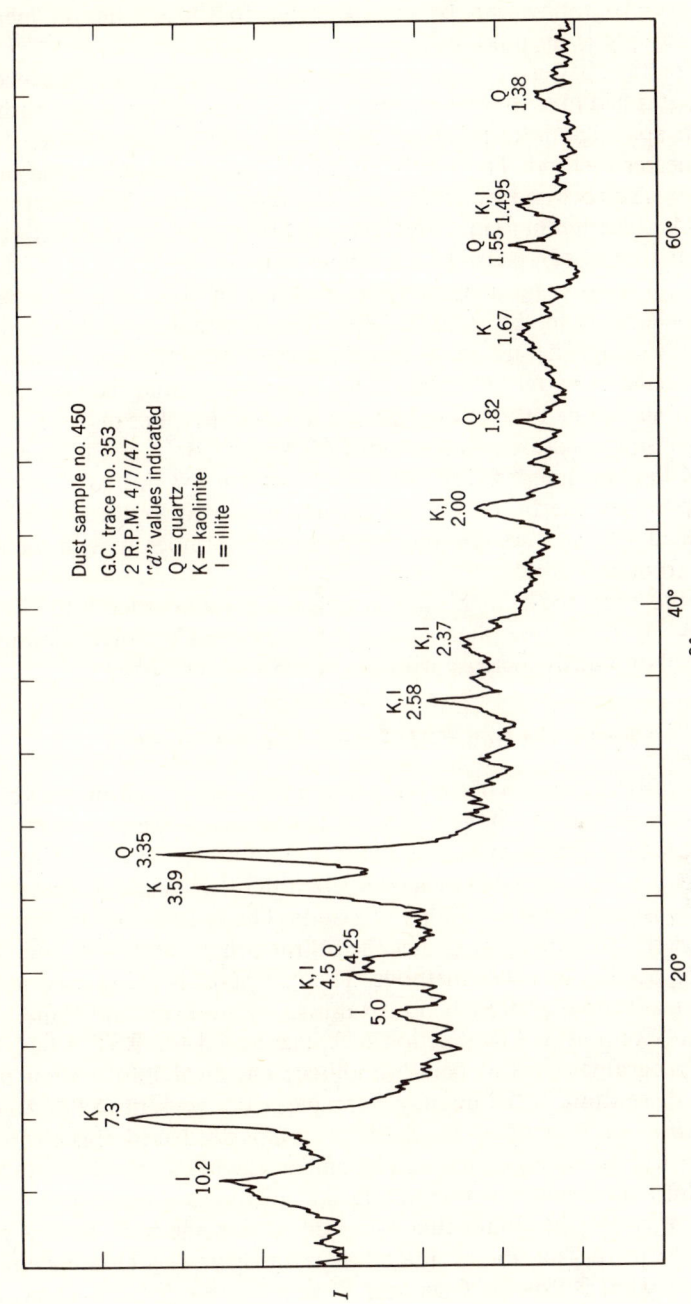

Fig. 6.17-1. Exploratory diffractometric trace by Klug and Alexander on Industrial Hygiene Foundation Sample 450, a siliceous dust of average particle diameter below 5μ. Fine dusts, other things equal, are the most serious silicosis hazards. After Klug and Alexander, Ref. 4, p. 431.

emission spectrography, can be indispensable to the diffraction method when the sample is an unknown.

The diffraction pattern of this sample, a mixture of two known powders, was photographed in a Guinier camera (see below) in order to establish the interplanar spacings more precisely than is possible via the routine Hull-Debye-Scherrer method. The results of measurements made on the developed film were processed by computer. The most intense line occurred at $d = 3.121$ Å; the attempt to identify the crystal that gave this line encounters the obstacle that *eleven* possible candidates exist in the group $d = 3.14$ to 3.10 Å. X-ray emission spectrography showed Ag, Zn, Br, S, and Cl to be the important elements in the sample. With this information at hand, the d values for the sample left no doubt that β-ZnS (sphalerite) was a constituent. With the sphalerite pattern eliminated, the film showed that the other phase was also cubic and had the lattice spacing $a = 5.637$ Å. Because this phase had the rock salt structure and contained Ag, Cl, and Br, it had to be a solid solution of the two halides. The theoretical pattern for this solid solution was calculated by use of a computer program due to Smith (see below), and the calculated d-values and intensities agreed completely with the relevant data from this film.

The need to know the elements in a sample if x-ray diffraction is to identify successfully the crystalline phases present and to establish their amount—this need will obviously increase with the number of such phases.

6.18 The Dow *ZRD* Search-Match Program

The illustrative examples of the foregoing section show that the risk of relying on x-ray diffraction alone increases as one leaves the necessary-and-sufficient category of determinations. This is but another way of saying that x-ray diffraction methods must then compete with other methods of analysis, both in the cost and in the reliability of results. The last determination cited was done in the traditional way: get the diffraction pattern first, and then seek help as needed by other methods. The computer is being used to expedite this traditional approach; for example, by Johnson and Vand [27] on the Joint Committee File (Section 6.3), also called the ASTM file. The computer programs are still being modified; chemical information now appears in the output. If the unknown sample is not present in the ASTM *Powder Diffraction File*, upon which the programs are based, this chemical information will be incorrect, but still helpful as a guide to listed substances that give diffraction patterns resembling that of the sample.

There is of course no doubt that x-ray diffraction methods have a permanent place in the scheme of analytical chemistry. The Dow Chemical Company, in what is now its Chemical Physics Research Laboratory, has played a major part in bringing this about. Late in 1968, powder diffraction samples were being done at an annual rate near 3000, and samples by x-ray

AN ASSESSMENT OF X-RAY DIFFRACTION METHODS

emission spectrography at an annual rate near 4200, in this laboratory at Midland, Mich. [28]. A reversal of the traditional approach in x-ray diffraction by this laboratory is a noteworthy step, which L. K. Frevel describes as follows:

> In view of the observation that any dependable powder diffraction analysis of a genuine unknown is supplemented or confirmed by a semiquantitative or quantitative spectroscopic analysis, it becomes expedient to reverse the procedure and obtain the element data first to facilitate finding the appropriate diffraction standards. Moreover, in most routine chemical identifications by the powder method, no formal quantitative comparison is made of the data in the manner illustrated [herein]. Usually the analyst relies on visual inspection of the d-values on one or more ASTM cards and concerns himself only with the approximate order of the three most intense lines for each phase identified. Direct comparison of films taken under equivalent conditions is convenient and rapid but suffers from subjectiveness and lack of numerical records. Expertise in this field is acquired by experience coupled with a reasonable knowledge of the optical principles of x-ray diffraction. Nonetheless, this type of qualitative analysis remains subjective and makes phase identification by x-rays as much of an art as a science. This situation will probably change with the accumulation of reliable quantitative powder patterns issued by the National Bureau of Standards (NBS Circular 539 and NBS Monograph 25) and with the availability of precision cameras with monochromators for Cu $K\alpha 1$. It is thus timely to consider computer programs to carry out the entire searching and matching process by fast digital computers.

This thinking led to the Dow *ZRD* Search-Match program [21], which can only be outlined here.

The *ZRD* Search-Match program is carried out by computer with an input that as a *maximum* includes the following three kinds of data:

1. $\{Z\}$, the atomic numbers of all elements in the sample in the range $Z = 3$ to 93, *excluding* C, N, O, and F. Hydrogen is of course excluded also. These five ubiquitous elements are difficult to identify by spectrographic methods.

2. $\{R\}$, the codified polyatomic groups (radicals, complex ions, ligands, clathrates) in the sample. Inclusion of these groups is desirable because specific information about them is easily obtained by other methods such as infrared absorption.

3. $\{D\}$, the powder diffraction data $\{v, d_v, I_v\}$, where v is the number that distinguishes a diffraction line.

The *minimum* input data are $\{Z_0\}$ and $\{D\}$, and they include only the atomic numbers of the principal elements in the sample. A vital question: What is an acceptable match between input data and data stored in the computer? The question is complex because its answer involves not only the reliability of the input $\{D\}$—about which the analyst has some knowledge and over which he can exercise some control—but also the reliability of the stored data, which is highly variable and often unknown. In conformity with the instructions given it, the computer *searches* its memory for data that *match* the input data within the limits imposed, and then presents as output the crystalline phases found in the sample.

To show what the *ZRD* Search-Match program can accomplish, we use Frevel's Sample U-3.6 [21, Table XI]:

1. Principal elements present by x-ray emission spectrography: Y, As, Ag, Pb, Se, and K.
2. Output of program: Y_2O_3, As_2O_3, Ag_3AsO_4, KH_2AsO_4, and $PbSeO_4$ were identified as crystalline phases present.
3. Presence of minor phases suggested by 16 weak unidentified lines.

The great advantage of the *ZRD* Search-Match program is that it properly puts the horse before the cart in the analysis of wholly unknown samples. Take Sample U-3.6. The data from x-ray emission spectrography might have been semiquantitative (easily obtained) or quantitative (more costly). With these data in hand, the need to know the crystalline phases in the sample could have been evaluated. If this need warranted diffraction work, such work would then be done, and it could be done with the assurance of minimum ambiguity in the interpretation of results.

To operate at full effectiveness, the *ZRD* Search-Match program needs certified standard patterns of greater precision than those now available. In the meantime, the ASTM file can be used with risks of accidental matches for mixtures that contain more than three crystalline components.

6.19 Automation and Computerization. Camera Method.

A camera method for powder diffraction is inherently more difficult to automate than is the diffractometric, in which exposure times are shorter and no film needs to be developed. Frevel [29] has however succeeded in automating the measurement of powder patterns on film to produce digital d and I computer output readings that could then be used, if desired, as input for the computerized identification of the crystalline phases present. Three pieces of equipment were used: the 24–3404 Jarrell-Ash microphotometer, which measures the displacements on the film that give d values for the sample; the Bristol Dynamaster potentiometer recorder, which measures the transmittances that fix line intensities; and the IBM 526 card punch that (after the necessary encoding) presents the output data.

The camera used by Frevel [30] is of greater interest here than are the details of the equipment just mentioned above. In this, an AEG Guinier double-cylinder camera (114.7 mm in diameter), the Cu $K\alpha1$ line selected by a bent-crystal monochromator (the Johansson arrangement; see Chapter 5) was used to obtain d-spacings and intensity data more precisely than is possible in the Hull-Debye-Scherrer method. See Figure 6.19–1. The

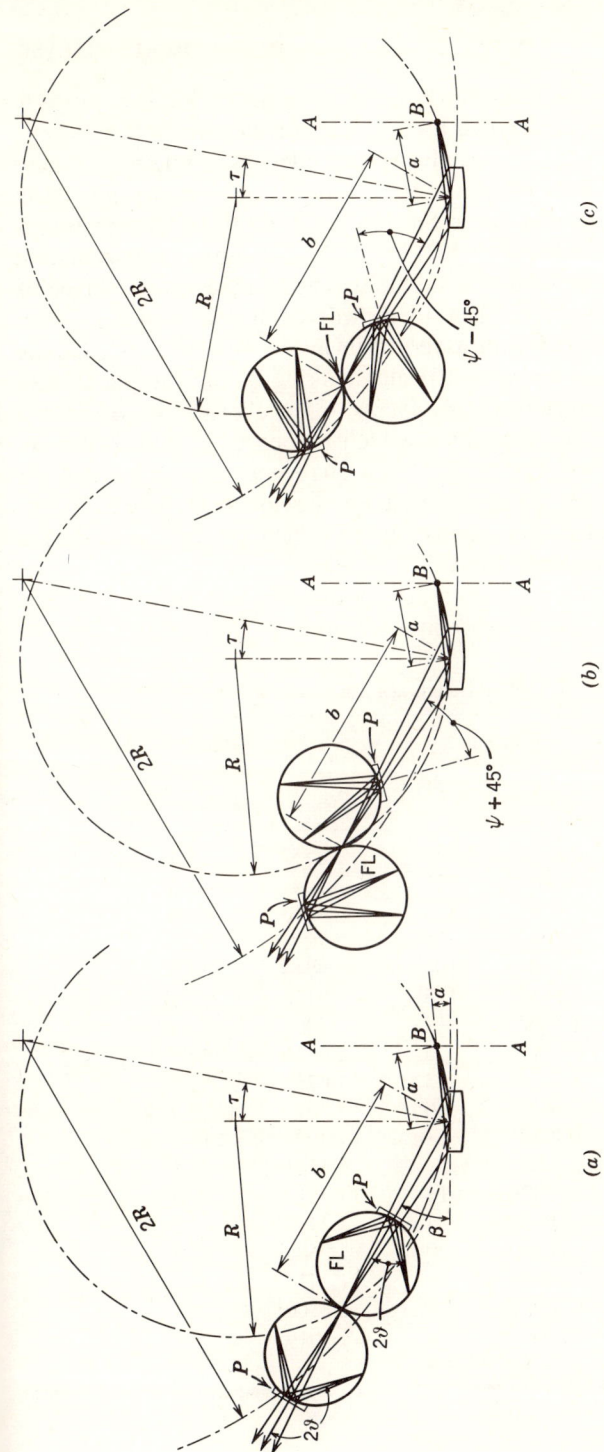

Fig. 6.19–1. Optical diagram for AEG double-chamber Guinier-type camera. A fine-focus x-ray tube (axis A–A) generates a beam from which the bent-crystal monochromator (Johansson arrangement) selects a line (e.g., Cu Kα1) that penetrates the first sample P as a convergent beam that comes to a focus at FL at the wall of first chamber, whereupon it diverges to produce back-reflected x-rays from the second sample P at the wall of the second chamber. In this way, diffraction lines are produced simultaneously in the first chamber (low θ) and in the second (high θ). The camera can operate with the chambers in the three positions shown. With this equipment, 0.01 to 0.02% Al_2O_3 can be detected in SiO_2, and 0.01% SiO_2 in Al_2O_3. See the discussion of parafocusing in Chapter 5. After Hofmann and Jagodzinski, Ref. 30, p. 603. (a) Symmetrical position (Beam axis coincides with normal to sample). (b) Asymmetric position (Beam axis 45° above normal to sample). (c) Asymmetric position (Beam axis 45° below normal to sample).

increased precision results because (1) Cu **K**α1 is better for this purpose than the Cu **K**α doublet even when the average wavelength of the doublet is used in the calculations, (2) the absorption effect is negligible because the sample is so thin; with parafocusing geometry, any absorption effect is independent of θ as pointed out in Section 6.6, and (3) the background is greatly reduced because only the Cu **K**α1 line enters the camera; in particular, the continuous x-ray spectrum is eliminated. The sample is introduced as follows. A small amount thereof as a fine powder is dusted onto aluminum foil 0.008-mm thick covered with a very thin layer of petrolatum. Gentle tapping of the sample holder and delicate stroking of the sample surface with cleansing tissue then remove any excess to leave a randomly oriented powder layer less than 0.01-mm thick, which is then oscillated during an exposure of several hours. Patterns of high quality are thus obtained for which the aluminum lines serve as reference; obviously it would be advantageous to have data of this quality on the substances that are listed in the file of the Joint Committee (Section 6.3).

Data for Case 2, Section 6.17, obtained by the best human comparisons at Dow and by the automated procedure are given in Table 6.19-1.

Table 6.19-1. Comparison of Diffraction Data for Case 2

Spacing Number v^a	Human		Automated	
	d_v, Å	I_v	d_v, Å	I_v
1	3.253	63	3.2552	66.7
2	3.121	175	3.1247	347.3
3	2.816	150	2.8203	304.6
4	2.704	75	2.7055	71.5
7	1.993	100	1.9944	113.5
8	1.9132	125	1.9136	237.7
9	1.7004	18	1.7001	11.1
10	1.6311	75	1.6316	112.1
11	1.6270	15	1.6258	17.8
12	1.5620	10	1.5617	8.9
14	1.4093	13	1.4096	10.0
15^b	—	—	1.3741	3.1
16	1.3532	15	1.3527	11.9
17	1.2924	3	1.2939	2.5
18	1.2610	20	1.2606	18.3
19	1.2416	40	1.2415	37.9
21	1.2097	6	1.2100	4.7
23	1.1504	10	1.1510	11.4

[a] Spacings v = 5, 6, 13, 20, and 22 are not listed because they belong to aluminum reference lines.
[b] The line at v = 15 is spurious.

AN ASSESSMENT OF X-RAY DIFFRACTION METHODS

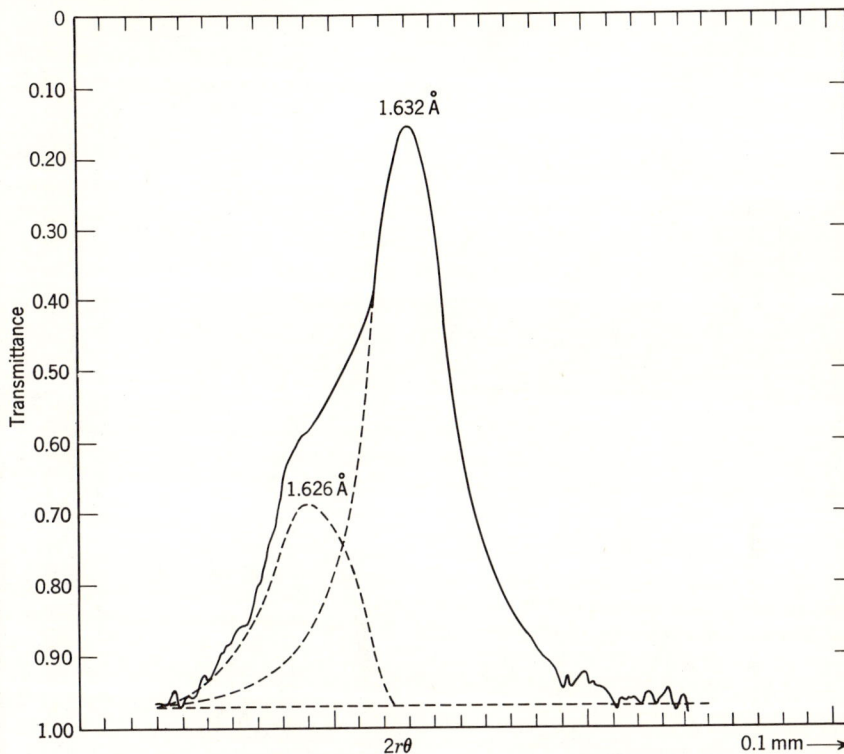

Fig. 6.19–2. A slow scan (0.25 mm/min) of two overlapping Bragg reflections for Case 2. θ is in radians. The two lines in the figure are listed as $v = 10$ and 11 in Table 6.19–1, from which the d spacings in the figure were taken. After Frevel, Ref. 29, Fig. 2.

The automated procedure is quicker, objective, and more reliable than human comparison. Figure 6.19–2 shows the resolving power achievable.

6.20 Automation and Computerization of the Diffractometer

Samples for the diffractometer are adapted to handling by machine; there is no need for lengthy exposures, and no film to develop; the detector gives digital (cps) values of the intensities—these are among the reasons why automation and computerization of a diffractometer system are natural developments. A successful system of this kind has resulted from the joint efforts of the Chevron Research Company, La Habra, Calif., and the Datex Corporation, Monrovia, Calif. [31]. The system is shown in Figure 6.20–1. The reader will recognize many of the components as having been discussed earlier (Chapters 1, 2; and the bent crystal monochromator—apparently

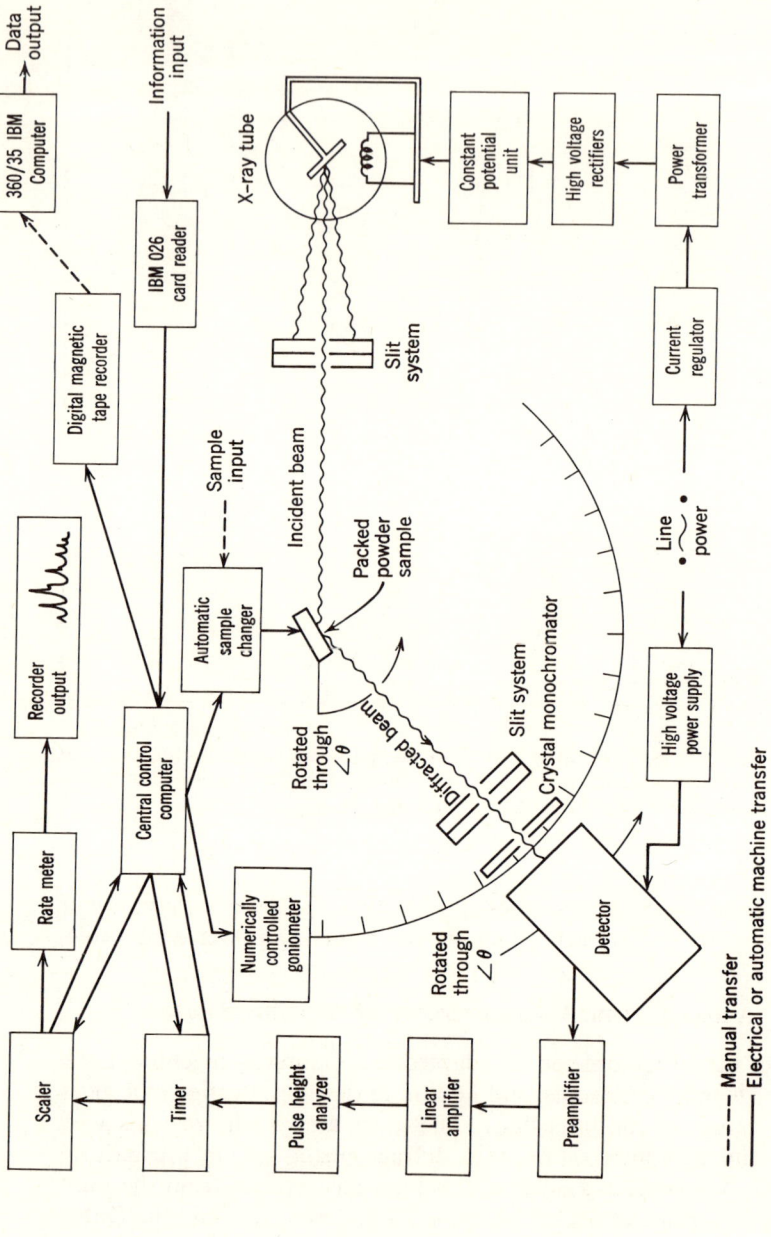

Fig. 6.20–1. Automated and computerized x-ray diffractometer system of the Chevron Research Company. Among the components: General Electric XRD-5 power supply; transistorized scaler and rate meter; high-intensity copper-target x-ray tube (40 mA; 50 kV); Tempres tube-mount; Datex central control computer. Encoderdyne drive motor, 2θ display system; Precision Instrument Co. incremental tape recorder; Leeds and Northrup analog recorder; D. and O. Machine, Inc, automatic sample changer; Tectronix oscilloscope for monitoring shape of pulses from proportional detector; modified IBM 026 card reader/punch, which transmits operating instruction for storage in the computer's memory and can be

of the Cauchois transmission type—in Chapter 5). The system is controlled by one computer and produces data for final processing by another.

The operating statistics are impressive. In the words of R. W. Rex [31]:

> All operating variables are numerically controlled by the input information on a command card. Once a deck of cards is loaded into the card reader, the samples inserted into the automatic sample changer, and the system started, no further human intervention is needed until the run has been completed. Uninterrupted runs have been as long as 18 days. Samples may be added to the sample changer and cards added to the deck in the card reader during operation, as desired, to permit runs of indefinite length.

After the system has established and subtracted background, it analyzes the net diffraction pattern to recognize peaks, shoulders, and the wing edges ("tails") of diffraction peaks. It next measures peak heights and areas, identifies phases, and calculates amounts present with reference to internal or mutual standards.

The diffractometer system can obviously control processes or materials. By August 1966 it had operated for nearly 6000 hours and had then an availability rating of 98%. The system is said to have sufficient flexibility for use in research and for determinations on unknowns. No analytical results are given.

6.21 Conclusion

The work in the Dow and Chevron laboratories shows that remarkable progress has been made in the generation of x-ray diffraction information for chemical analysis by use of either a camera or a diffractometer. The great need now is for improvement and standardization in information retrieval. The problem is formidable. What computer language to use? Should information about chemical composition be included in the programs? What standards of reliability should be adopted? Should one rely on peak intensities or integrated intensities, or have both available? How allow for differences in experimental method and in equipment? These questions have not yet been definitively answered. At Dow, Frevel and Adams [32] have supplemented the *ZRD* Search-and-Match program with a program based only on $\{d_v I_v\}$. Others, notably Johnson and Vand, who give a general discussion of the problem [27], have similar programs in different computer language. Kehl [33] describes action taken by the Joint Committee. What is done will determine how soon analytical chemistry benefits fully from the computerized retrieval of information.

We close with a mention of the successful computer simulation by Smith [34] and others of x-ray diffractometer traces (I versus 2θ) for powdered samples. That Frevel used one such program in his calculations was mentioned above. The simulation takes into account the factors listed in Section

6.7 as governing the intensities of diffracted lines. The simulated and actual patterns are, even in complex cases, nearly identical. Quartz, for example, has a cluster of five closely spaced lines that appear near $68°(2\theta)$ with Cu $K\alpha$; these are almost perfectly simulated by the computer. Patterns calculated by computer will be increasingly useful as standards.

The reader will appreciate that it has been necessary in this chapter to select for discussion a few papers from the many that deal with the automation and computerization of x-ray diffraction methods. The choice was made in the hope of indicating the significant trends; completeness and definitiveness are beyond reach. The articles by Merritt and Streib [22] are guides to further information.

MISCELLANEOUS METHODS

X-ray diffraction can yield information about topics other than crystal structure (not considered in this book) and composition. With two of these topics—the determination of crystallite size, and the nature of linear polymers and fibers—the analytical chemist needs at least a nodding acquaintance, which the following material is intended to provide. For further information, and for information on the topics not covered, see Refs. 4, 5, 19, and 22.

A substance need not be crystalline to give a diffraction pattern [35]. As one might guess from the nature of the scattering of x-rays by electrons and other scattering centers (Section 1.14), any sort of order—no matter how localized or vestigial—will lead to constructive and destructive interference. Diffraction patterns, even for gases and liquids, consequently result. Such patterns become sharper, the greater the order; ultimate sharpness (lines of minimum FWHM) is reached with a perfect crystal or crystals of greater than a minimum size. The FWHM (see Chapters 2 and 5) for diffraction lines is often expressed in terms of 2θ. For the discussion to follow, the considerations just outlined lead to these conclusions: (1) The amorphous state merges gradually into the crystalline. (2) The smaller the crystallites (small crystals), the broader their diffraction lines. (3) Preferred, as contrasted with random, orientation of ordered regions leads to non-uniform diffracted intensities, recognizable as "graininess" in the photographed diffraction pattern.

6.22 The Determination of Crystallite Size

It is true here as elsewhere that the FWHM of a line is due in part to the apparatus, so that the observed broadening B of a diffraction line reflects both *instrumental* and *pure* diffraction broadening. The latter was shown by Scherrer [36] to be governed by a relation formally identical with

MISCELLANEOUS METHODS

$$md = th = 0.89\lambda/B \cos\theta \qquad (6.22\text{–}1)$$

where m is the number of diffracting planes in a crystallite of thickness th in the direction normal to these planes. In accord with (6.22–1), the broadening B (or FWHM) becomes noticeable when th is near 1000 Å and becomes appreciable near 200 Å. It is given by

$$B = \sqrt{B_S^2 - B_A^2} \qquad (6.22\text{–}2)$$

where B_S is the measured width and B_A is the width of a line without broadening. Measured in radians, B_A is usually obtained from a powder-diffraction pattern of crystals larger than 1000 Å.

Particle-size measurements have a large range of useful applications—paint pigments, carbon for electrical brushes, and metal catalysts being some of the more common. An obvious extension of the method is the determination of particle shape. If all the lines are broadened uniformly, the crystals are more or less equidimensional in all directions. Any plane, or planes, with smaller broadening denote an extension of the crystal in the direction normal to the plane. A plane with more broadening indicates a short dimension. Care must be used in interpreting line broadenings because lattice defects or strains can produce somewhat similar effects.

6.23 X-Ray-Diffraction Studies of Linear Polymers and Fibers

Polymers consist of large molecules built up from one or more simple building blocks. The building process may produce a regular or random structure—linear, planar, or three-dimensional. A high degree of crystallinity may be associated with some of the modes of growth. A typical diffraction pattern obtained from an ordered polymer is shown in Figure 6.23–1 together with the pattern of a purely amorphous material. It is apparent that the ordered polymers are intermediate between crystalline and amorphous. The x-ray pattern provides a way to estimate the degree of crystallinity and to determine the structure in the ordered region.

The degree of order in polymers is not great enough, in general, to provide a pattern to very high values of θ. In most cases the interpretable pattern can all be recorded on a flat plate in the forward-beam direction from the sample. Standard powder equipment is often used and is very satisfactory.

The diffractometer is particularly valuable in following polymerization processes. A metal or glass plate can be dipped into a reaction vessel and immediately put into the sample position. If the equipment is set up to monitor a line that changes during the process, each step can be followed rapidly.

Many linear polymers are noncrystalline, but if they are stretched the molecules line up in the direction of stretch. The diffraction pattern can be used to follow the stretching. Simple devices can be set up directly in the x-ray equipment to control the tension and elongation.

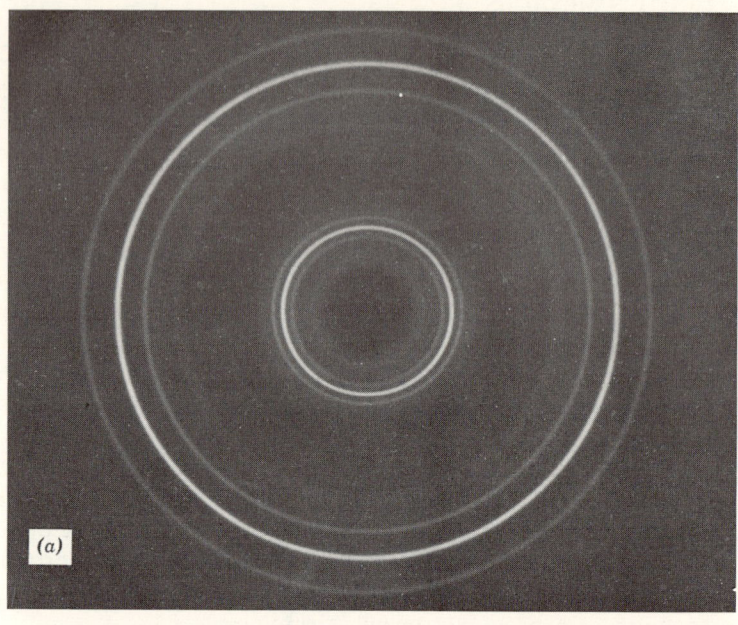

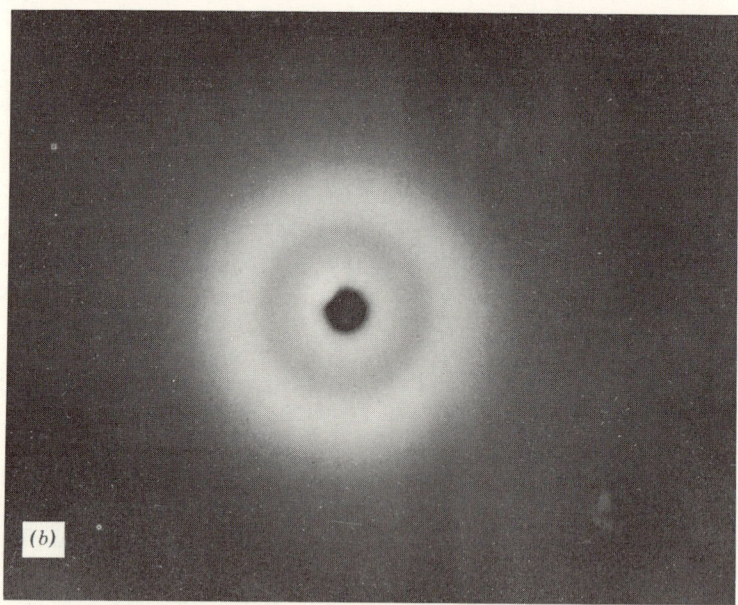

Fig. 6.23–1. Diffraction pattern from (*a*) an ordered (*b*) an amorphous polymer. The pattern from the ordered polymer evidently approaches that from a crystalline material.

The similarity of the techniques of polymer diffraction to those of crystalline-powder diffraction does not carry over to the interpretation of their diffraction patterns. Usually a preferred orientation is present in polymer patterns and the samples must be rotated to bring the planes into diffracting positions. Also, the number of crystal planes developed well enough to give diffraction tends to be small, with a consequent reduction in the number of reflections.

If a diffraction pattern is made of a fiber, or of an orientable sample stretched perpendicularly to the x-ray beam, the pattern will give reflections characteristic of the unit of repeat along the fiber axis. These reflections can be converted to d spacings by Bragg's law. The spacings so calculated agree well with those calculated from the bond lengths of the monomers (polyethylene, 2.5 vs 2.6 Å).

The reflections at right angles to the fiber (or stretch) axis are characteristic of the interfiber packing and have been found to vary regularly as different side groups are substituted on the polymer backbone.

A recent important book by L. E. Alexander in this field [37] is recommended to all readers seeking further information.

REFERENCES

1. A. W. Hull, *J. Am. Chem. Soc.*, **41**, 1168 (1919).
2. B. E. Warren, *J. Am. Ceram. Soc.*, **17**, 73 (1934).
3. *New York Times*, Sunday, August 17, 1969, p. 50.
4. H. P. Klug and L. E. Alexander, *X-Ray Diffraction Procedures*, Wiley, New York, 1954, 716 pp.
5. L. V. Azároff and M. J. Buerger, *The Powder Method in X-Ray Crystallography*, McGraw-Hill, New York, 1958.
6. (Mrs.) K. Y. Lonsdale, *Crystals and X-Rays*, Van Nostrand, New York, 1945.
7. L. K. Frevel, *Acta Cryst.*, **17**, 907 (1964).
8. G. G. Johnson, Jr., and V. Vand, *Advances in X-Ray Analysis*, Vol. 11, Plenum Press, New York, 1968, p. 376.
9. *International Tables of X-Ray Crystallography*, Kynoch Press, Birmingham, England; Vol. 1, 1952; Vol. 2, 1959; Vol. 3, 1962.
10. B. C. Giessen and G. E. Gordon, *Science*, **159**, 973 (1968).
11. A. W. Hull, *Phys. Rev.*, (a) **9**, 84; (b) **9**, 564 (1917); (c) **10**, 661 (1917).
12. P. Debye and P. Scherrer, *Phys. Z.*, (a) **17**, 277 (1916); (b) **18**, 291 (1917).
13. H. Friedman, *Electronics*, April 1945, p. 132.
14. J. D. Hanawalt, H. Rinn, and L. K. Frevel, *Ind. Eng. Chem., Anal. Ed.*, **10**, 457 (1938).
15. M. Straumanis and A. Ievinš, *Die Präzisionsbestimmung von Gitterkonstanten nach der asymmetrischen Methode*, Julius Springer, Berlin, 1940; reprinted by Edwards Brothers, Ann Arbor, Mich., 1948.
16. L. Alexander and H. P. Klug, *Anal. Chem.*, **20**, 886 (1948).

17. G. W. Brindley, *Phil. Mag.*, **36**, 347 (1945). The subject of microabsorption has a complex and confusing history.
18. G. L. Clark and D. H. Reynolds, *Ind. Eng. Chem., Anal. Ed.*, **8**, 36 (1940).
19. G. L. Clark, *Applied X-Rays*, 4th ed., McGraw-Hill, New York, 1955, especially pp. 429–435.
20. L. K. Frevel, *Ind. Eng. Chem. Anal. Ed.*, **16**, 209 (1944).
21. L. K. Frevel, *Anal. Chem.* **37**, 471 (1965).
22. L. L. Merritt, Jr., and W. E. Streib, (a) *Anal. Chem.*, **36**, 399R (1964); (b) **28**, 493R (1966); (c) **40**, 429R (1968).
23. L. D. Whittig, *Agronomy*, **9**, No. 1, 671 (1965).
24. G. W. Kunze, *Agronomy*, **9**, No. 1, 568 (1965).
25. G. W. Brindley, Ed., *X-Ray Identification and Crystal Structure of Clay Minerals*, The Mineralogical Society, London, 1951, 345 pp.
26. E. A. Meyers, E. J. Warwas, and C. K. Hancock, *J. Am. Chem. Soc.*, **89**, 3565 (1967).
27. G. G. Johnson, Jr., and V. Vand, (a) *Ind. Eng. Chem.*, **59**, 18(1967); (b) *Advances in X-Ray Analysis*, Vol. 11, Plenum Press, New York, 1968, p. 376.
28. L. K. Frevel, letter to H. A. L., dated October 9, 1968.
29. L. K. Frevel, *Anal. Chem.*, **38**, 1914 (1966).
30. E. Hofmann and H. Jagodzinski, *Z. Metallk.*, **46**, 601 (1955).
31. R. W. Rex, *Advances in X-Ray Analysis*, Vol. 10, Plenum Press, New York, 1967, p. 366.
32. L. K. Frevel and C. E. Adams, *Anal. Chem.*, **40**, 1335 (1968).
33. W. L. Kehl, *ACA Newsletter*, June 1966, p. 3.
34. D. K. Smith, *Norelco Reporter*, April–June 1968, p. 57. See also the following reports of the Lawrence Radiation Laboratory, Livermore, Calif.: UCRL-7196; URCL-50264; UCRL-70078; UCRL-70674.
35. P. Debye, *Ann. Physik*, **46**, 809 (1915).
36. P. Scherrer, *Göttinger Nachrichten*, **2**, 98 (1918).
37. L. E. Alexander, *X-Ray Diffraction Methods in Polymer Science*, Wiley, New York, 1969.

Chapter 7

Measurement of Film Thickness Simple Trace Determinations

7.1 Introduction

It is not immediately obvious why the measurement of film thickness may logically be treated at this point, where the description of absorptiometric methods is largely completed and the discussion of methods involving x-ray emission is about to begin. The measurement of film thickness can be done by methods of both kinds, and this chapter can therefore serve as a convenient bridge.

More than convenience is involved; the bridge is important because it leads gradually into the full complexity of x-ray emission spectrography. When film thickness is being determined, composition must be known; in *trace determinations*, thickness and composition may both be variable. We are concerned here only with *simple* trace determinations, by which we mean those in which *interelement effects* are absent. Such determinations are x-ray emission spectrography of the simplest kind.

Recapitulation and further explanation are needed. That the intensity of an analytical line in x-ray emission spectrography will depend on the amount present of the element sought (denoted by **E**) is a commonplace demonstrated in elementary fashion in Section 4.4. The sample S is composed of **E** in a matrix M, the composition of which can also influence the intensity of the analytical line. These influences depend in the first instance on how the sample S (**E** included) absorbs x-rays. Because M and **E** both contribute to this absorption, we henceforward call these influences *interelement* (*not* matrix) *effects*. There are three such effects, two of which have appeared in Chapter 6—namely, *positive* absorption effects, in which the relevant line-intensity is *increased* owing to a change in matrix composition (6.13–13); and *negative* absorption effects (6.13–14), in which that intensity is *decreased* by such a change. The third interelement effect is the *enhancement effect*, postponed for later discussion because it is not generally important here. What is meant by simple trace determinations should now be clear.

The films of this chapter rest upon substrates that may have nothing else in common with the film. We are thus not concerned with thickness gauging, as of steel strip, which was discussed in Chapter 3: the principles that govern x-ray absorption and the behavior of polychromatic beams are of course important here as there. The concept of *critical thickness* (Section 6.13) is important here and in x-ray emission spectrography further on.

As usual, the theoretical treatment here will be no more complex than is necessary. Fuller theoretical treatment, such as will be needed for x-ray emission spectrography, was carried through by Shiraiwa and Fujino [1] for the methods of this chapter, and the resulting equations were experimentally verified.

X-ray diffraction was the first x-ray method to be applied to the measurement of film thickness. Shortly after Clark, Pish, and Weeg [2] had obtained diffraction patterns from coatings as thin as $5(10^{-6})$ cm, three successful laboratory methods were developed for estimating thickness by use of diffraction techniques. Birks and Friedman [3, 4] used a Geiger detector to measure attenuation by a coating of a suitable x-ray line diffracted by the substrate. Gray [5] and Eisenstein [6] independently developed photographic methods in which thickness was determined by comparing the integrated intensity of a line diffracted by the coating with that of another line diffracted by the substrate. It is now certain that x-ray diffraction cannot generally compete in the measurement of film thickness with the methods described below.

THICKNESS OF PLATING

7.2 Attenuation of an Unresolved Beam from the Substrate. Tin Plate

Tin being costly, the thickness to which it is plated on steel must be rigorously and continuously controlled. That this is done by x-ray methods is important and of historical interest, for the success of the first such method in the 1940s may be regarded as having revived interest in x-ray absorption and emission in chemical analysis [7].

In 1950, Beeghly [8] published results to show that the thickness of tin plate could be measured by the attenuation of an unresolved x-ray beam resulting from the absorption of a polychromatic beam from a copper-target tube by the iron substrate (Figure 7.2–1). Two years later, Pellissier

THICKNESS OF PLATING

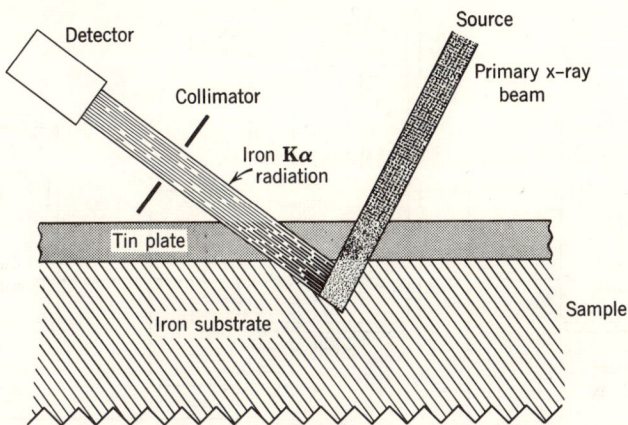

Fig. 7.2–1. Diagram of Beeghly's experiment. Note that the polychromatic x-ray beam is attenuated and filtered; and that the substrate beam, being largely iron $K\alpha$, is attenuated in the main. A manganese filter in the emergent beam will increase its spectral purity.

and Wicker [9] revealed that a similar method had been under investigation in the United States Steel Company since 1946.

This application may be described by saying that x-ray excitation converts the substrate into an x-ray source that yields a (nearly) monochromatic beam (Fe $K\alpha$, β) for absorptiometry of the tin plating. The substrate acts as a characteristic-line generator for the thickness determination. X-rays other than the characteristic lines of the substrate are undesirable if they are incident at appreciable intensity upon the detector. Among x-rays that could distort the results are diffraction peaks and characteristic lines of the film. The detector should be positioned so that the former do not enter, and the latter can be avoided in this case by choosing the x-ray tube voltage high enough to excite the characteristic lines of the substrate but low enough to prevent excitation of interfering lines from the film. The excitation potentials of interest are: iron **K** spectrum, 7.11 kV; tin **K** spectrum, 29.2 kV; voltage used, 20 kV. The **L** lines of tin (wavelength 3.0 Å and longer), which are excited in the sample at 4 kV and above, are absorbed strongly enough by the air in the optical path to be practically unimportant.

Philips Electronic Instruments has played an important role in the further development of the method, and Figure 7.2–2 is a schematic diagram of a recent Norelco Tin Coating Weight Gauge [10, 11]. Noteworthy features follow.

1. The *main unit* (one constant-potential x-ray generator and one recording console for each gauge head) supplies power up to 6 W (15 kV at 0.4 mA) and records tin-plate thickness directly in *pounds per base box* (lb/bb).

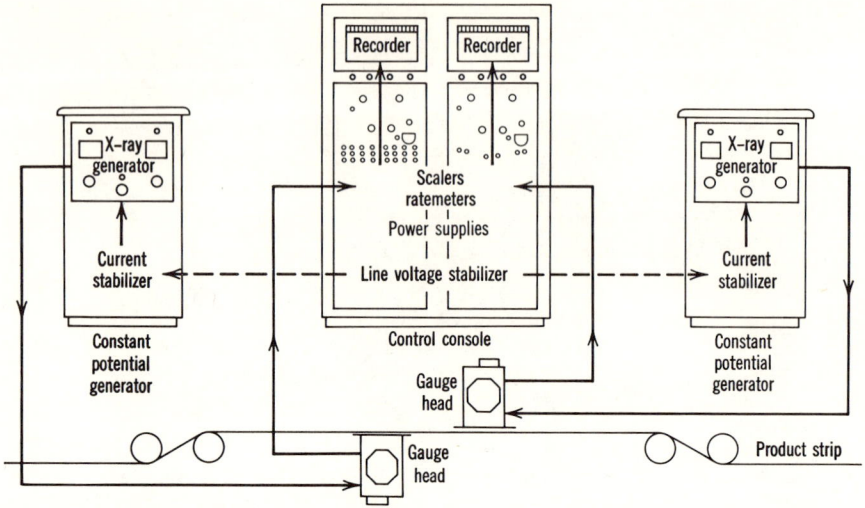

Fig. 7.2–2. Schematic diagram of Norelco Tin Coating Weight Gauge as shown in Philips Electronic Instruments Circular RC 389 3MF 762. The diagram may serve also for the Norelco Zinc/Aluminum Coating Weight Gauge, which differs in certain important respects from the gauge for tin. See text. Figure reprinted by permission.

(One pound per base box is about 11 g tin/m^2, or a plate about $1.8(10^{-4})$ cm thick.)

2. As shown, one gauge head scans the upper side of the strip while the other scans the lower. The strip may be up to 40 feet from the recording console. The heads can traverse strips up to 48 in. wide, and photoelectric sensors automatically reverse the scanning direction when the head arrives within 2 or 3 in. of the strip edge.

3. Each gauge head contains an air-cooled tungsten-target x-ray tube, two scintillation detectors, two preamplifiers, and a built-in reference standard. (Again, the comparative method.) It also contains a special optical system that compensates for fluttering ($\pm 1/8$ in. from the mean position) of the strip.

4. The output from each detector is fed into the recording console through a linear amplifier, pulse-height discriminator, three scaling circuits, and a rate meter. The output of the rate meters appears on a calibrated strip chart as pounds per base box per side.

5. Performance data: Over the entire range from 0.1 to 0.55 lb/bb/side, precision is ± 0.01 lb/bb/side, as is accuracy. The time constant is 1 sec. Recalibration required at intervals not to exceed 12 hours.

6. The satisfactory precision and accuracy are evidence that difficulties arising from the unavoidable oscillations of the tin-plate strip have been successfully overcome.

7.3 Attenuation of a Characteristic Line from the Substrate. Various Examples

ZINC OR ALUMINUM ON STEEL. In the method just described, the spectral purity of the emergent beam can of course be greatly improved by Bragg reflection. When interfering characteristic lines are unavoidably excited in both film and substrate, such wavelength resolution is desirable even though the concomitant reduction in cps requires increased intensity in the polychromatic beam from the x-ray tube.

Success in the x-ray control of tin-plate manufacture soon led to demands for similar control of the hot-dip galvanizing of steel and of its coating with aluminum [12]. The zinc problem is more difficult than that of tin-plate control because the characteristic lines of zinc are excited (see foregoing paragraph) and because x-ray absorbance for a given thickness is much lower than for tin. In the case of aluminum, the characteristic lines, though excited, are readily absorbed by air, but the absorbance of Fe Kα by the plating is even lower. High precision is thus more difficult to attain for zinc and for aluminum coatings than for tin plate, and the background (especially in the case of aluminum) makes matters worse.

Nevertheless, Philips offers a zinc/aluminum gauge of acceptable precision and accuracy—say, 2% of coating weight, but the equipment is more complex. The x-ray optical system in the gauge head contains a *water-cooled* tungsten-target tube (to give the higher beam intensity required), a Bragg crystal, and collimators, these being the most important differences from the tin-plate gauge, of which the zinc/aluminum gauge is a modification.

LABORATORY INVESTIGATION OF THE METHOD. In these applications, a polychromatic x-ray beam passes through a film of one metal, which it may excite; is attenuated and filtered; and excites a characteristic line in another metal (the substrate), which characteristic line may excite a characteristic line of the film, be attenuated, and perhaps diffracted on its way through the film to the detector. It is not obvious why the measured intensity under these complex conditions should be simply related to film thickness.

Accordingly, Zemany and Liebhafsky [13, 14] investigated the method on silver and on zirconium as substrate metal covered with a "plating" built up of iron foil 0.6-mil (0.00152-cm) thick. The substrate metals were chosen because their Kα lines, although differing considerably in wavelength, can yet excite the K lines of iron, such excitation being desired to make the problem complex. The use of foil made it easy to obtain "plating" thicknesses of high precision.

The times for 2^{14} counts were measured with a Geiger detector for each substrate bare and covered with different thicknesses of iron foil. One background count was determined for iron near the wavelength of each

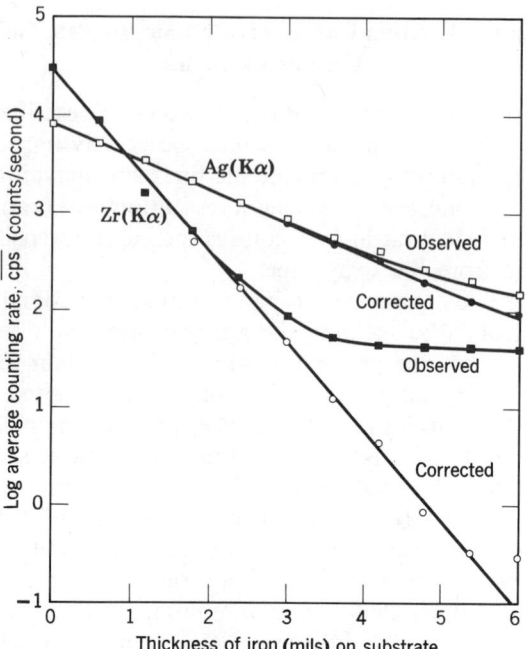

Fig. 7.3–1. Experimental results, as observed and after correction for background, for iron foil upon silver or zirconium. The ordinate is logarithmic, which means that the corrected results obey Beer's law. The tungsten-target tube was operated at 50 kV and 50 mA. Wavelength resolution was accomplished by Bragg reflection from a flat crystal of lithium fluoride. Note the large deviation of the corrected Zr Kα datum for greatest thickness.

substrate Kα line, and the corrected counting rate for each experiment was obtained by subtracting the background counting rate from the observed. The growing importance of the background at increasing thicknesses of iron foil is clear from Figure 7.3–1, where average counting rates, observed and corrected, are plotted as logarithmic ordinates against "plating" thicknesses.

Virtually all the corrected average counting rates lie close to a straight line for each substrate; this means that Beer's law is obeyed in each case over almost the entire thickness range. With zirconium, where the corrected counting rate varies through five powers of ten, the concordance of the results is especially impressive even though it is fortuitous at the two greatest thicknesses owing to the relatively large, unavoidable fluctuations in the background.

The experimental results in Figure 7.3–1 are clearly in accord with Beer's law as expressed in (1.10–1), (1.10–2), and (1.10–3); that is to say

THICKNESS OF PLATING

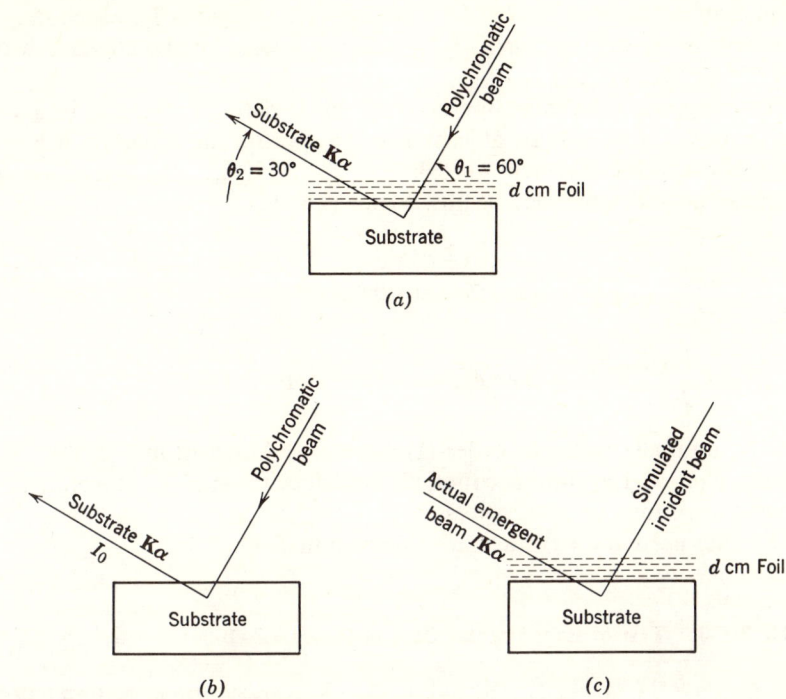

Fig. 7.3–2. Beer's law and Figure 7.3–1 *(a)* Experiment as carried out. *(b)* Determination of I_0 as carried out. *(c)* Experiment simulated as analog of *(a)* to rationalize conformity with Beer's law of corrected results in Figure 7.3–1. Corresponding angles are identical in *(a)*, *(b)*, and *(c)*. The thickness d of foil is measured vertically, whence the total beam path in foil is $d\,(\csc\theta_1 + \csc\theta_2)$. Beams are treated as rays, and the surfaces are plane. In *(c)*, both beams are the Kα line of the substrate.

$$\ln(I_0/I) = (\mu_1 \csc\theta_1 + \mu_2 \csc\theta_2)\rho d \qquad (7.3\text{–}1)$$

where $d\csc\theta$ is the path length of the beam in the foil, and each product $\mu\rho$ is a *linear* absorption coefficient. See Figure 7.3–2. Clearly, the simplest possible relationship connects intensity and film thickness d.

However, this is no ordinary case of Beer's law. The surprising validity of this law makes sense if one regards I_0 as that part of the incident beam capable of exciting the Kα lines of the substrate, and then assumes I_0 to be incident upon the iron foil at angle θ_1 as shown in Figure 7.3–2. This assumption is in accord with the way I_0 was established—that is, as a corrected counting rate for the base substrate. (We may ignore as trivial the inconsistency that the background counting rate used to obtain the observed counting rate was measured on iron; this procedure is correct when the substrate carries iron foil.) The intensity of the incident polychromatic beam does not appear in (7.3-1); what appears instead is I_0, which may be regarded as an *effective intensity*, a concept related to the effective wavelength (see Section 3.8).

The foregoing rationalization becomes more plausible if it can be shown that μ_1 and μ_2 in (7.3-1) are at least near the appropriate values for iron. For this purpose, it is convenient to replace μ_1 and μ_2 by μ_{avg}, an overall average for incident and emergent beams. From (7.3-1),

$$\mu_{avg} = \frac{2.303 \, Sl}{(3.155)(7.86)(0.00254)} = 36.56 \, Sl \quad (7.3-2)$$

where

$$Sl = \frac{\log(I_0/I)}{d \text{ (in mils)}} = \text{the negative of the slope in Figure 7.3-1} \quad (7.3-3)$$

and the numbers taken in order (1) convert the logarithms, (2) take path geometry (θ_1 and θ_2) into account, (3) introduce the density of iron, and (4) convert from mils to centimeters.

The data needed for the test are collected in Table 7.3-1.

Table 7.3-1. Test of Experimental Data in Figure 7.3-1

Substrate	Sl (7.3-3)	μ_{avg}	Kα, substrate, Å	K Edge, substrate, Å	Known μ (iron)
Zr	0.944	34.7	0.79	0.69	38.5 at 0.71 Å
Ag	0.333	12.2	0.56	0.49	14.1 at 0.50 Å

Our contention is that I_0 is comprised of wavelengths shorter than the critical wavelength of the absorption edge. If this is true, the value of μ_{avg} should be lower than that of the emitted Kα line of each substrate; see (1.13-2). The table confirms this.

Two final observations: The thickness range over which the method can be relied upon is determined largely by the value of μ_{avg}; the larger this coefficient, the smaller is this thickness range. It was found in this connection that a single layer of 0.6-mil iron foil on a nickel substrate absorbed the Kα line of nickel so effectively that the method was useless at this thickness. The mass absorption coefficient involved was near 400. Because iron Kα has a much lower excitation potential than the corresponding line of either silver or zirconium, it is not possible here to excite the substrate Kα without exciting iron Kα as well. Following are estimated Kα counting rates (counts per second) for the massive metals under the experimental conditions employed; iron, 200,000; zirconium, 32,000; silver, 8100. The thickness

THICKNESS OF PLATING

measurement could not succeed were the iron $K\alpha$ lines not eliminated by wavelength resolution.

DIVERSIFIED APPLICATIONS. AMP, Incorporated, of Harrisburg, Pa., is a manufacturer of small electrical components, which are odd in size and shape. These components must often be plated with precious metals. Accurate and precise measurements of plating thickness are needed not only for cost control, but also because plating-porosity, resistance to corrosion and wear, and electrical characteristics of the components all vary with thickness. Zimmerman [15] reports the outstanding success of x-ray methods at AMP on samples too diversified for description here. Masking and point-to-point exploration were often needed for obtaining localized information, especially on curved surfaces. Of particular importance now is the excellent agreement of his data with (7.3–1) over a hundred-fold thickness range for tin, silver, or gold on copper. He has prepared calibration curves, presumably satisfactory, also for tin on steel; and for nickel, palladium, or solder plate on copper. We shall return to his work.

PRODUCTION CONTROL AT IBM. Glade [16] reports on the successful production control of gold and rhodium plating thickness on the lever tip of IBM reed switches. The tip, which makes the electrical contact, is succes-

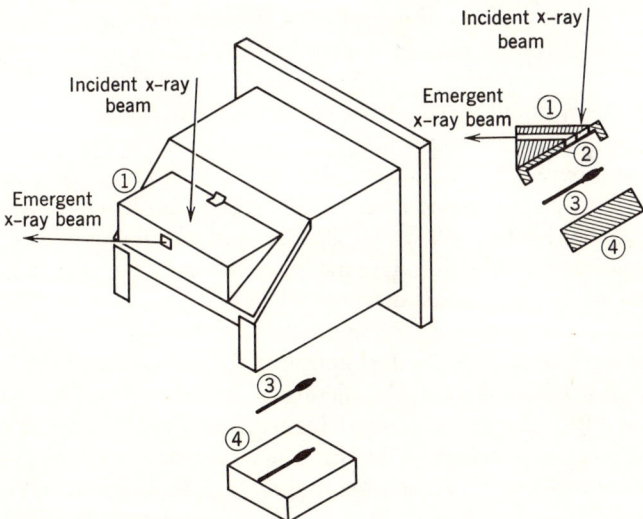

Fig. 7.3–3. Production control of plating thickness by use of a modified Lloyd aperture mounted on the sample drawer of a General Electric SPG-4 x-ray emission spectrograph. The x-ray beam is incident upon the sample through an opening (0.080 × 0.025 in.) in a copper mask 0.010 in. thick. The emergent beam reaches a scintillation detector after Bragg reflection from a flat lithium fluoride crystal. After Glade, Ref. 16, Fig. 1. (1) aperture; (2) mask; (3) sample; (4) fixture with cavity to hold sample.

sively plated with gold and with rhodium; close control of the plating is needed to ensure long life and satisfactory operation. Gold thickness is measured by the attenuation of nickel Kα. We return to the rhodium problem later.

The work is a good example of the differences between production and laboratory control methods. The need here was for a rugged device for reliable and reproducible operation by relatively untrained personnel under conditions where the thickness of gold could be obtained in a short time and read directly from a calibration curve of total counts (N_T) over a 40-sec interval against thickness in the range 0 to $100(10^{-6})$ in. Background corrections were not made. Figure 7.3–3 gives an idea of how the measurement was carried out. Earlier control by metallographic methods took 7 hours for measurement at a single point, which permitted sampling only once a shift. With the x-ray methods, samples from all plating lines are examined in 2 hours; precision is, if anything, slightly improved; and samples are taken thrice in each shift.

7.4 Film Thickness by X-Ray Emission Spectrography

In the method just described, the characteristic lines of the film were an interfering nuisance. In x-ray emission spectrography with x-ray excitation, a characteristic line is the analytical line and thus the source of information about the sample. When film thickness is to be measured, we then have a method that can satisfactorily complement the absorptiometric method of Section 7.3.

Let us begin qualitatively. Consider individual atoms of an element deposited on a thin substrate highly transparent to x-rays—say atoms of cobalt upon paper. Let a characteristic line (say cobalt Kα) be excited by a polychromatic beam, x-ray source and detector both being located above the sample. See Section 4.4. So long as the number of cobalt atoms is small, they will not noticeably attenuate the incident beam, nor will an x-ray quantum radiated by any cobalt atom be absorbed by any other. Under these conditions, the intensity of the characteristic line will be proportional to the number of cobalt atoms and hence to the thickness of the cobalt film.

As the film grows thicker, the metal will begin to filter and to attenuate the incident polychromatic beam, which will become shorter in wavelength and weaker as it penetrates. The intensity of cobalt Kα will now increase with thickness at a continuously decreasing rate. If the metal is thick enough, even the shortest-wavelength x-rays will fail at a certain depth to excite Kα quanta at a rate high enough to reinforce measurably the emergent beam, virtually all such "deep" quanta being absorbed by the cobalt on their way to the detector. The depth at which this first occurs is the *critical depth* for the experimental conditions. The critical thickness is the same as

THICKNESS OF PLATING

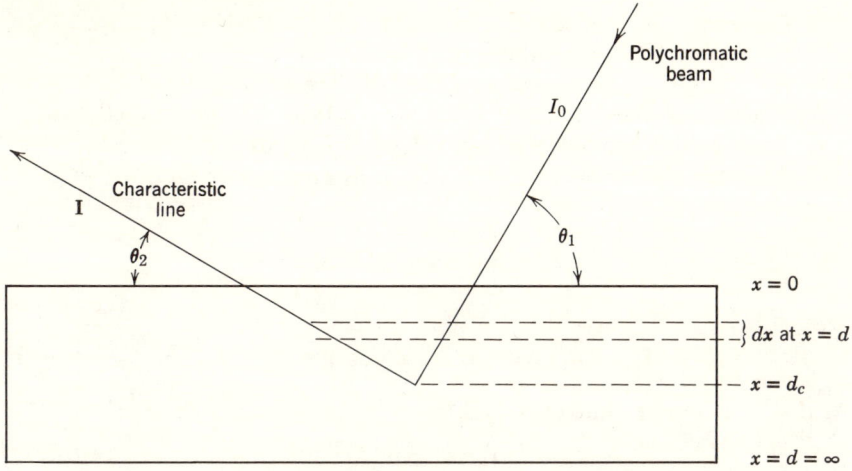

Fig. 7.4–1. Diagram for derivation of simplified relationship between intensity of emitted characteristic line and film thickness when the film is a pure element. Compare with Fig. 7.3–2.

this critical depth, and "critical" is used because an increase in thickness (or in depth) beyond the critical will not measurably increase the intensity of cobalt $K\alpha$, which is at its maximum value for the experimental conditions.

The significance of the relationship between sample thickness and emitted intensity was appreciated by Glocker and Schreiber [17]. Koh and Caugherty [18] proved experimentally that the intensity of a characteristic line emitted by a thin metallic film decreases below that for the massive metal as the film thickness decreases below about 0.003 cm, and they suggested that this variation in line intensity could be used to measure thickness. Calculation [19] showed 0.003 cm to be a reasonable value for the critical thickness.

As in the less complex case of Section 7.3, we shall develop here the simplest conceivable relationship between measured (this time, emitted) intensity and thickness. The treatment to be developed is modeled upon that in Sections 4.4 and 7.3, with both of which it is to be compared.

Consider, as in Figure 7.4–1, that a polychromatic beam (intensity I_0 measured in cps) is incident upon a film so thin that single values of θ_1 and θ_2 may be used for all values of x to evaluate the characteristic-line (here taken to be $K\alpha$) quanta measured by the detector. These quanta are generated and radiated isotropically in elements of volume dV, of altitude dx, and of unit area. Under these conditions:

1. Effective intensity: $k_1 I_0$ (k_1, fraction effective in exciting $K\alpha$)
2. Effective intensity at x: $k_1 I_0 e^{-a_1 x}$ ($a_1 = \csc \theta_1 \mu_1 \rho$)

3. Decrease in effective intensity in dV: $dI = |k_1 I_0 a_1 e^{-a_1 x} dx|$
   (| | means "absolute value")
4. $K\alpha$ quanta/sec generated in dV: $+\, d\mathbf{I} = |k_2 dI|$
5. Fraction of these quanta detected: $+\, d\mathbf{I} k_3 e^{-a_2 x}$ $(a_2 = \csc\theta_2 \mu_2 \rho)$
6. Total quanta/sec detected (cps): $\mathbf{I}_d = \int k_3 e^{-a_2 x} d\mathbf{I}$
7. After simplification by combining constants:

$$\mathbf{I}_d = \int_{x=0}^{x=d} k a_1 I_0 e^{-ax}\, dx$$

or,

$$\mathbf{I}_d = k a_1 I_0 (1 - e^{-ad})/a \qquad a = a_1 + a_2 \qquad (7.4\text{–}1)$$

At $d = \infty$ ("infinite thickness")

$$\mathbf{I}_\infty = k a_1 I_0 / a \qquad (7.4\text{–}2)$$

and, after dividing (7.4–1) by (7.4–2), we have

$$\mathbf{I}_d / \mathbf{I}_\infty = (1 - e^{-ad}) \qquad (7.4\text{–}3)$$

whence

$$\log(1 - \mathbf{I}_d/\mathbf{I}_\infty) = -0.4343 ad = \text{const.} \qquad (7.4\text{–}4)$$

ANALYSIS OF (7.4–1). Three important thickness regions are defined as follows by the value of the exponential term in (7.4–1); see Table 7.4–1.

Table 7.4–1. Thickness Regions According to (7.4–1)

e^{-ad}	Region	d or x	dI/dx
Negligible	"Infinite thickness"	Exceeds critical thickness	Approaches zero
Significant	Exponential	Intermediate	Variable
Near unity	Linear	Approaches zero	Approaches constancy

When d equals or exceeds the critical thickness, d_c, (7.4–1) is useless for the determination of thickness: all such values of d are "infinite" and yield $\mathbf{I}_\infty$ as measured intensity. This region, useless for thickness determinations, is the preferred region for x-ray emission spectrography of bulk samples. Although the region is beset by interelement effects, sample thickness is not a significant variable. To establish d_c, the critical thickness, one arbitrarily assigns a value near unity (e.g., 0.99) to the intensity ratio in (7.4–3) and solves for $d = d_c$.

THICKNESS OF PLATING

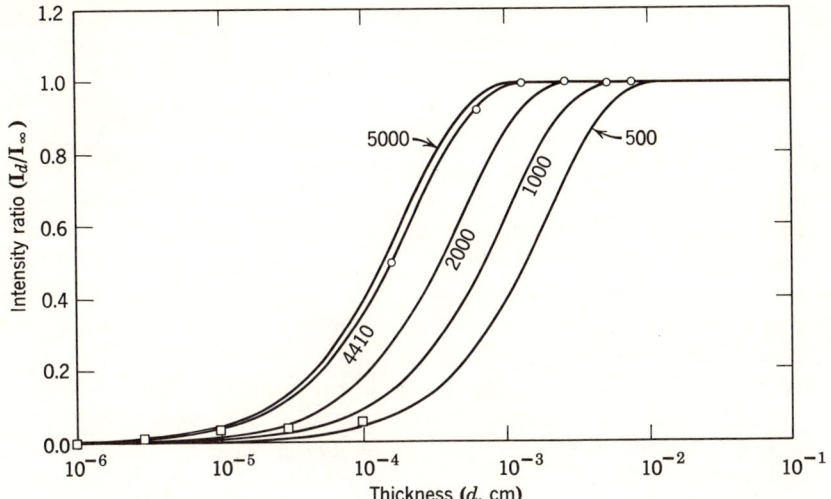

Fig. 7.4–2. Test of simple intensity-thickness relationship (7.4–3) over the entire thickness range for chromium-on-molybdenum. Circles: plated coatings; squares: evaporated coatings. Numbers are values of a/ρ for which the curves were calculated. From Liebhafsky and Zemany, Ref. 14, Fig. 3.

In the exponential region, useful thickness determinations are possible and intensity readings (cps) are conveniently plotted as logarithmic ordinates against known thicknesses to give linear calibration curves according to (7.4–4).

The linear region is of first importance in determinations of surface density and of composition. It permits determinations at minimum surface density. In determinations of composition, the region is important because interelement effects are absent. More later.

VERIFICATION OF (7.4–3). Liebhafsky and Zemany [14] undertook to verify (7.4–3), including the existence of the three regions specified in Table 7.4–1, by use of a single, practically important, example: namely, chromium-plated molybdenum, the plating being intended to protect the molybdenum against oxidation. Thin films were prepared by evaporating chromium in vacuum—and thick films by electroplating this metal—onto the molybdenum disks that served as substrate. The tungsten-target x-ray tube was operated at 50 kV and 50 mA except at counting rates above 3000 cps, for which the tube current was reduced to 5 mA and counting rates were adjusted. A multiple-chamber Geiger detector was used. The background corrections were complicated [14] and are not discussed here. The results are given in Figure 7.4–2.

The data in Figure 7.4–2 are intensity ratios according to (7.4–3) plotted against thickness alongside curves calculated for five values of a/ρ. The experimental points fit the curve for $a/\rho = 4410$ except in two cases where flaking of evaporated coatings was observed. Thicknesses are much less precisely known here than they were in the work with iron foil [13].

Figure 7.4–2 confirms Table 7.4–1. The region of "infinite thickness" extends upward from the critical thickness, which is near 10^{-3} cm; the exponential region begins there and stops near 10^{-5}; the linear region extends from there downward to zero.

Calculations based on known data [14] showed the value of 4410 to be somewhat high, though not unreasonably so in light of the uncertainty in the effective wavelength of the polychromatic beam and of the mass absorption coefficients. A high value could be due to the excitation of Cr **K**α by characteristic lines, notably Mo **K**α, from the substrate. Because such enhancement is an easy introduction to the kind of enhancement that is one of the interelement effects (Section 7.1), the matter was explored. The Cr **K**α count was taken for an adequate time at 50 kV and 5 mA for the molybdenum sample with the thickest chromium plate, and for massive chromium. The calculated cps values were: sample, 2576; Cr, 2513. Excitation by a substrate line does seem to have enhanced the intensity of a characteristic line in the film *above* that found for the massive metal under the same external excitation. The experiment is a convincing demonstration of enhancement, albeit under conditions seldom met with in x-ray emission spectrography.

The existence of the three regions of Table 7.4–1 being verified, we now look at experimental thickness determinations in the two promising regions. The region of "infinite thickness" belongs in the domain of determination of *composition* (not thickness) by the x-ray emission spectrography of bulk samples.

EXPONENTIAL REGION. Figure 7.4–3 contains the results of Zimmerman [15] on observed intensities for known gold and silver films. From the x-ray tube voltages given, it is clear that an **L** line was used for gold, and a **K** line for silver. The results confirm (7.4–4).

Zimmerman [15] also experimented with intensity *ratios* (α to β) of **L** and **K** lines as alternatives for thickness determinations in special cases. In accord with the modest wavelength differences between α and β, these alternative methods were of low sensitivity. These methods deserve to be kept in mind, however, because intensity *ratios* are immune to changes in certain experimental conditions (such as moderate fluctuations in x-ray tube voltage) that alter individual intensities.

THE LINEAR REGION. In differential form, (7.4–1) is

$$d\mathbf{I} = ka_1 I_0 e^{-ax} dx \qquad (7.4\text{–}5)$$

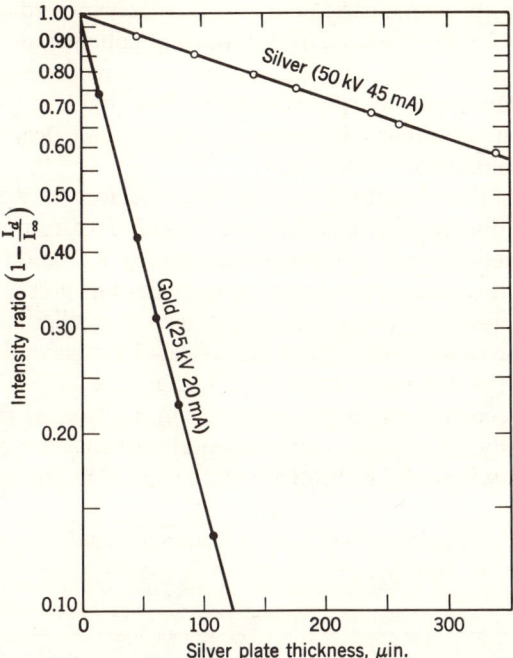

Fig. 7.4-3. Calibration curves for thickness determinations of gold and silver plate by x-ray emission spectrography. The ordinate scale is logarithmic. Note the pronounced difference in slope (hence, sensitivity). The method is more sensitive for gold because the **L** lines of gold are much more strongly *self-absorbed* than the **K** lines of silver. See (7.4-4). Relevant data in Appendices. After Zimmerman, Ref. 15, Fig. 5 and Fig. 6.

As ax approaches zero, it becomes permissible to write

$$\Delta I = ka_1 I_0 \Delta x \qquad \text{(compare with 4.4-2)} \qquad (7.4-6)$$

So long as (7.4-6) is valid, we are in the linear region and interelement effects are absent. The chief difficulty in testing (7.4-6) is the preparation of reliable standards (see, for example, Figure 7.4-2). Furthermore, the values of Δx will be nominal—not actual—values: they are averages over the sample area viewed by the detector, and they are based on bulk density. The actual density of thin films may be smaller than the bulk.

To illustrate these points, let us complete the cobalt story begun in Section 4.4. For the calculation there, $\Delta d (= \Delta x)$ for a monolayer of cobalt atoms was taken as 2.3 Å. For Co **Kα**, a value $I_\infty = 5(10^5)$ cps/cm² was measured under conditions near those of Chapter 5. Then for a monolayer by (7.4-2) and (7.4-6)

$$\Delta I = |aI_\infty \Delta x| = 4410 \times 5(10^5) \times 2.3(10^{-8}) = 51 \text{ cps/cm}^2 \qquad (7.4-7)$$

The agreement with the measured (15 cps) and the expected (30 cps) values from Table 5.14–1 is satisfactory. The important conclusions are that satisfactory order-of-magnitude calculations can be made from first principles for ordinary x-ray emission spectrographs; and that even such spectrographs, though not designed for high sensitivity, can deal with samples approaching monolayers.

Examination of (7.4–5) shows that the linear region extends from $x = 0$ to that value of x for which the exponential term differs from unity by more than the experimental error in the measurement of I. The differential quotient dI/dx, which relates I to x, hence also to amount present, depends on a_1 and (in exponential fashion) on $a = a_1 + a_2$: the smaller a, the larger the linear region. The dependence is too complicated for general discussion—especially when the sample contains more than one element so that the mass absorption coefficients depend upon composition in the manner of (1.10–5). Obviously, any rigorous experimental verification of (7.4–5) should be done with excitation of the characteristic line by a monochromatic x-ray beam.

An empirical verification of (7.4–6) is simpler. That this important re-

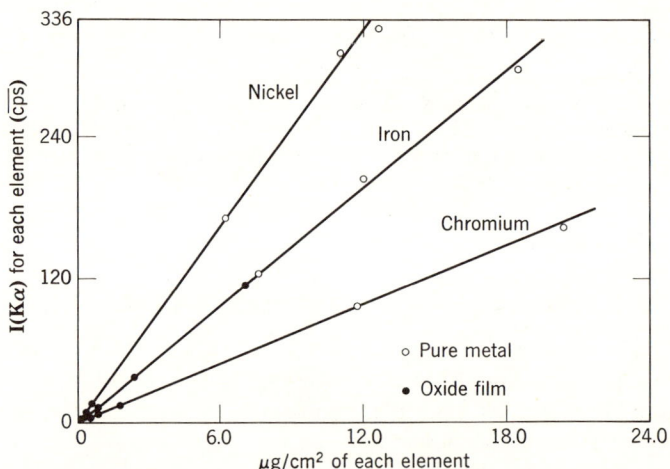

Fig. 7.4–4. Characteristic-line intensity and amount of metal per unit area for the components of Type 304 stainless steel. Note (1) that the abscissa is proportional to nominal thickness; (2) that data for the individual metals (open circles) agree with those for the oxide film (solid circles) from this stainless steel; (3) that the linear relationship (7.4–6) is valid for all three metals; (4) and, as an index of precision, that the average counting rates $[\overline{(cps}/\mu g) (cm^2)]$ determined from evaporated films of the *stainless steel* showed the following *percentage* differences from the values of the figure: − 2.6 (Ni); − 1.8 (Fe); + 1.2 (Cr). These differences are gratifyingly small. Values of $\overline{cps}/(\mu g) (cm^2)$ from the figure: for Ni, 26.9; for Fe, 16.4; for Cr, 85. After Rhodin, Ref. 21, Fig. 3 and Fig. 5.

lationship, according to which ΔI is proportional to the number of atoms (free or combined) present of the element identified by the characteristic line, is actually valid was established by Pfeiffer and Zemany [20] and confirmed by Rhodin [21] in a classic investigation. As the establishing of composition, not of thickness, was the principal objective in these investigations, we shall discuss them later and present here in Figure 7.4–4 only part of Rhodin's excellent results.

TIN PLATE. Electrolytic tin plate, its great industrial importance aside, is an excellent material on which to test thickness determinations by x-ray emission spectrography because it is uniform and because its low value of a gives it a large linear region (7.4–5).

To make the test, 24 rectangles (each 1×1.5 in.) were cut from each of six standard panels (each 6×6 in.) [14], supplied by the United States Steel Company. One such rectangle was selected at random for a thickness measurement on each face by the attenuation of Fe $K\alpha$ and by the emission of Sn $K\alpha$. The results for the second method appear in Figure 7.4–5. The linear region appears to extend to nearly $1.5(10^{-4})$ cm, which is further than one might have expected on the basis of (7.4–5), but this merely illustrates that experimental error makes it difficult to establish the upper limit of the linear region. The intercomparison of results by the two methods was satisfactory

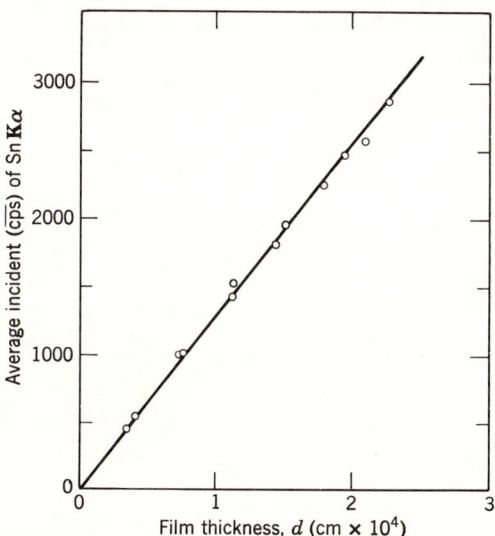

Fig. 7.4–5. Determination of tin-plate thickness by x-ray emission spectrography. The known film thicknesses (abscissas) were obtained by measuring the attenuation of Fe $K\alpha$ and were based ultimately on chemical determinations. After H. A. Liebhafsky and P. D. Zemany, Ref. 14, p. 458, Fig. 6.

[14, Fig. 5]: not only did the results at the higher thickness show the expected absorption effect attributable to the exponential term in (7.4–5); but the best straight line through the data intersected the ordinate axis, on which the attenuation results were plotted, precisely at the $\overline{\text{cps}}$ value for Fe $K\alpha$.

Inasmuch as each standard panel had been divided into 24 rectangles, only one of which was run, a test for uniformity was made by measuring the time for 2^{14} counts for Sn $K\alpha$ from the same side of each rectangle cut from the standard with the thinnest tin plate. A standard deviation of 2.1% was obtained, and the mapped data showed that plate thickness on the standard was not uniform. Furthermore, 12 replicate counting intervals on one rectangle gave a standard deviation of 0.7%, in close agreement with 0.8%, the predicted value for 2^{14} counts. For the full significance of these experiments, see Section 2.20 and Chapter 8.

X-ray emission spectrography was adapted to the control of tin-plate manufacture by Applied Research Laboratories, who developed the Quantrol [22] for this purpose. In this instrument, the intensity of tin $K\alpha$ is measured relative to that of scattered x-rays entering the detector from an analyzing crystal set for the reflection of x-rays 2.2 Å in wavelength. See Table 7.4–2.

Table 7.4–2. **Quantrol Reliability on Electrolytic Tin Plate**

Thickness (lb/bb)[a] Measured by	
Chemical Method	X-Ray Method,[b] Method II
0.245	0.240
0.49	0.492
0.51	0.512
0.72	0.710
0.74	0.727
0.92	0.913
0.94	0.938
0.97	0.983

[a] The bb signifies "base box." One pound per base box equals approximately 11g/m².
[b] Trace averaged over 1 minute.

These are impressively precise results for the continuous control of a manufacturing process, and they of course suggest that a servomechanism could be used to completely automate this and similarly controlled processes.

DIVERSE APPLICATIONS

The applications discussed above were selected to demonstrate the principles of the attenuation and emission x-ray methods for film thickness, and to show their suitability for industrial control. The tin-plate application is for these purposes the best and simplest example. More complex applications with features of particular interest are now sketched.

7.5 Films on Semiconducting Silicon

To provide for electrical contact and metallic conduction, silicon in semiconducting devices is often coated in vacuum with aluminum films, the thickness of which ranges up to 10,000 Å and must be carefully controlled. Cline and Schwarz [23a] proved that these thicknesses could be measured satisfactorily by getting corrected cps values for Al Kα (8.34 Å), and that this was more satisfactory than either optical interferometry or measuring the attenuation of Si Kα (7.13 Å). Their assessment that the interferometric method is usually inferior to the x-ray is supported by general experience. Attenuation of Si Kα gave lower sensitivity ($\pm$ 24 Å as compared with $\pm$ 6 Å for emitted Al Kα) for very thin films [23a], and encountered trouble from variable thickness of substrate oxide. Variable attenuation of Si Kα by this oxide is serious because oxygen has a mass absorption coefficient near 1000 at the wavelength of this line.

In this work, the minimum thickness of aluminum measured was 0.18 μ (1800 Å). In contrast to the tin-plate case, no linear region was established. To the extent that uncertainties in interferometry allowed, the results obeyed (7.4-4); that is, they lay in the exponential region. We have here just another example of the fundamental importance of x-ray absorption. As Figure 1.11-1 shows, an *element* will be relatively transparent to its characteristic lines in the wavelength region near the corresponding absorption edge, for the wavelength of these lines always exceeds somewhat the critical wavelength of the edge. But as wavelength increases, this transparency is reduced. Let us compare mass absorption coefficients for tin and for aluminum at wavelengths shorter and longer than the critical absorption wavelength. We have, approximately: Sn, 43 (0.4 Å) and 12 (0.5 Å); Al, 3300 (7 Å) and 320 (8 Å). These mass absorption coefficients help determine $a = a_1 + a_2$ in (7.4-1), and their influence here greatly exceeds that of the density. For aluminum, the mass absorption coefficient even on the "transparent" side of the edge is seven times that of tin on the short-wavelength side. It is not surprising, therefore, to find that aluminum (in spite of its lower density) has a much narrower linear region than tin.

The matter just discussed must be kept in mind when a film or sample contains more than one element. In that case, the mass absorption coefficients of the other elements will help determine the mass absorption coefficient of the sample, and the difference between a_1 and a_2 will be decreased below what it would have been for the element alone.

PHOSPHORUS IN SILICA-ON-SILICON. The introduction by thermal "driving-in" of phosphorus into silica films on a semiconducting silicon substrate stabilizes device behavior by reducing the movement of ions in the silica films. Pink and Lyn [24] point out that a non-destructive method of determining phosphorus would be helpful in understanding the stabilization and in controlling the manufacture of the devices. The problem is complex and not strictly one of measuring thickness, but it has enough in common with the aluminum-on-silicon determination to warrant mention here. In this case, the cps values for P Kα were found to be proportional to total phosphorus colorimetrically determined. Gaseous POCl$_3$ was the phosphorus source, and both x-ray and colorimetric values of phosphorus content were proportional to the time the silicon had been exposed to the gas.

7.6 Multiple Plating

As more than one plate is put on a metal, the opportunities for the generation and attenuation of characteristic lines increase, and it is possible in principle to determine the thickness of each coating by combining attenuation and emission data. In practice, the complexities multiply also: the intensity of the characteristic line emitted by a coating will depend on the thickness so long as this is below its critical value; and enhancement—witness the case of chromium on molybdenum in Section 7.4–2—can also occur. No general discussion can be given. It should be possible to judge the applicability of x-ray methods in most cases on the basis of fundamental data according to the considerations of this chapter. Two statements seem safe: If possible, determine thickness on a plating while it is still exposed. Do not attempt to determine thicknesses of too many platings at once.

Birks, Brooks, and Friedman [25a, b] pioneered in simultaneous thickness determinations of multiple platings. In the case of nickel on copper on steel (substrate), they determined the nickel from emission data, and the copper from attenuation of Fe Kα by use of a calibration curve (cps for Fe Kα against known thickness of copper) that took into account the attenuation of Fe Kα by the (now known) thickness of the outer layer of nickel. Emitted lines from coatings beneath the outer coating could also be used upon proper correction for attenuation [25a, b, c].

At IBM, Glade [16] reports that the lever tips (0.110 by 0.033 in.) of Section 7.3 are rhodium plated after the gold plate has been put on, the thickness of the gold having been measured on control samples (see Section

7.3). The thickness of the rhodium plating is established on control samples by counting the emitted rhodium Kα with elimination of W Kα (III) by pulse-height discrimination. Within the precision required for control, the rhodium results are independent of gold thickness over the working range; corrections for background are not made. "Broad-range" conditions have been chosen that make it possible for the operators, not technically trained, to get thickness data that are more precise for both metals than data by metallographic methods.

7.7 Radioactive Sources

We have met radioactive sources in Sections 2.25 and 3.2, and we meet them again later. Among them are true radioactive x-ray sources such as $_{26}Fe^{55}$, the K-capture isotope of Section 3.2, and there are others that serve only as sources of *energy* for the excitation of x-rays. This energy varies over a wide range from one source to another and is carried in different ways: for example, by gamma rays, electrons, alpha particles, and protons.

In an exploratory investigation, Cook, Mellish, and Payne [26a] pointed out the advantages of radioactive sources for thickness gauging and did enough experiments to show that these advantages are realizable. Advantages: (1) Favorable geometry when source and sample can be close neighbors, which ensures adequate, though low, cps values with sources weak enough to be safe. (2) Replacement of wavelength resolution by energy resolution of the simplest kind—namely, a proportional detector with single-channel pulse-height selection. (3) Sources available in enough variety to make the method reasonably flexible; attenuation of substrate K lines, and intensities of K and L lines emitted by the common platings are all useful.

The results obtained are in accord with what has been presented above. The sources are of limited value with a proportional detector when simple energy resolution is inadequate (as, for example, when $_{29}Cu$ and $_{28}Ni$ are both present); the lithium-drifted detectors (Sections 2.15 and 2.25) have a future here. One control application is mentioned in [26a] and discussed by Cameron and Rhodes [26b]: the use of a tritium source and a sealed proportional detector for gauging tin-plate by the attenuation of Fe Kα. Cameron and Rhodes [26b] discuss other examples also. There is an interesting complication involving the argon escape peak (Section 2.12) when emitted Sn L lines are used [26a]. The energies of peak and lines are so nearly identical that the measured Sn L intensity (near 3.5 keV) becomes virtually *independent* of tin thickness even though the critical thickness has not yet been reached. The reason: the escape peak is generated by the absorption in the detector of Fe K lines, the intensities of which *decrease* as the tin coating thickens and the Sn L intensity *increases*. The two changes tend to compensate, and this explains the observed result. A change of filler gas is indicated.

Zemany [26c] used $_{26}Fe^{55}$ for measuring the thickness of titanium plating on Kovar (Fe-Co-Ni). Because $_{26}Fe^{55}$ is a source of virtually monochromatic x-rays, emitting the Mn K lines, it excites *selectively*; it cannot generate K lines for elements above $_{25}Mn$ in atomic number, or L lines for elements above $_{61}Pm$. He discusses additional applications of this and other K-capture isotopes.

7.8 II-VI Compounds

The Roman numerals identify groups in the periodic table: thus cadmium sulfide is a II-VI compound that can serve as a qualitative x-ray detector [27]. The discovery by the Aerospace Research Laboratories [28] that cadmium sulfide and other II-VI compounds show a photovoltaic effect started extensive work on solar cells containing these as active materials. In such cells, they can be used as thin films—single crystals are not needed. To measure the thickness of these films, Chan [29] has tried an unusually wide variety of x-ray methods and found them satisfactory. As the films are *binary* compounds, the thickness measurement involves three considerations not present with elements: (1) stability of the film toward x-rays; (2) two mass absorption coefficients help determine a (Section 7.5); and (3) characteristic lines emitted by two elements are available for thickness measurement.

Chan points out that the films are difficult to prepare and that standards for calibrating the x-ray methods are a second problem. Satisfactory solutions for both seem to have been found. Satisfactory calibration curves were obtained for the following x-ray methods: (1) Electron excitation (at 10 kV and 0.2 mA) of S Kα with the Betaprobe (see below) in CdS film on Al substrate; wavelength resolution with pulse-height discrimination; flow proportional detector. (2) In (1), attenuation of Al Kα with necessary modification in equipment. (3) For CdS films, excitation of Cd Kα in conventional x-ray emission spectrograph with air path and in vacuum; no pulse-height selection. (4) For CdS films, excitation of Cd Kα by radioactive source; Si(Li) detector; pulse-height analysis (200 channels); counting interval, 240 sec. (5) For CdSe films, excitation of Cd Kα and Se Kα in conventional x-ray emission spectrograph; air path, no pulse-height selection. (6) For CdSe films, excitation of Cd Kα and Se Kα; other conditions as in (4). (7) and (8) For CdTe films, repetition of (5) and (6) with obvious modifications. (9) and (10) For HgSe films, repetition of (5) and (6) with obvious modifications, with excitation of Hg Lα, and with different radioactive source. This list shows the variety of conventional x-ray methods applicable to films.

Among the many interesting aspects of Chan's work, only the use of the (English) Telsec Betaprobe will be mentioned. In this device (see Chapter 10) electron excitation of x-ray spectra is accomplished by a defocused electron beam with an area near $\frac{1}{2}$ in.² As will appear in Chapter 9, electron excitation gives high x-ray intensity, penetrates very little below the surface, and is more likely to alter the sample than is x-ray excitation. That attenuation of Al Kα could be used to measure thickness means that significant excitation of the substrate occurred even when the film was 0.5 μ (5000 Å) thick; no doubt the characteristic lines of sulfur played some role in the excitation, which means that a_1 is an uncertain quantity. See Section 7.5.

An aside. The x-ray emission electron microprobe (electron microprobe for short) is a specialized x-ray emission spectrograph with electron excitation, full treatment of which is impossible in this book. It represents the extreme [square microns of area] in localized determinations. It was used by Cline [23b] for the simultaneous determination of the thickness of both coatings in the combination Au-Mo-Si (substrate) used to make integrated circuits. His work is mentioned here because the combination he investigated competes with the aluminum-coated silicon discussed above and because he treats the effectiveness of electron penetration for the generation of x-rays.

7.9 Permalloy Films. Composition and Thickness

These films, which have become critically important as a means for storing information on magnetic tapes, are binary alloys of nickel and iron, with the former in excess. Composition and thickness (usually, thousands of angstroms) are both important variables; occasionally mass replaces thickness. The two variables in question can be determined simultaneously by measuring the intensities of Ni Kα and Fe Kα, but elaborate calibration is needed. Bertin [30] gives an excellent discussion, with references, of this problem which is too specialized to be more than mentioned here.

7.10 Other Methods. Other Properties

In his important review, Bertin [30] says of films, "it is fair to state that a more nearly complete characterization is possible by x-ray methods than by any other group of related methods." In addition to thickness, or mass per unit area (already discussed) and composition (discussed later), information can be obtained by x-ray methods about surface smoothness; lattice parameters, degree and kind of crystallinity, and crystallite size and orientation; stress in crystalline films; and density of dislocations and stacking faults. In addition to methods regarded as principal methods in this book, total reflection of x-rays, divergent-beam (Kossel) photography and diffraction topography are described by Bertin.

SIMPLE TRACE DETERMINATIONS

7.11 Trace Determinations. General

Trace determinations, qualitative and quantitative, are conveniently subdivided into (1) traces as major constituents, a class in which the total sample is obviously minute; and (2) traces as minor constituents in samples not unusually small. It is occasionally expedient to isolate the constituents of interest from a large sample so that they can be determined under simpler conditions; this amounts to shifting the determination from the second class to the first. Partly for this reason, most trace determinations carried out with ordinary x-ray emission spectrographs are of the first class.

The usefulness of x-ray emission spectrography for trace determinations is clearly foreshadowed in the work of Laby [31], von Hámos [32, 33], and Engström [33, 34]. The method cannot reach into the micromicrogram range with any assurance when ordinary equipment is used, nor can it reveal chemical constitution—both objectives that are often within reach of the classical microchemical methods that are growing continually more powerful as the result of work such as that performed by Yoe and his collaborators [35]. But with special equipment, fractions of monolayers can be determined. The reader will remember the simplified calculations for the monolayer of cobalt, for which (7.4-7) gave 51 cps for 1 cm^2.

Such a sample weighs 0.2 μg, but the calculation is no guarantee that such amounts can always be detected, much less determined. The success of the determination will obviously depend upon the size and the reproducibility of the difference $N_T - N_B$, N_T being the total count at the 2θ value for the peak of the analytical line and N_B the count for the background at that value. But this is not a simple matter. Among the factors to be considered are: (1) Proper choice of analytical line (**K** or **L**) and of x-ray tube target. (2) Voltage and current for x-ray tube, which should be near their maximum stable values. (3) High spectral purity of incident beam so as to give minimum interference with analytical lines and low background. Increase in wavelength owing to Compton scattering can lead to interference. (4) Resolution good enough to resolve peaks at satisfactory intensity—but no better (Chapter 5). (5) Background low enough so that counting error and statistically established limit of detection are satisfactory (Chapter 8). (6) Interelement effects to be kept at a minimum.

As the object here is to present only *simple* trace determinations, in which interelement effects are negligible (or deliberately neglected), these six factors will be discussed elsewhere, mainly when x-ray emission spectrography proper is treated. An inkling of their importance is given by Figure 7.11-1 [36].

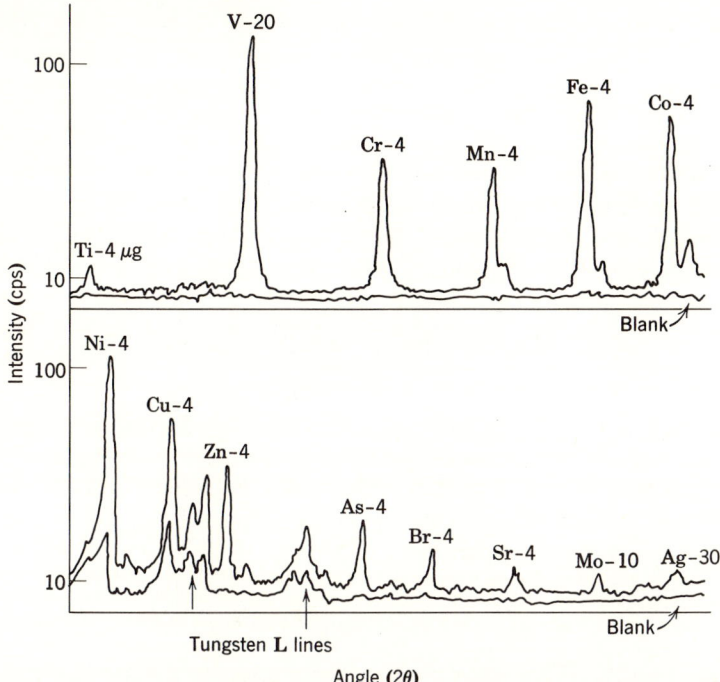

Fig. 7.11–1. X-ray spectrographic chart recording obtained from a composite spot prepared on Mylar film and containing 14 elements. The number of micrograms of each element is given near its Kα line. The lower chart is a continuation of the upper one. Note the tungsten, copper, and nickel in the background spectrogram. The copper and nickel are impurities in the tungsten x-ray tube. The increase in intensity of the tungsten L lines and background in the sample spectrogram (in comparison with the blank spectrogram) is due largely to the use of helium when the sample spectrogram was run. From Ref. 36, Fig. 8–6.

Figure 7.11–1 is the chart recording for a composite sample containing 14 elements in microgram amounts—a Class 1 determination. The number of micrograms of each element is given near its Kα peak. To reduce background due to scattered x-rays, thin Mylar was used as substrate as an improvement over filter paper. The determinations were made with helium in the x-ray path, but a "blank" (Mylar only) was recorded in air to emphasize that this is an unacceptable procedure for precise work in which the difference $N_T - N_B$ is small. (A comparison of the recorded W L intensities makes this point; the actual intensities must have been almost identical.) The heights of the copper and nickel peaks were enhanced by excitation of these metals as impurities in the x-ray tube. The maximum sensitivity (cps/microgram) occurs in the middle of the range of atomic numbers represented in the figure. It would be futile to attempt a detailed explanation of Figure 7.11–1, which serves the purpose of indicating what results may be expected in simple Class 1 trace determinations done without maximum optimization. Needless to say, these results are encouraging: identification of each element is positive, and semiquantitative results would be easy to obtain—always provided a proper blank is run.

The expectation that trace determinations would become an increasingly important part of x-ray emission spectrography [36] has been borne out, as an examination of three recent books [37–39] will show. The excellent review by Campbell and Thatcher [39] is particularly to be recommended.

7.12 Kinds of Simple Trace Determinations

The simplest of trace determinations are those for which (7.4–6) is valid: sample thickness is in the linear region; interelement effects are absent; and Δx and ΔI in (7.4–6) are proportional to the number of atoms present (for unit area) of the element sought. So long as all the sample is in the beam, Δx need not rigorously be constant over the entire sample area so long as (7.4–6) is obeyed in the form

$$\Delta I = k' a_1 I_0 m = k'' a_1 I_0 n_a \tag{7.12–1}$$

where m is the mass of the sample and n_a the number of atoms present in the sample of the element sought. Further discussion below.

Spot tests are an important way of applying (7.12–1) to Class 1 determinations. The technique is well known [40] and has proved very valuable in analytical chemistry. As often carried out, a reagent (specific if possible) is made to react in or on filter paper with the element sought, usually present as a trace. The results are normally qualitative or semiquantitative, it often being difficult to make them quantitative by methods other than x-ray emission spectrography [41]. With x-rays, however, not only is it possible to get quantitative results, but also there is no need of a reagent to develop a color!

Pfeiffer and Zemany [42] in 1954 proved that the advantages just mentioned are easily realizable in practice. Figure 7.12–1 shows data obtained by the x-ray emission spectrography of spots formed by evaporating drops of a known zinc salt solution on filter paper. The gratifying linearity is proof that the conditions under discussion are practically met. The slight deviation from linearity at the highest amount of zinc shows that a negative absorption effect enters with increasing sample size. But it is unlikely that conditions are ideal, for the solution saturates the paper, and the zinc salt residue is neither upon the surface as a thin film nor uniformly distributed. That linearity is nevertheless achieved illustrates another advantage of working with small samples: so long as all the sample intercepts the incident beam and interelement effects are negligible, there can be considerable

SIMPLE TRACE DETERMINATIONS

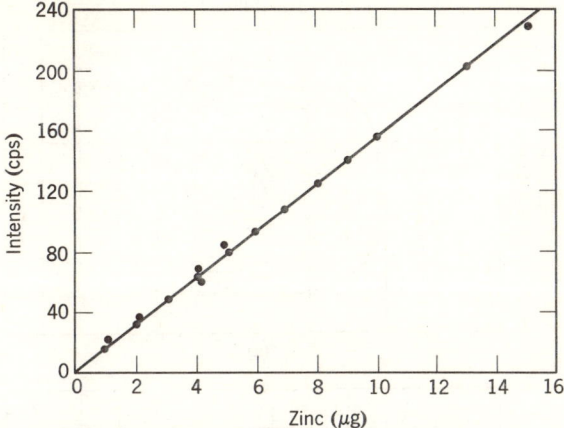

Fig. 7.12–1. Counting rate corrected for background of zinc Kα line for various amounts of zinc on Schleicher and Schuell No. 740E $\frac{1}{2}$-in. disks. Samples were prepared by placing measured amounts of zinc sulfate solution on the paper and drying to remove the water. From Pfeiffer and Zemany, Ref. 42.

departure from ideal conditions without essential loss of simplicity.

The point just made leads to a suggestion worthy of wider exploitation than it has yet received. Why not deliberately carry out Class 1 determinations on *large* samples so as to avoid interelement effects? If the spot-test technique is to be used, one might dissolve a representative part of the sample, dilute as needed, and proceed. We have here a "trade-off" situation so typical of analytical chemistry that detailed discussion is superfluous. An important factor in such a "trade-off" is whether the x-ray spot-test method routinely carried out has the minimum reliability (accuracy and precision) needed. This depends, among other things (see Chapter 8), not only on the uniformity with which the sample is distributed but also upon the uniformity of the x-ray beam.

In a qualitative way, the uniformity of the x-ray beam in a spectrograph can be examined by having it darken a glass slide. Quantitative evidence of nonuniformity appears in Figure 7.12–2, which records the results of an experiment on a Philips spectrograph. The sample, an iron foil 0.1 × 0.1 in., was used as a probe and placed in contiguous positions over the area of the sample cavity. Readings of iron Kα were made on a Geiger detector, and the approximate contour maps in Figure 7.12–2 were constructed from those readings. If a sample fills the sample cavity, therefore, the detector response is a response integrated over regions in which the analytical-line intensity varies markedly. This variation will change with the x-ray optical system and with the goniometer setting. The important thing to remember is that such variations do exist and must be taken into account in the positioning

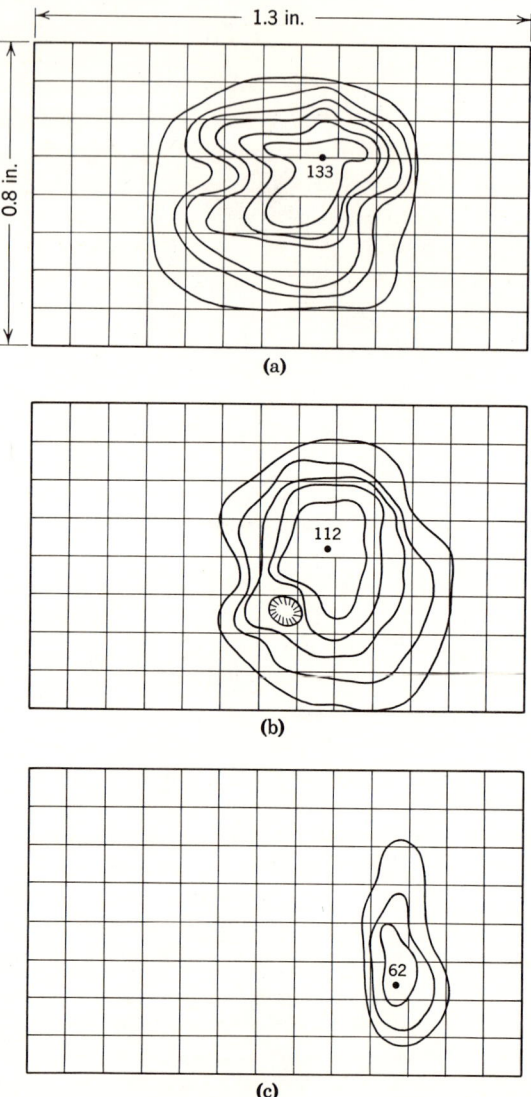

Fig. 7.12–2. Contour maps showing spectrograph sensitivities for the iron Kα line at various positions of the sample. (*a*) At surface of sample holder. (*b*) 0.16 in. below surface of sample holder. (*c*) 0.32 in. below surface of sample holder. The sensitivity changes with the x-ray optical system, with the goniometer setting, and with the distance of the sample below the surface of the sample holder. The contour interval is 20 cps, and the numbers are cps readings corresponding to the points in the figure. Authors' unpublished results 1958.

SIMPLE TRACE DETERMINATIONS

of samples small in area if quantitative, or even semiquantitative, results are sought. Even if positioning is properly done, a marked change in the area of a sample is liable to lead to error even when interelement effects are absent; clearly, samples of different areas may have different integrated intensities for unit area of incident beam.

Lack of uniformity in the x-ray beam will clearly be much less serious than in Figure 7.12–2 when the spot-test technique is used, and this technique is consequently to be recommended when the sample need not be examined non-destructively. *Rotation* of the paper or other substrate in the plane perpendicular to the axis of the beam will go far to eliminate the effect of nonuniformity both in the beam and in the distribution of the sample on the substrate. The elimination of interelement effects by dissolving the sample for the spot-test technique can be important even when the sample is small: the thickness of the undissolved sample will often be great enough to make (7.12–1) invalid—particularly is this true for nonuniform samples in which a light element (analytical line strongly absorbed) is to be determined.

The spot-test technique is of course not the only way of carrying out simple trace determinations; examples of other ways appear below. Class 2 determinations in which the matrix is virtually transparent to x-rays are one example: in the determination of tetraethyllead in gasoline with $L\alpha$ as analytical line, (7.12–1) should be obeyed at low concentrations; as the concentration is increased, one should encounter the exponential region and finally reach a critical concentration corresponding to the critical thickness of Table 7.4–1. A related example is a Class 2 determination in which interelement effects due mainly to the matrix are appreciable, but constant: as long ago as 1930, Laby [31], with a photographic plate as detector, proved that copper or iron in zinc could be detected in concentrations near 1 part per million by weight. To be sure, he used electron excitation, which means that absorption effects were minimized. Another trace determination is expediently included here: the qualitative or semiqualitative, in which high precision is not sought so that interelement effects are deliberately ignored. The important point here is this: If the sample is properly presented to an x-ray emission spectrograph for trace determinations, results at least semiquantitative can be obtained easily by making simple calibrations.

7.13 Examination of Bank Notes

Figures 7.13–1 and 7.13–2 are examples of the kind of trace determinations just discussed. A genuine bank note was placed in an x-ray emission spectrograph so that a chosen area, known as to extent and location, was exposed to the x-ray beam. The chart recording in Figure 7.13–1 was quickly made. Chart recordings for the identical area were made on several counter-

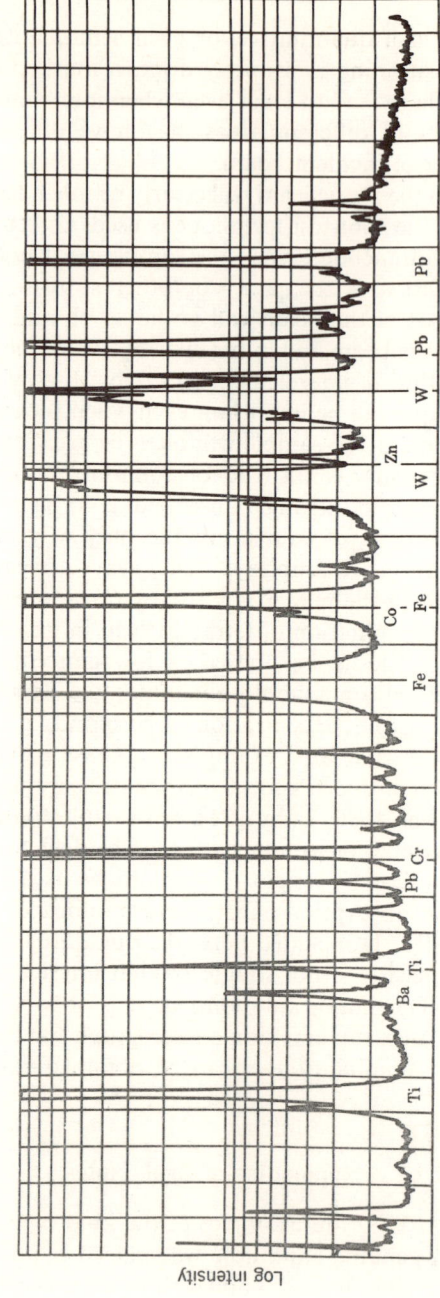

Fig. 7.13-1. Chart recording of the emission spectrum from a genuine bank note. From Ref. 36, p. 163, Fig. 7-1.

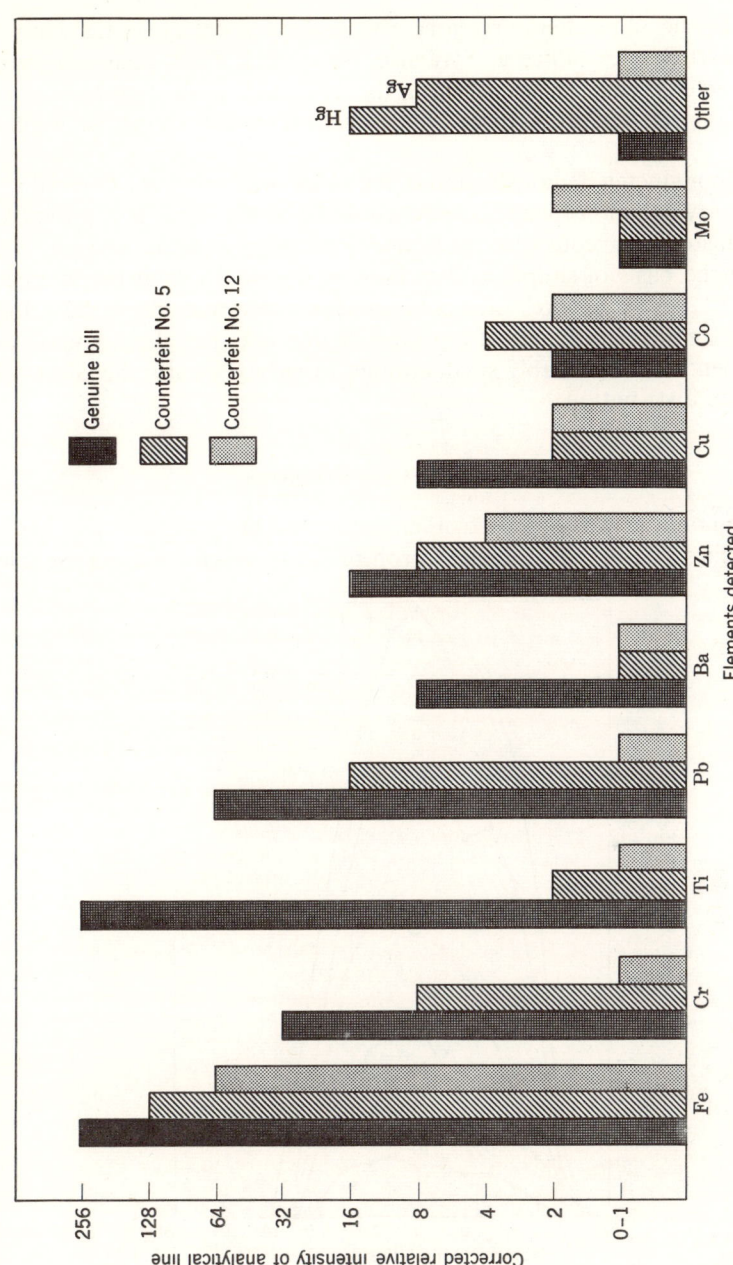

Fig. 7.13-2. Histograms made from x-ray spectrographic chart recordings for one genuine and two counterfeit bank notes. These counterfeit bills were obtained through the courtesy of the United States Secret Service Office at Syracuse, N.Y., with whose permission these data are presented. From Ref. 36, p. 227, Fig. 8-4.

feit bills of the same denomination, obtained by courtesy of the United States Secret Service Office at Syracuse, New York. Each such recording was characteristically different from the rest and from Figure 7.13–1. Histograms prepared from three of the recordings are shown in Figure 7.13–2.

A more convincing demonstration of the value of simple trace determinations by x-ray emission spectrography would be hard to find. It is probable that the histograms could be translated into fairly reliable quantitative results on the basis of simple calibrations on known spots in the manner of Figure 7.12–1. To be sure, some of the elements detected were in the sizing of the bank-note paper, and others were in the ink; but the amounts of these elements were probably small enough to reduce greatly the effects of nonuniform distribution.

7.14 Vacuum Tube Problems

Anyone familiar with the electron-emission and life problems that beset vacuum tubes knows that it is often prohibitive and sometimes impossible

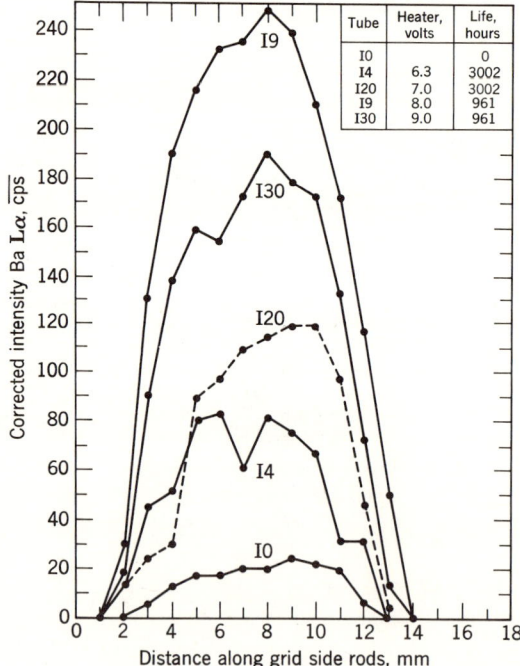

Fig. 7.14–1. Distribution of barium along side rods of grids in vacuum tubes. From Bertin and Longobucco, Ref. 43, p. 582, Fig. 9.

SIMPLE TRACE DETERMINATIONS

to obtain by classical analytical methods the chemical information needed for the solving of these problems. Bertin and Longobucco [43] have shown that trace determinations by x-ray emission spectrography are much more useful here than classical methods can ever be. There is room here for only one of the several ingenious techniques they used to obtain the different kinds of information needed. Having made provision in the sample drawer of a General Electric XRD-3 spectrograph for rotating the sample in two mutually perpendicular directions, they could scan a sample and register the intensity of analytical lines from restricted areas by use of a General Electric Heinrich milliprobe [44].

During tube operation, barium is volatilized from the oxide-coated cathode and deposited on various parts—for example, on the nickel grids, where it can sometimes lead to objectionable secondary electron emission. Figure 7.14–1 shows the distribution of deposited barium (amounts in micrograms) along grid side-rods as obtained by x-ray emission spectrography for tubes of different *effective* life (lower heater voltage makes for longer life). The position of Curve I 9 in the figure cannot be explained, but there is no doubt it is correct. Comparison is worth making of the data in Ref. 43 with those laboriously obtained on a related problem by conventional methods [45].

7.15 Films Near the Monolayer Level

The films now to be discussed stand in sharp contrast to those in the practical applications of the two preceding sections. We move here to the absolute lower limit of the linear region, where the limit of detection can be a few per cent of a monolayer. We are thus in the region of the cobalt example used previously [36] and cited above; but the work about to be described is much more difficult to do because it involves the ultimate in surface and sample preparation, and because the elements dealt with are *light* elements (carbon, oxygen), the determination of which presents special problems treated in a later chapter.

For an understanding of the interactions of metals and gases, familiar examples of which are various corrosion reactions, Cohen and his colleagues [46] at the National Research Council of Canada have proved it necessary to study the formation of clean surfaces on metals, and the initiation and growth of thin films on such surfaces. Of primary concern here is their linking of high-energy *electron* diffraction (HEED) and x-ray emission spectrography, which makes it possible to obtain information simultaneously about the structure and the composition of films in vacua better than 10^{-9} torr. During a HEED experiment, the sample not only diffracts the electron beam, but becomes a line source of characteristic x-rays that can be used to establish film composition. The addition of a Philips spectrometer makes the HEED

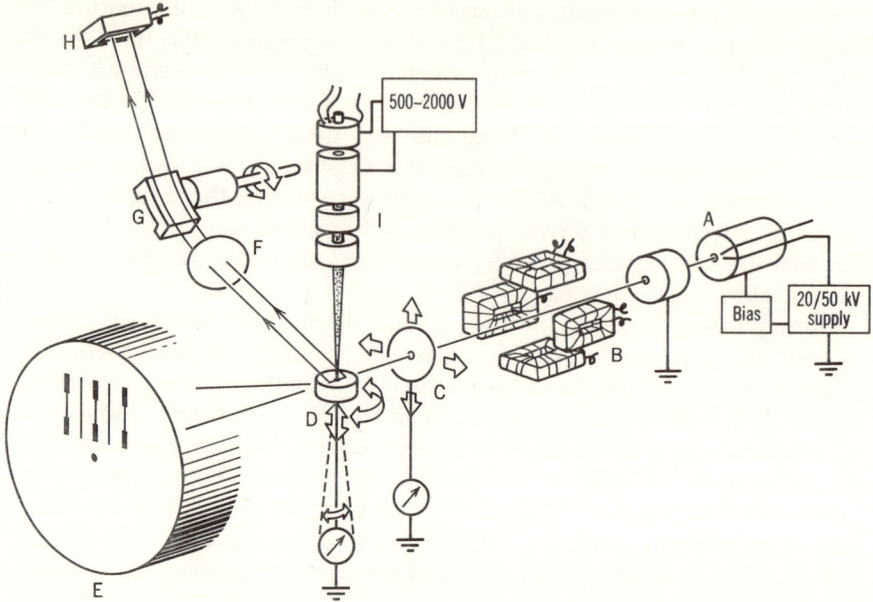

Fig. 7.15–1. Present experimental arrangement at the National Research Council of Canada for simultaneous HEED and x-ray emission determinations on thin films in high vacuum. Legend: A, electron gun, which produces a slightly divergent beam of 50-kV electrons; B, deflection coils for alignment of beam with C (diameter, about 100 μ), an aperture that limits the beam and collects about 99% thereof to provide a signal for the stabilization of A; D, sample mounted upon an adjustable manipulator; E, phosphor screen for display of HEED pattern; F, G, H, components of Philips x-ray spectrometer for light elements; F, thin window separating two vacua (10^{-3} torr in spectrometer; 10^{-9} torr in sample chamber); G, Henke-type *curved* Bragg reflector, *mica with lead stearate overlay*; H, flow proportional detector; I, source of lower-energy electrons for excitation of characteristic lines: normal incidence shown, other positions possible. Broad arrows in figure show directions of possible mechanical movements. See Chapter 5. From P. B. Sewell, Ref. 47.

equipment into a combined diffractometer-emission spectrograph, to which an additional electron source (I in Figure 7.15–1) has been added to increase sensitivity [46d; 47].

Analytical lines in samples can generally be excited by means other than x-rays or electrons: films are no exception. Saylor and Marks [48] used protons and ions of heavier elements (argon, nitrogen) on oxide films of thicknesses up to 5000 Å on nickel, aluminum, and their alloys; energy resolution was used. Hart, Olson, and Smith [49] used 100-keV protons to excite oxygen $K\alpha$ from known thicknesses of oxide anodically produced on high-purity aluminum and found a limit of detection corresponding to about 0.1 monolayer of oxygen.

SIMPLE TRACE DETERMINATIONS

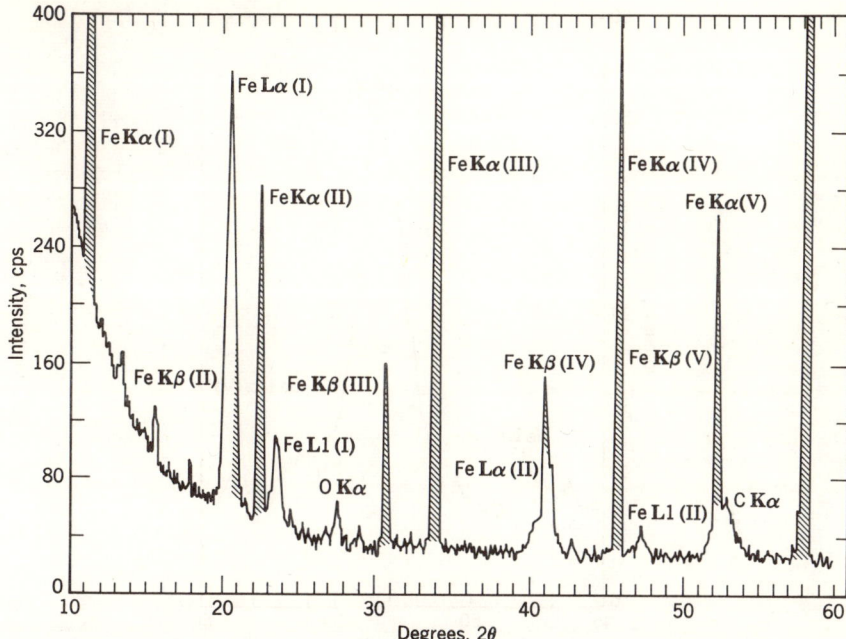

Fig. 7.15–2. X-ray emission spectrum from electropolished surface of high-purity iron recorded as excited during HEED experiment under these conditions: Electron beam, 40 kV and $3(10^{-5})$ mA; grazing angle of incidence, about $1°$; take-off angle for x-rays, $15°$. All cross-hatched lines were reflected from mica substrate of curved Bragg reflector and can be eliminated by pulse-height selection. All other lines were reflected from lead stearate film on mica substrate. Roman numerals show order of Bragg reflection in either case. From Sewell, Mitchell, and Cohen, Ref. 46d, p. 71, Fig. 6.

We take up again the electron excitation of analytical lines, not only because of its importance to the analytical chemist, but also because the experiments done on the apparatus of Figure 7.15–1 are interesting analytical chemistry. Figure 7.15–2 is an x-ray emission spectrum excited by the HEED technique from the electropolished surface of high-purity iron and recorded without pulse-height selection or discrimination. The rising background with decreasing 2θ is normal. Particularly noteworthy features: (1) The presence of iron lines at high intensities and of higher orders in Bragg reflection (Roman numerals). (2) The fact that Fe **K** lines, of whatever order, all reflected from mica, can be removed by pulse-height selection, which is indicated by the cross-hatching in the figure. As will appear, it is possible to isolate C **K**α in this way. The figure is a classic illustration of discussions in Chapter 2. (3) The C **K**α peak arises from carbon introduced when the surface was cut by spark erosion under oil. (4) The O **K**α peak is

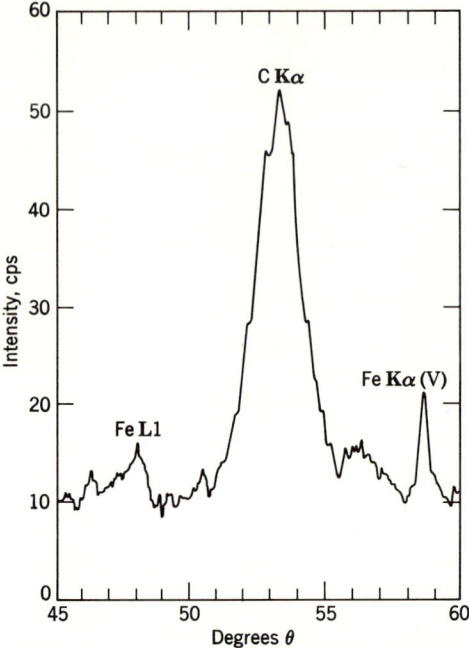

Fig. 7.15–3. C Kα peak from monolayer of stearic acid on electropolished iron. Conditions approximately as in Fig. 7.15–2 except that pulse-height discrimination was used: Fe Kβ (V) was removed completely, but Fe Kα (V) was not. From Sewell, Mitchell, and Cohen, Ref. 46d, p. 73, Fig. 7.

indicative of a thin oxide film and can be measured without serious interference by iron lines.

Scientific interest apart, the determination of carbon on "clean" metal surfaces is a matter of first importance because carbon contamination can be a great nuisance in all equipment (such as the electron microprobe) that offers electrons the opportunity to crack stray hydrocarbon vapors in high vacuum. Sewell, Mitchell, and Cohen accordingly carried out a carbon determination on a carefully prepared monolayer of stearic acid on the best clean iron surface they could make. The C Kα intensity, shown in Figure 7.15–3, was used to establish the sensitivity of the method for carbon [26d], the area of a stearic acid molecule and the structure of the iron (100) surface being known [46d]. For the C Kα peak, Figure 7.15–3 gives 4.1 ± 0.3 as the ratio $(N_T - N_B)/N_B$. The calculations for the stearic acid monolayer indicate a value of 0.52 for this ratio when the iron surface is covered by a monolayer of carbon. For a counting interval of 200 sec, the estimated limit of detection for carbon on iron is about 0.1 monolayer, or $2.5(10^{-9})\,\text{g/cm}^2$.

SIMPLE TRACE DETERMINATIONS

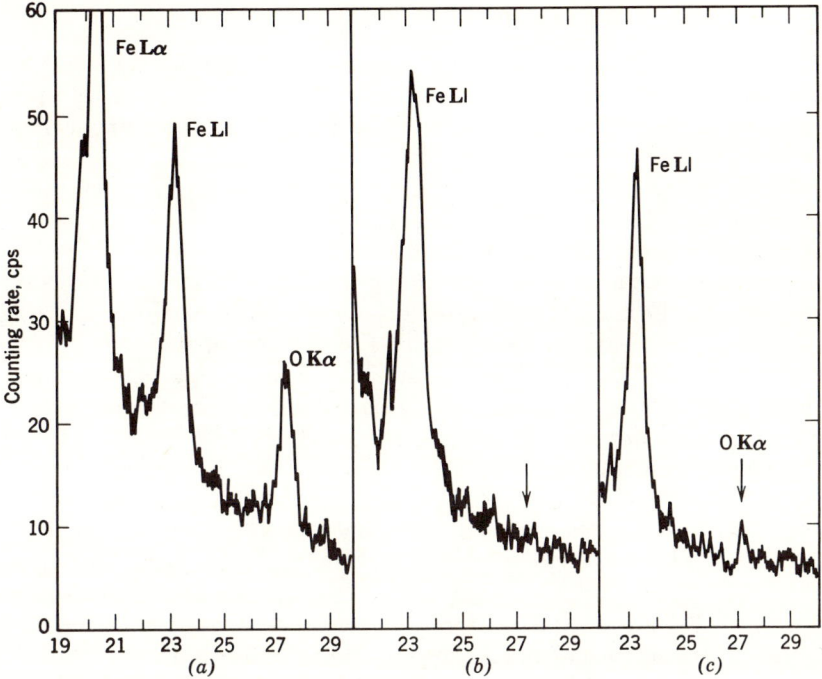

Fig. 7.15–4. The O $K\alpha$ peak from oxygen on iron: *(a)* air-formed oxide film; *(b)* after reduction by H_2; *(c)* after exposure for 1 hr to water vapor at 10^{-7} torr. From Sewell, Mitchell, and Cohen, Ref. 46d, p. 76, Fig. 9.

The work is being continued with various substrates (Si, Fe, Pt, Ta) and various overgrowths (O, F, B, C) to obtain comparative sensitivity data for three methods of electron excitation (40 kV and 1800 V with grazing incidence; 1800 V with normal incidence). Sensitivities are expressed in g/cm^2 for unit value of the ratio $(N_T - N_B)/N_B$: thus values near $6(10^{-8})$ g/cm^2 are found for oxygen on these substrates by the third method of excitation. This method gives fivefold lower limits of detection (see Chapter 8) than the first [47].

Figure 7.15–4 shows that the method will be useful for studying the kinetics of thin-film formation. The strongest O $K\alpha$ peak, for which $(N_T - N_B)/N_B$ was 1.5, was given by a film of cubic oxide ($Fe_3O_4 - \gamma Fe_2O_3$) known to be of mean thickness 17 ± 2Å, equivalent to about ten layers of oxygen in the close-packed plane of Fe_3O_4. The effectiveness of the method can be judged on this basis.

7.16 Thin Films and Auger-Electron Spectrography

Chapter 1 spoke of the interplay between x-rays and electrons and its importance to analytical chemistry. A comparison of this section with that preceding makes the point as well as it can be made. In the preceding section, films were studied by measuring intensities of characteristic lines produced by electron excitation and the accompanying generation of (Auger) electrons was neglected; here this situation is reversed.

Section 1.23 explained that Auger electrons are ejected when radiationless transitions occur in atoms containing holes formed by the ejection of photoelectrons. Both kinds of electrons can serve to characterize the emitting atoms; in the case of photoelectrons, the Einstein equivalence law must be taken into account (Section 4.5). The emission of Auger electrons competes with the generation of characteristic lines by either x-ray or electron excitation; data in an appendix show that the production of Auger electrons easily wins the competition at low atomic numbers. Accordingly, determinations of light elements—difficult by x-ray emission spectrography—offer Auger electrons their best chance of being useful in chemical analysis.

The possibility that Auger electrons might be useful in chemical analysis was neglected until recently, when Harris [50] in valuable investigations proved that they are well adapted to *qualitative* determinations on surfaces. His apparatus consisted of an electron gun to excite the sample in its holder, an electron-energy analyzer, and an electron multiplier. The Auger-electron energies range up to about 2000 eV, and the experimental data are electron-multiplier output (response) recorded against Auger-electron energy (analogous to x-ray intensity as a function of wavelength). The characteristic Auger peaks are difficult to isolate from the large and complex background. Harris made an important contribution by using the *derivative* of the response curve, automatically obtained by electronic differentiation, to accomplish this isolation. Figure 7.16–1 shows results by the Auger-electron method for an application [51] related to those of the previous section.

However, the relationship between the material covered by these two sections is much closer than this. Weber and Pina [52a] showed that the usual three-grid low-energy electron diffraction (LEED) system could be used to give the differentiated response proved necessary by Harris to make Auger-electron spectrography useful. It is therefore possible to link LEED and Auger-electron spectrography for the examination of thin films. In Section 7.15, HEED and x-ray emission spectrography were linked to the same end.

Weber and Pina [52a] reported a sensitivity of 0.1 monolayer Cs on Si, a sensitivity since improved by them, according to Palmberg and Rhodin [52b], who have used Auger-electron spectrography to study the surfaces of metals of face-centered-cubic structure.

SIMPLE TRACE DETERMINATIONS

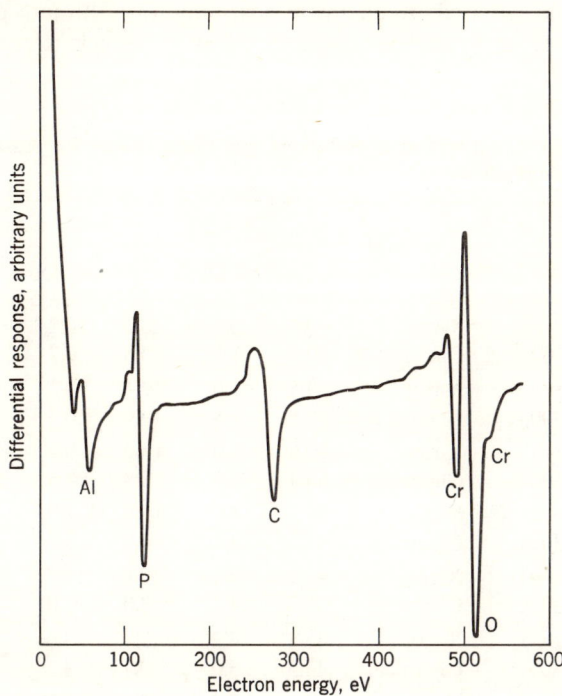

Fig. 7.16-1. Auger-electron examination of electropolished aluminum left for 30 minutes in a phosphoric-chromic acid bath at 90° and then thoroughly washed with water. Differentiation gives maxima and minima characteristic of the elements and permits much more certain evaluation of the (qualitative) results than is possible for the undifferentiated response. From Dunn and Harris, Ref. 51, p. 81, Fig. 2.

Because we are concerned primarily with x-ray methods, Auger-electron spectrography will have to be dismissed here. Sections 7.15 and 7.16 show that x-rays and electrons must be regarded as interconvertible and interchangeable in analytical chemistry.

7.17 Alloy and Oxide Films

Rhodin's objective [21] in the investigation that produced Figure 7.4–4 was to establish by x-ray emission spectrography the composition of oxide films formed on stainless steel, such films being of first importance in connection with the passivity and with the atmospheric oxidation of these alloys. Experimental details: (1) Metals and alloys deposited by evaporation on 0.25-mil Mylar (polyester) film to reduce scattered background below what it would have been with filter paper. (2) Oxide films formed by wet oxidation and isolated by chemical stripping. (3) Ultimate calibration by microbalance; colorimetric methods used to check x-ray results. See Table 7.17–1.

Table 7.17–1. Composition of Substrate and Oxide Films for Austenitic Stainless Steels by Two Methods

Sample	Method	Fe		Cr		Ni	
		%[a]	Std. Dev.	%	Std. Dev.	%	Std. Dev.
Type 304 alloy	X-ray	70.5	0.5	20.2	0.2	8.7	0.1
thickness, 300 Å	Chemical[b]	69.6	1.2	20.6	0.7	9.0	0.1
Type 316 alloy	X-ray	69.5	1.0	19.4	0.2	11.3	0.1
thickness, 300 Å	Chemical	68.5	1.8	19.6	0.3	10.9	0.2
Type 347 alloy	X-ray	70.6	2.0	18.2	0.2	9.8	0.1
thickness, 300 Å	Chemical	69.4	2.1	18.8	0.6	10.2	0.2
Type 304 oxide-1	X-ray	37.2	4.0	13.4	2.0	9.4	1.2
thickness, 30 Å	Chemical	34.0	3.1	14.0	3.4	10.7	4.0
Type 316 oxide-1	X-ray	42.1	3.2	12.6	0.7	4.9	0.6
thickness, 30 Å	Chemical	45.0	4.2	13.0	2.0	5.4	2.1
Type 347 oxide-1	X-ray	45.4	4.2	17.7	3.0	7.7	0.8
thickness, 30 Å	Chemical	44.5	5.0	20.7	3.2	8.2	2.6
Type 304 oxide-2	X-ray	15.0	2.0	45.5	3.7	3.8	0.2
thickness, 300 Å	Chemical	15.1	2.6	46.6	5.0	4.0	0.5
Type 316 oxide-2	X-ray	25.0	3.2	36.8	3.1	5.0	0.3
thickness, 300 Å	Chemical	25.2	4.0	36.1	4.3	5.2	0.8
Type 347 oxide-2	X-ray	24.8	2.8	36.1	2.1	2.2	0.1
thickness, 300 Å	Chemical	25.0	3.2	35.2	4.1	2.2	0.1

[a] Percentage by weight. Sample weight 25 to 100 μg.
[b] The chemical method was microcolorimetric.

These data are a classic example of what can be accomplished on thin films by use of the microbalance, of microcolorimetry, and of x-ray emission spectrography with conventional (in this case, Philips) equipment. All x-ray measurements were in the linear region; interelement effects were not detected. In general, data by the two analytical methods are in excellent agreement with precision somewhat higher for the x-ray. The striking differences in composition between the 300 and 30-Å oxide films are of course important as regards the corrosion of these important steels. Discussion of this matter requires consideration of trace elements (Si, Nb, Mo) not determined by the x-ray method, wherefore the interested reader is asked to consult Rhodin's articles.

In the electronics industry, low-power resistors are made by evaporation of resistance-heated nichrome alloys in vacuum. As chromium is more volatile than nickel, the condensate becomes poorer in chromium as evaporation proceeds. Control of resistivity therefore requires determination of composition, and Spielberg and Abowitz [53] used a conventional Philips x-ray emission spectrograph to that end. The estimated precision, largely reflecting the effect of counting error, was 3%, or comparable with Rhodin's; likewise comparable was the experience with interelement effects. About 10 minutes is needed to determine the Ni/Cr ratio in a thin film.

Spielberg and Abowitz [53] took unusual care with the calibration of their x-ray emission method. In addition to optical interferometry and microcolorimetry, which we have met before, they used x-ray absorptiometry with a monochromatic beam at a sufficiently low angle of incidence (6°) to ensure adequate attenuation. (The substrate for the film must be adequately transparent.) The measurements were done in a Philips powder diffractometer with Cu $K\beta$ reflected from a silicon (111) Bragg crystal before it was incident on the film. These calibrations and the microcolorimetric were considered more reliable than the interferometric, which suffered because (as often happens) the density of the evaporated films was less (up to 40% less) than the densities of the metals in bulk. Finally, Spielberg and Abowitz [53] found that beryllium behaved differently as substrate for condensing the metal vapors than did glass or quartz—an indication that x-ray emission spectrography is an ideal way to measure "sticking coefficients" of relatively nonvolatile species.

7.18 Class 1 Determinations with Prior Concentration

Chief among the reasons for concentrating a trace to make possible a Class 1 determination are (1) to increase sensitivity; (2) to eliminate elements that emit interfering lines; (3) to eliminate (or reduce) interelement effects.

Three of the methods of concentrating the trace are among the orthodox methods of separation in analytical chemistry.

1. A liquid in which the trace is dissolved or suspended is evaporated (aqueous solution) or chemically removed (burning, digesting with oxidizing acids); and the residual trace, redissolved if necessary, is transferred onto a suitable substrate (usually Mylar or paper) before being placed in the spectrograph. Examples: Addink [54] determined 3 to 10 ppm of zinc in blood by dry-ashing 5-ml samples at 490°C. Davis and Hoeck [55] determined vanadium (e.g., 0.4 ppm $\pm$ 10%) and nickel (e.g., 1.3 ppm $\pm$ 2.3%) in 5 to 20 g of residual fuels and charging stocks after concentration by digestion. Gunn [56] determined nickel, vanadium, and iron in feed-stock oils after

charring 10 g of sample with sulfuric acid, igniting, dissolving the ash in hydrochloric acid, adding cobalt ion as internal standard, and finally measuring intensity ratios after transfer to a circle (diameter, $1\frac{3}{8}$ in.) of filter paper.

2. An example of trace separation by precipitation (and subsequent ashing) is the determination of uranium in oilfield waters by Kehl and Russell [57] at levels as low as 0.01 ppm in the water.

3. The mercury cathode was used by Cavanagh [58] to separate high-purity iron from traces of niobium, tantalum, hafnium, zirconium, uranium, and thorium, which were eventually determined after transfer to a Mylar substrate. These determinations convincingly demonstrate the comparative advantages of x-ray emission spectrography.

Examples of two unusual methods of concentrating traces: (1) Campbell and Leon [59] found that selective oxidation of lead concentrated arsenic and antimony in the surface layers, where the two impurities could be determined. (2) Hirt, Doughman, and Gisclard [60] concentrated by impingement the heavy elements in airborne dust from a known volume of air on a glass-fiber filter-disk. Because of the present concern about air pollution and because of the improvement in x-ray equipment since 1956, this technique will no doubt become more practiced and better known.

As regards trace determinations, ion-exchange membranes (or papers) and x-ray emission spectrography are made for each other: such membranes not only separate and concentrate trace elements in solution, but also serve as a substrate for direct insertion in the spectrograph. Grubb and Zemany [61a] early appreciated these advantages and demonstrated that 0.001 ppm cobalt in solution could be determined in this way; a later application from this laboratory [61b] was the determination of potassium ion liberated from the surface of mica that had been ground. Campbell and Carl [62] discussed various applications of the technique in 1956; Campbell and Thatcher [39] studied it carefully at the Bureau of Mines.

Ion-exchange-membranes present the problems (background, inter-element effects, x-ray absorption by covering films) usual in Class 1 determinations. In addition, they have problems of their own. The usual exchange is that of a dissolved cation with hydrogen ion of the membrane. The cation to be determined must compete in this process with other cations, and especially with hydrogen ion, present in the solution. Also, the exchange process requires time and may never be complete [39, 63]. The rate of exchange can be increased by using powdered or liquid resins [64], by ordinary stirring, and by ultrasonic agitation.

Table 7.18–1 shows what can be done [39]. Campbell and Thatcher used an Amberplex C-1 cation-exchange membrane held in the spectrograph by the device in Figure 7.18–1 after it had collected zinc during 24 hours

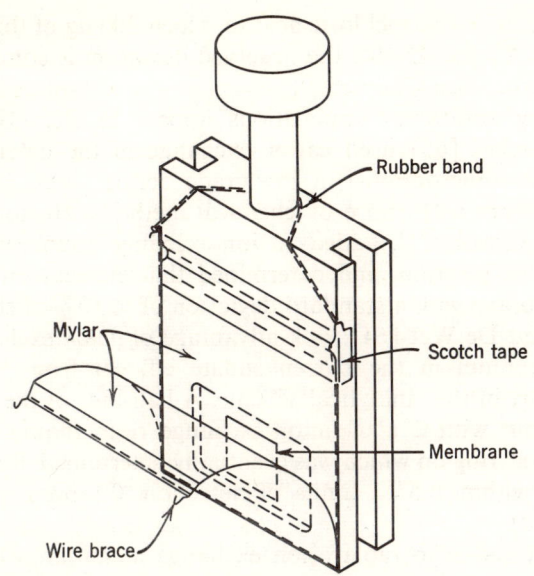

Fig. 7.18–1. Holder for ion-exchange membrane carrying sample for Class 1 determination. From Campbell and Thatcher, Ref. 39, p. 51, Fig. 1.

Table 7.18–1. Results of Zinc Determinations on Ion-Exchange Membranes [39]

Membrane	Side	Zinc, μg	Deviation, %
1	a	199	− 0.5
	b	196	− 2.0
2	a	203	+ 1.5
	b	202	+ 1.0
3	a	197	− 1.5
	b	197	− 1.5
4	a	200	0.0
	b	201	+ 0.5
5	a	203	+ 1.5
	b	201	+ 0.5
6	a	199	− 0.5
	b	204	+ 2.0

from 1 liter of 0.01 N hydrochloric acid to which 200 µg of this cation had been added. As N_T was 25,600, the observed deviation is comparable with the counting error. (See Chapter 8.)

The following additional applications appear in Ref. 39. Campbell, Leon, and Thatcher [65] used cation-exchange in the determination of iron and copper (micrograms) in low-grade copper ores, and obtained results in agreement with those by chemical methods. Horton and Moak [66] dissolved Zircalloy II, added an ion-exchange membrane to collect thorium from the solution, and determined that element on the washed and dried membrane with a standard deviation of ± 10% at the 5 µg level. Van Niekerk and De Wet [64] took advantage of *anion* exchange to concentrate the uranium in the barren sulfate effluent from ion-exchange columns not part of this analytical scheme. A half-liter of the effluent was shaken for 5 min with 2 g of anion-exchange resin (liquid or powder), the uranium appearing on which was eventually determined. Results: 1 ppm determined to within ± 5%; limits of detection, 0.1 ppm (powder) and 0.2 ppm (liquid).

The present trend is to replace ion exchange membranes by papers in which the ion-exchange resins are incorporated.

7.19 Class 2 Determinations

Detailed discussion of Class 2 determinations is best postponed for this reason: Because the matrix, often a solvent, is never entirely transparent to x-rays, the critical depth (which corresponds to critical thickness in a film) is established by the matrix and by the element sought, which means that this critical depth usually decreases with increasing concentration of the element. Consequently, Class 2 determinations are virtually all made on samples of depths exceeding the critical.

7.20 Conclusion

Thickness and trace determinations, and studies of surfaces, are among the most interesting and important uses for x-ray emission spectrography. Such studies often need to be done on localized areas; when extreme localization (sq microns) is needed, the highly specialized spectrograph called the electron microprobe must be used.

The advantages of joining simple trace determinations to thickness measurements in this chapter seemed to outweigh the disadvantages of discussing such determinations in advance of x-ray emission spectrography proper. In general, the requirements for success there are more easily met than for the applications of this chapter, in which intensities are likely to be low, counting times long, reliable standards difficult to prepare, and careful handling of the sample a major consideration. Being able to look at monolayers is worth the price!

A final word. If a trace is to be determined, reacting it chemically is sometimes advantageous, especially if the element sought is light. As procedures of this kind are routine in classical analytical chemistry (witness the determination of phosphorous as phosphomolybdate), no discussion is needed. The trick is to convert the trace into a compound in which at least one element is easy to determine by x-ray emission spectrography. The same trick will occasionally be useful when the amount of sample is not limited.

REFERENCES

1. T. Shiraiwa and N. Fujino, in *Advances in X-Ray Analysis*, Vol. 12, Plenum, New York, 1969, p. 446.
2. G. L. Clark, G. Pish, and L. E. Weeg, *J. Appl. Phys.*, **15**, 193 (1944).
3. L. S. Birks and H. Friedman, *Phys. Rev.*, **69**, 49 (1946).
4. H. Friedman and L. S. Birks, *Rev. Sci. Instr.*, **17**, 99 (1946).
5. R. B. Gray, *Phys. Rev.*, **69**, 49 (1946).
6. A. Eisenstein, *J. Appl. Phys.*, **17**, 874 (1946).
7. L. S. Birks, *X-Ray Spectrochemical Analysis*, Interscience, New York, 1959, p. 80.
8. H. F. Beeghly, *J. Electrochem. Soc.*, **97**, 152 (1950).
9. G. E. Pellissier and E. W. Wicker, *Elec. Mfg.*, **49**, 124 (1952).
10. *Norelco Reptr.*, **3**, 58 (1956).
11. *Norelco Coating Weight Gauges*, Philips Electronic Instruments, Circular RC 389 3 MF 762.
12. J. A. Dunne, in *Advances in X-Ray Analysis*, Vol. 6, Plenum, New York, 1963, p. 345.
13. P. D. Zemany and H. A. Liebhafsky, *J. Electrochem. Soc.*, **103**, 157 (1956).
14. H. A. Liebhafsky and P. D. Zemany, *Anal. Chem.*, **28**, 455 (1956).
15. R. H. Zimmerman, in *Advances in X-Ray Analysis*, Vol. 4, Plenum, New York, 1961, p. 335.
16. G. H. Glade, in *Advances in X-Ray Analysis*, Vol. 11, Plenum, New York, 1968, p. 185.
17. R. Glocker and H. Schreiber, *Ann. Phys.*, **85**, 1089 (1928).
18. P. K. Koh and B. Caugherty, *J. Appl. Phys.*, **23**, 427 (1952).
19. R. M. Brissey, H. A. Liebhafsky, and H. G. Pfeiffer, *Am. Soc. Testing Materials Spec. Tech. Publ.*, **No. 157**, 43, 1954.
20. H. G. Pfeiffer and P. D. Zemany, *Nature*, **174**, 397 (1954).
21. T. N. Rhodin, *Anal. Chem.*, **27**, 1857 (1955).
22. Applied Research Laboratories, *Spectrographer's News Letter*, **X**, No. 1 (1957).
23. (a) J. E. Cline and S. Schwartz, *J. Electrochem. Soc.*, **114**, 605 (1967); (b) J. E. Cline, in *Progress in Analytical Chemistry*, Vol. 2, Plenum, New York, 1969, p. 83.
24. F.X. Pink and V. Lyn, *Electrochem. Tech.*, **6**, 258 (1968).
25. (a) L. S. Birks, E. J. Brooks, and H. Friedman, *Anal. Chem.*, **25**, 692 (1953); (b) Ref. 7, pp. 80–82; (c) W. C. Keesaer, in *Advances in X-Ray Analysis*, Vol. 3, Plenum, New York, 1959, p. 77.

26. (a) G. B. Cook, C. E. Mellish, and J. A. Payne, *Anal. Chem.*, **32**, 590 (1960); (b) J. F. Cameron and J. R. Rhodes, *Brit. J. Appl. Phys.*, **11**, 49 (1960); (c) P. D. Zemany, *Rev. Sci. Instr.*, **30**, 292 (1959).
27. (a) H. A. Liebhafsky, *Anal. Chem.* **26**, 26 (1954); (b) **28**, 583 (1956) summarizes the earlier literature on cadmium sulfide as x-ray detector.
28. D. C. Reynolds, G. Leies, L. L. Antes, and R. E. Marburger, *Phys. Rev.* **96**, 533 (1954).
29. F. L. Chan, *Developments in Applied Spectroscopy*, Vol. 7A, Plenum, New York, 1969, p. 3. This article contains references to extensive earlier x-ray work by the author.
30. E. P. Bertin, in *Progress in Analytical Chemistry*, Vol. 2, Plenum, New York, 1969, p. 35.
31. T. H. Laby, *Trans. Faraday Soc.*, **26**, 497 (1930).
32. L. v. Hámos, *Arkiv Mat., Astron. Fysik,* **31A**, No. 25 (1945).
33. L. v. Hámos and A. Engström, *Acta Radiol.*, **25**, 325 (1944).
34. A. Engström, *Acta Radiol. Suppl.*, **63** (1946).
35. J. H. Yoe, *Anal. Chem.*, **29**, 1246 (1957).
36. H. A. Liebhafsky, H. G. Pfeiffer, E. H. Winslow, and P. D. Zemany, *X-Ray Absorption and Emission in Analytical Chemistry*, Wiley, New York, 1960, pp. 225–237.
37. I. Adler, *X-Ray Emission Spectrography in Geology*, Elsevier, New York, 1966, pp. 160–163.
38. R. Jenkins and J. L. de Vries, *Practical X-Ray Spectrometry*, Springer-Verlag, New York, 1967, pp. 132–133, 151–156, 172–173.
39. W. J. Campbell and J. W. Thatcher, *Developments in Applied Spectroscopy*, Vol. 1, Plenum, New York, 1962, pp. 31–62 with 60 references.
40. F. Feigl, *Spot Tests, Vol. I, Inorganic Applications* (translated into English by R. E. Oesper), Elsevier, New York, 1954.
41. E. H. Winslow and H. A. Liebhafsky, *Anal. Chem.*, **21**, 1338 (1949).
42. H. G. Pfeiffer and P. D. Zemany, *Nature,* **174**, 397 (1954).
43. E. P. Bertin and R. J. Longobucco, in *Advances in X-Ray Analysis*, Vol. 7, Plenum, New York, 1964, p. 566.
44. K. F. J. Heinrich, in *Advances in X-Ray Analysis*, Vol. 5, Plenum, New York, 1962, p. 516.
45. J. P. Blewett, H. A. Liebhafsky, and E. F. Hennelly, *J. Chem. Phys.*, **7**, 478 (1939).
46. (a) P. B. Sewell, C. D. Stockbridge, and M. Cohen, *Can. J. Chem.*, **37**, 1813 (1959); (b) P. B. Sewell and M. Cohen, *Appl. Phys. Lett.*, **7**, 32 (1965); (c) P. B. Sewell and M. Cohen, *Appl. Phys. Lett.*, **11**, 298 (1967); (d) P. B. Sewell, D. F. Mitchell, and M. Cohen, in *Developments in Applied Spectroscopy*, Vol. 7A, Plenum, New York, 1969, p. 61.
47. P. B. Sewell, letter to H. A. L., January 13, 1970.
48. W. P. Saylor and C. L. Marks, in *Advances in X-Ray Analysis,* Vol. 12, Plenum, New York, 1969, p. 457; and earlier references there given.
49. R. R. Hart, N. T. Olson, and H. P. Smith, Jr., *J. Appl. Phys.*, **39**, 5538 (1968).
50. (a) L. A. Harris, *J. Appl. Phys.*, **39**, 1419 (1968); (b) **39**, 1428 (1968).
51. C. G. Dunn and L. A. Harris, *J. Electrochem. Soc.*, **117**, 81 (1970).
52. (a) R. E. Weber and W. T. Pina, *J. Appl. Phys.*, **38**, 4355 (1967); (b) P. W. Palmberg and T. N. Rhodin, *J. Appl. Phys.*, **39**, 2425 (1968).
53. N. Spielberg and G. Abowitz, *Anal. Chem.*, **38**, 200 (1966).
54. N. W. H. Addink, *J. Iron and Steel Inst.*, **1960**, 199.
55. E. N. Davis and B. C. Hoeck, *Anal. Chem.*, **27**, 1880 (1955).
56. E. L. Gunn, *Anal. Chem.*, **33**, 921 (1961).

REFERENCES

57. W. L. Kehl and R. G. Russell, *Anal. Chem.*, **28**, 1350 (1956).
58. M. B. Cavanagh, *The Application of X-Ray Fluorescence to Trace Analysis,* Naval Res. Lab. Rept. 4528, 1955, 4 pp.
59. W. J. Campbell and M. Leon, *Fluorescent X-ray Spectrograph for Dynamic Selective Oxidation Rate Studies: Design and Principles,* Bureau of Mines Rept. of Investigations 5739, 1961, 21 pp.
60. R. C. Hirt, W. R. Doughman, and J. B. Gisclard, *Anal. Chem.,* **28**, 1649 (1956).
61. (a) W. T. Grubb and P. D. Zemany, *Nature,* **176**, 221 (1955). (b) P. D. Zemany, W. W. Welbon, and G. L. Gaines, *Anal. Chem.,* **30**, 299 (1958).
62. W. J. Campbell and H. F. Carl, *Pittsburgh Conf. on Analytical Chemistry and Applied Spectroscopy*, Pittsburgh, March 1956, paper 92.
63. F. W. Lytle, *Determination of Trace Elements in Plant Material by Fluorescent X-Ray Analysis,* University of Nevada, Reno, Nev., M. S. thesis, 1958.
64. J. N. Van Niekerk and J. F. DeWet, *Nature,* **186**, 380 (1960).
65. W. J. Campbell, M. Leon, and J. W. Thatcher, *Solution Techniques in Fluorescent X-Ray Spectrography*, Bureau of Mines Rept. of Investigations 5497, 1959, 24 pp.
66. W. S. Horton and W. D. Moak, *Determination of Microgram Quantities of Thorium in Zircalloy II by X-Ray Fluorescence Spectroscopy with Ion-Exchange Membranes,* Knolls Atomic Power Lab. Rept. KAPL-M-WSH-4, 1959, 11 pp.

Chapter 8

Reliability of X-Ray Emission Spectrography Statistical Considerations

> Anything that will discourage men from believing general propositions I welcome only less than anything that will encourage them to make them.
> O. W. Holmes, Jr. to H. L. Laski, January 7, 1924.

8.1 Reliability of X-Ray Methods. General

The quotation above fits the theory of errors, which rests on generalizations that need to be questioned and tested. Fortunately, the testing is easy with x-ray emission spectrography as it can readily provide statistical information *en masse*.

Reliability in general is determined by both *accuracy* and *precision;* and accuracy implies the existence of a "true" value. To illustrate: Suppose that an alloy prepared to contain the element sought at weight-fraction a is analyzed for this element by two different wet methods, numerous replicate determinations being made by each. The analytical results $(x'_1, x'_2, \ldots, x'_n)$ by the first method, and those by the second $(x''_1, x''_2, \ldots, x''_n)$, will each cover a range of values, and each set of results will be distributed about its mean ($\bar{x}'$, mean for first method; $\bar{x}''$, mean for second method). To arrive at the form of the distribution, we list the number of results falling within various intervals measured from the mean; that is, within the contiguous intervals each Δx wide and located on both sides of the mean $\bar{x}'$—and within similar intervals measured from $\bar{x}''$. The numbers of results in the various

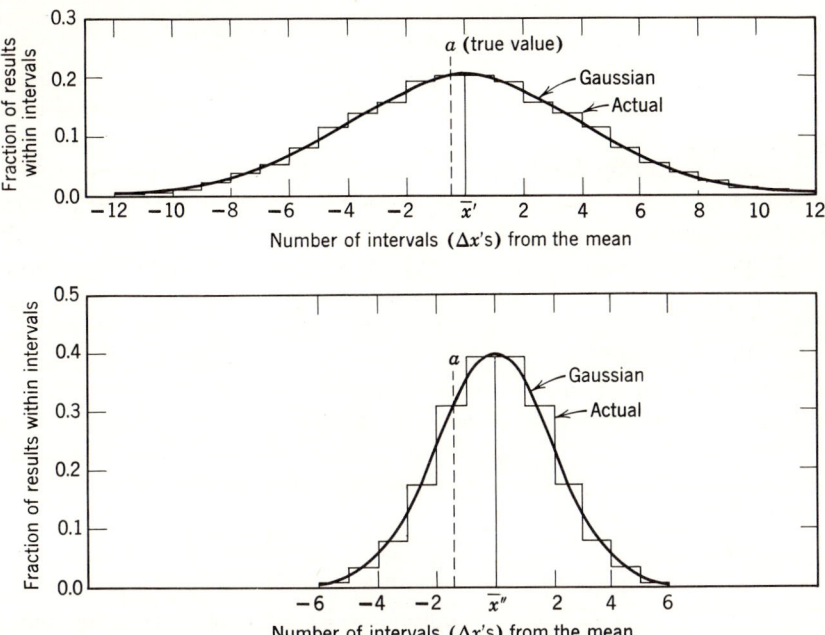

Fig. 8.1–1. Comparisons of Gaussian with assumed "actual" frequency distributions. Enough experimental results for such comparisons are seldom available, nor is the agreement likely to be as good as in the figure.

intervals establish the *frequency* distributions for the two methods. Two illustrative distributions (the "actual" distributions) are shown as stepped curves in Figure 8.1–1. In this figure, true values and means are plotted along the abscissas, which have been subdivided into intervals Δx, the mean being taken as the origin. For each such interval, there is plotted as ordinate the fraction of all measurements that falls within the interval. (This fraction is a more convenient parameter than the number of results within an interval.) The "true" value for present purposes is taken to be a, on the assumption that the preparation of the alloy was a more reliable procedure than its analysis.

Figure 8.1–1 illustrates two statements that experience has shown to be generally valid for analytical results obtained by wet methods: (1) The true value a and the mean $\bar{x}$ are different quantities, and one cannot be predicted from the other. (2) No conclusions about the frequency distribution can be drawn from a or from $\bar{x}$. One more generalization applies to *comparative* x-ray methods: if the comparison is properly carried out, questions of accuracy will never arise; "properly" presupposes a standard so reliable that it may be considered to contain the element sought at the "true" weight-fraction *a*. The reliability of a comparative method thus carried out is measured by its precision. We are justified in dismissing accuracy from further consideration here because methods such as x-ray emission spectrography are not *absolute* methods of determining the amount present of the element sought.

To make further progress, it is necessary to idealize the experimental results in Figure 8.1–1 by drawing continuous distribution curves through the stepped curves plotted from the analytical results. In principle, such continuity could be attained by decreasing the size of the intervals, Δx, until the steps disappear and a smooth curve results. But what is to be the shape of this curve? Here we must face the fact that the data available for ordinary analytical methods do not suffice to justify the assumption that one type of distribution curve fits all analytical methods; they do not even suffice to justify the less sweeping assumption that one type of curve fits one analytical method under the various conditions (different days, different operators) met with in the best practice. Note the emphasis on *type*: not one *curve*, but one type of curve.

A BRIEF DIGRESSION. In the language of statistics, either of the *stepped* distributions in Figure 8.1–1 records all measured values of x for a (statistical) *sample* of a *population* large enough to be represented by a *continuous* frequency distribution. All simple frequency distributions are characterized by a mean and a variance. The square root of the variance is the standard deviation. For the population, the mean is μ (*not* a mass absorption coefficient) and the variance is σ^2. For any sample smaller than the population, the mean is $\bar{x}$ and the (estimate of) variance is s^2. Now, $\bar{x}$ and s^2 for any sample can never be as reliable as μ and σ^2 because $\bar{x}$ and s^2 are only the experimental estimates of μ and σ^2. We are concerned in practical applications with these estimates, which must not be confused with μ and σ^2. Nevertheless, for simplicity's sake, we shall call s^2 the variance. For further discussion, see Ref. 1.

Let us return to the problem of what *type* of continuous distribution to choose for the results of an analytical method. The classical choice appears in Figure 8.1–1, where Gaussian curves have been drawn. The equation for this type of curve is

$$n_x = \left(\frac{n}{s\sqrt{2\pi}} \exp\left[\frac{-(x - \bar{x})^2}{2s^2} \right] \right) dx \qquad (8.1-1)$$

which gives the number of determinations n_x in an interval dx in the region of x, and hence for any value of $x - \bar{x}$. The number n_x is thus proportional to the ordinate of Figure 8.1–1. The variance s^2 in (8.1–1) is defined by

$$s^2 = \frac{\sum_i (x_i - \bar{x})^2}{n - 1} \qquad (8.1-2)$$

The standard deviation s is the square root of the variance; graphically, it is the horizontal distance from the mean to the point of inflection of the distribution curve. The standard deviation is thus an experimental measure of precision; the larger s, the flatter the distribution curve; the greater the range of replicate analytical results; and the less precise the method. In Figure 8.1–1, method 1 is less precise but more nearly accurate than method 2. One hopes that a and $\bar{x}$ will coincide, and that s will be small; but this happy state of affairs need not exist: in general, $\bar{x}$ and s cannot be predicted from a, nor from each other.

An analytical method is usually subject to more than one error. The standard deviation for the method will therefore be a composite of individual standard deviations. So long as these errors are independent, the standard deviations should combine as follows to give the overall standard deviation s:

$$s = \sqrt{s_a^2 + s_b^2 + \cdots + s_m^2} \qquad (8.1–3)$$

In (8.1–3), the subscripts identify the different individual standard deviations.

The conclusions reached above have already been used in Chapters 2 and 5. If, as was done in Figure 8.1–1, the *fractions* of determinations in intervals dx are plotted instead of their numbers, then (8.1–1) may be regarded as giving the *probability* that a determination will fall in interval dx. Probabilities being multiplicative, comparison of (8.1–3) with (8.1–1) shows that the overall standard deviation s may be regarded as that of an overall Gaussian made up of the individual Gaussians corresponding to subscripts $a, b, \ldots m$. This simple treatment may fail when the errors are not "well-behaved."

8.2 The Standard Counting Error

Lacking better information, we usually assume errors in x-ray emission spectrography to be independent and random. (Drift caused by changes in the electronic system is definitely not random.) Before we consider errors in general, we shall examine the one that is not only important and unavoidable, but that also sets x-ray emission spectrography apart from all methods that do not depend upon the counting of quanta. This is the *standard counting error*, s_C.

Under the simplest conditions, the standard counting error is *approximately* equal to the square root of the total number of counts; or

$$s_C = \sqrt{N} \qquad \text{(approximately)} \qquad (8.2–1)$$

The proof of this relationship is one of the triumphs of probability theory. The underlying considerations are most obviously applicable to radioactive systems, and it was to these that they were first applied [2].

Consider a system of one kind of radioactive atoms in which on the average $\bar{N}$ atoms are observed to decompose in a given time, Δt, and $\bar{N}$ particles are emitted to be counted, the number of radioactive atoms *remaining sensibly constant*. When such counts are made, a series of values $N_1, N_2, \ldots, N_{n-1}, N_n$ is obtained. The question now is whether the frequency distribution of these counts can be quantitatively explained on the assumption that radioactive decay is a random process. The affirmative answer was obtained first by von Schweidler [2]. The treatment below follows that of Beers [3a].

Subdivide the counting interval Δt into b equal subintervals, each so small that the chance of two atoms decomposing in the same subinterval may be neglected. The probability that an atom will decay in a given interval is then $\bar{N}/b$. The probability of having one atom decay in, for example, each of the first N subintervals is $(\bar{N}/b)^N$, and the probability of having none decompose in the remaining $(b - N)$ subintervals is $(1 - \bar{N}/b)^{b-N}$. The probability of having the N atoms decompose in this *particular* way, in time Δt, is the product of these two probabilities; or

$$(\bar{N}/b)^N (1 - \bar{N}/b)^{b-N}$$

This problem is identical in principle with that of drawing b balls out of a bag containing a very large number of white and of black balls in the ratio $\bar{N}$ to $b - \bar{N}$. The probability above is that of drawing N white balls and $b - N$ black balls in any one given order. (White balls represent particles emitted during the counting interval Δt.) But there are other ways of getting N white balls in b draws of one ball at a time. The much larger probability P_N of drawing N white balls *irrespective of order* from such a bag was first computed by Bernoulli according to what we now call the binomial distribution law; that is,

$$P_N = \frac{b(b-1)\ldots(b-N+1)}{N!} \left(\frac{\bar{N}}{b}\right)^N \left(1 - \frac{\bar{N}}{b}\right)^{b-N} \qquad (8.2\text{--}2)$$

The numerator of the first term is the number of ways N white balls could appear in b draws, and the denominator $N!$ is the number of ways these same N white balls could be interchanged. (Division by $N!$ in the first term reflects the fact that the order in which any specific white ball is drawn is unimportant, since this division by $N!$ in effect makes individual white balls indistinguishable.) If the decomposition of radioactive atoms and the resultant emission of charged particles really follow the laws of chance that govern the drawing of balls from a bag, then radioactivity must be a random process.

The matter can be subjected to experimental test if (8.2–2) can be made more tractable. This equation, though generally applicable, is hopelessly cumbersome when b is large, as it will always be in actual cases. Fortunately, as b becomes very large, (8.2–2) simplifies to

$$P_N = \frac{b^N \bar{N}^N e^{-\bar{N}}}{N! b^N} \quad \text{(identity of factors preserved)}$$

or

$$P_N = \frac{\bar{N}^N e^{-\bar{N}}}{N!} \quad \text{(factors combined)} \quad (8.2\text{–}3)$$

which is the Poisson distribution. It is important whenever the number of counts taken is low enough to make a count of zero fairly probable. The analytical chemist, except occasionally in trace determinations, will deal with counts so large that he need not concern himself with the Poisson distribution.

What distribution does concern him? It turns out to be the Gaussian, for the Poisson distribution as $\bar{N}$ becomes larger approaches more and more closely to the Gaussian. It may be shown analytically that, for large $\bar{N}$,

$$P_N = \frac{1}{\sqrt{2\pi\bar{N}}} \exp \frac{-(N - \bar{N})^2}{2\bar{N}} \quad (8.2\text{–}4)$$

Comparison of (8.2–4) with (8.1–1) results in two important conclusions, of which the second is the less obvious. First, if one equation represents a Gaussian distribution, so does the other. Second, the standard deviation in (8.2–4), which we have chosen to call the standard counting error, must be

$$s_C = \sqrt{\bar{N}} \quad (8.2\text{–}5)$$

The second conclusion is more important than might seem at first glance, as the following shows. (1) Owing to (8.2–5), the Gaussian distribution of (8.2–4) is *unique* for each $\bar{N}$. Because s_C is the horizontal distance from the mean to the point of inflection of the distribution curve, $\bar{N}$ alone—not average *and* standard deviation as in the case of (8.1–1)—suffices to define the ideal distribution curve. (2) From what has just been said, it follows that s_C is *predictable* from the total count and *controllable* through it. We have here a situation highly unusual in analytical chemistry, for (if everything works out favorably) we have in x-ray emission spectrography the possibility of predicting and controlling the best precision attainable so long as the analytical-line intensity is established by counting.

As a graphic summary of the properties of the three distributions, illustrative binomial, Poisson, and Gaussian distributions (the latter unique for the $\bar{N}$ chosen) have been plotted in Figure 8.2–1.

The next thing, of course, is to see whether radioactive counts in fact distribute themselves according to the unique Gaussian defined by $\bar{N}$ and $s_C = \sqrt{\bar{N}}$. As Figure 8.2–2 shows, Rutherford and Geiger [4] proved in 1910 that they do.

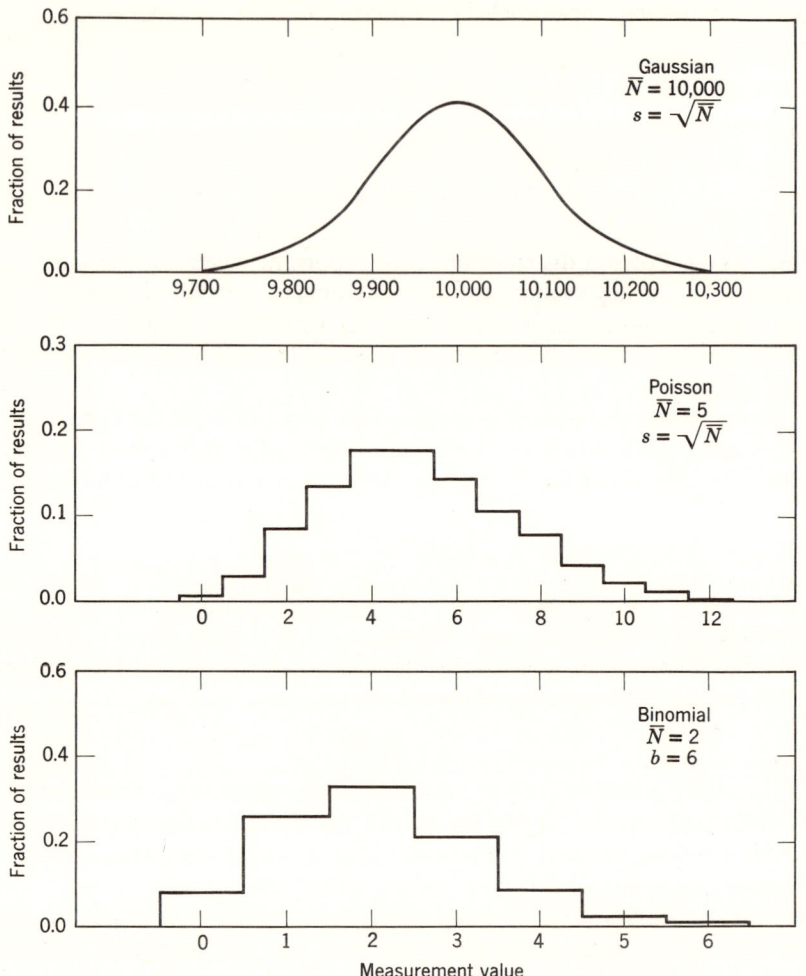

Fig. 8.2–1. Three frequency distributions for analytical results obtained by counting.

We need a similar proof for x-ray emission spectrography [5]. That the emission of quanta ought to resemble radioactivity in being a random process was pointed out by Einstein [6]. But the emission of x-rays is not x-ray emission spectrography, even though the conclusion $s_C = \sqrt{N}$ is usually applied to it without comment or misgiving. The conclusion thus applied can be strictly valid only when operating conditions are ideal. This statement becomes clear if the spectrograph is regarded as a means of maintaining in the sample a sensibly constant and extremely large number

RELIABILITY OF X-RAY EMISSION SPECTROGRAPHY

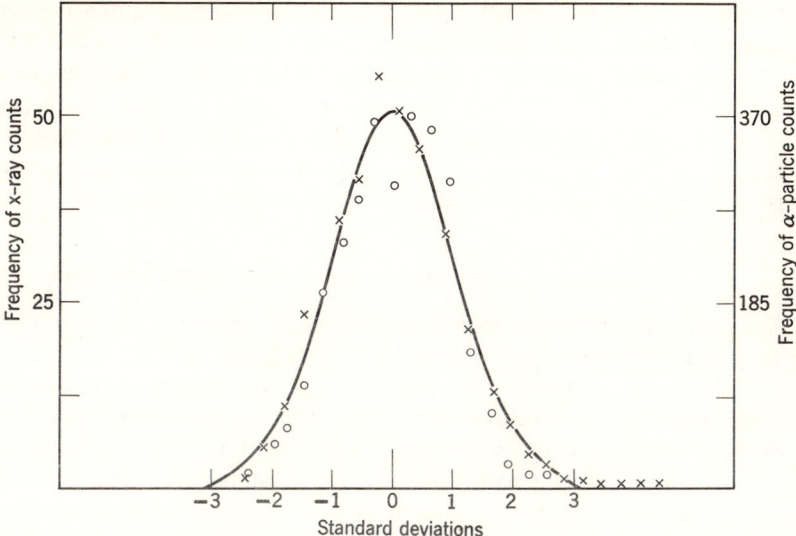

Fig. 8.2–2. Experimental proof that x-ray emission spectrography and radioactivity both conform to the unique Gaussian fluctuation curve based on $\bar{N}$ alone. Crosses = data of Rutherford and Geiger; circles = x-ray emission data; solid line = theoretical Gaussian curve. From H. A. Liebhafsky, H. G. Pfeiffer, and P. D. Zemany, Ref. 5.

N_0 of virtually identical excited atoms that emit the x-ray quanta being counted by the detector. The emission of such a quantum by any one of these N_0 atoms is a spontaneous process. This system of N_0 excited atoms maintained in the sample by the action of the exciting beam is thus analogous to a radioactive sample. The conclusion $s_c = \sqrt{\bar{N}}$ therefore ought to hold in the former case, *provided always* that the spectrograph system is functioning so well that other errors are not significant in comparison with the standard counting error under consideration.

This reasoning was tested experimentally [5] as follows: with a tungsten sample in the spectrograph, the goniometer was adjusted until a counting rate near 100 cps was obtained. This counting rate is high enough to give a convenient counting interval and low enough to eliminate significant coincidence errors in the Geiger detector. The time required to reach 1024 counts was then measured 393 times in succession. For each individual counting interval, the number of counts recorded for 10 sec was calculated by simple proportion. In this way, a body of data was obtained for which $t = 10$ sec; $n = 393$; and $\bar{N} = 1018$.

These data are plotted in Figure 8.2–2 about the Gaussian curve for which the standard deviation is the square root of the mean. The data of

Rutherford and Geiger, which were obtained by counting alpha-particles, are plotted about the same curve.

The radioactivity and the x-ray data lie satisfactorily about the *unique Gaussian* prescribed by (8.2–4). This conclusion was reached by applying the chi-square goodness-of-fit test according to Ref. 1, p. 620. The theoretical treatment by von Schweidler applies thus to radioactivity and to x-ray emission spectrography under satisfactory operating conditions.

8.3 The Standard Counting Error as an Operating Criterion

The standard counting error is an invaluable criterion for judging operating conditions in x-ray emission spectrography.

For the usual accurate analytical method, the mean, $\bar{x}$, is assumed identical with the true value, and observed errors are attributed to an indefinitely *large number* of small causes operating at random. The standard deviation, s, depends on these small causes and may assume any value; mean and standard deviation are wholly independent, so that an infinite number of distribution curves is conceivable. Under ideal conditions, x-ray emission spectrography differs sharply from such a usual case, because the only uncertainty present results from the random emission of x-ray quanta. We then have only a *single* small cause of error operating at random, and the individual counts must lie on the *unique* Gaussian curve for which the standard deviation is the square root of the mean. This unique Gaussian is a *fluctuation* curve, not an error curve in the strictest sense; there is no true value of N such as that presumably corresponding to a of Section 8.1—there is only a most probable value $\bar{N}$.

Inasmuch as s_C results from fluctuations that cannot be eliminated so long as quanta are counted, this standard deviation is the irreducible minimum for x-ray emission spectrography. Not only is it a minimum, but it is also a predictable minimum. When the standard deviation, s, significantly exceeds the standard counting error, s_C, it is likely that errors resembling those the analytical chemist usually encounters are superimposed upon the random fluctuations associated with the emission process.

A comparison of standard deviation s to standard counting error s_C is thus a useful criterion for the reliability of analytical results obtained by x-ray emission. To illustrate the simplest possible application of this criterion, consider again the x-ray data plotted in Figure 8.2–2, which are given in Table 8.3–1. The individual N's summarized in the 8.3–1 table could, in x-ray emission spectrography, appear eventually as analytical results; that is, as the x's of Figure 8.1–1 with $\bar{x}$ as their mean. For these 393 individual N's, the standard deviation is

$$s = \sqrt{\sum_i (N_i - \bar{N})^2 / 393 - 1} = 30 \text{ counts} \qquad (8.3\text{–}1)$$

RELIABILITY OF X-RAY EMISSION SPECTROGRAPHY

which is to be compared with the standard counting error

$$s_C = \sqrt{\overline{N}} = \sqrt{1018} = 32 \text{ counts} \qquad (8.3-2)$$

Table 8.3–1. Random Fluctuations in 393 Groups of N Counts Each

Midpoint of Subinterval[a]	Frequency	Midpoint of Subinterval[a]	Frequency
1100	2	1010	49
1090	2	1000	39
1080	3	990	33
1070	10	980	26
1060	18	970	14
1050	41	960	8
1040	48	950	6
1030	50	940	2
1020	41		

[a] Subinterval defined by N (as listed) $\pm$ 5 counts, except that lowest and highest subintervals include all extreme values. Boundary values were included in the higher subinterval.

These two standard errors are to be considered identical, for comparisons of this kind can never be highly refined. The identity of these errors is a welcome indication that the spectrograph system was satisfactorily stable over the rather long period (nearly a working day) required for taking the data in Table 8.3–1.

Let us risk repetition by pointing out that the agreement of s and s_C is not trivial. The former quantity is a *standard deviation* whose value is determined by the dispersion of the individual results about their mean. The standard counting error, s_C, is simply the square root of the mean. The two will be identical in x-ray emission spectrography if the process is truly random and subject to no error except the standard counting error. We shall attempt to point up the distinction between s and s_C by always calling the former the *standard deviation* (8.3–1); and the latter, the *standard counting error* (8.3–2). Because a mean, $\overline{N}$, is not ordinarily established, it is customary to approximate s_C by taking the square root of a single count N: compare (8.2–1) with (8.2–5). The standard deviation s can, of course, be calculated according to (8.1–2) for any experimental quantity.

When s significantly exceeds s_C, other errors are present, and these may be in the equipment, in manipulation, or in the sample (either unknown or standard).

One simple example of poor operating conditions is that of a volatile sample evaporating rapidly enough to change the optical path during the counting period. Owing to this change, the count registered during t seconds decreases with time if the spectrograph is not readjusted. This decrease is superimposed upon the random fluctuations discussed with the result that s exceeds s_C.

A series of experiments paralleling those of Table 8.3–1 was carried out on an open cell containing toluene, with the goniometer set to give a counting rate near 100 cps at the midpoint of the series. The results are given in Table 8.3–2.

Table 8.3–2. Effect of Drift Due to Evaporation Superimposed on Random Fluctuations[a]

Midpoint of Subinterval[b]	Frequency	Midpoint of Subinterval[b]	Frequency
635	28	1035	33
735	31	1135	34
835	28	1235	16
935	26	1335	4

[a] 200 groups of N counts each; $t = 10$ sec.
[b] Subinterval defined by N (as listed) $\pm$ 50 counts, except that lowest and highest subintervals include all extreme values.

Inspection of Table 8.3–2 shows that the distribution is far from Gaussian. Calculation of the standard deviation and standard counting error gives

$$s = 198 \text{ counts} \gg s_C = 31.9 \text{ counts} \qquad (8.3\text{–}3)$$

Whenever s thus exceeds s_C, operating conditions should be improved immediately. Such an excess of s over s_C calls for an immediate investigation, in which it is usually best to begin with the most probable cause of error, and to continue with others in order of decreasing probability until the discrepancy can be eliminated. In the example just cited, covering the cell is all that would be required. In other cases, positioning of the sample could be at fault; the difficulty might be heterogeneity of the sample, or drift in the electronic components, or something else. In such cases an analysis of variance (Section 8.7) is advisable.

8.4 The Practical Standard Counting Error in More Complex Cases

According to the law for the combination of errors, the standard counting error will be larger and more complex if more than one count is needed to establish an analytical result. Several *practical* examples follow. Being practical, they involve N's—not $\bar{N}$'s. Compare again (8.2–1) with (8.2–5).

SIMPLE CORRECTION FOR BACKGROUND. Let N_T be the total count, and N_B the proper background count over the counting interval Δt for N_T; the counting error for the difference $N_T - N_B$ is

$$s_C = \sqrt{(s_{CT})^2 + (s_{CB})^2} = \sqrt{N_T + N_B} \qquad (8.4\text{--}1)$$

because each individual counting error (s_{CT} for N_T; s_{CB} for N_B) is the square root of the corresponding count. The addition of squares under the first radical follows from the rule for the combination of independent errors to give the error of a difference.

Note that N_B may be approximated as the product of Δt and a counting rate measured for the background over a different counting interval; for example, over an interval longer than Δt if one wishes to establish the background with greater precision. Whether such higher precision is worth while depends upon the relative contribution of N_B in (8.4–1).

The analytical chemist often wishes to express the counting error *relative to the amount present* of the element sought; in a simple case we have

$$\mathbf{s}_C = \sqrt{N_T + N_B}\,/(N_T - N_B) \qquad (8.4\text{--}2)$$

This relative standard counting error ($\mathbf{s}_C$) is analogous to a relative standard deviation. Either may be expressed as a fraction or in percentage; (8.4–2) is particularly useful in emphasizing the importance of the background in trace determinations. Note **boldface s.** See also Table 8.4–1.

2. SIMPLE RATIO. COMPARATIVE METHOD. Here and in 3 below we deal with *relative* standard counting errors of ratios. Though these cases are more complex than (8.4–2), we shall use **s** to represent them.

The determination of a major constituent might be done by comparison without correcting for background. Such a comparison entails an increase in standard counting error—no matter whether the comparison is of an analytical line in an unknown with the same line in a standard, with a line of an added internal standard, or with a scattered line in the background.

For a simple quotient and large values of N, the rule for the combination of errors leads to

$$\mathbf{s}_C = \sqrt{(\mathbf{s}_{CSt})^2 + (\mathbf{s}_{CU})^2} = \sqrt{\left(\frac{\sqrt{N_{St}}}{N_{St}}\right)^2 + \left(\frac{\sqrt{N_U}}{N_U}\right)^2}$$

or

$$\mathbf{s}_C = \sqrt{\frac{1}{N_{St}} + \frac{1}{N_U}} \qquad (8.4\text{--}3)$$

where the subscripts St and U denote standard and unknown.

3. RATIO CORRECTED FOR BACKGROUND. COMPARATIVE METHOD. This case is a combination of the two just described. Numerator and denominator are each a difference, and each has the standard counting error prescribed by (8.4–1). The relative standard counting error of the quotient is

$$s_C = \sqrt{\frac{(N_T + N_B)_{St}}{[(N_T - N_B)_{St}]^2} + \frac{(N_T + N_B)_U}{[(N_T - N_B)_U]^2}} \qquad (8.4\text{–}4)$$

4. ILLUSTRATIVE CALCULATIONS. Several simple examples, in all of which a count of 2000 is assumed to measure the amount present of the element sought, are summarized in Table 8.4–1.

Table 8.4–1. Illustrative Examples of Counting Errors

Case	$(N_T)_U$, counts	$(N_B)_U$, counts	$(N_T)_{St}$, counts	$(N_B)_{St}$, counts	s_C, counts	Ratio to Case 1	s_C, parts per 100
1	2000	0	No standard used		~45	1	2.2
2	5000	3000	No standard used		~90	2	4.5
3	2000	0	2000	0	~63	$\sqrt{2}$	3.2
4	5000	3000	5000	3000	~126	$2\sqrt{2}$	6.3

The sixth and seventh columns of the table show how the standard counting error pyramids with the number of terms it contains. The last column, calculated according to (8.4–4), is of particular interest because it relates the standard counting error to the amount present of the element sought (on the simplifying assumption, of course, that this amount is proportional to $N_T - N_B$).

It is well to remember that the equations of this section deal with cases that differ fundamentally from that of (8.1–3). This equation deals with the *different* errors in the result of a *single* measurement (i.e., N or x), and the others are combinations of standard counting errors of *different* quantities that go to make up a *complex* datum that usually cannot be obtained in a single measurement.

8.5 Measurement of Standard Counting Error in Complex Cases

To test the calculated results in Table 8.4–1, measurements were made in the authors' laboratory on a General Electric XRD-5 D/S spectrograph under conditions that minimized all errors except the counting error. Instead of comparing standard and unknown, analytical lines for two elements, cobalt and iron, were compared because this could be done on a single sample.

A composite spot containing about 2 µg each of cobalt and of iron was

RELIABILITY OF X-RAY EMISSION SPECTROGRAPHY

prepared by evaporating aqueous nitrate solutions on filter paper. The instrument was set to count cobalt $K\alpha$, and twenty successive values of N_T were obtained, each over a counting interval of 40 sec. The instrument was then set for iron $K\alpha$, and the counting program repeated. Finally, at a setting appropriate for the background, the same counting program was carried through. The individual counts are given in Table 8.5–1.

The results in Table 8.5–1 show that calculations, such as those underlying Table 8.4–1, provide useful guidelines for x-ray emission spectrography.

Table 8.5–1. Evaluation of Standard Counting Error
A. Basic Data

	Range of the 20 Values[a] of N	Mean, $\bar{N}$
$(N_T)_{Co}$	5760 to 6040	5867
$(N_T)_{Fe}$	4830 to 5080	5005
N_B	2140 to 2300	2240

B. Comparison of Standard Counting Error and Standard Deviation

Determination	Equation for s_C or $\bar{s}_C$	s_C or $\bar{s}_C$	s (8.1–2)
Co (no background correction)	(8.2–1)	77	71
Co (background correction)	(8.4–1)	90	90
Ratio[b]—Co/Fe (no background correction)	(8.4–3)	0.024[d]	0.019
Ratio[c]—Co/Fe (background correction)	(8.4–4)	0.052[e]	0.045

[a] Individual values rounded to nearest ten.
[b] Value of ratio, 1.172.
[c] Value of ratio, 1.312.
[d] Or about 2% of ratio.
[e] Or about 4% of ratio.

8.6 The Goniometer Setting as a Source of Error

When analytical line intensities are measured repeatedly on a laboratory spectrograph, the following two steps are of interest as possible sources of error. (1) Establishment of the goniometer position (2θ value) corresponding to maximum intensity. (2) Resetting of the goniometer to this position. The second step is absent in comparative intensity measurements provided standard and unknown can be counted without disturbing the goniometer.

In the General Electric XRD-5 D/S, the goniometer circle is graduated in degrees, and a setting of 2θ is made by means of a goniometer adjustment

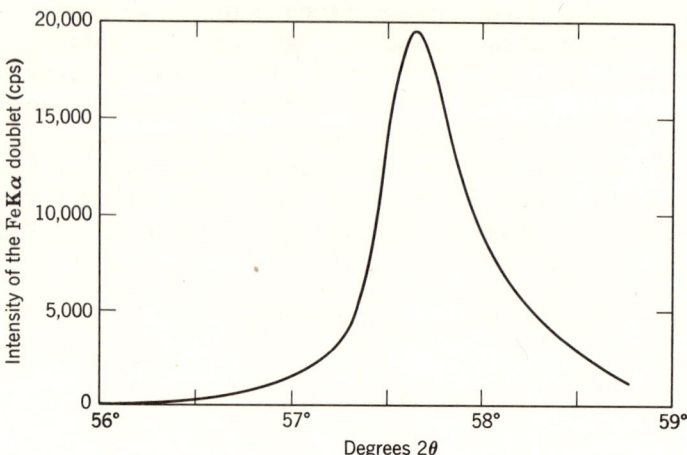

Fig. 8.6–1. X-ray-line shape from the XRD-5 D/S optical system with a 10-mil collimator.

drum. One revolution of the drum displaces the circle by 1 degree. The circumference of the drum is subsivided into 100 parts, each of which corresponds therefore to 0.01° in 2θ. In precise work, the goniometer setting for maximum intensity is located by systematically taking counting rates at properly selected positions of the drum. The position of greatest counting rate fixes the goniometer setting; that is, this setting is made to the nearest 0.01 degree. Obviously, this setting may be inaccurate by as much as 0.005 degree. How will this influence the analytical results? Because this question relates to accuracy, we are safe in saying that there will be no error if a comparative intensity measurement is being made. Resetting the goniometer by means of the drum is a different matter, for this is a question of precision.

The magnitude of this reset error, if measured in cps, will be determined not only by the error in 2θ, but also by the counting rate at the goniometer setting actually made. The variation of counting rate with goniometer setting for an analytical line will depend upon the shape of the peak generated by the line. Accordingly, this shape was measured for a typical analytical line, iron $K\alpha$. The measurement consisted in taking counting rates at selected drum positions that covered the 2θ-range (about 3 degrees) of the line. Resolution was moderate, and a Soller slit 3.5 in. long with 0.010-in. spacing provided collimation. The results are given in Figure 8.6–1.

To facilitate estimation of the reset error in cps, the curve in Figure 8.6–2 was obtained by differentiating graphically the curve in Figure 8.6–1. As the calculus requires, the differential curve has the value zero at the maximum intensity ($2\theta = 57.65°$) in Figure 8.6–1. Also, the two points of inflection in Figure 8.6–1 appear as a maximum and a minimum in Figure 8.6–2: note how rapidly cps changes with 2θ at these points.

RELIABILITY OF X-RAY EMISSION SPECTROGRAPHY

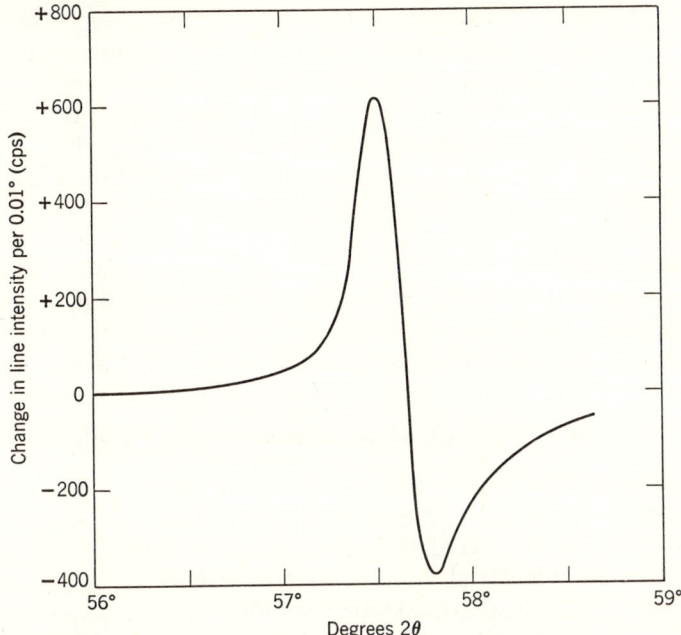

Fig. 8.6–2. The rate of change of intensity for the x-ray line of Fig. 8.6–1.

Figure 8.6–2 shows the advantage of locating the intensity maximum with high precision. Near this maximum, an error of even 0.01 degree in positioning the adjusting drum corresponds to a change of only about 0.1 % in the counting rate, which is usually a negligible change in N_T. The reset error is much greater at goniometer settings near either point of inflection in Figure 8.6–1. A marked advantage of doing x-ray emission spectrography near the intensity maximum of the analytical line is the virtual elimination of the reset error. In many cases, this advantage will be more important than the obtaining of a higher counting rate at the maximum.

The analysis just concluded applies in simple cases. When high resolution is necessary, as when interfering lines are present, the reset error may be much more important.

8.7 Analysis of Variance

Up to this point, we have dealt with known individual errors, and with the standard counting error s_C as an operating criterion. So long as s is comparable with s_C (hence s^2 with s_C^2), and operating conditions are satisfactory, nothing more is required.

When s^2 significantly exceeds $s_C{}^2$, there may be several important sources of variation. It is then advisable to discover what these sources are. Analysis of variance is a systematic procedure for making this discovery.

In analysis of variance, the variance due to each source of variation is systematically isolated. A test of significance, the F-test, is then applied to indicate the importance of each source. The interested reader is urged to consult books on statistics [7] for discussions of this valuable statistical method.

Several examples are given to show how analysis of variance can be used in x-ray emission spectrography. The data are summarized in tabular form under headings already given except for "sums of squares," under which are listed data that correspond to the numerator of the fraction in (8.1-2).

8.8 Higher Standard Counting Errors. Drift

For an analysis of variance at counting rates above those at which a Geiger detector is satisfactory, data were taken under the following conditions: General Electric XRD-5 D/S spectrograph with SPG-4 (flow proportional) detector; analytical lines, iron $K\alpha$; counting interval, 40 sec; approximate counting rate, 40,000 cps. For more than a thousand equal consecutive counting intervals, values of N_T (one to each interval) were printed out, the mean $\bar{N}_T$ being just under 1,700,000 counts. A thousand consecutive N_T values from within the data were used in the analysis of variance. These thousand values were subdivided sequentially into 100 sets of 10 each. The analysis was designed to show the short-term variation within sets and the long-term variation among sets. The results are summarized in Table 8.8–1.

Table 8.8–1. Analysis of Variance. Drift and Counting Error

Source of Variation	Sum of Squares	Degrees of Freedom	Variance
Short-term (within sets)	1,810,561,500	900	2,011,735
Long-term (among sets)	14,340,024,200	99	144,848,729
Counting error ($s_C{}^2 = \bar{N}_T$)	—	—	1,696,370

The last column of Table 8.8–1 shows (1) that the long-term source of variation clearly overshadows the short-term, the ratio of variances exceeding 70; and (2) that the short-term variance is comparable with $\bar{N}_T$, which is, of course, the standard counting error squared.

The long-term source of variation is most probably a drift arising in the detector system, perhaps caused by changes in room temperature during the long experiment. The analysis of variance cannot give the cause of the drift nor can it, as carried out above, reveal how the drift changes with time. But the analysis does warn of instability in the complex electronic system, and it emphasizes again the importance of comparing the unknown quickly with a standard.

The short-term source of variation is in large part the counting error. By taking square roots, we arrive at $s = 1418$ and $s_C = 1302$ counts from the last column of Table 8.8–1. When we consider that the long-term drift must have increased the short-term variance, we are justified in concluding that the virtual identity of s and s_C has been established at the higher counting rates (40,000 cps) of the present experiment. This is a gratifying accomplishment.

8.9 Standard Counting Error Comparable with Manipulative Errors

Two manipulative errors that occur in laboratory x-ray emission spectrographs need special attention. The first is the error in setting the goniometer—the reset error; see Section 8.6. The second is that traceable to the placing of samples in the spectrograph, either to the repeated placement of a single sample, or to the replacement of one sample by one or more other, geometrically and chemically identical samples. The second error is called the placement error.

Data for an analysis of variance were collected under the following conditions: spectrograph and detector, as above; sample, stainless steel plate, $1\frac{3}{4} \times 1\frac{1}{4} \times \frac{1}{4}$ inches; analytical line, iron $K\alpha$ (peak at 57.58 degrees); counting interval, 10 sec. The measurements were made at the inflection point, $2\theta = 57.05$ degrees, where the change of counting rate with angle was 10,000 cps for 0.01 degree.

The following groups of experiments were carried out, there being 20 counting intervals (20 consecutive experiments) in each group, with each counting interval yielding one value of N_T. (1) With goniometer and sample undisturbed, a group of N_T values was taken to assess drift in the manner of Table 8.8–1. (2) In each experiment, with the sample always in place, the inflection point was approached from the side of *higher* values of 2θ; that is, by turning the goniometer drum in the direction of *smaller* 2θ. This is the group of DOWN experiments. (3) A group of UP experiments was done, identical with the DOWN experiments except that the inflection point was approached by turning the drum in the direction of *larger* 2θ. (4) With the goniometer remaining undisturbed at the setting for the inflection point, the sample was inserted, counted, removed, and replaced until the data for the 20 intervals had been accumulated.

The first group of experiments showed that no significant drift was present; the standard deviation ($s = 468$ counts) and the standard counting error ($s_C = 491$ counts) were virtually identical. An analysis of variance for the second and third groups of experiments is summarized in Table 8.9-1.

Table 8.9-1. Analysis of Variance. Reset Errors and Counting Error

Source of Variation	Sums of Squares	Degrees of Freedom	Variance
Reset errors and counting error	17,431,500	38	458,724
Counting error ($s_C^2 = \bar{N}_T$)	—	—	244,305
Combined reset errors by difference	—	—	214,419
Difference between reset errors	85,264	1	85,264

The last column of Table 8.9-1 shows the combined reset errors to be comparable with the counting error. (The subtraction of variances is justified because long-term drift had been proved absent in the first group of experiments). By use of data given above, the combined reset error, which is the mean of UP and DOWN errors, can be expressed as an angle:

$$\sqrt{214,000} \times 0.01/10,000 = 0.0005 \text{ degree} \quad \text{(approximately)} \quad (8.9\text{-}1)$$

A significant difference between the UP and DOWN errors would point to backlash in the adjustment drum. Although there is a difference (Table 8.9-1), the F-test [8] shows that this difference has a low level of significance.

Equation 8.9-1 gives 0.0005 degree as the standard deviation of the reset error. Near the peak, where the change of counting rate with angle is small (Section 8.6), this may be neglected. The goniometer adjustment mechanism in this spectrograph is satisfactory.

It should be mentioned that the results in Table 8.9-1 were obtained only after experience had taught that the adjustment drum must be pressed inward for the most precise results. Early trials in which such pressure was not exerted gave reset errors *ten* times as large.

The fourth group of experiments gave the results in Table 8.9-2.

The variances show that the standard placement error is comparable with the standard counting error. Hence an increase in N_T might produce a worthwhile increase in the precision of the analytical result.

To summarize: The standard counting error will include all random errors

Table 8.9-2. Analysis of Variance. Placement and Counting Errors

Source of Variation	Sums of Squares	Degrees of Freedom	Variance
Placement and counting errors	7,902,100	19	415,900
Counting error ($s_C^2 = \bar{N}_T$)	—	—	245,108
Placement error	—	—	170,792

in the spectrograph. If only random errors exist, values of N will have a Gaussian distribution. Systematic errors of any kind will distort and/or shift the Gaussian curve; but such effects, if small, may escape detection. A comprehensive analysis of variance will give the magnitude of each significant systematic error, and the magnitude of the combined random errors.

8.10 Counting Strategy. Guaranteed Reliability

Good counting strategies are based on common sense and the fundamental relations $s_C = \sqrt{N} = \sqrt{\bar{r}\Delta t}$, $\bar{r}$ being the mean counting rate and Δt the counting interval. In the simplest case, these relations apply for total count N_T and for background count N_B. The counting strategy in general fixes s_C, N_T, N_B, Δt_T, and Δt_B for a single determination on the basis of $\bar{r}_T$ and $\bar{r}_B$ if these are known or can be approximated. In this way s_C is also fixed; see Section 8.4.

The following procedure is recommended for the most difficult case, that of a sample about which nothing is known. (1) Place the sample in the spectrograph and estimate r_T at the 2θ setting for the analytical line and r_B near that setting. (2) If $r_T > r_B$, proceed as follows. If r_T and r_B are comparable, proceed on the basis of Section 8.11. Because r_T for pure elements can be near 500,000 cps, the former situation is the more likely. (3) Decide on the minimum acceptable s_C. This quantity is defined by (8.4–2), but the following *approximation*

$$s_C = \sqrt{N_T}/N_T \quad \text{(approximately)} \quad (8.10\text{–}1)$$

is often useful. Obviously, if $s_C = 0.01$ (or 1%) is adequate, N_T must be 10^4 counts. (4) Calculate the corresponding Δt_T. (5) Take $\Delta t_B = \Delta t_T$. (6) When the determination has been made, calculate a better s_C from (8.4–2). (7) If this s_C is unacceptably high, increase $\Delta t_B = \Delta t_T$ as needed for subsequent determinations. Alternatively, prior to the first determination, increase Δt_T above that calculated in step 4 as judgment dictates.

This simple, common-sense procedure seems preferable to using more elaborate mathematical methods. Consider an extreme case for which $N_T = 10^4$ and $N_B = 5000$ counts. The s_C according to (8.10–1) is 1%; the s_C according to (8.4–2) is about 2.5%, and in this case the preliminary r_B would have forewarned the analytical chemist to increase Δt_B according to step 7.

Many counting strategies are conceivable. Two—namely "fixed count" and "fixed (counting) time"—deserve special mention because they are often built into spectrograph systems. The first might be accomplished by pressing a button to cause counting to continue until a predetermined number of counts, N, has been accumulated *no matter what is being counted*. That this can be a straitjacket if blindly followed is made clear by two extreme cases: (1) the element sought is absent; and (2) the background being counted is negligible. The second strategy might be accomplished by pressing a button that causes counting to continue until a predetermined time (the counting interval) has elapsed, again no matter what is being counted. Examination of (8.4–1) shows this to be a better procedure, but note the paragraph that follows the equation. Also, see Section 8.11.

How reliable are results obtained by x-ray emission spectrography? That depends upon s for the complete determination, which includes *all* errors, many of which (such as those deriving from sample preparation) are *not* included in s_C. The point is implicit in the discussion of analysis of variance, but it warrants explicit statement.

Two ways of looking at reliability will be discussed on the basis of Figure 8.2–2, which involves only s_C. No similar mass of s data is available. Accuracy does not enter the discussion. Background is assumed negligible.

1. Suppose that the $n = 393$ counts were taken to establish μ, the mean of a normal (Gaussian) universe that contains all possible values of N. We may then say that μ lies in the range $\overline{N} \pm 3\sigma/\sqrt{n}$, the boundaries of the range being a pair of *confidence limits* for the determination. The shape of the Gaussian leads us to expect that this statement will be correct for every 997 $\overline{N}$ values out of 1000; or, if we are willing to paint the lily, for every 9973 $\overline{N}$ values out of 10,000. (Narrower limits include predictably smaller fractions analogous to 99.73% for 3σ.) The *experimental* value of s justifies the statement

$$s = s_C = \sigma = 32 \text{ counts} \qquad \text{(very nearly)} \qquad (8.10\text{–}2)$$

so that the limits for μ are $1018 \pm (3 \times 32)/20$ or 1013 and 1023 counts.

2. Consider the second, more common case. A single N is the basis for a reported value of W, the weight fraction of an element in a submitted sample, the "unknown." How reliable is W? In the present discussion, we must assume that (8.10–2) is valid, and that a calibration curve exists which relates N to W. We may take for granted that 393 determinations on standards have given us a reliable value of s. For the unknown, the single value of N will very probably lie between 940 and 1100 counts (Table 8.3–1); more extreme values may (very rarely) occur. (Note in Figure 8.2–2 the five highest radioactivity results.) For cases such as this—namely, single determination on an unknown by a method for which s had been reliably established—the authors believe it best to drop confidence limits and to *guarantee* the result to within $3s$; that is, to report $N \pm 96$ counts in the present case [9]. If this procedure proves unsatisfactory, make replicate determinations and/or modify the guarantee. Use standard works [1, 7] as guides. The proof of the pudding is in the eating.

Clearly, the large amount of work needed to establish s reliably is justified only if unknowns few in number are extremely important, or if many

RELIABILITY OF X-RAY EMISSION SPECTROGRAPHY

unknowns are to be run, as in process control. Otherwise, the best guarantee possible has to be an informed guess that should not be *less conservative* than $\pm 3s_C$. When background is appreciable in the simplest case, this guarantee will be $(N_T - N_B) \pm 3s_C$ with s_C calculated according to (8.4–1). For other cases, see below.

8.11 Qualitative Trace Determinations. Statistical Considerations

In the trace determinations of Chapter 7, the element **E** to be determined was usually known to be present. When this is not the case, a decision about the presence of **E** must be made on the basis of N_T and N_B to see whether a quantitative determination is warranted. This decision involves two inherent risks: the *producer risk* (reporting **E** absent though present—think of a producer delivering goods for which he is not paid) and the *consumer risk* (reporting **E** present though absent—think of a consumer paying for goods not delivered). The risks become serious when $\overline{N}_T - \overline{N}_B$ approaches zero, and negative values for differences of individual counts become possible. Several such values are indicated by arrows in region III, Figure 8.11–1, which illustrates the problem. The figure should be compared with Figure 7.12–1, which has no region III.

Statistically, we approach the problem in two steps: (1) Is a Gaussian curve applicable to the results of trace determinations on a single sample? (2) What is to be the basis for the needed decision? The first question was answered affirmatively by Zemany, Pfeiffer, and Liebhafsky [10] when they showed that 91 values of $N_T - N_B$ for a "spot" containing zinc in microgram amounts conformed to a Gaussian frequency distribution, and that this was likewise true for 216 analogous values for strontium. Agreement with the Gaussian was about like that in Figure 8.2–2, and the standard deviation closely equaled s_C, which for the zinc-spot determinations was

$$s_C = \sqrt{\overline{N}_T + \overline{N}_B} = 35 \text{ counts} \qquad (8.11\text{–}1)$$

and

$$s_C = \sqrt{\overline{N}_T + 3\overline{N}_B} = 274 \text{ counts} \qquad (8.11\text{–}2)$$

for the strontium-spot determinations because 3 kinds of counts were needed here to establish the background. The second question was answered by applying as follows the 3s guarantee [9] introduced above. Negative amounts present are physically impossible. One therefore begins by allowing only for the consumer risk (reporting present when absent), which is determined by fluctuations in N_B. According to the guarantee, this risk is ignored when

$$N_T > \overline{N}_B + 3s_{CB} \qquad (8.11\text{–}3)$$

The position just outlined was tested on manganese spots [11], for which it meant a report of "present" if $N_T > 4870 + 3(70) = 5080$ counts and a report of "presence not proved" if $N_T < 5081$ counts. The tests were painstakingly made by a careful operator to whom the presence or absence of manganese was not known. The reports made appear in Table 8.11–1.

Table 8.11–1. Results on the Detection of Manganese

Mn Added, μg	Number of Runs	Report "Present"	Report "Not Proved"
None	7	None	7
0.001	22	9	13
0.002	14	13	1
0.003	6	6	0

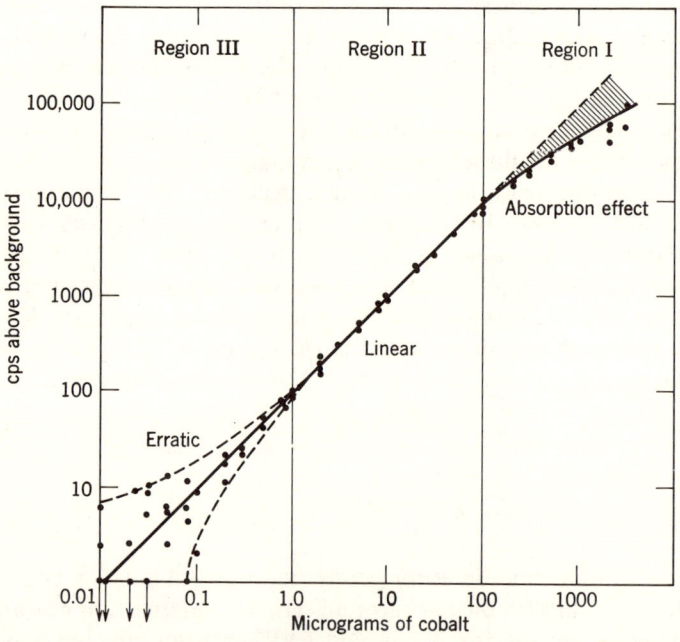

Fig. 8.11–1. Cobalt determinations that demonstrate occurrence of three regions in trace determinations by x-ray emission spectrography under simplest conditions. Uneven distribution of cobalt in sample probably a source of error in region I. Erratic distribution in region III has a statistical basis; the envelope (broken curve) is two standard deviations away from the solid line in this, the region of *qualitative* determinations. For meaning of arrows below region III, see text. After H. A. Liebhafsky, H. G. Pfeiffer, and P. D. Zemany, Ref. 11, p. 322, Fig. 1.

Two conclusions follow from the table: (1) The consumer risk has been reduced or eliminated: manganese was never reported present when absent. (2) If one wishes to be sure of finding manganese whenever it is present, the amount present should exceed 0.002 μg. But, by how much?

To attempt an answer, one must consider also the fluctuations in N_T. According to [11], one then arrives at a logical definition of the *minimum amount guaranteed detectable* (MAGD) as

$$\text{MAGD} = 3k\sqrt{\overline{N}_T + \overline{N}_B} \simeq 6k\sqrt{\overline{N}_B} \; \mu g \qquad (8.11\text{–}4)$$

where $\overline{N}_T$ is the count for MAGD present. In this case, with a counting interval of 1000 sec and a counting rate for manganese of $r = 170$ cps/μg, k is 1/170,000. The more exact solution of (8.11–4) gives

$$\text{MAGD} = 430/170{,}000 = 0.00253 \; \mu g \; \text{Mn} \qquad (8.11\text{–}5)$$

and the approximate

$$\text{MAGD} = (6 \times 70)/170{,}000 = 0.00247 \; \mu g \; \text{Mn} \qquad (8.11\text{–}6)$$

Obviously, MAGD = 0.003 μg should be taken as the minimum amount of manganese guaranteed detectable as higher precision is neither warranted nor needed. Agreement with Table 8.11–1 is thus complete, and (8.11–4) is preferable to the less conservative (8.11–3). The MAGD is a logical boundary between regions II and III, Figure 8.11–1.

8.12 Regression Methods

In 1886, Sir Francis Galton described as a "regression toward mediocrity" his discovery that deviations in stature of sons from the mean stature of all men were smaller than (though in the same direction as) corresponding deviations for their fathers from the mean of all fathers [12]. As regards stature, the sons were "regressing toward mediocrity" because they were becoming more like all other men.

Somewhat illogically, regression is now used to describe many and diverse relationships among variables, those between sample composition and interelement effects in x-ray emission spectrography being a recent example. Analytical chemists may be pardoned for feeling toward regression as Molière's M. Jourdain did toward the speaking of prose.

A *simple linear regression* of y on x is represented algebraically by

$$y = a + bx \qquad (8.12\text{–}1)$$

with y the dependent, and x the independent variable. Similarly

$$y = a + b_1 x_1 + b_2 x_2 + \cdots + b_n x_n \qquad (8.12\text{–}2)$$

represents a *multiple linear regression* of y on the variables $x_1 \ldots x_n$. More complex regression equations containing higher terms (e.g., $x_1 x_2$ or $x_1{}^2$), though often useful, will not concern us. It is usually desirable that regression treatments provide for obtaining values of the coefficients a, $b_1, \ldots, b_n$ by the method of least squares. A computer is helpful.

In x-ray emission spectrography, a general regression treatment must assume that the analytical-line intensity of any element depends on the kind and weight-fraction of *every* element in the sample, and that all these relationships affect the relationship between **I** and **W** (boldface type is used for quantities pertaining to the element being determined). In this situation, (8.12–1) is inadequate, but there is hope that a treatment according to (8.12–2) will suffice. This equation will then represent a multiple linear regression of **I** on the n weight-fractions [**W** and $(n-1)$ W's] in a sample containing n elements. As a W for any such element decreases, it becomes permissible at some point to disregard the presence of that element. The intensity **I** must have been properly corrected for background.

The problem subdivides itself into: (1) *calibration*, or establishment of the values of interelement coefficients, which are represented by the b's in (8.12–2); (2) *determination*, or establishment of the needed weight-fractions in an unknown sample.

Calibration is simplest when least-squaring is not needed. The **I**'s will then be measured under these conditions: (1) $s = s_C$; (2) $\mathbf{s}_C$ at least as small as needed; (3) no significant inaccuracies in W's for standard samples; and (4) number of standards needed equals number of elements present in significant amount. The last condition gives the minimum required number of standards and follows from the other three.

How many interelement coefficients will be needed? No one can tell in advance. A rational beginning, in accord with (8.12–2), can be made as follows. Consider a sample containing only elements A and B and represent the interelement coefficients by α's with the appropriate subscripts. We begin by considering as proportional to W_A the change produced in the interelement effect due to A by changing W_A, and use as proportionality constant the coefficient α_{AA}; correspondingly, α_{AB} applies to the change in the interelement effect due to A produced by changing W_B; and so on for α_{BB} and α_{BA}—altogether a total of four linear coefficients, all assumed constant. By logical extension, an n-element sample needs n^2 coefficients, the establishment of which in turn requires n^2 equations based on information from standard samples. One standard sample can contribute one equation modeled on (8.12–2) for each element it contains: that is, it contributes as many equations as it has **I**'s to be measured. The minimum number of standard samples therefore is n as was stated above. We do not ask whether this scheme is physically logical—only whether it works!

To proceed with an actual calibration, it is expedient to use a multiple regression equation differing in form, though not in substance, from (8.12–2). We use illustratively the form given by Criss and Birks [13] with weight-fractions W replacing "mass concentrations" C. We take standard samples 1, 2, and 3 each containing elements A, B, and C at different weight-fractions. Instead of I^S, for the sample we shall use an intensity ratio, or normalized intensity, $\mathbf{R} = I^S/I^E$, where the denominator is the intensity for the appropriate pure element. We shall neglect all instrumental factors. We then write for sample 1:

when $\mathbf{I}$ is for A: $\mathbf{W}^1/\mathbf{R}^1 = \alpha_{AA}\mathbf{W}^1 + \alpha_{AB}W_B^1 + \alpha_{AC}W_C^1$
when $\mathbf{I}$ is for B: $\mathbf{W}^1/\mathbf{R}^1 = \alpha_{BB}\mathbf{W}^1 + \alpha_{BA}W_A^1 + \alpha_{BC}W_C^1$ (8.12–3)
when $\mathbf{I}$ is for C: $\mathbf{W}^1/\mathbf{R}^1 = \alpha_{CC}\mathbf{W}^1 + \alpha_{CA}W_A^1 + \alpha_{CB}W_B^1$

The equations for samples 2 and 3 are identical except that the identifying superscript is 2 or 3. The quantities in boldface type need no identification as to element as they are for the element being determined. From these $n^2 = 9$ equations, values of the nine interelement coefficients can be calculated. Each of the nine equations may be regarded as a variant of (8.12–2).

In the simple calibration example just discussed, it was assumed that no "least squaring" was needed. If errors in the standards or in the intensity measurements make least-square-fitting necessary, or if it is desired to take instrumental factors (differences in spectrographs [14]) into account, then the number of different standard samples must exceed the number of elements present.

Determination is the *inverse* of calibration and consequently requires a different form of regression equation; e.g.,

A being determined: $(\mathbf{R}\alpha_{AA} - 1)\mathbf{W} + \mathbf{R}(\alpha_{AB}W_B + \alpha_{AC}W_C) = 0$
B being determined: $(\mathbf{R}\alpha_{BB} - 1)\mathbf{W} + \mathbf{R}(\alpha_{BA}W_A + \alpha_{BC}W_C) = 0$ (8.12–4)
C being determined: $(\mathbf{R}\alpha_{CC} - 1)\mathbf{W} + \mathbf{R}(\alpha_{CA}W_A + \alpha_{CB}W_B) = 0$

Equations 8.12–4 and 8.12–3 differ only in form. As only a single sample (the unknown) is involved, superscripts are not needed.

However, regression equations alone do not suffice for determination as they did for calibration. In calibration, known weight-fractions, different for each sample, were inserted into the regression equations and constant coefficients were sought. Here the weight-fractions are unknown, and each may vary from zero to unity. Mathematically, the equations 8.12–4 are always satisfied when weight-fraction is zero. In this situation, these equations are useless for practical determinations without additional information; namely,

$$W_A + W_B + W_C = 1 \qquad (8.12\text{–}5)$$

or

$$W_A + W_B + W_C = K < 1 \qquad (8.12\text{–}5a)$$

As was implied above, (8.12–5) was not needed in the calibration experiments because known values for all weight-fractions were available. When elements (other than A, B, or C) are present in weight-fractions so small that these need not be considered, (8.12–5a) replaces (8.12–5).

How trustworthy is such a multiple linear regression treatment? It is most reliable when unknowns are the same *kind* of samples as the standards; when the two have a common composition range over all of which the treatment is adequate; and when operating conditions are always the same. Clearly, the fewer the unknowns, the greater the advantage of comparing standard and unknown in a way that involves no regression treatment.

When many unknowns are to be run, the obvious advantages of the regression treatment outlined above are reinforced by others. Change of operating conditions, including change of spectrograph, can be taken into account by including more than the minimum number of standards to give the needed "transfer coefficients." In the same way, least-square values of all coefficients can be obtained to increase reliability. More complex regression treatments might be justified. The treatment might prove satisfactory also for unknowns considerably different from any standard.

This chapter has attempted to show that certain statistical generalizations useful in analytical chemistry are particularly applicable to x-ray emission spectrography, by means of which they can readily be tested.

REFERENCES

1. C. A. Bennett and N. L. Franklin, *Statistical Analysis in Chemistry and the Chemical Industry*, Wiley, New York, 1954; see index.
2. E. von Schweidler, Premier Congrès International Radiologie, Liège, 1905.
3. (a) Y. Beers, *Introduction to the Theory of Error*, Addison-Wesley, Cambridge, Mass., 1953; (b) J. L. Doob, *Stochastic Processes*, Wiley, New York, 1953; (c) R. D. Evans, *The Atomic Nucleus*, McGraw-Hill, New York, 1955.
4. E. Rutherford and H. Geiger, *Phil. Mag.* [6], **20**, 698 (1910).
5. H. A. Liebhafsky, H. G. Pfeiffer, and P. D. Zemany, *Anal. Chem.*, **27**, 1257 (1955).
6. A. Einstein, *Phys. Z.*, **18**, 121 (1917).
7. O. L. Davies, *Statistical Methods in Research and Production with Special Reference to the Chemical Industry*, Oliver and Boyd, London; and Hafner, New York, 2nd ed. revised, 1954; especially Chapter 5. See also Ref. 1, Chap. 7.
8. Ref. 7, pp. 60 and 70.
9. H. A. Liebhafsky, E. W. Balis, and H. G. Pfeiffer, *Anal. Chem.*, **23**, 1531 (1951).
10. P. D. Zemany, H. G. Pfeiffer, and H. A. Liebhafsky, *Anal. Chem.*, **31**, 1776 (1959).
11. H. A. Liebhafsky, H. G. Pfeiffer, and P. D. Zemany, in *X-Ray Microscopy and X-Ray Microanalysis, Proc. 2nd Int. Symp.*, A. Engström, V. Cosslet, and H. Pattee, Eds., Elsevier, New York, 1960. p. 321.
12. Ref. 1, p. 37.
13. J. W. Criss and L. S. Birks, *Anal. Chem.*, **40**, 1081 (1968).
14. J. Lucas-Tooth and C. Pyne, in *Advances in X-Ray Analysis*, Vol. 7, Plenum, New York, 1964, p. 523.

Chapter 9

X-Ray Emission Spectrography. General

9.1 Introduction

In modern analytical chemistry, defined to include both characterization and control [1], many excellent methods, some old, more new, compete unremittingly for the work to be done. The value of x-ray emission spectrography, electron microprobe techniques included, will mainly determine the eventual position of x-ray methods in this competition.

Earlier chapters have introduced the reader to x-ray emission spectrography for specialized applications, generally under simple conditions. The authors believe that an x-ray unit should be the nucleus of every new, modern laboratory in which *ultimate* analysis plays an important role. Such a unit should be capable of the determinations (absorption, diffraction, thickness) discussed in earlier chapters, but x-ray emission spectrography will be its main activity. This chapter shows that, for determinations of all but the lightest elements, the method can serve usually as court of original jurisdiction and sometimes as court of last resort.

X-ray emission spectrography is an *operational* name. For each element sought, a characteristic x-ray line (the analytical line) is emitted, is identified, and has its intensity measured. This intensity, even after correction for background, is not usually proportional to the amount present of the element sought: most samples with which the analytical chemist must deal do not show this proportionality to the extent demonstrated in Chapter 7. The x-ray emission spectrography of *things as they are* has to be the principal concern. The emphasis is on x-ray *emission*: not included is spectrography in which x-rays excite optical lines, as for the determination of rare earths in phosphors [2].

For the early history of x-ray emission see Section 1.21, in which the quotations from Moseley proved him to be its founder. He said further [3] that it "may even lead to the discovery of missing elements, as it will be possible to predict the position of their characteristic lines." A great landmark, the discovery of hafnium by Coster and Hevesy [4], once more showed Moseley to be a true prophet.

The discovery of new elements has now become commonplace. However, the story of Hevesy and hafnium [5, pp. 177–252] repays study, not only because it shows x-ray emission spectrography to great advantage over classical methods of analysis and over optical emission spectrography, but also because $_{72}$Hf occupies a strategic position in the periodic table, as one discovers upon completing Table 4.7–3 to include all 32 N electrons. But let Hevesy [5] speak:

The presence of the so-called "rare earths" group is, in view of [Bohr's electron arrangement for the periodic system]...no longer an unexplained anomaly... but a natural consequence.... As the N group has 32 electrons and 18 were already present in xenon and the preceding elements, the number of rare-earth elements *cannot exceed 14* including cerium. Cerium having the atomic number 58, the last rare-earth element must have the atomic number 71, element 72 must belong in the titanium...group. *Contrary to this conclusion, the literature records the existence of 15 rare-earth elements,* ...

Professor Bohr consulted with Dr. Coster...and with the writer [Hevesy] who was attached to his [Bohr's] laboratory, asking if we considered the evidence of the existence of the rare-earth element of atomic number 72 as conclusive. We were all three unanimous that it was not. However, in those days none of us thought to look for the proper element 72,...

While waiting for the necessary equipment [for research unrelated to hafnium], *and more as a pastime,* I [Hevesy] suggested to Dr. Coster that we look for the proper element 72.... *The first exposure...showed at once the presence* of the element looked for.... However, we did not consider this evidence as conclusive because, through a queer coincidence the zirconium $K\alpha$ doublet of the second order nearly coincides with the hafnium $L\alpha$ doublet. [No pulse-height selection in those days! See Figure 9.1–1.]

...It was only after we had succeeded in the chemical separation of hafnium from zirconium that we announced the discovery of the new element... [pp. 180–183].

We soon recognized that zirconium and hafnium were chemically so closely related that there was little hope of finding a characteristic chemical reaction for hafnium... [p. 185].

Hauser made attempts to find new elements in zirconium by investigating the optical spectrum for samples of different origin. The failure of his efforts was due to the intricate nature of the zirconium spectrum and furthermore to the fact that there was not sufficient variation in the intensity of the hafnium lines, which were all rather weak, to permit the detection of the presence of a new element in his samples.

X-RAY EMISSION SPECTROGRAPHY. GENERAL

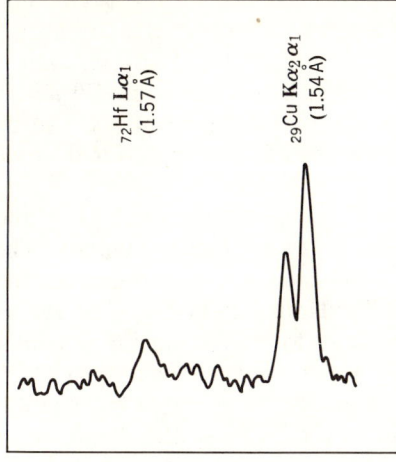

Fig. 9.1-1. The first photometer curve made in the search for element 72. Norwegian zircon was treated with boiling acids to remove soluble constituents, and the powdered residue was then fixed on the copper target of an x-ray tube. Note that the $_{72}$Hf Lα1 peak is clearly identifiable even above the high background produced by electron excitation. After Ref. 5, p. 183, Fig. 73.

That it could have been possible to discover the element in the optical spectrum can be best seen from the fact that the spectrum taken by Rowland contained only one hafnium line... in contrast to the spectrum taken by all other investigators. X-ray spectroscopy facilitated enormously that task, and therefore *the discovery of hafnium may be considered to be one of the most beautiful examples of Röntgen-ray analysis.* [Emphasis and bracketed material supplied.]

Thus the new elements "found" in zirconium since its discovery in 1789—among them, *norium* (1845); *jargonium* (1869); *nigrium* (1869); *nipponium* (1908); and *euxene earth* (1901)—were laid decently and finally to rest. The advantages that x-ray emission spectrography, even with primitive equipment, holds over wet methods and optical spectra led Hevesy to develop this x-ray method for chemical analysis when he went to Freiburg in 1926 [6]. The hafnium story is a landmark in the history of atomic structure, of the periodic table, and of x-ray emission spectrography.

To terminate this brief history, we mention that Hadding seems to have been the first to use x-ray emission specifically for analysis, and it is appropriate that his first published data [7] should have been for a metal (impure platinum) because x-ray emission spectrography is particularly suited to the examination of metallic materials. Eddy and Laby [8] carried out admirable work, the results of which compare favorably with the best obtained today. Three investigators, Friedman, Birks, and Brooks [9, 10] at the Naval Research Laboratory ushered in the new era of x-ray emission spectrography by proving that the use of modern detectors makes the method too attractive for the analytical chemist to overlook. Progress since that time has been—in the current idiom—"just too much" as the *Analytical Chemistry Annual Reviews* [11] show.

Earlier chapters contain brief descriptions of enough applications to show in detail what the methods there discussed can do. With x-ray emission spectrography, this is no longer possible. This chapter can cover only matters of general importance for most applications. Fortunately, the original literature can be reached easily through the *Analytical Chemistry Annual Reviews*, which give information not only by element but also by application. The Appendices to this book include an abbreviated guide that will fill most needs.

In the important *Analytical Chemistry* series, fundamental [11] and applied reviews appear in alternate years. They are complementary. The reader who wants information about x-ray emission spectrography and related methods should begin with the fundamental reviews. The reader who wants to choose among different methods for determining a particular element should turn to the applications reviews, specific reference to which will not usually be made in this book. One must be prepared for a shock. The applied reviews are arranged by field: in the 1969 issue, 2568 numbered entries appear in the bibliography on pharmaceutical and related drugs. The references on x-ray emission spectrography are somewhat less numerous, but they appear in many fields, particularly in metallurgy—ferrous; light metals; and Zr, Hf, V, Nb, Ta, Cr, Mo and W being the 1969 subdivisions. Such is modern analytical chemistry!

With methods and applications so diverse and numerous, critical comparison, though badly needed, becomes almost impossible. Clearly, activation analysis and atomic absorption spectroscopy are formidable rivals of x-ray emission spectrography among the newer methods, and old methods "never die; they just fade away."

Though adequate critical comparison of the different methods for ultimate analysis is lacking, the authors stand by an x-ray unit as the best nucleus of a *new* modern laboratory of analytical chemistry. The problem of older laboratories is of course more complex: capital investment already made must be considered, as must training and experience already available. The future seems to belong to large laboratories, some of which sell their services to small users. The diversity of instrumental methods, the cost of equipment, and the pervasive, irresistible influence of the computer, all favor the growth of large laboratories [1, 12].

In addition to the reviews just discussed and the *Advances in X-Ray Analysis* often referred to, the authors strongly recommend to the x-ray spectrographer several books [13–16].

THE EXCITATION OF ANALYTICAL LINES

9.2 General

Characteristic lines were first excited by x-rays and by electrons (Section 1.20), and these are the principal agents used for excitation in x-ray emission

THE EXCITATION OF ANALYTICAL LINES

spectrography today. As the knowledge of x-ray spectra (Chapter 4) and of atomic structure increased, it became clear that such excitation could be accomplished by any means of supplying to an atom the energy needed to eject an inner electron. Bombardment by protons and by ions of other elements can do this (Section 7.15). More to the point as regards analytical chemistry, so can the various emanations from radioactive atoms (Sections 2.25 and 7.7). The reader will remember the distinction between those atoms that emit characteristic x-rays (radioactive *x-ray* sources, such as $_{26}Fe^{55}$) and those that supply the excitation energy in other ways (radioactive *energy* sources). The discussion that follows assumes samples of greater than critical thickness, a parameter that varies sharply with the method of excitation. It is understood also that, especially with radioactive sources, excitation can often be mixed; for example, x-rays plus electrons.

Excitation by proton bombardment, long studied by nuclear physicists, deserves special mention. Birks and co-workers [17] compared proton excitation with electron and x-ray excitation as regards yield of analytical-line quanta and peak-to-background ratio, N_T/N_B. Interesting results of this work appear in Table 9.2–1.

Table 9.2–1. Excitation of Analytical Lines by Protons and by Electrons [17a]

Analytical Line	N_T/N_B, 15 keV electrons	N_T/N_B, 34 keV electrons	N_T/N_B, 0.7 to 2 meV protons
Ti Kα1	9.62(10^3)	20.8(10^3)	5(10^4)
Cu Kα1	2.24(10^3)	5.72(10^3)	5(10^4)
Ge Kα1	0.88(10^3)	3.97(10^3)	1(10^5)

Notes. (1) Proton excitation requires many times as much energy as electron excitation. (2) The N_T/N_B values are more favorable for protons. (3) The N_T/N_B values in the table, having been corrected back to natural line width, are much more favorable than those important to the analytical chemist, who must use Bragg reflectors that give a much greater FWHM to get the intensities he needs (Chapter 5). Values of N_T/N_B for the electron microprobe, for example, can range from 300 to 1000 [17a]. At present, protons of the high energy needed to excite analytical lines are best produced by a van de Graaf generator, which usually makes the cost prohibitive.

The intensity of the resulting analytical line is determined not only by the energy of the agent that excites it, but also by how this agent interacts with the sample. This interaction in turn depends upon the agent (α-particles, for example, tend to interact strongly with the atomic nucleus) and upon the samples (the electron shells outside the nucleus change with atomic number). Mathematical treatments of this intensity problem are complex, generally do not give results useful to the analytical chemists, and are not emphasized below.

9.3 Electron Excitation

Before 1928, analytical lines were excited mainly by electron bombardment of the sample, which could be regarded as replacing the target in a Coolidge tube and had the drawbacks—severer then than now—that attend determinations on samples in high vacuum. After the publication by Glocker and Schreiber [18] of an important paper that showed the advantages of x-ray excitation for analytical chemistry, electron excitation rapidly lost ground. The success of the electron microprobe has reversed the trend. The reversal is being accelerated by the use of radioactive energy sources, and by the development of spectrographs, such as the Betaprobe Direct Electron Excitation X-Ray Spectrometer (Telsec Probe for short) of the British Telsec Instruments Limited, in which average results can be obtained over a sample area of 10×10 mm (see Chapter 10).

Electrons can of course be taken out through thin metal windows, as from a Coolidge tube that has no target, and arrangements of this kind were in early use for electron excitation of analytical lines [19]. Their future usefulness in analytical chemistry is doubtful.

The early experience with electron excitation (see Chapter 1; especially Figure 1.20–1) revealed its main features, which are (1) rapid transfer of energy to the sample, hence low critical thickness, 10^5 Å being a good value for the depth to which 50-keV electrons penetrate aluminum [20]; (2) much higher overall efficiency of x-ray production than in x-ray excitation, the x-rays for which are produced by electron bombardment; (3) high background; (4) relatively low absorption and enhancement effects; and (5) greater risk of altering the sample, especially if it happens to be plant or animal tissue. The need to work in vacuum is not the disadvantage it once was. Not only is improved equipment now available, some of which permits multiple loading of samples, but the increasing need for light-element determinations has forced much work with x-ray excitation into vacuum also.

As the "sphere of influence" of the electron beam used to excite analytical lines in the electron microprobe extends only about a micron, or 10^4 Å, in each of the three directions, internal standards—either added to, or present in the sample—are not generally useful for determinations with this instrument. This disadvantage has made it necessary to establish by fundamental calculations the relationship between analytical-line intensity and the amount present of the element sought. These calculations, which have been remarkably successful, rest upon more sophisticated views of the interaction between electrons and matter than those alluded to in Chapter 1.

THE EXCITATION OF ANALYTICAL LINES

Duncumb and Shields [21] describe this interaction as follows. Electrons entering the sample are scattered at random in a complicated way that produces a distribution, varying with depth of penetration, in their number and energy. This distribution governs the intensity of the emitted analytical line. Calculations of the intensity can be based either on a model of the distribution that assumes some form for the average electron scattering, as does theirs; or on a model in which trajectories of individual electrons are followed [22].

An electron entering a sample interacts with the constituent atoms in two principal ways: (1) by losing energy as it ejects an orbital electron, and (2) by undergoing elastic scattering (without significant loss of energy) to experience either a large change of direction (greater than, say, 90°) via Rutherford scattering or a change much smaller. Repeated small changes of direction (multiple scattering) lead eventually to loss of the original direction as depth of penetration increases: *diffusion* of the electrons has then begun. Rutherford scattering can cause electrons to reemerge from the sample (*back scattering*). The picture is that of an electron continuing to

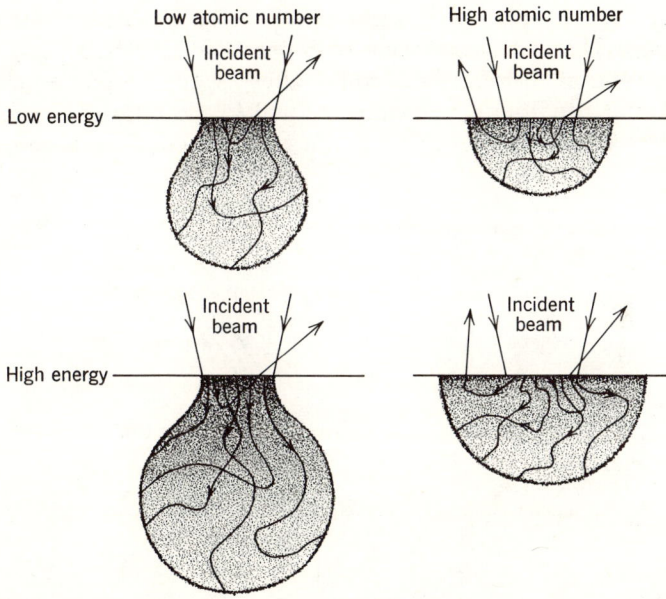

Fig. 9.3-1. Qualitative representation of the interaction between incident electron beam (diameter, say, 1 μ) and sample in an electron microprobe. Situation at low atomic number (e.g., Al) described in text. As Z increases, so does the chance of interaction between incident electron and sample atom. Result: electron diffusion begins much nearer the sample surface with a marked modification in the "sphere of influence" of the incident beam. Note that shape of this "sphere" is independent of electron energy. Back-scattered electrons are indicated by emergent arrows. From Duncumb and Shields, Ref. 21, p. 618, Fig. 1.

lose energy so long as it can eject orbital electrons from atoms, being slowed down as a result, gradually assuming random direction owing to multiple scattering, but occasionally undergoing large changes of direction when Rutherford scattering occurs.

How does this complex interaction change with atomic number? In the *light* elements (low density, low atomic number), electrons will on the average penetrate further before diffusion begins, and the chance of their being back-scattered is consequently small. Increasing the energy of the incident electrons of course increases their path length, but the depth of complete diffusion remains an approximately constant fraction of the electron penetration. For the electron microprobe, the situation is illustrated in Figure 9.3–1; for electron excitation of a larger sample area, Figure 9.3–2 is more nearly applicable.

The comprehensive investigation by Eddy and Laby [8] may serve as the classic example of what can be accomplished with electron excitation and a photographic plate as detector. Their results on a copper-zinc alloy are remarkable for the high precision attained. To be sure, this case is among the most favorable that can be imagined. No calibration was required and the percentage of only one element needed to be established, for the alloy was binary. The atomic numbers of copper and zinc being adjacent, the intensity ratio of their Kα lines could, after an appropriate adjustment of experimental conditions, be assumed equal to the ratio of the number of atoms present of each metal. Under these simple conditions, compositions

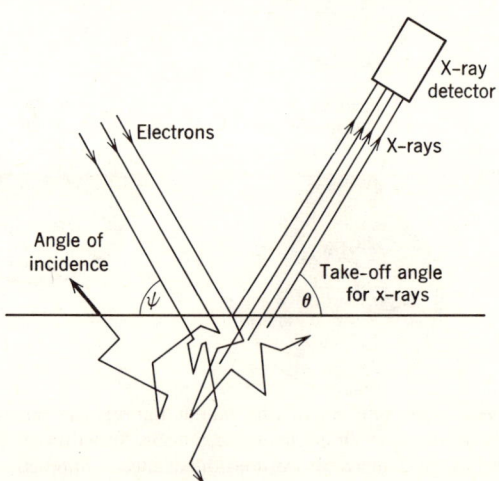

Fig. 9.3–2. Qualitative representation of interaction between individual electrons and sample unrestricted as to area. A single back-scattered electron is indicated. See Fig. 9.3–1. From Archard and Mulvey, Ref. 22, p. 626, Fig. 1.

could be calculated satisfactorily from intensity ratios, as is shown by the following results for a series of 16 x-ray determinations on such an alloy found by chemical methods (details not given) to contain 73.00% copper: average copper content, 73.16%; standard deviation for a single determination, 0.27% copper. There is no better practical illustration of advantages inherent in the simplicity of x-ray spectra, and no better example of a useful built-in standard.

9.4 X-Ray Excitation

Discussions of the Coolidge tube and of the interaction of x-rays with matter began in Chapter 1, and the topics have reappeared frequently enough to make detailed reference impractical. Most of the work is old. One would think, therefore, that little remains to be said about the x-ray excitation of analytical lines. One would be wrong. As analytical lines are still most commonly excited in this way, the continuing growth of x-ray emission spectrography has led to important changes in x-ray excitation, and the end is not in sight.

Conventionally, such excitation was carried out with a Coolidge tube containing a tungsten target, and the x-rays traveled an air path, an arrangement at once highly flexible and most convenient. The tube was commonly operated at 60 Hz, at 50-kV peak, and with half-wave rectification. The needs in x-ray emission spectrography for greater sensitivity, greater precision, and greater capability for the light elements have changed this simple situation. The changes, usually successful, have their price. Modern x-ray excitation tends to be more costly and less convenient than the conventional; in filling specialized needs, it meets strong competition from alternative ways of getting the work done. Equipment changes belong in a later chapter; here we shall sketch only their significance as regards analytical results.

Whether a change from the conventional is worthwhile depends here as always upon the laboratory for which it is being considered. For example: Should the excitation voltage available be raised from 50 to 100-kV peak? The answer depends on (1) how important heavy-element determinations are in the laboratory's work; (2) whether determinations with L lines are so unsatisfactory that K lines must be used; (3) whether the actual results obtained with K lines will be sufficiently better, which involves questions such as the intensity available after Bragg reflection at short wavelengths (~ 0.2 Å), where (Chapter 5) satisfactory crystals are hard to find because the d values needed are small; (4) whether 75-kV peak excitation is adequate; and (5) what kinds of voltage excitation are involved —see Figure 9.4–1. This by no means complete summary for a single proposed change (for one thing, cost and alternative methods were not mentioned) shows the complexities that can make an embarrassment of riches out of progress in equipment.

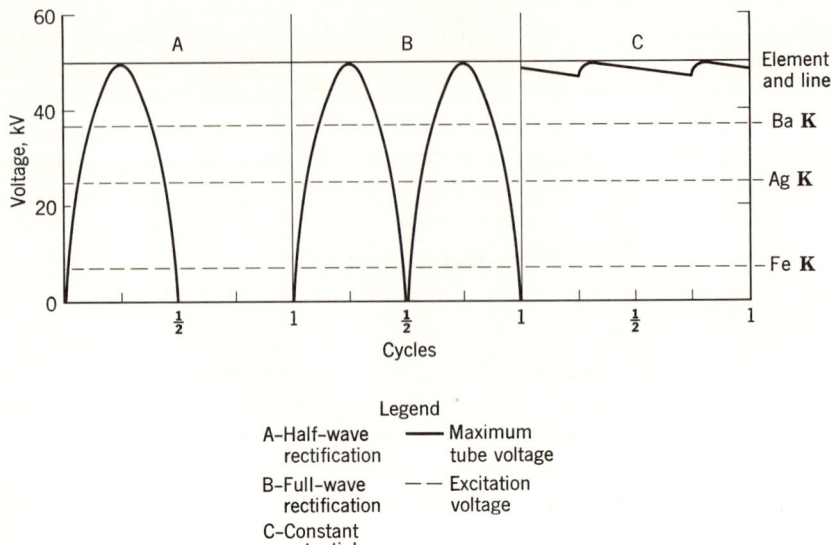

Fig. 9.4-1. Excitation efficiency for three common kinds of excitation voltage. The time interval for excitation of the sample is evidently proportional in each case to the total length of broken line under the voltage curves during one cycle, which leads to the relative intensities quoted in the text. After I. Adler, Ref. 14, p. 36, Fig. 19.

The principal advantages and disadvantages of x-ray excitation can be inferred from a comparison of this with electron excitation (Section 9.3). Convenience, flexibility, and low background (see Figure 1.20–1) bear repetition, as does the statement that some of the changes away from the conventional decrease convenience and (sometimes) flexibility: thus liquid samples are awkward to handle in vacuum.

EXCITATION VOLTAGE. We condense here Adler's description [23] of recent changes in excitation voltage that make Coolidge tubes better x-ray sources for the excitation of analytical lines. Minimum requirements for the tube: power in the range 3–5 kW; output, full-wave rectified (Figure 9.4–1); voltage, variable in the range 15–50 kV; current, variable in the range 10–50 mA; power supply, stabilized against line fluctuations and with output regulated. Preferably and usually, the tube target is grounded, which makes necessary a high negative voltage on the filament, ensures easily against shock hazard, and permits cooling of the target with flowing tap water.

Excitation of x-rays at constant voltage is preferable even to excitation with full-wave rectification. Adler [14, p. 38] calculates the following relative intensities for **K**-lines in samples excited by x-rays produced at 50 kV with

THE EXCITATION OF ANALYTICAL LINES

the three common kinds of voltage; the intensities are listed in the order half-wave rectification, full-wave rectification, and constant voltage: Fe as sample—43, 86, 100; Ba as sample—13, 26, 100. See Figure 9.4–1.

Coolidge tubes operating at 75 kV and 100 kV are now available. The advisability of installing excitation equipment for higher than 50 kV was discussed briefly above. In the determination of uranium, the **K** lines will not soon replace the **L**!

SPECTRAL PURITY. The only characteristic lines leaving a Coolidge tube should be those of the target. The increasing use of x-ray emission spectrography for trace determinations has made spectral purity more important than ever. Spectral purity should be assayed periodically by recording the tube spectrum after it has been scattered in the spectrograph by a solid polymer (e.g., lucite) as sample.

THE EMERGENT SPECTRUM. To satisfy modern needs, notably for the calculation of interelement effects and for the excitation of light elements, it is desirable to know in detail how the (integrated) intensity varies over wavelength range in the spectrum as it emerges from the Coolidge tube. Gilfrich and Birks [24] satisfied these needs for tubes with W, Mo, Cu, and Cr targets over the range 15–50 kV peak for tungsten and at 45-kV constant voltage for all targets. Their experimental arrangement appears in Figure 9.4–2. They found three features for all targets: (1) an unexplained "secondary peaking" that appears as a slight shoulder in the relative intensity-wavelength plot for the continuous spectrum; (2) substantial contributions

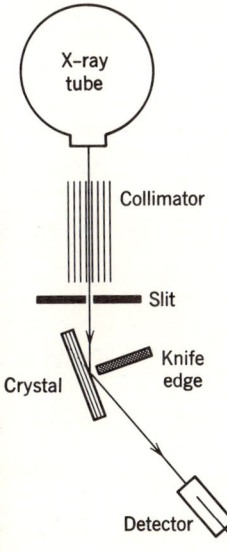

Fig. 9.4–2. Sketch of experimental arrangement for establishing the variation of integrated intensity with wavelength for x-ray beams from Coolidge tubes with different targets. The single-crystal (LiF) spectrometer has a slit to limit the beam width, which ensures that the crystal will intercept all the beam even at the smallest Bragg angle ($\theta = 4.5°$) used. The knife edge shields the flow proportional detector from rays scattered by the edges of the slit. After J. V. Gilfrich and L. S. Birks, Ref. 24, p. 1077, Fig. 1.

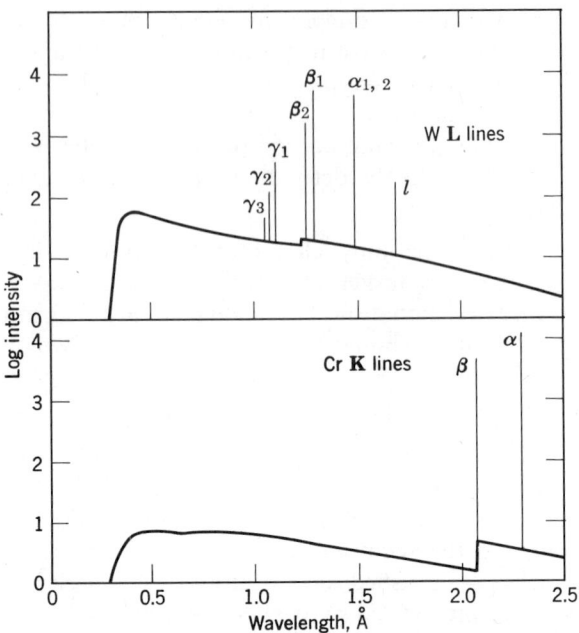

Fig. 9.4–3. Relative intensities, on a logarithmic scale, for x-ray spectra emerging from two Machlett OEG-50 tubes, one with a chromium, the other with a tungsten target. The tubes were operated at 45 kV (constant voltage). The figure cannot be used to estimate the contribution of the line spectrum because that would require too large an increase in the ordinate axis. In a table in the reference cited, the measured integrated intensities of each line have been corrected to the proper natural line breadth in order to eliminate characteristics of the spectrometer and give the integrated spectral distribution. This correction is large enough so that the figure is only a general guide to the excitational distribution. After Gilfrich and Birks, Ref. 24, p. 1078, Fig. 3.

to the total intensity by the characteristic lines at 45 kV (constant potential): 75% by Cr **K**, 60% by Cu **K**, and 24% by W **L**; (3) an influence of absorption jumps at critical absorption wavelengths of the targets. Relative intensities of beams emerging from tubes with tungsten and chromium targets can be estimated from Figure 9.4–3.

INFLUENCE OF PHOTOELECTRIC ABSORPTION (of absorption edges or jumps; see Section 1.11 *et seq.*). This influence causes the expected discontinuity in the continuous spectrum, as was discovered by Kulenkampff [25] in experiments like those of Figure 9.4–4. What this means for the excitation of analytical lines from light elements is shown by Figure 9.4–5.

UNWANTED X-RAY EXCITATION. In general, unwanted x-ray excitation falsifies the intensity of the analytical line by raising background or by exciting characteristic lines that change this intensity. Unwanted excitation of characteristic lines in the Bragg crystal was discussed in Section 2.23 as a

THE EXCITATION OF ANALYTICAL LINES

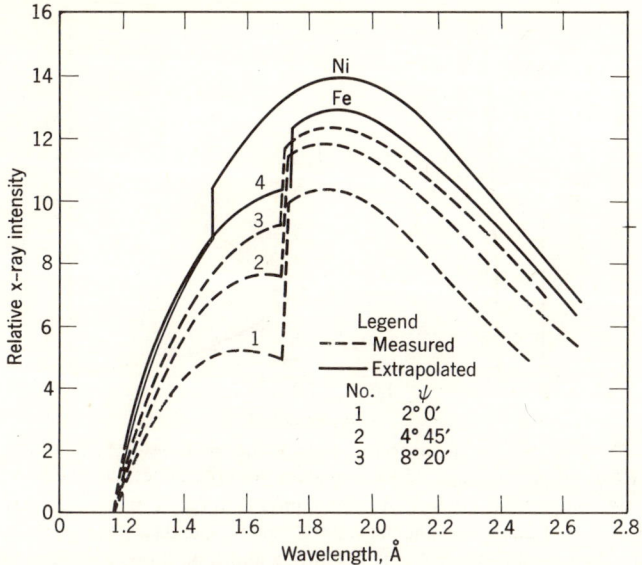

Fig. 9.4–4. Discovery of the influence of absorption jumps on continuous x-ray spectra produced by electron excitation at 10.470 kV. The electron beam is incident on the target at angle ψ (variable); the relative x-ray intensity is measured at 90° (fixed). Relative intensity (measured over 20 sec with an ionization detector) decreases with ψ mainly because the excitation process becomes less efficient in accord with discussion in Section 9.3. The solid lines are for maximum relative intensity as obtained by extrapolation. Critical absorption wavelengths are known to be 1.74 Å (Fe) and 1.49 Å (Ni), in excellent agreement with the figure; this proves that the continuous spectra jump in intensity because an absorption edge is crossed in the direction of increasing wavelength. The characteristic lines generated are not shown. From H. Kulenkampff, Ref. 25, p. 580, Fig. 12.

nuisance in the determination of phosphorus. Another nuisance is peculiar to x-ray diffraction. If the characteristic line being used for the determination of d excites characteristic lines in the sample, the latter will be radiated uniformly in all directions and may enhance the background enough to spoil the determination. A common example is the use of Cu Kα for diffraction work on samples rich in iron. An appropriate change of target is the obvious remedy.

CHOICE OF TARGET. X-ray tubes are now available, not only with a greater variety of targets, but also with increased power ratings for most of these. For the excitation of a single analytical line, choice of the best target is a simple matter when data such as those of Gilfrich and Birks [24] are available for the emergent x-ray beam generated by the target. Such data are needed for every kind of target important to the analytical chemist.

In the usual situation, the analytical chemist would like being prepared to determine elements ranging upward from, say, fluorine in atomic number. He is therefore willing to compromise as regards excitation efficiency in

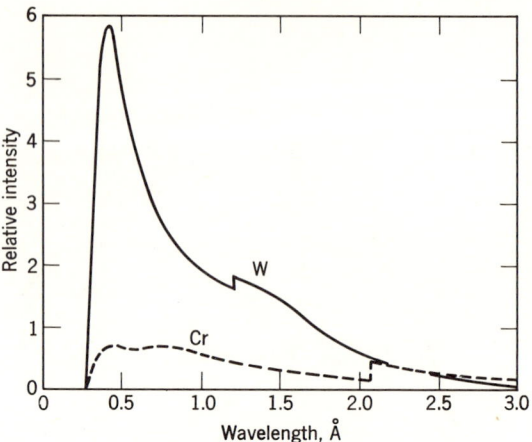

Fig. 9.4–5. Continuous spectra emerging from Coolidge tubes (Machlett OEG-50) with tungsten and chromium targets at 45 kV (constant voltage). Note that the chromium target gives at 3 Å several times the intensity from the tungsten target. For this situation, important for x-ray excitation of analytical lines of light elements, the absorption jump in chromium (just beyond 2 Å) is responsible. After J. V. Gilfrich and L. S. Birks, Ref. 24, p. 1078, Fig. 4.

order to reduce capital investment and to escape frequent changing of x-ray tubes. One useful compromise consists of the General Electric tube that contains a tungsten and a chromium target so installed as to permit simple external switching from one target to the other. What this means as regards excitation efficiency over the range of atomic numbers is foreshadowed by Figures 9.4–3 and 9.4–5, and amplified by Figure 9.4–6.

Whether a dual-target tube is a better bargain than the two equivalent single tubes depends again upon the work of the laboratory. If determinations of elements heavier than $_{22}$Ti are comparable in frequency with those of the lighter elements, the dual-target tube certainly seems a good buy. It can be a good buy also in many other situations, even when its life does not exceed that of a single-target tube.

X-RAY TUBES FOR EXCITATION OF ULTRASOFT ANALYTICAL LINES. Somewhere in the region of atomic numbers below $_{11}$Na to $_9$F, sealed Coolidge tubes cannot usefully excite even the (ultrasoft) analytical lines of shortest wavelength, which are F **K**(18.3 Å), O **K**(23.6 Å), N **K**(31.6 Å), C **K**(44.7 Å), B **K**(67.6 Å), and Be **K**(114 Å). Demountable tubes developed by Henke [26] fortunately have this capability.

The problems attending x-ray excitation in the ultrasoft region (wavelengths of 10 Å and more) are formidable. The characteristic-photon yield is low (Section 1.23). Mass absorption coefficients are huge. Consequently, an electron beam of high power must strike a target of large area in a tube with very thin (and therefore fragile) windows. Targets must be demountable so that the wavelength range can be covered. Contamination of these

THE EXCITATION OF ANALYTICAL LINES

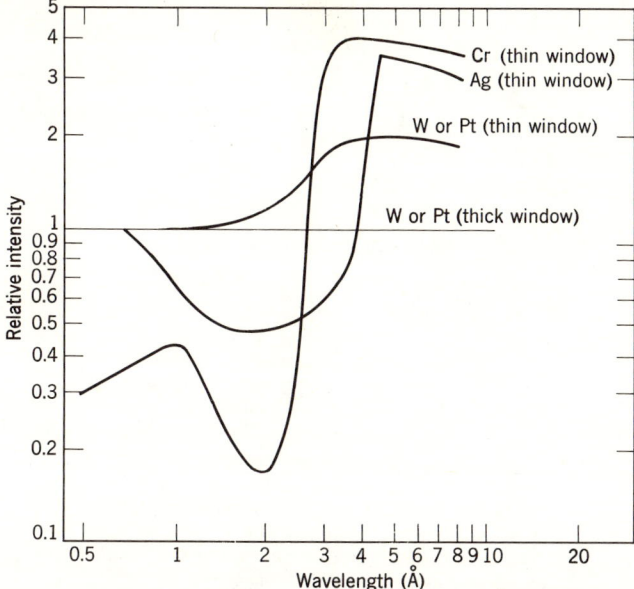

Fig. 9.4–6. Relative measured intensity as a function of wavelength in the emergent spectra of General Electric Coolidge tubes with three different targets and with thin and thick beryllium windows. In general, the chromium target is better suited than the tungsten for the excitation of analytical lines in elements below $_{22}$Ti. Courtesy General Electric Company.

targets, as by tungsten from the cathode, cannot be tolerated because the mass absorption coefficients in the ultrasoft region are so high. Deposition of carbonaceous materials on the target must also be prevented.

Henke [26] has managed to solve these formidable problems by developing the continuously pumped tube shown schematically in Figure 9.4–7. Powerful electron beams are electrostatically constrained to impinge on two large ($\frac{1}{4} \times 1$ in.) lateral focal spots on an interchangeable target of triangular cross section. The tube operates near 10^{-5} torr and it is isolated from the sample chamber by a sliding gate that covers the tube port with a thin, highly transmitting window until the sample chamber has been pumped down, usually to about 10^{-3} torr. Tube housing and anode are water-cooled. The anode is operated at maximum power, and the cooling water is then converted to steam. The output x-ray intensity is enhanced by having the anode very near the large, thin window. Ultrasoft x-ray sources that are interchangeable with standard Coolidge tubes have been built. Such sources will no doubt soon be on the market, and it will be interesting to see whether they or large-scale electron excitation (as in the Telsec probe mentioned above) will prove more effective in the determination of light elements.

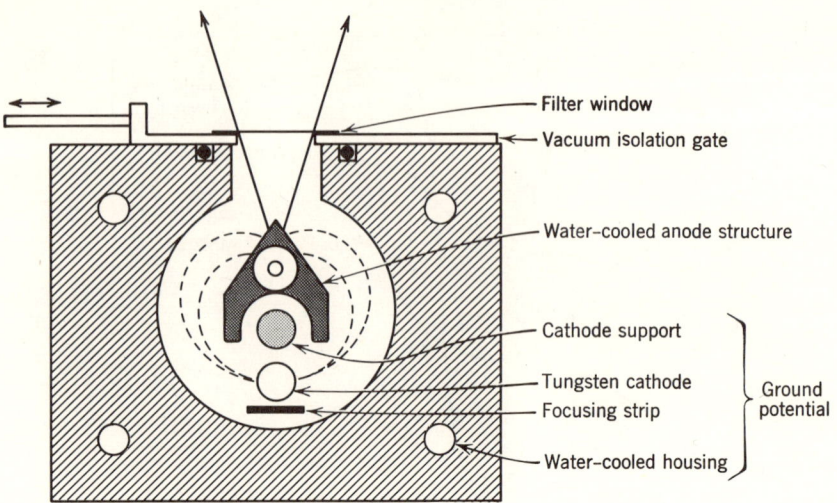

Fig. 9.4-7. Schematic diagram of a Henke demountable ultrasoft x-ray source. Operation at anode voltages from a few tenths to 20 kV at 2 to 4 kW, as dictated by the target, provides x-ray beams of large cross section and high intensity in the wavelength region above 10 Å. Short description of tube in text. Note ingenious "concealment" of cathode to prevent contamination of target by tungsten. The anode is operated at several hundred degrees Centigrade or higher to minimize the risk of contamination by carbonaceous matter, which exists even at high vacuum. From B. L. Henke, Ref. 26, p. 296, Fig. 8.

9.5 Comparison of Electron and X-Ray Excitation

Zemany [27] has made the illuminating comparison in Table 9.5-1 of electron excitation at one extreme, namely in the electron microprobe, with x-ray excitation at the other, namely in an emission spectrograph. The estimates of optical efficiency are based on reasonable assumptions about losses in the two devices.

Table 9.5-1. Comparative Data for the Excitation of Analytical X-Ray Lines

	Emission Spectrograph	Electron Microprobe
Power of electron beam, W	$2.5(10^3)$	$5(10^{-3})$
Excited sample area, cm²	1	10^{-8}
Power of excited x-ray beam, W	$5(10^{-2})$	$2.5(10^{-5})$
Power per unit sample area, W/cm²	$5(10^{-2})$	$2.5(10^{+3})$
Optical efficiency	$\sim 10^{-5}$	$\sim 2(10^{-4})$
X-rays detected, W	$5(10^{-7})$	$5(10^{-9})$

THE EXCITATION OF ANALYTICAL LINES 371

Note how much greater is the x-ray power per unit area for the microprobe. The table shows the great range of x-ray emission spectrography.

9.6 Excitation by Radioactive Sources

Radioactive sources provide energy that can be used for excitation. This energy is not often predominantly in x-ray form so that not many radioactive *energy* sources qualify as radioactive *x-ray* sources, $_{26}Fe^{55}$ (Section 3.2) being a conspicuous exception. Generally, then, we deal here with *mixed excitation*. Spectral purity in excitation can be achieved by use of a "characteristic-line generator" such as was introduced by Engström for absorptiometry (Section 3.7): that is to say, the radioactive energy is used to excite a suitable characteristic line in a substance intermediate between radioactive source and sample, the analytical line in which is thus produced by x-ray excitation.

Practical radioactive sources are weak for reasons of cost and safety, and they inevitably become weaker with age. Bragg resolution cannot be used in conjunction with them—the attendant intensity loss is prohibitive. But necessity here becomes a virtue. Radioactive sources can be placed so close to sample and detector that favorable geometry largely compensates for their weakness and makes possible the building of portable, primitive x-ray emission spectrographs useful, for example, in mineralogical exploration. See Sections 2.25, 7.7, and 10.7.

The intensities of analytical lines excited by radioactive sources must therefore be established by using a suitable detector and by employing either the balanced filter method of Ross (Chapter 5) or energy resolution in the form of pulse-height discrimination or analysis. See Chapter 2. Worthy of a second mention is the overriding importance that the new solid-state detectors (Section 2.15) have for radioactive sources: used along with multichannel pulse-height analyzers, these detectors make possible laboratory spectrographs with radioactive sources that are much more sophisticated than the portable instruments mentioned above. See Section 2.25.

For detailed information about radioactive sources for the excitation of analytical lines, see the excellent review by Rhodes [28], our mainstay for this section and especially for Table 9.6-1.

The *Bremsstrahlung*, listed above for the two sources that contain an ordinary element along with the radioactive isotope, is analogous to, but more complex than, the continuous x-ray spectrum produced, say, by a thick tungsten target. In the radioactive case, the electrons of course are ejected by the nucleus and give rise to *internal* Bremsstrahlung before they leave the nucleus, and *external* thereafter. The latter is of greater practical importance [29]. The sources that rely on Bremsstrahlung have

Table 9.6–1. Radioactive Energy Sources Useful for Exciting Analytical Lines

Source	Half-Life, years	Useful Energies (Kind and amount)	Heaviest Element Usefully Excited
$_{26}Fe^{55}$	2.7	Mn **K**, 5.9 keV	$_{24}Cr$
Tritium—Zr	12.3	Br'ung, 2–12 keV; Zr **L**, 2 keV[a]	$_{30}Zn$
$_{48}Cd^{109}$	1.3	Ag **K**, 22 keV; γ-ray, 88 keV	$_{43}Tc$
$_{61}Pm^{147}$—Al	2.6	Br'ung, 10–100 keV	$_{60}Nd$
$_{95}Am^{241}$	470	γ-rays, 26 and 59.6 keV; Np **L**, 11 to 22 keV	$_{69}Tm$
$_{64}Gd^{153}$	0.65	γ-rays, 97 and 103 keV; Eu **K**, 42 keV	$_{88}Ra$
$_{27}Co^{57}$	0.74	γ-rays, 14, 122, and 136 keV; Fe **K**, 6.4 keV	$_{98}Cf$

[a] Br'ung is *Bremsstrahlung*; see text.

practical emission efficiencies (10^{-5} to 10^{-3} photons per disintegration) several orders of magnitude lower than do radioisotopes alone.

With a radioactive source that generates about 10^7 photons/sec, the radiation hazard is small even from unshielded sources. To achieve cps values in the range 10^3 to 10^5, which are needed if counting intervals below 100 sec are to suffice for reasonable counting errors, the geometrical efficiency for excitation must be a few per cent inasmuch as the characteristic-photon yield may also be only a few per cent. Geometries of adequate efficiencies are shown in Figure 9.6–1.

Radioactive sources, such as $_{84}Po^{210}$ that emit α-particles are of particular interest for light elements [30]. The characteristic-photon yield with elements near carbon is relatively very favorable when excitation is by

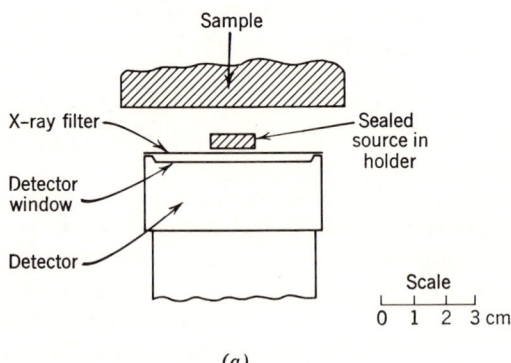

(a)

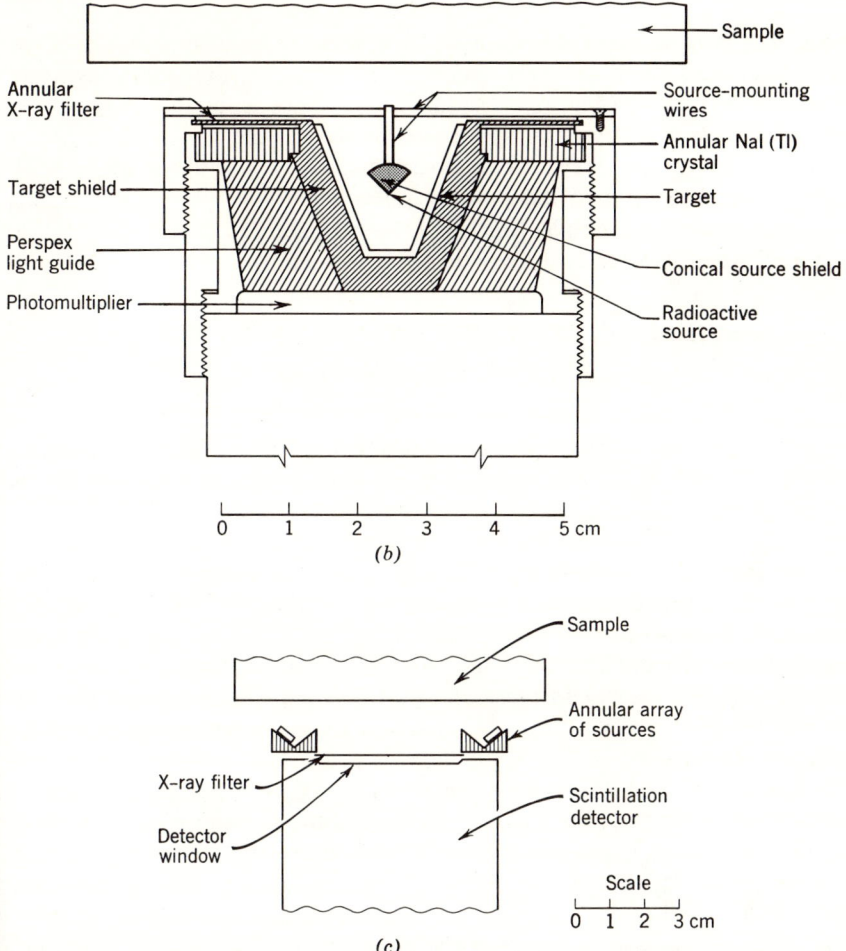

Fig. 9.6-1. Three geometrical arrangements of radioactive sources that give analytical-line intensities adequate for many trace determinations. (*a*) "Central source" arrangement of sealed source, sample, and scintillation detector. (*b*) Same arrangement (in principle) with characteristic-line generator ("target") added. The spectral purity of the characteristic line generated to excite the analytical line can exceed 90%. A simple "target" is often made of a powdered oxide with a minimum amount of binder (epoxy resin). If more than one characteristic line is needed, other oxides are mixed in. "Targets" are easy to change. (*c*) Annular arrangement for use with a tritium-zirconium source and a scintillation detector—*annular* for increased intensity. Here the radioactive source is not sealed but consists of a thin layer of tritiated zirconium on tungsten. From J. R. Rhodes, Ref. 28, p. 684, Fig. 1; p. 686, Fig. 2; p. 690, Fig. 6.

α-particles; furthermore, back-scattering of α-particles is negligible as is the Bremsstrahlung they generate. In a test of α-particle excitation, Watson, Sturseth, and Howard used a beam of α-particles from the 88-in. Texas A&M variable cyclotron, tightly focused on the sample, which was a Mylar film carrying μg/cm^2 of Cu, Sn, or Pb and inclined at 45° to the beam. With a Si(Li) detector and a 900-channel pulse-height analyzer, good results were obtained; for example, four Pb **L** lines ranging in energy from 9.2 to 14.8 keV were detected and resolved [31].

PRINCIPAL DIFFICULTIES AND COMMON REMEDIES
9.7 General Considerations

A sample (S) may be either *standard* (St) or *unknown* (U) and either *original* (as brought to the laboratory) or *prepared* (as placed in the spectrograph); the second distinction is often so obvious that it need not be explicitly made.

A common assignment for the analytical chemist is to determine the *amount present* of an element in a sample, or in *unit weight* of sample. A more general, less common, assignment is to establish sample *composition*; that is, all elements present and the amount of each. In the kind of work for which x-ray emission spectrography competes, determination of *concentration*, or amount present in *unit volume* of sample, is infrequent. Only the establishment of sample composition is an *analysis*; anything less is one or more *determinations*.

Because the sample for x-ray emission spectrography takes so many forms (films, evaporated residues, powders, solutions, slurries, solid articles as made or as found), composition units need discussion. No matter how an analytical line is excited, the prime considerations are the number and kinds of atoms that are affected in the excitation process, and the number and kinds of atoms that influence the emergent intensity of the analytical line after excitation. For samples of greater than critical thickness (and no others concern us in this chapter), the volume of sample affected is usually unknown and need not be known. Units of weight (or mass) are therefore the logical way to express composition, a statement corroborated by the fundamental importance of the mass absorption coefficient. The weight-fraction, W or **W** (see below), is a simple, logical, and generally satisfactory unit to express composition.

Surely most analytical chemists have misgivings about the increasing and unjustified use of *concentration* (usual symbol C) as the unit of composition in x-ray emission spectrography; for example, in the case of alloys. The (self-contradictory) *concentration in weight per cent* is coming into use, although concentration is a *volume* unit, and *weight per cent* needs no help—it can stand alone!

PRINCIPAL DIFFICULTIES AND COMMON REMEDIES

We do not mean to be pedantic. Calibration curves, as we soon see, often deviate from a linear relationship between analytical-line intensity and amount of corresponding element in the sample. It is always permissible to choose composition units that give the most nearly linear calibration curves. When such curves are based on standards, the composition units are not important in themselves, always provided the same unit is used for standard and for unknown.

X-ray emission spectrography, semiquantitative or quantitative, is concerned with the relationship between the intensity $\mathbf{I}$ of the analytical line and $\mathbf{W}$, the weight-fraction, if this unit of composition is chosen. (These quantities appear boldface when they are for an element being determined.) Under the simplest conditions,

$$\mathbf{I}^S = \mathbf{W}^S \mathbf{I}^E \qquad (9.7\text{--}1)$$

where the superscripts denote sample and (pure) element, both being of "infinite" (i.e., greater than critical) thickness.

The reader will have formed the (correct) impression by now that it is almost impossible to avoid *identifying* elements heavier than fluorine by x-ray emission spectrography when their weight-fractions exceed, say, 0.001–0.01. As Anater [32] has convincingly demonstrated, and others have found, it is easy to apply (9.7–1) and obtain *semiquantitative* results that establish $\mathbf{W}$ for many elements within about $\pm 50\%$. Truly *quantitative* results, $\mathbf{W}$ within $\pm 1\%$ or better, can be obtained if certain difficulties are remedied.

The difficulties of concern here are deviations from the proportionality of (9.7–1). The principal difficulties are of four kinds. They are discussed below, as are the applicable remedies.

9.8 Difficulties Traceable to Equipment

No generally applicable discussion of equipment difficulties can be detailed or precise. The situation differs with the degree of automation and with the extent to which solid-state circuitry is used. The best that can be done is to list the most common causes of equipment difficulties.

Loss of spectral purity in x-ray tubes and *problems with gas-filled detectors* were discussed in earlier chapters.

BRAGG REFLECTORS. These, especially those with high coefficients of *thermal expansion*, will cause difficulties if their temperatures change enough to alter significantly the lattice spacing.

FAULTY GONIOMETER SETTING. The best instruments can be set easily to $0.005°$ (2θ), which ordinarily corresponds to an error of less than 0.1% in $\mathbf{I}$; see Chapter 8.

MISALIGNMENT OR IMPROPER POSITIONING OF SAMPLE. See Figure 7.12–2. Related problem: change in level of a liquid sample; see Chapter 8.

INSUFFICIENT WARM-UP. The usual prescription is a minimum of 30 minutes for the x-ray components and less for the electronic circuitry (detectors and counting circuits). Because an appreciable part of the useful life of an x-ray tube may be spent in warming up if most determinations are made on equipment that has been shut down, it seems worthwhile to establish whether 30 minutes is longer than is needed for warming up modern equipment containing a maximum of solid-state circuitry.

UNSTABLE X-RAY GENERATOR. Indications of progress: Ashby and Proctor [33] found x-ray tube current and voltage to remain within $\pm$ 0.02 and $\pm 0.05\%$ of their mean values over almost 24 hours during which the line voltage fluctuated $\pm 15\%$. Unplanned intensity fluctuations in the x-ray beam that were caused by changing absorbance of the air path could be detected, so constant was the x-ray source. Yee and Deslattes [34] found that a solid-state current stabilizer could keep x-ray tube current constant to $\pm 0.1\%$ for at least 30 minutes in the range 20–1000 mA. Older equipment cannot be expected to perform this well; furthermore, with any equipment, the intensity fluctuations in the x-ray beam will depend somewhat on fluctuations in the line voltage. In general, fluctuations in x-ray tube current are less serious than fluctuations in x-ray tube voltage.

Not all causes of equipment difficulty have been listed. Why not? (1) As stated above, they are not usually serious and are becoming less so. (2) They depend on the equipment—not only on kind, but also on age, history, and environment. (3) Some of the important difficulties may be beyond what the analytical chemist can remedy or conveniently get remedied. What can the analytical chemist do? *Get to know his system by carrying out an analysis of variance* in the manner of Chapter 8. Find out what contribution a given kind of equipment difficulty makes to the standard deviation and use the standard counting error as a yardstick for establishing its seriousness. Remember that short-term intensity changes (fluctuations) and long-term changes (drifts) present different problems, the relative seriousness of which depends upon W and upon I: trace determinations differ from the determination of major constituents. Whether it is desirable to alternate getting N_T and N_B depends upon what kind of uncontrollable intensity change is likely to occur over the counting intervals needed to establish N_T and N_B. It comes to this: Nowhere in analytical chemistry is the statistical approach easier to apply and more certain to prove rewarding than with x-ray emission spectrography. Ways in which intensity comparisons can remedy equipment difficulties will appear in Section 9.18.

Extensive data on equipment difficulties are not easy to come by. Every one interested in x-ray emission spectrography should study the history Eastman [35] has given of a Norelco x-ray emission spectrograph used

PRINCIPAL DIFFICULTIES AND COMMON REMEDIES

almost daily for 3 years in the Kaiser Aluminum and Chemical Corporation Ravenswood (West Virginia) Reduction Laboratory by operators without formal analytical training. Determinations routinely done with minimal supervision are (1) Ca in the electrolyte of the aluminum cells, (2) Si, P, S, and Mn in cast iron, and (3) S, K, Ca, V, Mn, Fe, and Ni in carbon materials. Eastman cites extensive data to support his conclusion that "routine [x-ray] results...are more reliable than chemical results by the same personnel...[because] chemical procedures inherently contain more potential sources of error." (Words in brackets supplied.) Table 9.8–1 is a 3-year maintenance log. The troubles were neither esoteric nor epidemic and most could have occurred with any spectrograph system!

Table 9.8–1. Maintenance History of Norelco Spectrograph in Kaiser Laboratory[a]

Date		Problem	Symptom
July	1959	Leaky water hose	Shorted battery
July	1959	Poor connection in P-10 filler gas line	Gas wasted
Aug.	1959	Slipping goniometer gear	Unreliable 2θ values
Sept.	1959	Fumes from plastic in room	Cps low and erratic
Oct.	1959	Exit collimator dented during installation of flow proportional detector	
Jan.	1960	Shorted high-voltage cable	Smoke
June	1960	Dust from samples in collimator	Reduced cps for S
March	1961	Oscillating current and line controls	Synchronous oscillations in regulators
June	1961	Leaky hose to x-ray tube	Wet samples
July	1961	Bragg crystal wrongly positioned	Broken crystal
Aug.	1961	Detector needs repair	High N_B
Aug.	1961	Main power supply needs repair	No power
Sept.	1961	Impure helium supplied	Reduced cps for Ca
June	1962	Short in high-voltage cable	Smoke
July	1962	(Repetition of Aug. 1959)	
July	1962	Dust in system	Reduced cps for S
July	1962	Entrance collimator washed with acetone	Fell apart

(Helium path bellows changed four times in these three years. X-ray tube replacements, being expected and unavoidable, are not listed.)

[a] From Eastman [35].

9.9 Difficulties Caused by Unwanted Contributions to Intensity. Background

Our x-ray spectra consist of peaks superimposed on the background that is always with us. Not only the background, but certain kinds of peaks as well, qualify as difficulties because they are unwanted contributions to intensity. Before the background difficulty is discussed in detail, a brief summary of the situation is in order. Unwanted peaks receive further attention later.

We list six kinds of peaks: analytical-line peaks (wanted); diffraction peaks caused by Bragg reflections *from the sample*; higher-order reflections from the Bragg reflector; peaks that result from incoherent scattering by the sample of a characteristic line of the target; Compton peaks; peaks of characteristic lines (other than the analytical line) from the sample—all five unwanted. When these peaks merge with the wanted peak, spectral interference results.

Any peak will merge into the background as the intensity of the peak decreases. Figure 8.11–1 is a notable example: here the analytical peak disappears into the background as the amount of the element to be determined decreases to zero.

The background problem is complex. When light elements are being determined, the problem is almost too complex for detailed discussion. We therefore "zero in" on the heavier elements and assume that pulse-height selection has eliminated cosmic rays, radioactivity, higher-order Bragg reflections, and electronic noise. This leaves us with a background composed principally of scattered x-rays that can come from the x-ray tube, the sample, the sample holder, gas in the optical path, the collimator, and from other parts of the spectrograph. Possible contributions by characteristic lines of the Bragg reflector and by abnormal Bragg reflections were dealt with in Section 5.8. Fortunately one can cope with the background without knowing of what it consists. Background is higher with electron- than with x-ray excitation.

The following symbols are used: N_T is the count for analytical line plus background over the counting interval Δt; N_B is the background count, defined above, for the same interval; N_B', N_B'', and so on are counts taken to establish N_B, and they may be taken (for example) at goniometer settings different from that for N_B, or over a different counting interval, or in the presence of a material (perhaps a substrate or solvent) other than the sample. The corrected count $N_T - N_B$ normally is used to arrive at the amount present of the element sought. Usually $N_T - N_B$ derives from a peak intensity, though such a difference can be formed for integrated intensities also (Section 6.7).

Must a background correction always be made? By no means. Sometimes satisfactory working curves are obtained without correction. The ideal such curve is a straight line intersecting the origin in a plot with analytical-line intensity as ordinate and weight-fraction of element sought as abscissa. If N_B is constant and not corrected for, the straight line in this plot will intersect the ordinate axis, and the intercept will measure N_B. Incorporating a correction for background does not necessarily improve the reliability of a working curve.

In fact, the making of a background correction entails disadvantages. It is often necessary to establish N'_B to make such a correction. This entails extending the counting time, makes a determination more costly, and increases the chance of equipment difficulties. Finally, as was shown in Chapter 8, the use of N_B in computing analytical results actually increases the counting error; for example, according to the law by which errors combine, the statistical error in the difference $N_T - N_B$ is greater than the statistical error in N_T alone.

On the other hand, a determination of the background has the following advantages. The background may serve as an internal standard (see below). Even when it is not so used, the obtaining of a reasonable background count is moderately good assurance that the spectrograph is functioning properly. Finally, when a single determination is carried out for which no calibration curve is available, the corrected count $N_T - N_B$ is a better basis for comparing the unknown with a single standard than is N_T alone. In summary: background corrections should be made only if they lead to more reliable analytical results, and opportunities to reduce the background —if possible, to the point where it may be neglected—are usually worth exploiting.

9.10 Remedies for Background Difficulties

Further discussion of (9.7–1), now needed, was postponed because the effects of equipment difficulties, being often unpredictable, cannot be simply formulated. In (9.7–1), the proportionality constant relating $\mathbf{W}^S$ to $\mathbf{I}^S$ is $\mathbf{I}^E$, the intensity for the pure element under the experimental conditions. Obviously, this relationship is idealized and can be expected to hold only under conditions as simple as the thin-film conditions of Chapter 7. With samples exceeding critical thickness, the best that can generally be expected is to find that

$$\mathbf{I}^S = k'\mathbf{W}^S \qquad (9.10\text{–}1)$$

Any difficulties in the $\mathbf{I}^S - \mathbf{W}^S$ relationship will have found adequate remedies if k' is constant. For example, the background difficulty will have been remedied if

$$\mathbf{I}^S = k'\mathbf{W}^S = k(N_T - N_B) \qquad (9.10\text{–}2)$$

Figure 7.12–1 contains data for such a case, the counting interval being 1 sec. Equation 9.10–2 is the slope-intercept relationship mentioned above as ideal for a plot of N_T against $\mathbf{W}^S$. Even if no background difficulty exists, this linear relationship will not hold if other difficulties (e.g., absorption effects) have not been remedied.

The remedy for a background difficulty is different in the different situations, which may be summarized as follows:

1. The background is negligible. No comment needed.
2. The background, though appreciable, need not be corrected for.
3. Background correction is required, but the situation is simple enough that straightforward correction suffices.
4. Background correction is required, but the correction varies with wavelength.
5. The background is so large that statistical fluctuations in N_B may vitiate a determination of the element sought. See Chapter 8. No comment needed here.
6. Light elements are being determined.

The first situation is most frequently met in the determination of major constituents. An extreme example is the counting of an analytical line for a pure metal when N_T may be 10^5 counts and N_B only 50 over the same short counting interval so that $N_T - N_B$, which measures the intensity of the analytical line, is sensibly identical with N_T.

The second was mentioned above, where it was indicated that background determination and correction could be omitted if this gives a more serviceable working curve. For an example, see Ref. 36.

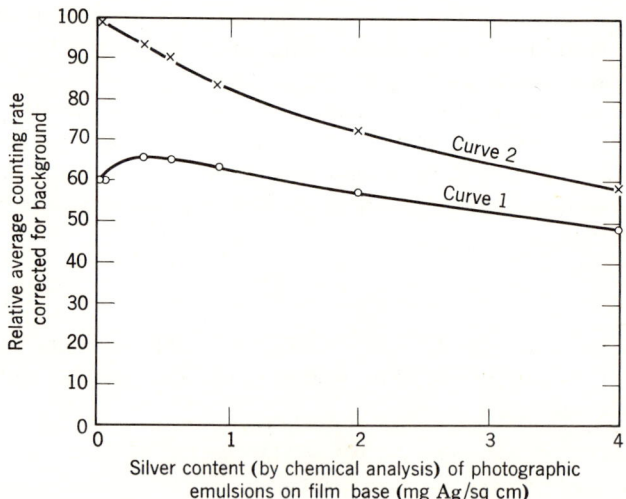

Fig. 9.10–1. The effect of a nonlinear background on the analysis for silver. Simple subtraction of adjacent background gave inversion observable on curve 1 near ordinate axis. Allowance for variation of background with wavelength gave the more reliable curve 2. The ordinate scale is arbitrary. We are indebted to the Eastman Kodak Company for the films and the results.

PRINCIPAL DIFFICULTIES AND COMMON REMEDIES

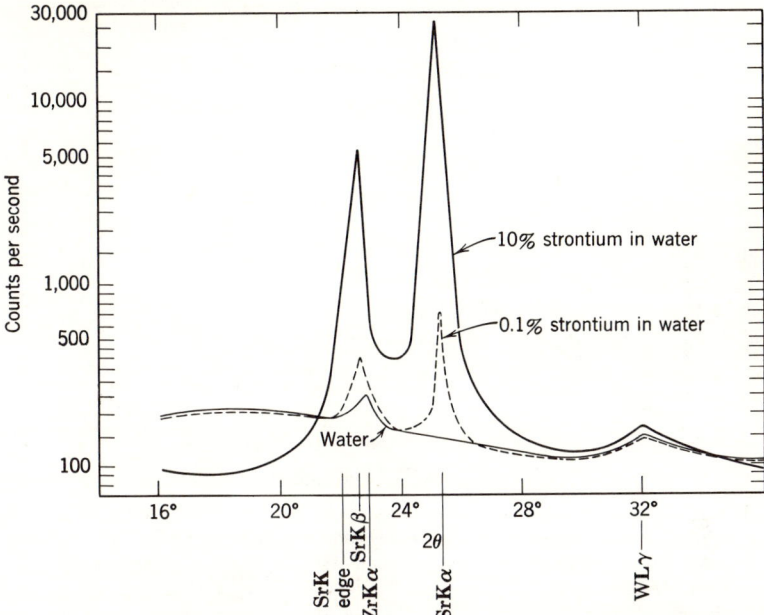

Fig. 9.10-2. Complex background in the determination of strontium in aqueous nitrate solutions. Note: (1) Gradually decreasing background with increasing 2θ as shown by water. (2) Unexpected presence of zirconium $K\alpha$ owing to impurity in glass cell. (3) Pronounced decrease in background at low 2θ for 10% solution. Cause: High absorbance on short-wavelength side of strontium absorption edge. (4) Slight increase in background with increasing strontium on long-wavelength side of edge. Cause: increased Compton scattering of Sr $K\alpha$ by sample. (5) Peak (almost merged with background) at 30° (2θ) results from scattering of target line by sample. Procedure: Use Sr $K\alpha$ as the analytical line. Establish background correction $F = N_B/N'_B$ on water, taking N'_B near $2\theta = 30°$ on unknown solutions. Authors' unpublished results, 1958.

By a straightforward correction of the background, as in the third situation, we mean the following. A background count N'_B is taken at a single goniometer setting far enough removed from that of the analytical line so that the line does not contribute measurably to the count. If taken over the counting interval for the line, N'_B is assumed equal to N_B and is therefore subtracted from N_T. If N'_B is taken over a different counting interval, the calculation is made by adjusting N'_B for the difference on the basis of an assumed constant background counting rate.

The fourth includes cases in which interfering (sometimes unresolved) lines and diffraction peaks enhance the background, and absorption edges affect it. Unless the interfering lines are satellites, they are usually easy to recognize even if reflected in orders higher than the first. (Satellites are not likely to be present at high intensity, but they may occasionally need to be considered, especially when light elements are being determined.) To

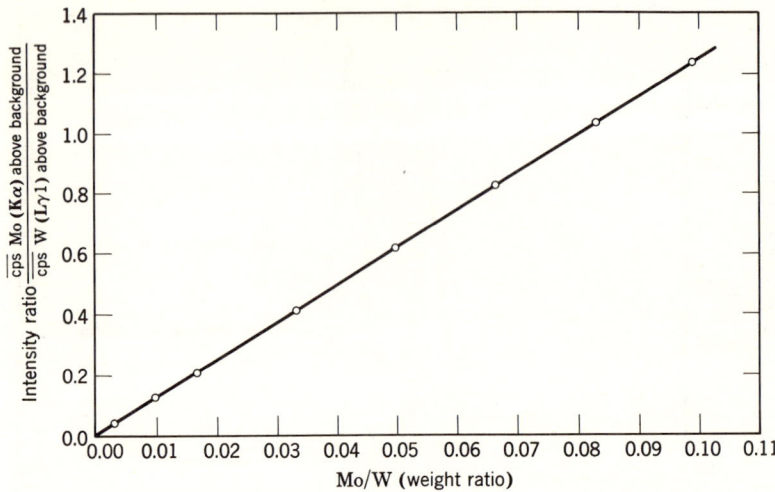

Fig. 9.10-3. Working curve for the determination of the molybdenum-tungsten ratio in aqueous solution. The background varies with wavelength, and the analytical line (tungsten $L\gamma 1$) was superimposed upon the scattered characteristic lines of the tungsten target. Nevertheless, a satisfactory linear working curve (see figure) was obtained when tungsten and molybdenum background corrections were calculated and applied. Background correction factors were established and used. See J. E. Fagel, Jr., H. A. Liebhafsky, and P. D. Zemany, Ref. 36.

illustrate the diversity possible here, three examples are given in Figures 9.10–1, –2, and –3. It is clearly impossible to cover all eventualities: special cases need individual treatment.

The case of Figure 9.10–3 is worth noting. The background difficulty in the determination of tungsten was made more serious because the sample scattered the characteristic lines from the tungsten target. Ordinarily one does not choose as target material the element being determined in a sample. That a successful determination is possible under these adverse conditions was worth demonstrating because such a choice of target is often dictated by other considerations.

The sixth situation is important because rapid progress in the determination of light elements is being made in spite of serious difficulties, among which the background difficulty is prominent. The discussion here supplements that in Section 2.23.

The upper curve of Figure 9.10–4 shows that the background problem can be of crucial importance when light elements are determined with a flow proportional detector. With a scintillation detector, the problem can be even worse because there is photomultiplier noise to contend with also. The principal x-ray components of the background are scattered x-rays, interfering lines, and characteristic lines of shorter wavelength that appear by higher-order Bragg reflection. Satellites may sometimes be present.

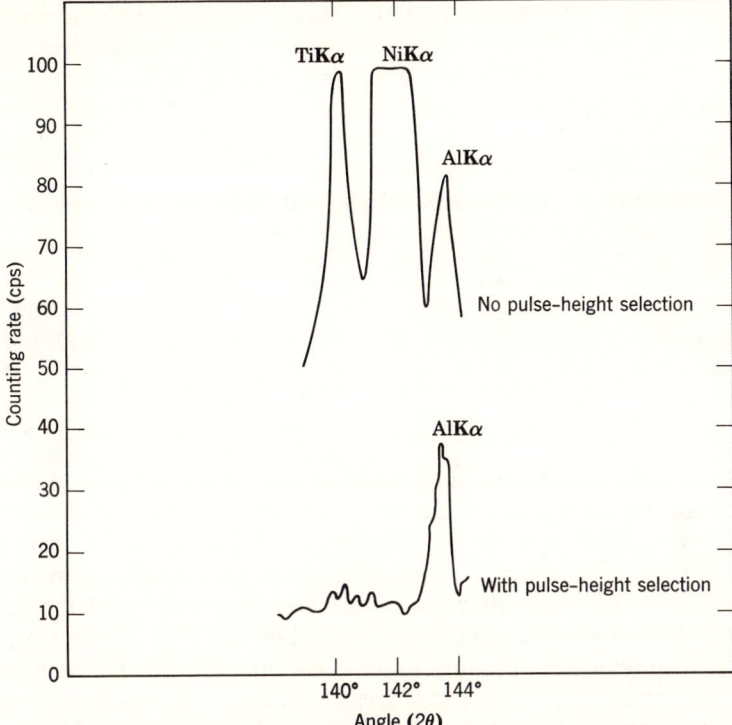

Fig. 9.10-4. Background in the determination of aluminum in Alnico. Pulse-height selection facilitates the determination by reducing in importance the background and the lines (in 5th- and 3rd-order reflection) of the matrix elements, nickel and titanium. After Norelco Reptr., 3, 78.

Higher-order Bragg reflections are the more disturbing when the analytical line is weak and more likely when its wavelength is long. Measurements in the authors' laboratory have given the following relative intensities for the reflection of Mo Kα from a lithium fluoride crystal at the order shown in parentheses: 100 (1); 35 (2); 10 (3); 2.5 (4); 1.1 (5). Unfortunately, such monotonic decrease in intensity with order cannot always be counted on: Moseley [37] did his work with a potassium ferrocyanide crystal, "a fine specimen," for which the third-order reflection was the most intense among the first three orders, all of which were strong.

Calculations were made for aluminum to illustrate the unfortunate situation of the light elements as regards interference with their analytical lines by lines undergoing higher-order Bragg reflections. The results appear in Table 9.10-1. Only the strong lines of each element were considered. The table gives values of the product $n\lambda$. The nearer such a value is to the wavelength of aluminum Kα, the more serious could be the contribution to background. The effective elimination of such contributions by pulse-height selection is one of the great services electronics has rendered x-ray emission spectrography.

Table 9.10–1. Higher-Order Bragg Reflections and Aluminum Kα

	Al, Kα	Ti, Kα	Mn, Kα	Ag, Lα	Rh, L$\beta 2$	Ba, Lα	Eu, Lα	Gd, Lα
Wavelength, Å	8.34	2.75	2.10	4.15	4.13	2.78	2.12	2.05
Order (n)	1	3	4	2	2	3	4	4
Product ($n\lambda$)	8.34	8.25	8.40	8.30	8.26	8.34	8.48	8.20

Pulse-height selection also filters out scattered x-rays and photomultiplier noise when the energy differences are large enough. The case of scattered x-rays is straightforward, but the noise problem is complex. To begin with, noise pulses are distributed so that they are increasingly numerous at lower energies; consequently, noise is not a serious problem with the heavier elements. With the light elements, noise pulses not only increase N_B, but they can also interfere with pulses due to the analytical line. The noisiness of different photomultipliers varies greatly, so that it is important to select carefully a tube for use with a scintillation detector in the determination of light elements. Finally, as the pulses due to noise are predominantly of low energy, pulse-height selection will become progressively less effective in coping with noise as the x-ray wavelength increases.

Table 9.10–2 contains unpublished results from the authors' laboratory that illustrate the need for pulse-height selection in the determination of light elements. In the case of silicon, the background was due mainly to scattered x-rays. In the case of sulfur, photomultiplier noise was also present. The counting interval was 10 sec for N_T (total count) and for N_B (background). The excellent results for sulfur could not have been obtained had there not been careful and fortunate selection of the photomultiplier.

We repeat finally that V_L, the base of the window in pulse-height selection, may have to be adjusted if the background is to be computed from a value N_B' obtained at $\theta' - $ a goniometer setting different than that for the analytical line. See Section 2.23.

Reducing background was mentioned above. To be worthwhile, this measure must lead to an increase in the ratio of N_T to N_B: reducing N_B alone, which can be done by changing excitation conditions or by filtering so as to decrease the intensity of scattered x-rays, can be self-defeating. On the other hand, such measures as reducing scattering by having a vacuum

Table 9.10-2. Need for Pulse-Height Selection (PHS) in Reducing Background

	Detector	Sample	Wavelength (Å)	N_T	N_B
No PHS	Flow proportional[a]	Silicon	7.11	2,449	17
With PHS	Flow proportional[a]	Silicon	7.11	2,445	13
No PHS	Scintillation	Sulfur	5.36	23,280	20,410
With PHS	Scintillation	Sulfur	5.36	7,500	40

[a] P-10 filler gas, consisting of 90% argon, 10% methane.

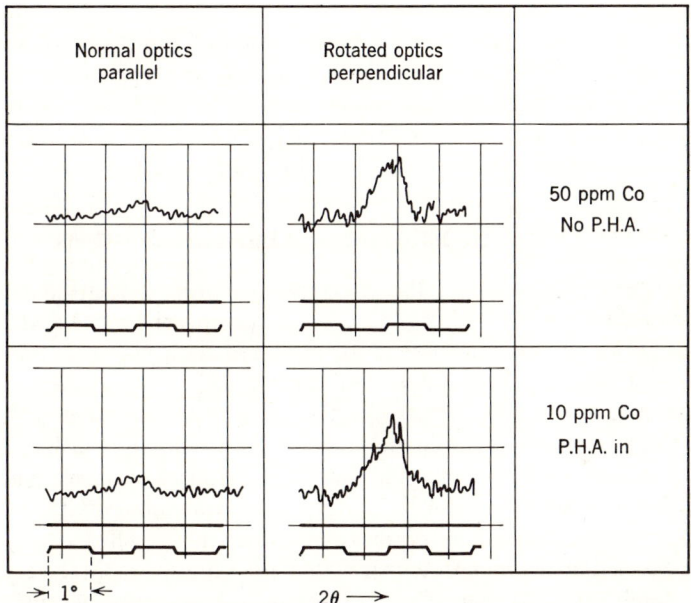

Fig. 9.10-5. "Normal" and "rotated" optics in the determination of trace amounts of cobalt in aqueous solution to demonstrate reduction of scattered background owing to its linear polarization: Co Kα; topaz crystal; 82.5 (2θ) background levels made identical by adjusting ratemeter. "Normal" means the geometrical arrangement in the Philips *PW*1520 spectrograph, which has a vertical goniometer plane, so that planes through sample and Bragg crystal are parallel at 90° (2θ). "Rotated" means the geometrical arrangement in which the intensity of polarized x-rays entering the detector has been minimized by rotating x-ray tube and sample holder approximately 90° about the axis of the collimator. P.H.A. means "pulse-height analysis": note its effectiveness. After K. P. Champion and R. N. Whitten, Ref. 38, p. 1082, Fig. 2. Compare with Fig. 1.14-2.

optical path, by selecting cell windows and sample mountings (Mylar scatters less than filter paper!), and by choosing the proper collimator are generally helpful. The detector and electronic circuitry also need careful consideration. Perhaps the best procedure for an analytical chemist is this: Make sure a reduction in N_B is worthwhile. Try in order of decreasing effectiveness and convenience the various ways of bringing it about. Measure N_T/N_B to see what has been accomplished.

One way of reducing background deserves special mention as it is built into spectrographs with a horizontal goniometer plane and with the direction of excitation perpendicular to that plane. Scattered x-rays are highly plane-polarized (Section 1.14). Characteristic lines are not polarized at all. Consequently, a spectrograph geometry that reduces the chance for plane-polarized x-rays to enter the detector ought to increase the ratio of N_T to N_B. The reader may remember the experiment of Compton and Hagenow, Figure 1.14–2: here the detector D_1 corresponds to a detector in a spectrograph with a vertical goniometer plane, and D_2 to a detector in one with a horizontal. Champion and Whitten [38] appreciated and tested these considerations with the results shown in Figure 9.10–5.

9.11 Qualitative Introduction to Interelement Effects

The ensuing discussion of the effects in question is restricted to x-ray emission spectrography with *x-ray excitation*. Sections 9.2, 9.3 and 9.4 show that difficulties of this kind are to be expected with every method of excitation, but that they will differ with the method.

In the x-ray emission spectrography of multicomponent samples of greater than critical thickness, the intensity of any analytical line must be expected to vary with the kind and weight-fraction of every element in the sample. We have seen (Sections 6.13 and 6.14) that results obtained by x-ray diffraction are influenced by *absorption effects*, there called *matrix effects* because exerted by the matrix, which is all of the sample except the crystalline species sought. We have seen further (Section 7.4) that the absorption effect (often called *self-absorption*) in a pure element must be allowed for in thickness determinations that use the intensities of emitted lines. In Chapter 9, we must consider not only the effects just mentioned, but *enhancement effects* also. To establish a distinction between x-ray diffraction and x-ray emission spectrography, the sum of all such effects is called *interelement effects* here, even though this step departs from previous practice [39]. The matrix continues to be all of the sample except the constituent sought, which of course is an element (not necessarily a crystalline species) in x-ray emission spectrography.

The only kind of enhancement under discussion is the tendency toward increased intensity of an analytical line that results from the photoelectric

PRINCIPAL DIFFICULTIES AND COMMON REMEDIES

absorption of a characteristic line produced in the matrix of a sample. Changes in the scattering of x-rays that accompany changes in composition can also lead to intensity changes in analytical lines, but these are not singled out for separate discussion.

Interelement effects are a difficulty in x-ray emission spectrography because they militate against proportionality in (9.7–1) and (9.10–2). Fortunately, these effects are amenable enough to calculation so that this difficulty can often be remedied *within certain limits* by calculations based either on intensity values for known standards (multiple regression methods) or on fundamental parameters [40]. More later.

Chapter 6 made clear that absorption effects can be *positive* (lead to increased **I**) or *negative* (lead to decreased **I**). Enhancement effects can lead only to increased **I**. The simplest enhancement occurs when the analytical line λ_E of element E is excited within the sample by the characteristic line λ_F of some element in the matrix. More complex enhancements are possible. Suppose a wavelength λ_0 in the beam incident on the sample can excite the characteristic lines λ_E, λ_F, and λ_G of the elements identified by subscript, no element being singled out for determination; and that the absorption edges are at wavelengths that permit excitation of λ_F and λ_G by λ_E, and of λ_G by λ_F. As Sherman [41] pointed out, the situation is then complex and representable as in Table 9.11–1.

Table 9.11–1. Enhancement Effects of Graduated Complexity

Line	Excited By				Contributions to Intensity
λ_E	λ_0	—	—	—	1
λ_F	λ_0	$\lambda_0\lambda_E$	—	—	2
λ_G	λ_0	$\lambda_0\lambda_E$	$\lambda_0\lambda_F$	$\lambda_0\lambda_E\lambda_F$	4[a]

[a] $\lambda_0\lambda_E\lambda_F$ means that λ_0 excites λ_E, which excites λ_F, which in turn excites λ_G.

Excitation by λ_0 involves absorption effects. These can be treated successfully (as we saw in Chapter 7) by narrow, parallel beam geometry, which is applicable to the well-collimated beam of a good spectrograph with a detector of small aperture. Unfortunately, this simple geometry does not apply to the enhancement components. As Sherman [41] emphasized, the enhancement contributions are generated within the sample and radiated with spherical symmetry, so that it is necessary to integrate at every point over the entire solid angle in order to obtain their intensities. In other words, one must deal here with broad, divergent beams [42, 43], and this leads to expressions not integrable in closed form.

The complexity of interelement effects is conveniently demonstrated with reference to (9.7–1). Such a demonstration, with iron as the element determined, appears in Table 9.11–2 and Figure 9.11–1, the five samples being assumed identical (except in composition) and subjected to monochromatic x-ray excitation.

Table 9.11–2. Qualitative Interelement Effects

Case	Sample	Intensity of Iron Kα	Net Effect	Comment
A	Fe	I^{Fe}	None	Reference standard
B	Fe-Al	$>W^S I^{Fe}$	Positive absorption effect	μ_{Al} less than μ_{Fe}
C	Fe-Pb	$<W^S I^{Fe}$	Negative absorption effect	μ_{Pb} greater than μ_{Fe}
D	Fe-Co	$W^S I^{Fe}$ (approx.)	No pronounced effect[a]	Mass absorption coefficients comparable; cobalt Kα cannot[a] excite iron Kα
E	Fe-Ni	$>W^S I^{Fe}$	Predominantly enhancement effect	Mass absorption coefficients comparable; nickel K excites iron K

[a] Note that cobalt Kβ, which is of shorter wavelength than cobalt Kα, can excite iron K. This small enhancement effect was disregarded for the sake of simplicity.

When a sample contains more than two elements, the relationship of I^S to that expected from (9.7–1) becomes more complex and can sometimes seem surprising. Consider a ternary alloy of $_{12}$Mg, $_{13}$Al, $_{14}$Si, and take 8Å as the wavelength of Al Kα, the analytical line assumed to be excited by a monochromatic beam of somewhat shorter wavelength. In the same order, the relevant mass absorption coefficients are (approximately): 3900, 280, 400. In this situation, a simultaneous *increase* in the weight-fractions of magnesium and aluminum can result in a *decreased* I^S because intensity reduction owing to increased absorption by the matrix more than compensates the increase to be expected from an increasing **W**. We have here a simple situation involving only absorption, and the results are quantitatively predictable. As the number of elements in the sample increases, and enhancement effects become possible (see Table 9.11–1), quantitative pre-

PRINCIPAL DIFFICULTIES AND COMMON REMEDIES

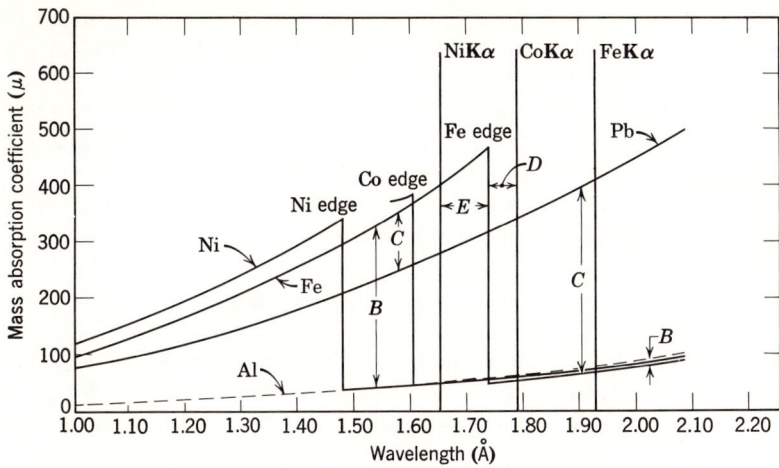

Fig. 9.11-1. Spectral data to illustrate interelement effects for three transition elements. (To avoid crowding, only part of the cobalt absorption curve is shown.) See Table 9.11-2. *Case B.* Substitution of Al for Fe decreases absorption of incident beam and has little effect on analytical line. Net positive absorption effect. *Case C.* Substitution of Pb for Fe decreases absorption of primary beam but greatly increases absorption of analytical line. Net negative absorption effect. *Case D.* Note wavelength relationship indicated in figure. Enhancement impossible. *Case E.* Note wavelength relationship in figure. Enhancement occurs.

dictability may disappear, but an informed guess should remain in the cards. Birks and Harris [44] discovered that the intensity of Fe $K\alpha$ as the analytical line in the chromite-olivine system *decreases* when the weight-fraction of chromium *increases* more rapidly than W^S, in this case the weight-fraction of iron. Mitchell and Kellam [45] have made a thorough study of several such cases with emphasis on Fe $K\alpha$, Cr $K\alpha$, and Ni $K\alpha$ as analytical lines. In general, the expected complexities were found and explained. We remind the reader that even the most difficult kinds of interelement effects in x-ray emission spectrography are far simpler and more nearly predictable than corresponding effects in the ultraviolet and visible regions, where the behavior of valence electrons is involved.

Absorption effects make x-ray emission spectrography almost useless for the determination of major constituents in one situation where classical analytical chemistry serves routinely and satisfactorily. Consider the series Fe, FeO, Fe_3O_4, Fe_2O_3. In x-ray emission spectrography of the solids, the four substances will all give virtually identical iron contents as measured by the intensity of an analytical line of iron excited in them. The reason why is easy to see. Assume the oxygen to be completely transparent (as it nearly is) to the incident x-rays capable of exciting iron $K\alpha$, and to iron $K\alpha$ itself. Then the intensity of iron $K\alpha$ will be determined by the number of iron atoms encountered by the incident beam and by the number encountered by quanta of the analytical line. For samples above the critical thickness, these numbers will be virtually identical in all four materials; the critical depth would, of course, be least for Fe and greatest for Fe_2O_3.

Let us return briefly to Sections 7.3 and 7.4. The mass absorption coefficients for the incident and emergent beams will have the form

$$\mu = \mu_{Fe} W_{Fe} + \mu_O (1 - W_{Fe}) \qquad (9.11-1)$$

where the subscript O represents oxygen. Values of a have been calculated for the iron series, as have critical depths, reasonable values being taken for the parameters involved. The results are summarized in Table 9.11–3.

Table 9.11–3. Calculated Results for Various Iron Materials

	Material			
	Fe	FeO	Fe_3O_4	Fe_2O_3
W_{Fe}, %	100	77.8	72.5	70.0
a	431	348	327	318
I, relative[a]	100	95.7	95.5	95.0
Critical depth, cm[b]	$1.36(10^{-3})$	$2.31(10^{-3})$	$2.72(10^{-3})$	$2.77(10^{-3})$

Assumed wavelengths: incident, 1.39 Å; analytical line, 1.937 Å

	μ_{Fe}	250 (1.39 Å)	71 (1.937 Å)
	μ_O	8.1 (1.39 Å)	22 (1.937 Å)

[a] Based on the relative amount of iron in the critical depth.
[b] The critical depth is taken as that beyond which only 1% of the quanta in the emergent beam is produced. For explanation of calculations, see Sections 7.3 and 7.4.

Inspection of the table shows that the quotient a/W_{Fe} is in fact nearly constant; that I changes much less rapidly than W_{Fe}; and that the critical depth has doubled when the highest oxide is reached. All three conditions are reflections of the (positive) absorption effect that occurs in this binary system when iron is replaced by oxygen, which has a lower mass absorption coefficient.

The determination of iron by x-ray emission spectrography can be carried out on these materials by eliminating the absorption effect (thin-film

PRINCIPAL DIFFICULTIES AND COMMON REMEDIES

methods of Chapter 7), or by keeping it constant (as by working with dilute solutions in a relatively transparent solvent).

We now give examples to show how nearly difficulties due to interelement effects can be remedied by calculation.

9.12 The Estimation of Absorption Effects. Aqueous Sodium Tungstate Solution as Example

The interelement effects in the determination of tungsten in aqueous sodium tungstate solutions are absorption effects only; enhancement does not occur. The determination is of interest for several reasons: (1) It is only moderately more complex than the determination of plating thickness (Section 7.4). (2) Because the analytical line W Lγ1 was superimposed on the same line emitted by the target and scattered by the sample, the background problem was unusually serious, as was mentioned in Section 9.10. (3) There was reason to hope that an oversimplified treatment, based on the use of an effective wavelength for the incident polychromatic beam, might give useful results even though such a treatment could not cope with interelement effects in more complex cases. (4) This determination is one in which critical depth changes with composition. For these reasons, the determination was studied in the authors' laboratory [46].

The treatment of film thickness in Chapter 7 must be changed to accommodate two differences between that case and this: here the depth (thickness) always exceeds the critical, and the mass absorption coefficients can change greatly as composition changes. Because the film in Figure 7.4–1 was a *pure element*, only two mass absorption coefficients (those in a_1 and a_2) were needed; a *single* mass absorption coefficient could be used to describe both the attenuation of the incident beam *on its way to* the volume element (height dx) in the figure, and the excitation of the analytical line *within* this volume element. Here, the absorption effect is described by using μ_1, the mass absorption coefficient for the solution, and the excitation effect must be described by using μ_W, the absorption coefficient of tungsten. Accordingly, the treatment that includes (7.4–1)–(7.4–4) must be changed as follows (please refer to earlier treatment, the numbering in which is retained below):

3. Decrease in effective intensity in dV:

$$dI = |k_1 I_0 \mathbf{a} e^{-a_1 x} dx| \qquad (\mathbf{a} = \csc \theta_1 \mu_W \rho)$$

(Note that boldface **a** is for the element being determined.)

4. Lγ1 quanta per second generated in dV: $+ \, dI = |k_2 dI|$

7. After simplification by combining constants:

$$I_d^S = \int_{x=0}^{x=d} kaI_0 e^{-ax}\, dx$$

or

$$I_d^S = kaI_0(1 - e^{-ad})/a \qquad (9.12\text{--}1)$$

and at $d = \infty$ (the case for all samples here) with $a^S = a$

$$I_\infty^S = \frac{kaI_0}{a^S} \qquad (9.12\text{--}2)$$

where the superscript refers to a sample of solution containing tungsten as the element sought, **a** being independent of the composition while the value of a^S depends upon that of μ^S, which changes with composition according to

$$\mu^S = \mu_W W + \mu_M W_M \qquad (9.12\text{--}3)$$

where M is the matrix in the sample to which the weight-fractions refer.

From (9.7–1), in the absence of absorption effects,

$$\frac{I}{W} = \text{constant} \qquad (9.12\text{--}4)$$

and, in the presence of absorption effects that vary from sample to sample, if these effects are given by (9.12–2):

$$\frac{aI}{W} = \text{constant (superscripts and subscripts omitted)} \qquad (9.12\text{--}5)$$

To test (9.12–5), aqueous sodium tungstate solutions containing tungsten over a wide range of weight-fractions were excited by a polychromatic x-ray beam of constant, independently determined effective wavelength; the intensities of W $L\gamma 1$ were measured; and the background cps was determined as for Figure 9.10–3. Values of a' were calculated for all samples according to

$$a' = \mu_1 \csc \theta_1 + \mu_2 \csc \theta_2 \qquad (a = a'\rho) \qquad (9.12\text{--}6)$$

with the mass absorption coefficients from (9.12–3). As the solutions vary but little in density, a' may be used to replace a in (9.12–5). The results appear in Table 9.12–1.

The constancy of the ratio in the lowest line verifies (9.12–5). It is gratifying that the oversimplified treatment of absorption effects should be adequate to deal with a threefold variation of the ratio in the line above the last, especially when a large and variable background correction had to be made.

The critical depth, calculated with 0.99 as the intensity ratio in (7.4–3), was 0.1 cm for the strongest, and 0.4 cm for the weakest, solution.

Table 9.12–1. Calculation of Absorption Effects for Sodium Tungstate Solutions

Beam	Wavelength	Mass Absorption Coefficients		
		Tungsten	Sodium Tungstate	Water
Incident polychromatic[a]	1.02 Å[a]	260	165	2.85
Tungsten Lγ1	1.10 Å	191	122	3.64
$\theta_1, 60°; \theta_2, 30°$				
Results of calculations				
$[10^3]$ W[b]	2.87 5.70	8.49 11.28	16.79 22.2	32.85 43.15
Background, $\overline{cps}$[b]	254.3 220.5	195.8 182.5	157.9 141.4	118.9 105.4
I, $\overline{cps}$[b]	184.0 320.6	416.0 488.7	617.5 697.9	788.8 874.5
10^{-3}[I/W]	64.1 56.2	49.0 43.3	36.8 31.4	24.0 20.3
10^{-4}[I a'/W]	80.3 81.3	80.2 79.2	81.1 80.9	79.5 81.3

[a] Effective wavelength, chosen with due regard to the position of the L_{II} edge (1.07 Å) of tungsten.
[b] $\overline{cps}$ means an average counting rate in counts per second. W is g tungsten per g sol'n.

No brief is made here for the use of this oversimplified treatment as a replacement for the more precise and powerful methods of calculating interelement effects. In cases where absorption effects predominate, and these are in the majority, (9.12–5) can serve as the basis for estimating differences in interelement effects between samples S and S'; that is,

$$\frac{aI^S}{W^S} = \frac{a'I'^{S'}}{W^{S'}} \qquad (9.12-5a)$$

If a and a' are unknown, their values can be approximated on the basis of simple absorbance measurements on S and S'. Also, as is clear from the definitions of the a's—see the text preceding (7.4–1)—measured or estimated mass absorption coefficients can be used in relations analogous to (9.12–5a), when ρ, csc θ_1 and csc θ_2 are *constant or nearly so*.

9.13 Regression Treatments of Interelement Effects

Regression treatments (see Chapter 8) to remedy interelement effects were used long ago by Sherman [47], who called them a "practical correlation of intensity and composition" in 1953 and concluded that "a close linear approximation" could be made "for limited ranges in composition." His equations follow for a three-element system (symbols and subscripts somewhat altered):

$$\left. \begin{array}{r} (a_{11} - t_1)W + a_{12}W_2 + a_{13}W_3 = 0 \\ (a_{22} - t_2)W + a_{21}W_1 + a_{23}W_3 = 0 \\ (a_{33} - t_3)W + a_{31}W_1 + a_{32}W_2 = 0 \\ W_1 + W_2 + W_3 \quad\quad\quad = 1 \end{array} \right\} \quad (9.13-1)$$

In place of I's, Sherman uses t's, which are times to fixed counts for each analytical line. Comparison with Section 8.12 shows immediately that Sherman's was a true regression treatment with a_{ij} as the interelement coefficients, which he calls "parameters to be computed from samples of known composition." He goes on to say that the computations involved in calibration and determination are "not inconsiderable, but no simplification is apparent for wide ranges in major compositions." Nothing is fundamentally changed today except that the advent of the computer has eliminated the labor of computation. Sherman tested (9.13-1) on known samples of chromium, iron, and nickel oxides with good results; see Ref. 47, Table III.

The calculation of interelement effects has been done in too many ways to report here. The interested reader should consult Ref. 40 for a good bibliography that also includes articles on the electron microprobe. Only one regression treatment will be thoroughly examined here.

Lucas-Tooth and Pyne [48] used the multiple regression method under conditions that gave it an excellent chance to prove itself: that is, they used it not in designed experiments on prepared standards, but upon what might be considered a "manufactured product"—sixty high-alloy steel samples, the chemical composition of which had been precisely established by conventional methods in several independent laboratories. They concentrated on interelement effects in the determination of chromium in these alloys by x-ray emission spectrography; the wealth of information available on these 60 samples made possible the obtaining of highly reliable least-squares solutions via the computer.

The seriousness of the interelement-effect difficulty is clear from Figure 9.13-1 and from Table 9.13-1, which also contains data compiled from Ref. 48 for discussion below.

Lucas-Tooth and Pyne use three weight-percentages in treating their results; these are $P_{n,m}^{\text{Chem}}$, chemically determined for element n in sample m; $P_{n,m}^{\text{x-ray}}$, corresponding, completely corrected x-ray value; and $P_{n,m}^{\text{ap}}$, an *apparent* corresponding value obtained by adjusting (normalizing) $\mathbf{I}$ so that

$$P_{n,m}^{\text{ap}} = \kappa \mathbf{I} \simeq P_{n,m}^{\text{chem}} \quad (9.13-2)$$

This definition of an apparent percentage proportional to $\mathbf{I}$ and to the

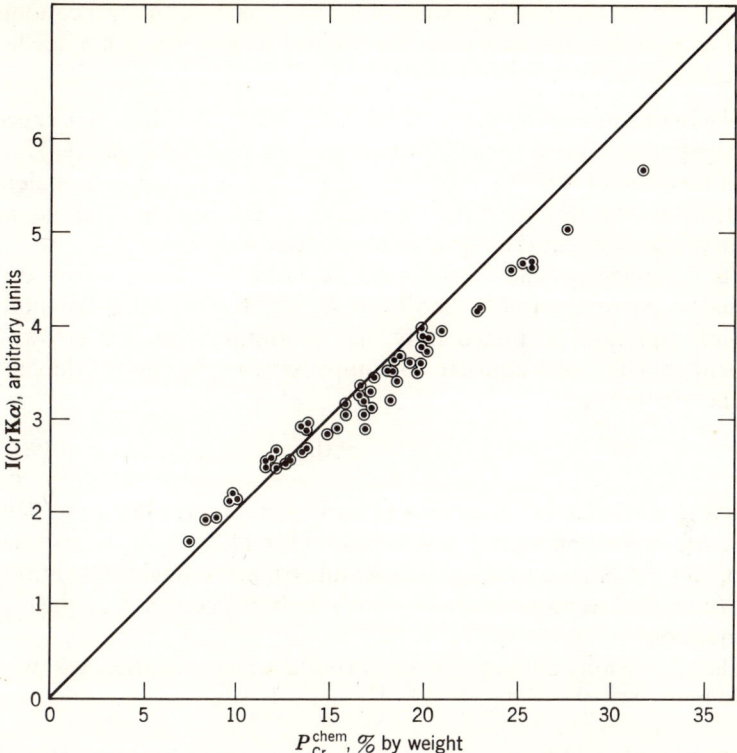

Fig. 9.13–1. The interelement-effect difficulty in chromium determinations on the 60 high-alloy steels used by Lucas-Tooth and Pyne. The divergences of points from the line are *not* due to uncertainties in the compositions, which are known with high precision, but (mainly) to interelement effects. The seriousness of these effects for a sample is measured by the horizontal distance of its chemically determined chromium content from the straight line. After Lucas-Tooth and Pyne, Ref. 48, p. 528, Fig. 2.

chemical percentage is a useful device that makes it possible to obtain, by adjusting κ, identical P^{ap} values for the same sample run on different instruments, or on the same instrument at different times. Such adjustment makes it possible to obtain and to use constant values of interelement coefficients even in the face of intensity fluctuations caused by equipment difficulties. Of course, more known samples must be measured under these conditions than would be needed were equipment difficulties absent. (See analysis of variance, Chapter 8.) Table 9.13–1 shows the usefulness of $P^{ap}_{n,m}$.

Table 9.13–1 shows that the four molybdenum-bearing steels have the greatest interelement effects, and that these effects are adequately compensated by calculation in all six cases. (Please compare the two columns of Δ's.) The multiple regression treatment by which this was accomplished is now outlined.

1. The regression equation contained one additive constant (α_0), expected to be negative as it must correct for background, one coefficient (k_0) for the element for which **I** is being measured, and—as a maximum—interelement coefficients to the number (x) of elements in the sample. The *maximum* number of constants in the regression equation was thus $x + 2$.

2. The interelement effect (coefficient k_{AC}) of C on A was assumed proportional to the product of P_C and **I** (for A), which means that the multiple regression equation contained only one quadratic term (see below), the coefficient in which was adjusted to compensate for equipment difficulties.

3. The relationship

$$\sum_1^n P_{n,m}^{\text{x-ray}} = 100 \tag{9.13-3}$$

was used to eliminate the interelement coefficient of a troublesome element (Fe). In the regression equation (see below) for $P_{\text{Cr},m}^{\text{x-ray}}$, $k_{\text{Cr,Fe}}$ was set equal to zero, and the other coefficients were adjusted accordingly. The introduction of (9.13–3) thus reduces to $x + 1$ the number of constants in the regression equation.

4. The regression equation for the chromium determination was written on this basis as

$$\begin{aligned}P_{\text{Cr},m}^{\text{x-ray}} = \alpha_0 &+ P_{\text{Cr},m}^{\text{ap}}(k_0 + k_{\text{Cr,Cr}}P_{\text{Cr},m}^{\text{ap}} + k_{\text{Cr,Ni}}P_{\text{Ni},m}^{\text{ap}} + k_{\text{Cr,Mn}}P_{\text{Mn},m}^{\text{ap}} \\&+ k_{\text{Cr,Mo}}P_{\text{Mo},m}^{\text{ap}} + k_{\text{Cr,Ti}}P_{\text{Ti},m}^{\text{x-ray}} + k_{\text{Cr,Si}}P_{\text{Si},m}^{\text{chem}} \\&+ k_{\text{Cr,Al}}P_{\text{Al},m}^{\text{chem}} + k_{\text{Cr,V}}P_{\text{V},m}^{\text{x-ray}} + k_{\text{Cr,W}}P_{\text{W},m}^{\text{x-ray}} \\&+ k_{\text{Cr,Cu}}P_{\text{Cu},m}^{\text{x-ray}} + k_{\text{Cr,Nb}}P_{\text{Nb},m}^{\text{x-ray}})\end{aligned} \tag{9.13-4}$$

5. The best possible values of the $x + 1$ constants in the regression equation were obtained via a least-squares computer program with the maximum amount of the best available information as input. Of the three kinds of P's available as input, the P^{chem}'s are the most, and the P^{ap}'s the least reliable. Seemingly not all the former were available when the x-ray study was done. Values thus obtained for the thirteen constants are given in Table 9.13–2.

The data in Table 9.13–2 were then used to calculate x-ray values for the chromium contents of each of the 60 samples. The root-mean-square difference between $P_{\text{Cr}}^{\text{chem}}$ and $P_{\text{Cr}}^{\text{x-ray}}$ was 0.07%. The largest single difference was 0.26%; here the sample contained 31.67% Cr by weight. That calculation was in this case an effective remedy for interelement difficulties is obvious from a comparison of Figures 9.13–1 and 9.13–2.

It would be difficult to find a more convincing example of the flexibility

Table 9.13–1. Selected Data from Lucas-Tooth and Pyne [48]; Information for Six Samples with Chromium Contents near 17% by Weight

Number [48]	Chromium Data					P^{chem} for Other Elements[d]									
	P_{Cr}^{ap}	$P_{Cr}^{x\text{-ray}}$	P_{Cr}^{chem}	Δ_1^a	Δ_2^b	Al	Si	Cu	W	V	Nb(Cb)	Mo	Ti	Mn	Ni	Fe
7415	17.17	16.74	16.81	+0.36	−0.07		0.27							0.58	0.62	81.5
7417	16.63	16.68	16.70	−0.07	−0.02		0.58							1.98	10.15	70.4
7518	15.86	16.82	16.81	−0.95	+0.01		1.10					0.49	0.93	2.48	11.89	66.0
7527	15.48	17.09	17.14	−1.66	−0.05		1.07					2.98	0.46	2.25	12.16	63.7
7542	14.46	16.99	16.93	−2.47	+0.06		1.12	3.57				4.38	0.56	2.19	19.87	51.1
8194	16.46	17.07	17.04	−0.58	−0.03		0.72				0.55	1.30		0.71	4.95	73.7
			60[c]			5	60	6	5	3	9	21	13	60	60	60

[a] $\Delta_1 = P_{Cr}^{ap} - P_{Cr}^{chem}$ (measure of interelement effect).
[b] $\Delta_2 = P_{Cr}^{x\text{-ray}} - P_{Cr}^{chem}$ (measure of effectiveness of interelement-correction).
[c] Numbers in *italics* show on how many of the 60 samples the element in question was chemically determined.
[d] For the six samples in the table, as for 54 of the 60 samples, the residual percentage by weight was 0.25. This residual never exceeded 0.61%. A blank space indicates an absent element.

Table 9.13-2. The Thirteen Constants for the Chromium Regression Equation [48]; Additive Constant, $\alpha_0 = -0.2173$; Element Coefficient $k_0 = 0.73835$

Eleven Interelement Coefficients

$k_{Cr,Cr} = 0.01403$ $k_{Cr,Ni} = 0.00304$
$k_{Cr,Mn} = 0.00462$ $k_{Cr,Mo} = 0.02878$
$k_{Cr,Ti} = 0.04640$ $k_{Cr,Si} = 0.00707$
$k_{Cr,Al} = 0.00208$ $k_{Cr,V} = 0.00794$
$k_{Cr,W} = 0.02876$ $k_{Cr,Cu} = 0.00328$
$k_{Cr,Nb} = 0.02928$

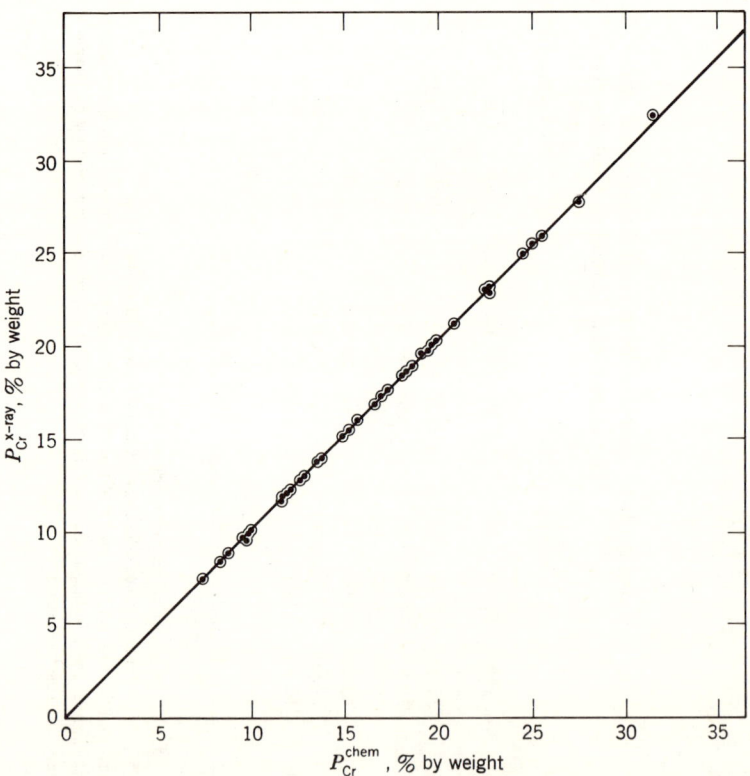

Fig. 9.13-2. Chromium percentages (ordinate) obtained by correcting for interelement effects by (9.13-4). Compare with Figure 9.13-1. After Lucas-Tooth and Pyne, Ref. 48, p. 534, Fig. 5.

and the usefulness of multiple regression methods applied to x-ray emission spectrography. Note that three measures of composition (the three P's) are available for use as appropriate. Note that their use makes (9.13-4) into a higher-order multiple regression equation (one linear, one quadratic, and ten cross-product terms containing P's) in contrast with the simpler (9.13-1). Note that (9.13-4) has a term (coefficient $\mathbf{k}_0$) specifically for the element being determined, and that it includes provision for equipment difficulties and (through normalization via P^{ap}) for transferring the calibration data to other spectrograph systems. It is interesting to compare this investigation with the earlier work of Lucas-Tooth and Price [49] on the copper-zinc-tin system.

Multiple regression methods are in essence empirical. The proof of the pudding is in the eating, and the meal will not be satisfactory if the regression equation is unsuitable. Such unsuitability limits the range of satisfactory applicability; the unsuitability can be so great as to make the method useless. The remedy is to improve the regression equation. One example of this is the introduction by Lucas-Tooth and Pyne [48] of a (negative) background-correction coefficient α_1 into a regression equation for $P_{\mathrm{Mn}}^{\mathrm{x-ray}}$ to compensate for an increase, due to the presence of chromium, in the background count for the determination of manganese. Their paper contains other evidence to show the dangers of following blindly the multiple regression approach. We mention only one: the range of nickel percentages for the 60 alloys proved too wide, with many samples at the lower end of the scale. Lucas-Tooth and Pyne say that in these circumstances another computer program would have been more suitable. A related question is this: as no one alloy had over 8 elements to be determined, is (9.13-4) too general? To answer questions such as this, it would be valuable to have a report of the subsequent usefulness of the Lucas-Tooth and Pyne approach for different spectrographs under different operating conditions.

There is not room for discussion of other interesting multiple regression treatments such as that by Beattie and Brissey [50], Alley and Myers [51], and Mitchell and Hopper [52]. We return to Criss and Birks [40] below. All these multiple regression treatments are an *extension of the comparative method of analysis* (unknown vs standard) *by mathematical means.*

9.14 Interelement Effects from Fundamental Parameters

When there is no enhancement, the calculation of interelement effects from fundamental parameters is simple. Glocker and Schreiber [43] did it successfully years ago, and Section 9.12 presents a recent example. Section 9.11 has already implied that the presence of enhancement effects raises the problem of calculation to a higher order of complexity—by no means an overstatement. It is therefore gratifying that Criss and Birks [40] were able to solve the problem satisfactorily.

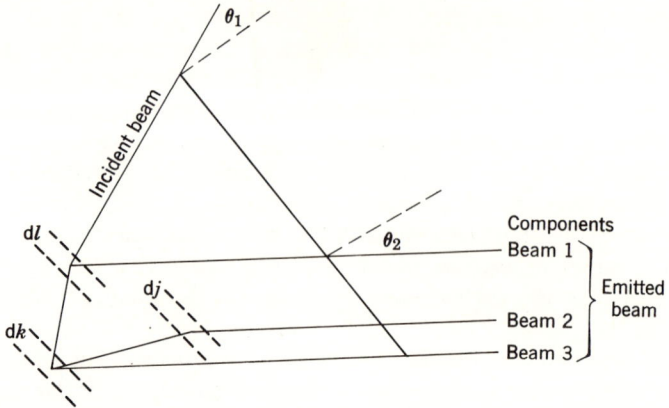

Fig. 9.14-1. Sherman's model for mathematical treatment of interelement effects for a sample containing only $_{24}$Cr, $_{26}$Fe, and $_{28}$Ni. A sample monochromatic beam enters the sample at θ_1 and the analytical lines are emitted at θ_2. The emitted beam is made up of 3 component beams originating in thickness elements at depths j, k, and l in which analytical lines are generated by the absorption of the quanta shown in parentheses below (only Kα quanta were considered):

Thickness Element	Significant Absorbed Quanta	Emitted Quanta	Component Beam
dl	(Incident Beam)	Cr Kα Fe Kα Ni Kα	1
dk	(Fe Kα) (Ni Kα)	Fe Kα from (Ni Kα) Cr Kα from (Fe Kα) and from (Ni Kα)	3
dj	(Fe Kα)	Cr Kα from (Fe Kα)	2

The quanta absorbed in dk originated in dl; those absorbed in dj originated in dk. Component beams 2 and 3 result from *enhancement effects*. See Table 9.11–1. After Sherman, Ref. 41, p. 284, Fig. 1.

First, a brief return to the pioneering work of Sherman. He treated the interelement-effect problem by making simplifying assumptions [41, pp. 284, 285]. His approach is shown in Figure 9.14–1 for $_{24}$Cr, $_{26}$Fe, and $_{28}$Ni as sole sample components. The reader will remember that incident and emitted beams are narrow and well collimated; hence they need only the simple "narrow beam" treatment in which the intensity from the total thickness of sample is obtained by integrating over thickness (represented in Figure 9.14–1 by l)—from zero to infinite thickness for the samples under consideration here. This is the kind of integration done in Chapter 7. For the beams generated by enhancement within the sample, the "broad beam" treatment is required, in which integration must extend over all space angles; this will involve functions of $|l - k|$ and $|j - k|$, and entail the frequent use of integrals of the form

$$\int \frac{\log(a + bx)}{\alpha} dx$$

which can only be integrated by numerical methods. For further details, see Ref. 41; see also [8–8] below.

To calculate intensities for excitation of the Cr-Fe-Ni sample by a 50-kW-peak beam from a tungsten-target Coolidge tube, Sherman used 1.2 Å (a value near the wavelength of W L lines) as the effective wavelength. Simplifying assumptions [41] had of course to be made. The calculated intensities were in good accord with the experimental results. He found, in terms of Table 9.11–1, that components excited by λ_E and λ_F are always important and that the contribution to the intensity of λ_G by $\lambda_E \lambda_F$ excitation may need to be considered in precise work.

Sherman appreciated that the use of an effective wavelength left something to be desired. He concluded further that the *calibration* problem could be satisfactorily solved, but that its inverse, the *determination* problem, had no rigorous solution. (See Section 9.13 for nomenclature.) Shiraiwa and Fujino [53] improved matters by developing an improved way of getting the effective wavelength, and calculated intensities in excellent agreement with experiment.

To facilitate reference and avoid confusion, we shall present the important work of Criss and Birks [40] almost as it is given in Ref. 15, pp. 90–93; three equations (*numbers bracketed*) will be copied from the book and a translation to our nomenclature will be included.

Criss and Birks made two important advances: (1) by using the experimental data of Gilfrich and Birks [24], they escaped the limitation imposed by having an effective wavelength represent excitation by a polychromatic beam. (2) By using a computer, they succeeded in employing an iterative procedure to solve practically the *determination problem* that Sherman concluded could not be rigorously solved.

Their three equations (their numbers) are

$$R_i = \frac{P_{Mi} + S_{Mi}}{P_{ii}} \quad [8\text{-}6]$$

where

$$P_{Mi} = g_i W_i \sum \frac{D_{i\lambda} \mu_i I_\lambda \Delta\lambda}{\mu_{M\lambda} \csc \phi_1 + \mu_{Mi} \csc \phi_2} \quad [8\text{-}7]$$

The P_{ii} is a similar equation except that $W_i = 1$ and $\mu_{i\lambda}$ and μ_{ii} are substituted for $\mu_{M\lambda}$ and μ_{Mi} in the denominator. The secondary "fluorescence" contribution S_{Mi} is given by

$$S_{Mi} = g_i W_i \sum \left\{ \frac{1}{\mu_{i\lambda}} \sum_j \left(D_{j\lambda} W_j K_j \mu_{ij} \mu_{j\lambda} \right. \right.$$

$$\times \left[\frac{1}{\mu_{M\lambda} \csc \phi_1} \ln\left(1 + \frac{\mu_{M\lambda} \csc \phi_1}{\mu M_j}\right) \right.$$

$$\left. \left. \left. + \frac{1}{\mu M_i \csc \phi_2} \ln\left(1 + \frac{\mu M_i \csc \phi_2}{\mu M_j}\right) \right] \right) \right\} \quad [8\text{--}8]$$

The symbols in these equations were given the following meanings [15].

R_i = ratio of the two intensities next defined.

P_{Mi} = intensity contribution from "element i in matrix M" resulting directly from excitation by polychromatic incident beam.

P_{ii} = analogous intensity for pure element i.

S_{Mi} = intensity contribution from element i in matrix M resulting from excitation by characteristic lines j generated in the sample. See Comment 2 below.

g_i = proportionality factor not present in R_i.

W_i = weight-fraction of element i.

$D_{i\lambda}$ = value zero at wavelengths λ (of the polychromatic beam) too long to excite charactcristic line of element i; is unity for all other wavelengths.

$D_{j\lambda}$ = similar constant for element j for characteristic wavelengths λ generated in the sample.

μ's = mass absorption coefficients (three) identified by their subscripts.

$I_\lambda \Delta\lambda$ = integrated intensity in polychromatic beam for interval $\Delta\lambda$ [24].

ϕ_1, ϕ_2 = angles made by incident (ϕ_1) and emitted beams with surface of sample.

K_j = product of two constants that enter into the "emission coefficient" (see below).

The following notes are intended to expedite translation to our symbols.

1. "Intensity from element i in matrix M" means the intensity of the analytical line for element i in sample S, or I^S. (S is *not* S_{Mi}!) S is infinitely thick.

2. In the manner of Table 9.11–1 for element G

$$I^S = I_{\lambda_0} + I_{\lambda_E} + I_{\lambda_F} + I_{\lambda_E \lambda_F} \tag{9.14--1}$$

P_{Mi} clearly equals I_{λ_0}; the next two contributions are examples of P_{ij}. Birks' [8–8] seems not to include contributions like the last term, for his equations are much less complex than Sherman's. (In Ref. 41, see (V) on p. 289 and the material that follows.)

3. The summation of $I_\lambda \Delta\lambda$ in [8–7] replaces I_0 in (9.12–1) and (9.12–2).
4. In Chapter 7 and in Section 9.12,

$$\mu'\rho' \csc \theta' = a' \quad \text{(with } \mathbf{a}, a_1 \text{ and } a_2 \text{ involved)} \quad (9.14\text{–}2)$$

The three mass absorption coefficients in these a's are in [8–7]; a fourth necessarily appears in [8–8] for each value of j. The θ's are easy to identify. The ρ has cancelled out in [8–6], as has the $\csc \theta$ in $\mathbf{a}$ in (9.12–2). The constant g_i has been likewise disposed of.

5. For an explanation of K_i, see Section 4.4–1, which—especially (4.4–6)—will make clear that:

$$\text{emission coefficient } \varepsilon = \mu K_i \quad (9.14\text{–}3)$$

The computer can solve for the composition of unknowns as follows. The initial assumption is made that the weight-percentage of each element equals the relative intensity of its analytical line, relative intensities being normalized to make them sum to 100%, or somewhat less if minor constituents have not been determined. According to a program based on the three Birks equations, the computer then calculates what intensities should have been observed for the assumed composition, compares these with those observed, adjusts composition and "iterates"—begins again. When iteration has succeeded in producing satisfactory intensities—say, such that the last assumed composition gives intensities within 0.1% of those observed—the computer prints out the last composition as the analytical result. Birks finds that 3 or 4 iterations suffice; that the calculations take less than 1 minute on fast computers; and that the cost is nominal.

Criss and Birks [40] carried out a valuable comparative study of six stainless steels, compositions being calculated by two regression methods and by the fundamental-parameter method. The samples apparently were flat, homogeneous, and "infinitely" thick. (For the importance of sample preparation, see below.) In regression method 1, the interelement coefficients were determined on the standards in Table 9.14–1.

Table 9.14–1. Weight Percentages in Standards for Method 1 [40]

	Hastelloy B	1187	Inconel X	410	301
Cr	0.00	21.62	14.00	12.60	17.90
Fe	6.00	27.40	8.50	86.09	72.67
Co	0.00	20.80	0.00	0.00	0.00
Ni	62.00	20.26	78.10	0.16	7.23
Mo	32.00	3.41	0.00	0.05	0.21
Total	100.00	93.49	100.60	98.90	98.01

Table 9.14–2. Comparative Results for Major Constituents in Each of Six Alloys

Stainless Steel No.	Element	Chemical	X-ray Results Calculated According to					
			Method 1		Method 2		Fundamental Parameters	
		% by Weight	% by Weight	Δ^a	% by Weight	Δ^a	% by Weight	Δ^a
303	Cr	17.2	17.8	0.6	Used as standard		18.5	1.3
	Mn	1.3					1.3	0.0
	Fe	71.2	73.8	2.6			71.5	0.3
	Ni	8.7	8.4	−0.3			8.7	0.0
	Total	98.4						
304	Cr	18.6	18.8	0.2	19.0	0.4	19.8	1.2
	Mn	1.4					1.5	0.1
	Fe	69.5	71.0	1.5	71.7	2.2	69.6	0.1
	Ni	9.4	9.7	0.3	9.3	−0.1	9.1	−0.3
	Total	98.9						
316	Cr	17.7	17.8	0.1	Used as standard		19.0	1.3
	Mn	1.8					1.8	0.0
	Fe	64.8	66.0	1.2			66.5	1.7
	Ni	12.8	13.2	0.4			12.7	−0.1
	Total	97.1						
321	Cr	17.8	18.5	0.7	18.4	0.6	19.0	1.2
	Mn	1.6					1.7	0.1
	Fe	68.2	71.7	3.5	70.6	2.4	68.7	0.5
	Ni	10.8	10.9	0.1	11.0	0.2	10.6	−0.2
	Total	98.4						
347	Cr	17.7	18.5	0.8	18.4	0.7	18.9	1.2
	Mn	1.6					1.7	0.1
	Fe	67.9	71.7	3.8	70.2	2.3	68.5	0.6
	Ni	10.7	11.7	1.0	11.4	0.7	10.9	0.2
	Total	97.9						
430	Cr	17.5	17.6	0.1	17.5	0.0	18.8	1.3
	Fe	81.3	79.3	−2.0	81.4	0.1	79.6	−1.7
	Total	98.8						

^{a} Δ = weight % (x-ray) − weight % (chemical).

PRINCIPAL DIFFICULTIES AND COMMON REMEDIES

For regression method 2, the standards were alloy 301 (Table 9.14–1) and alloy 303 (Table 9.14–2). The results appear in Table 9.14–2.

No definitive conclusions can be drawn from Table 9.14–2, but the following comments are offered because further work of this kind would be useful and welcome.

1. The standards in Table 9.14–1 do not seem well suited to determinations on the stainless steels: note the high Mo content of the Hastelloy B; the high Co content of alloy 1187, the absence of Mo and Co in Table 9.14–2; and the absence of Cr in the Hastelloy B. That good results are nevertheless obtained by method 1 does not necessarily mean that almost any standard can be used; there might have been compensating influences on the interelement coefficients.

2. A rough comparison of the various Δ's is perhaps permissible. The following is a reasonable, though arbitrary, method of so doing. Use the Δ's of method 1 for reference. Consider as equal to zero all Δ's that do not exceed 0.3 in absolute value. If among the other Δ's, Δ for method 2 exceeds that for method 1 in absolute value, record a "minus"; if the reverse is true, record a "plus." Treat the fundamental-parameter method the same way. On this basis, method 2 gets 6 (+'s) and 2 (−'s); the fundamental-parameter method gets 7 of each.

3. In connection with the effectiveness of method 2, see the comment about (9.13–4) toward the close of the last section.

4. Overall, the results in Table 9.14–2 are encouraging. There is no proof that the Δ's in the fundamental-parameter method result mainly from uncertainties in the parameter.

9.15 Comparison of Calculated Interelement-Effect Corrections

The principal limitation in any calculated correction as a remedy for difficulties due to interelement effects is precisely this: if nothing else can be calculated, *no other difficulty is remedied*. That is, such calculated correction will not remedy equipment difficulties, background difficulties, or difficulties with sample preparation. How far impurities may be disregarded is a moot question. If an empirical comparison of standard and unknown is needed for reliable results, a calculation of interelement effects is supererogatory.

Regression methods are less susceptible to the weakness just stated. We saw above that Lucas-Tooth and Pyne managed to remedy other than interelement difficulties by increasing the number of observations on which the regression equation is based. Owing to least-squaring, such an increase also leads to higher precision. The principal weakness of regression methods is that they are valid only for limited composition ranges, the extent of which cannot be predicted.

The great virtue of the fundamental-parameter method is that it escapes—at least in principle—the composition limitations just mentioned. It is at the mercy of uncertainties in the fundamental parameters. Sometimes at least, it will have to include $\lambda_E \lambda_F$ excitation of G (see above); this will complicate the equations, but should not daunt the computer. The question of suitable values for the fundamental parameters is probably trickier than it appears. Light elements, in which scattering rises relative to the photoelectric effect, and in which the x-ray spectra are affected by chemical binding, seem certain to give trouble. The effect of scattered x-rays on the intensity of the analytical line will depend upon the matrix. Finally, the method is dependent on the measured values of $I_\lambda \Delta \lambda$, for which the D's equal unity (see [8–7] and [8–8]) in the energy distribution; actual experience is needed to prove that the variations in this distribution from one tube to another (same target element, of course) cause little difficulty.

It seems to come to this: without the computer, the calculation of sample composition with interelement effects only taken into account would have limited usefulness. With the computer, the prospects for such methods are much brighter. Only time can tell whether they can compete with comparison of standard and unknown in a way that takes all difficulties into account.

9.16 Spectral Interference. Unfolding of Complex Pulse-Height Distributions

With modern equipment, spectral interference is no longer a major difficulty in most x-ray emission spectrography. Spectral interference not caused by characteristic lines of elements being determined has been discussed at various places, notably in Chapter 5 and in Section 9.10. The object here is to supplement earlier discussion of the characteristic-line case by presenting illustrative examples. The pulse-height distribution curve obtained without Bragg reflection from a multichannel analyzer will also be treated below as a case of spectral interference resulting from inadequate energy resolution.

A satisfactory solution of the problem of determining niobium and tantalum in the same ore was solved *without the use of pulse-height selection* by a happy marriage of chemical and x-ray methods. The chemical isolation from the ore of the mixed oxides of this pair of metals is a simple procedure, but the determination of the niobium-tantalum ratio in such a mixture by wet methods is difficult and time-consuming. The most serious obstacle to determining this ratio by x-ray emission spectrography was insufficient resolution at workable intensities of the second-order Kα line of niobium and the first-order Lα1 line of tantalum, this being the best analytical line for the tantalum determination [54]. With a tungsten-target tube, the characteristic lines of that element complicate the spectrum further [55].

PRINCIPAL DIFFICULTIES AND COMMON REMEDIES

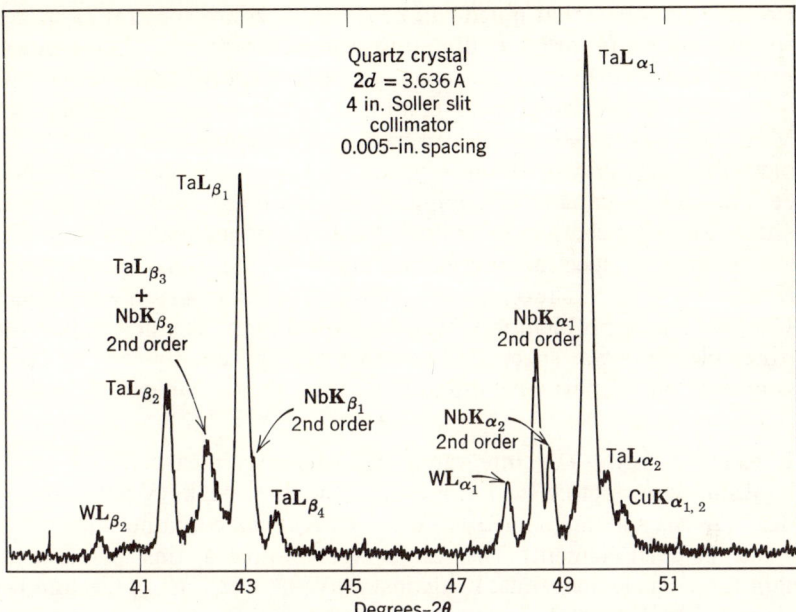

Fig. 9.16-1. X-ray spectra of niobium and tantalum as obtained on an ordinary emission spectrograph without pulse-height selection, but with other experimental conditions carefully chosen. From W. J. Campbell and H. F. Carl, Ref. 54.

Operating the x-ray source at 18 kV eliminates the niobium K lines (excitation potential, 18.986 kV), but at the price of low analytical-line intensity for tantalum. In spite of this resolving power-intensity problem, satisfactory methods were worked out to determine the niobium-tantalum ratio in the mixed oxides [56, 57].

Satisfactory resolution of tantalum $L\alpha 1$ without prohibitive loss of intensity was accomplished by Campbell and Carl [54] in the following way. To achieve more efficient excitation of the tantalum line and to reduce background and interference, the tungsten-target source was replaced by one with a molybdenum target. In the Norelco spectrograph, a Soller slit collimator (4 in. long with 0.005-in. spacing) was inserted between the Bragg crystal and detector, and an open-tube collimator with five evenly spaced thin plates was used between the crystal and the sample. The Bragg crystal was quartz ($2d = 3.636$ Å), a good choice because it gives sharp lines and improves resolution. An argon-filled Geiger detector was used because it is more efficient for the tantalum L than for the niobium K lines. The effectiveness of the optical system and the complexity of the spectra are clear from Figure 9.16-1. This example [39, pp. 201 and 202] is repeated here for two reasons: (1) it shows that much can be done to reduce spectral interference by proper choice of target, collimating system, Bragg crystal, and detector filler-gas; and (2) it illustrates a problem easier to solve today, when pulse-height selection is available to eliminate higher-order reflections.

Modern x-ray emission spectrographs, adequately collimated, with a choice of targets and of Bragg crystals, and with automatic pulse-height selection are thus capable of coping directly with almost all spectral interferences involving characteristic lines. Cases do occasionally arise in which the correction of spectral interference requires additional measurements. One such, the determination of $_{23}$V, $_{24}$Cr, and $_{25}$Mn in a steel was discussed in Chapter 5 as a problem in wavelength resolution. The determination of all three elements was successfully carried out by Zemany [58] by the following method, of particular interest because it uses *integrated intensities* (not cps, or $N_T - N_B$):

1. CALIBRATION. The integrated intensity was measured on each of several standards (actual steels) for each of the three peaks (V Kα, unresolved Cr Kα − V Kβ, and unresolved Mn Kα − Cr Kβ) that contained an analytical line (Kα for each element). The results were plotted against percentage by weight for each element thus: $\mathbf{I}'_V$ against % V; $\mathbf{I}'_{Cr-V}$ against % Cr; and $\mathbf{I}'_{Mn-Cr}$ against % Mn. Extrapolation to zero per cent gave the corresponding background intensity I_B for each element so that net integrated intensities (unprimed, e.g., $\mathbf{I}_V$) could be obtained by subtraction. Net integrated intensities for $W = 1\%$ by weight of each element were thus obtained; e.g., $\mathbf{I}_V^{1\%}$.

To arrive at % Cr and % Mn from the net integrated intensities of the unresolved peaks, Kβ/Kα intensity ratios are needed for each of the two elements; call these R_{Cr} and R_{Mn}. These were obtained from measurements on pure V_2O_5 and pure $K_2Cr_2O_7$.

2. DETERMINATION. On any unknown, U, the three net integrated intensities $\mathbf{I}_V^U$, $\mathbf{I}_{Cr-V}^U$, and $\mathbf{I}_{Mn-Cr}^U$ were obtained as they had been on the standards.

3. CALCULATION OF RESULTS.

$$V(\% \text{ by weight}) = \frac{\mathbf{I}_V^U}{\mathbf{I}_V^{1\%}} \qquad (9.16\text{–}1)$$

$$Cr\,(\% \text{ by weight}) = \frac{\mathbf{I}_{Cr-V}^U - R_{Cr}\mathbf{I}_V^U}{\mathbf{I}_{Cr}^{1\%}} \qquad (9.16\text{–}2)$$

The subtractive term in the numerator of the second equation is the net integrated intensity of V Kβ, and the subtraction corrects for its spectral interference with Cr Kα. The manganese content is calculated in analogous fashion to the chromium content. Satisfactory results were obtained.

Let us see now what might be done with spectral interference problems

PRINCIPAL DIFFICULTIES AND COMMON REMEDIES

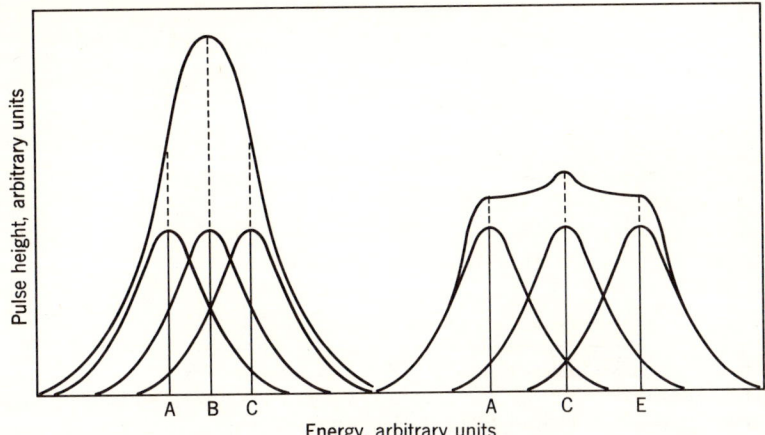

Fig. 9.16–2. Idealized composite pulse-height distributions from a multichannel analyzer unaided by wavelength resolution. The taller distribution is for neighboring elements A, B, and C; unfolded, it would give the three individual Gaussian distributions (one for each element) shown. The broader distribution is for elements A, C, and E, for which the atomic-number interval is 2 between neighbors; individual contributions again are shown.

when energy resolution alone (no Bragg crystal, multichannel analyzer only—see Chapter 2) is used. As energy resolution generally falls short of wavelength resolution, it seems unlikely that unaided energy resolution could deliver useful information in the $_{23}$V, $_{24}$Cr, $_{25}$Mn, $_{26}$Fe case, with which even wavelength resolution cannot cope simply and directly. A case such as $_{24}$Cr, $_{26}$Fe, $_{28}$Ni is more promising for energy resolution provided that one can forego distinguishing between the **K**α and **K**β lines of an element—and one probably can because the intensity ratio **K**α/**K**β is fairly large (near 5) and will not vary much from element to element in this small range of atomic numbers. The information from the multichannel analyzer in such a case is a single broad pulse-height distribution, which one assumes is the sum of three individual pulse-height distributions, one for each element. None of the four distributions is likely to be Gaussian: the **K**α and **K**β distributions for a single element are not resolved, and the distributions may be distorted for reasons given in Chapter 2; see especially Figures 2.12–1, 2.12–3, and the (idealized) Figure 9.16–2. Clearly, such procedure can *at best* give correct relative intensities for the elements present *when these elements are known*. It cannot remedy difficulties such as those under discussion in this chapter.

The difficulties attending wavelength resolution of analytical lines for the light elements led Dolby [59] in 1958 to develop three methods of *unfolding* composite pulse-height distributions obtained by use of a propor-

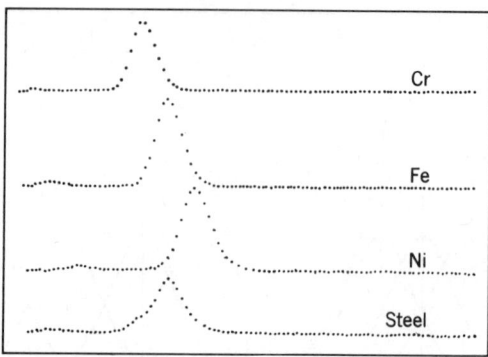

Fig. 9.16–3. Composite pulse-height distribution for a stainless steel and component distributions for three major constituents, all as displayed on a 400-channel pulse-height analyzer. Note relative positions of intensity peaks. A Machlett OEG-50 tungsten-target x-ray tube was operated at 15 kV and 2 mA. The proportional detector was side-window, xenon-filled. After L. S. Birks, R. J. Labrie, and J. W. Criss, Ref. 60, p. 702, Fig. 2.

tional detector. He began by postulating that the composite curve consisted of superimposed and unresolved Gaussian pulse-height distributions as components, these being of known positions, known FWHMs, but of unknown peak heights A, B, C, and so on. Then ordinate (intensity) values (M_1, M_2, M_3, etc.) of the composite curve obey the relations

$$\left. \begin{array}{l} M_1 = K_1 A + K_2 B + K_3 C \ldots \\ M_2 = K_4 A + K_5 B + K_6 C \ldots \\ M_3 = K_7 A + K_8 B + K_9 C \ldots \end{array} \right\} \quad (9.16\text{–}3)$$

If each component distribution is represented by an equation that is linearly independent of the others, (9.16–3) will have solutions of the form

$$\left. \begin{array}{l} A = k_1 M_1 + k_2 M_2 + k_3 M_3 \ldots \\ B = k_4 M_1 + k_5 M_2 + k_6 M_3 \ldots \\ C = k_7 M_1 + k_8 M_2 + k_9 M_3 \ldots \end{array} \right\} \quad (9.16\text{–}4)$$

in which the k's are functions only of the K's in (9.16–3). This is clearly another example of a multiple regression treatment (Chapter 8 and Section 9.13)—though, as applied by Dolby, it was not a wholly *empirical* regression treatment, as he assumed Gaussian shapes for the component distributions.

Dolby based a second method on the shift in the peak intensity of the composite pulse-height distribution that occurs when composition is changed. See Figure 9.16–3 [60]. His third method—"wave form analysis" —is not discussed here.

Dolby tested his two unfolding methods in the following ingenious way. In a scanning instrument resembling an electron microprobe, he had the

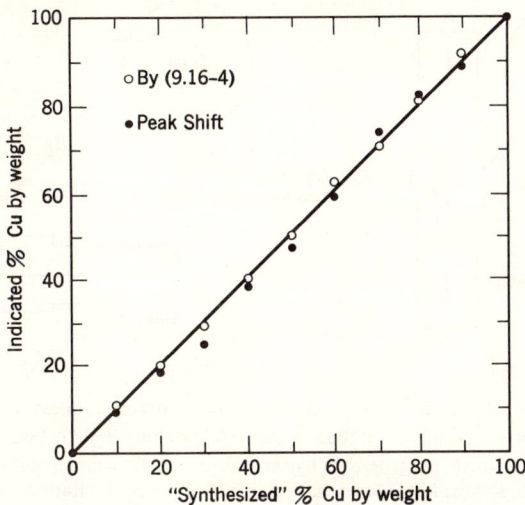

Fig. 9.16-4. Test of two ways of unfolding a synthesized composite pulse-height distribution curve with copper and zinc pulse-height distributions as components. After R. M. Dolby, Ref. 59, p. 85, Fig. 1.

electron beam move in linear sawtooth fashion across the edge of a thin copper sheet on a nickel background. The relative time the electron beam spent on each metal then made possible the computation of a "synthesized" % Cu by weight for comparison with percentages obtained by each of the two unfolding methods. Figure 9.16-4 shows his results. The same instrument gave these results for a brass: wavelength resolution, Cu $60 \pm 1\%$ by weight, Zn $40 \pm 1\%$ by weight; from composite distribution unfolded by (9.16-4), Cu $58\% \pm 4\%$ by weight, Zn $40 \pm 4\%$ by weight.

Consider now a composite distribution curve that results when determinations have to be made on actual samples. Such a curve is not likely to be Gaussian, for it will reflect all the complicating factors—some of them specifically described here as difficulties; others, such as the characteristics of the electronic circuitry, the presence of lines such as $K\beta$, pulse amplitude distortion, and shifts in pulse amplitude with counting rate having been mentioned earlier. In addition, there is the effect of the presence of possible unknown elements in the sample. Enough has been said to show that the regression treatment, which need not be based on equations formally identical to (9.16-3) and (9.16-4), in an actual case had best be a wholly empirical treatment.

Such a treatment was successfully carried out by Birks, Labrie, and Criss [60] on several stainless steels and other alloys. As the schematic diagram (Figure 9.16-5) shows, they were principally concerned, not

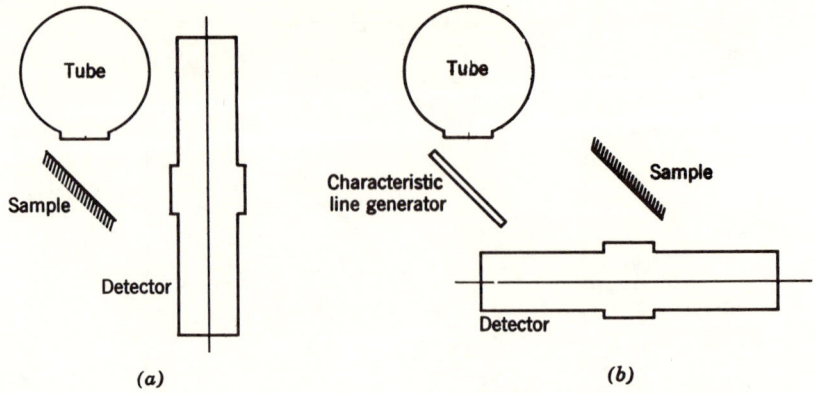

Fig. 9.16–5. Two schemes for testing unfolding procedures on stainless steels. Note favorable geometry. In scheme *b*, excitation of the sample is by the characteristic lines of $_{30}$Zn (Kα wavelength near 1.44 Å), which are generated by the polychromatic beam from the tube. For either scheme, the output is of the kind shown in Figure 9.16–3, where excitation conditions are given. After L. S. Birks, R. J. Labrie, and J. W. Criss, Ref. 60, p. 701, Fig. 1.

with long-wavelength x-rays as was Dolby, but with testing unfolding procedures on practical samples with analytical lines of shorter wavelength, and with the effective transfer of analytical-line quanta from sample to detector, which makes possible the use of x-ray tubes at low power and of radioactive sources, these being advantages that attend the elimination of Bragg reflection. We discuss only some of their work on stainless steels *possibly* containing manganese.

Scheme *a* of Figure 9.16–5 needs no comment. Scheme *b* is only one of several in which characteristic-line generators are used to good advantage. Their use would be impossible in conjunction with Bragg reflection because intensities would be too low; here, however, counting intervals near 0.4 minute sufficed even though the x-ray tube was operated at only 25 W. The characteristic-line generators can (as in Figure 9.16–5) give virtually monochromatic excitation, which simplifies fundamental-parameter calculations (Section 9.14), or they can be used to give selective excitation of one or more elements in the sample. Results by both schemes appear in Table 9.16–1.

The first thing to say about these results is that they are excellent. The weight percentages were estimated by simple proportionality with Type 316 stainless steel as a standard; that is, for Fe as example,

$$\frac{\mathbf{W}_{\text{Fe}}^{301}}{\mathbf{W}_{\text{Fe}}^{316}} = \frac{\mathbf{R}_{\text{Fe}}^{301}}{\mathbf{R}_{\text{Fe}}^{316}} \qquad (9.16\text{–}5)$$

where **R** is the ratio of the intensity for the element in the sample to that of the pure element. In all experiments, the standard deviation observed was very near the standard counting error (Chapter 8), which shows good control

PRINCIPAL DIFFICULTIES AND COMMON REMEDIES

Table 9.16–1 [60]. Composition of Stainless Steels from Unfolding of Composite Pulse-Height Distributions

Steel	% by Weight, Chemical	% by Weight[a], Scheme a	% by Weight[b], Scheme b
301	Cr, 17.9 Fe, 72.7 Ni, 7.2	18.3 70.8 7.9	18.6 72.0 7.2
303	Cr, 17.2 Fe, 71.2 Ni, 8.7	17.3 70.3 9.4	17.7 71.0 8.7
304	Cr, 18.6 Fe, 69.5 Ni, 9.4	18.7 68.2 9.6	18.8 69.0 9.2
321	Cr, 17.8 Fe, 68.2 Ni, 10.8	18.0 68.0 11.0	18.2 68.1 10.7
347	Cr, 17.7 Fe, 67.9 Ni, 10.7	17.9 67.1 11.2	18.0 67.8 10.9

[a] Average of 9 runs.
[b] Average of 10 runs.

of conditions. As the composition range of all samples was near that of the standard, no serious difference in interelement effects could have existed. The failure to consider the possible presence of manganese had no damaging influence; in fact, experiments in which the possible presence of this element was not neglected in the computer calculations did not give results quite as good—an indication that the manganese was present at a weight-fraction low enough so that an attempt to include it was self-defeating owing to increased statistical errors. Finally, a fundamental-parameter calculation (results not included above) proved useful in the case of scheme b, where such calculations could be made with increased confidence because excitation was virtually monochromatic.

The work included ingenious experiments (Figure 9.16–6) to establish the limit of detectability for Mn. How this is done is explained in the caption. Interelement effects are not—and need not—be taken into account in this simple experiment. Results: by the unfolding as practiced, the percentage by weight of $_{25}$Mn would have to be 2 or 3 before it could be detected in the presence of $_{23}$V and $_{27}$Co; and near 5% in the presence of $_{24}$Cr and $_{26}$Fe.

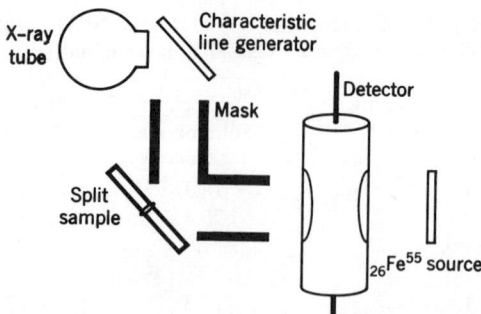

Fig. 9.16-6. Experiment to measure the limit of detectability of Mn in the presence of varying weight fractions of V and Co: (1) Characteristic lines of Mn enter the detector from the $_{26}Fe^{55}$ source, in which they are generated by **K** capture. Their intensity is varied by moving the source. (2) The ratio of the characteristic-line intensity of V to that of Co can be varied by changing the position of the split-sample containing the two elements as shown. (3) The presumed weight-fraction of Mn in the split-sample is obtained from intensity measurements in which this sample is replaced by pure Mn. After L. S. Birks, R. J. Labrie, and J. W. Criss, Ref. 60, p. 705, Fig. 7.

Birks, Labrie, and Criss [60] also show how useful information can be extracted from the composite distribution curves by schemes simpler than the computer-regression treatment. As the composite curve contains all the x-ray information available for a sample run as prescribed above, it is of interest to consider whether a direct mathematical comparison of such curves for standard and unknown might be feasible and worthwhile. In any case, the use of composite pulse-height distribution curves, long established in nuclear physics, promises to be worthwhile in analytical chemistry for establishing composition under special conditions without paying the intensity tribute demanded by Bragg reflection.

DIFFICULTIES ATTENDING SAMPLE PREPARATION. REMEDIES AND COMPENSATIONS

9.17 Importance of Sample Preparation

Chapter 8 has shown that the standard counting error sets the limit for the precision (measured as relative standard deviation in the analytical result) ultimately attainable in x-ray emission spectrography. For reasonable counting rates and counting intervals, this limit lies somewhere below one part per thousand in the weight-fraction of a major constituent. Only in exceptional cases does the relative standard deviation realized for actual unknowns fall much below one part per hundred. General experience indicates that much of this shortfall usually results from difficulties attending sample preparation: it is consequently minimized when routine determina-

SAMPLE PREPARATION. REMEDIES AND COMPENSATIONS

tions are done frequently on bulk metallic samples with reproducible surfaces. The diversity of samples is so great that significant improvement will not come easily in less favorable cases. Calculational methods generally will not help.

Samples are either standards or unknowns. As the comparative method, introduced for absorptiometry in Chapter 3, is the mainstay of x-ray emission spectrography also, it follows that unknowns and standards must be made truly comparable. In absorptiometry, this requirement was easy to meet if the two were alike, or nearly alike, in composition; ready commutation between standard and unknown was an added advantage. In x-ray emission spectrography, however, where the critical depth is usually small, unknowns and standards must be alike in other essential respects: for example, in their surfaces; in particle size if they are powders; in degree of heterogeneity (if any); and in the way they are presented to the x-ray beam. Such factors come into play mainly because the x-ray beam does not travel through the sample in x-ray emission spectrography as it does in absorptiometry; the fate of the spectrographic beam is usually far more complex.

It usually makes sense to run the *original* sample if the needed answer can thus be found [61]. If not, the sample must be *prepared* for spectrography. The preparative methods must produce truly comparable standards and unknowns even though the time and effort required, which can approach being prohibitive, far exceed those of the final x-ray determination.

Samples *heterogeneous throughout* are usually the most troublesome. They subdivide into samples where heterogeneity is a hindrance because average composition is sought, and samples in which the existence of heterogeneity is to be established or its nature studied.

Slow rotation in its plane of the sample in the spectrograph, sometimes called "spinning" the sample, is often recommended as compensating for heterogeneity when average composition is sought. Such compensation can be less than perfect, especially when the surface of the sample is not uniform and interelement effects are pronounced. In such case, rotation will lead to a result that is an *average* of values, each systematically in error. If a comparison with a standard is properly carried out, such an average may be more reliable than a single result obtained without rotation would have been. Such rotation is beneficial to the extent that it averages out the intensity distribution of the x-ray beam that excites a sample not uniformly distributed (e.g., a residue on filter paper or, worse yet, Mylar).

An interpolation about diffraction peaks. Such peaks result from Bragg reflection *by the sample of a characteristic line from the target.* They will not appear if the voltage across a Coolidge tube is too low to excite such lines, or if sample excitation is by electrons. To reach the detector, diffraction peaks must pass through the collimator and through the voltage window of the pulse-height selector. When present, they will be found at the 2θ-value for the target line; that is, at or very near 2θ-values for peaks from characteristic target lines scattered incoherently by the sample (see Figure 9.10–2). The effect on diffraction peaks of rotating the sample depends upon its crystallinity. Amorphous samples merely contribute scattered x-rays to the background. Samples that would give grainless powder diffraction patterns (Chapter 6) cannot generate strong diffraction peaks, and the intensity of any such peaks will not change when the sample is rotated. As the crystals in a sample become large enough to give grainy diffraction patterns, the intensity of the diffraction peaks may change on rotation of the sample; everything depends upon how the larger crystals are oriented. For example, a single-crystal sample could give a peak of maximum intensity when it is in the position most favorable with respect to the collimators. The peak would disappear when the crystal is rotated away from this position. Less extreme changes could occur with samples containing macrocrystals.

When it is necessary either to prove uniformity, or to establish the heterogeneity of a sample, *localized determinations* are needed. The ultimate among such determinations are those done by the electron microprobe.

There is of course heterogeneity of another kind, examples of which have appeared in Chapter 7. In these, the surface may be uniform but different in composition from the bulk of the sample, as in oxide films on metals or plated metals. This kind of heterogeneity is closely allied to the case of samples *supported on a substrate* and needs no separate discussion.

In summary: heterogeneity may be physical (particle size and bulk-density variation) or chemical or both. Heterogeneity below the critical depth will go unnoticed by the spectrograph.

Faulty sample preparation or handling is most surely detected by making a chemical analysis of a suspected unknown to obtain results for comparison with those from the spectrograph. This is often not feasible, and it is unnecessary if the fault can be detected by one or more of these measures: (1) leaving the sample in the spectrograph and making several independent counts, (2) resetting the goniometer between independent counts, (3) removing and returning the sample to see whether the counting rate changes, (4) scanning different portions of the sample, (5) counting after altering the surface; for example, a powdered sample can sometimes be stirred or shaken, and (6) making replicate x-ray determination on different portions of a sample.

The diversity of samples amenable to x-ray emission spectrography makes them difficult to classify. We recognize the following kinds of samples: (1) *diluted*, (2) *bulk solids*, (3) *supported*, (4) *powdered*, (5) *converted*, and (6) *radioactive*. As the way in which the comparative method is best applied differs with the kind of sample, we discuss this method before we return to the sample problems.

9.18 The Comparative Method in X-Ray Emission Spectrography

Except perhaps for the spectrograph called the electron microprobe, x-ray emission spectrography will for the foreseeable future rely mainly upon the (empirical) comparative method. Difficulties in sample preparation are best remedied in this way, and—as a welcome bonus—so are other difficulties already discussed, how many depending upon how the method is practiced. The method can be the closest thing there is to a panacea for x-ray-emission-spectrographic ills when the sample is homogeneous.

In the preceding section, it was tacitly assumed that an *unknown sample* was being compared with a *standard sample*, for each of which (9.7–1) or (9.10–2) is valid: we shall call this an *intersample comparison*. Of course, what is actually being compared is not one sample with another, but one *intensity* (e.g., I^U with I^{St}) or one *intensity ratio* (R^U with R^{St}) with another. This comparison need not be direct; it may be via a *calibration* or *working curve*; it may even be via a multiple regression treatment (Section 9.13). There seems no doubt that comparisons of properly chosen intensity ratios, I^S/I, should yield more reliable analytical results—no matter what the nature of I—than similar comparisons involving I^S.

Intrasample comparisons are useful in x-ray emission spectrography as well. These involve only *one* sample, the *unknown*, and the classic work of Eddy and Laby [8 and Section 9.3] may serve as prototype. They obtained quantitative results for $_{29}$Cu and $_{30}$Zn in brass on the assumptions that

$$\frac{I^U_{Cu}}{I^U_{Zn}} = \frac{W^U_{Cu}}{W^U_{Zn}} = \frac{W^U_{Cu}}{1 - W^U_{Cu}} \qquad (9.18-1)$$

the intensities being the net intensities of the $K\alpha$ lines. The validity of (9.18–1) shows that enhancement of Cu $K\alpha$ by Zn $K\beta$ could not have been appreciable. For comparable results obtained by the x-ray excitation of similar samples, see Ref. 48, Table I.

By contrast, Figure 9.10–3 represents an intersample comparison even though the Mo/W weight ratio is being determined by use of an intensity ratio: solvent and other substances are present. The figure contains a calibration curve for the determination of this weight ratio on unknowns.

An intersample comparison in a related case gave more complex results. The advantages of determining molybdenum on a powdered ore of known tungsten content by measuring the molybdenum-tungsten ratio are attractive enough to outweigh the difficulties often present. Accordingly, the working curve in Figure 9.18–1 was prepared by measuring the intensity ratio of molybdenum $K\alpha$ to tungsten $L\gamma 1$ on powders poured onto Mylar film. Table 9.18–1 gives data for the samples used. The synthetic (standard)

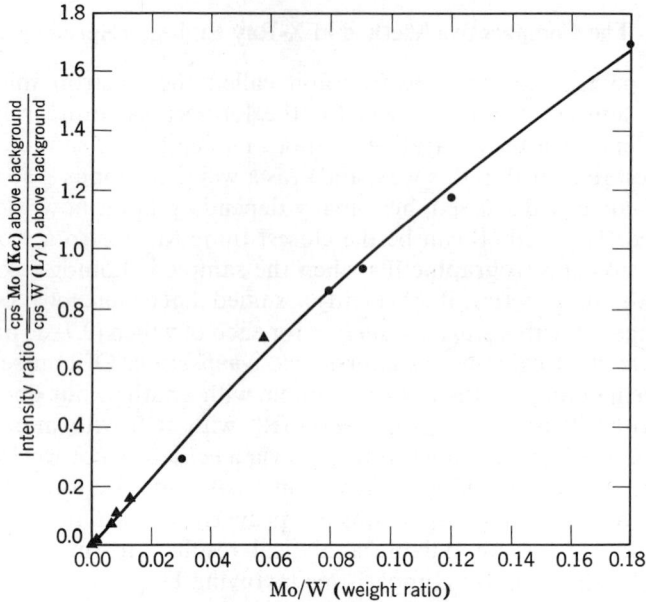

Fig. 9.18-1. Comparison of x-ray and chemical results for scheelite and synthetic samples. Triangles = scheelite ores (unknowns); dots = synthetic samples (standards). See J. E. Fagel, Jr., H. A. Liebhafsky, and P. D. Zemany, Ref. 36.

Table 9.18-1. Composition of Samples for Figure 9.18-1

Weight ratio × 100[a]	0.0177	0.069	0.652	0.829	1.37	3.00	5.76
Type of sample	Ore	Ore	Ore	Ore	Ore	Synthetic	Ore
% W by weight	56.2	56.3	56.7	60.3	48.3	53.5	53.8
Weight ratio × 100[a]		7.91		9.03		11.96	17.94
Type of sample		Synthetic		Synthetic		Synthetic	Synthetic
% W by weight		50.2		49.5		47.7	44.5

[a] Mo/W. See Figure 9.18-1. Compositions of ores by replicate wet analyses. The molybdenum content can be calculated from tungsten content and Mo/W ratio.

samples were prepared by evaporating solutions containing known amounts of sodium molybdate and sodium tungstate.

Points for both ores (unknown samples) and synthetic samples are satisfactorily close to the curve in Figure 9.18–1. Each point is the mean of several determinations. Curvature of the kind observed is to be expected from the changing absorption effects brought about by the decrease in tungsten content along the curve.

Standard deviations were calculated for Figure 9.18–1 for the five samples on each of which at least five determinations had been made. These standard deviations range roughly from 3 to 10% of the weight ratios. The high standard deviations are attributable to the uncertainties associated with powdered samples and with residues formed by evaporating solutions.

The reliability of the weight-ratio determination was greater for Figure 9.10–3 because (1) the tungsten content of the dissolved samples was chosen high enough (0.03 g/g solution) to keep the absorption effect sensibly constant; and (2) uncertainties associated with sample preparation were least for the solutions, the relative standard deviations being only 1/100 for Mo/W = 0.1 and 2/100 for Mo/W = 0.01. The extension of this kind of comparison by Bertin is noteworthy [62].

Internal standards also require intersample comparisons. For the *ideal* such standard

$$\mathbf{R} = \frac{\mathbf{I}^S}{I_{i\text{-}St}} = \frac{\mathbf{W}}{W_{i\text{-}St}} \qquad (9.18\text{–}2)$$

Usually proportionality of the two last terms, not equality, is the best that can be hoped for. To introduce the important internal-standard technique, which was fully developed by Hevesy [5] many years ago, we present the results of a study of the extraction of tungsten ores by sodium hydroxide, greatly facilitated by x-ray emission spectrography, which made it possible to determine tungsten rapidly and precisely in hundreds of sodium tungstate solutions containing various concentrations of sodium hydroxide. Because of the pronounced negative absorption effect due to sodium tungstate itself (Table 9.12–1), and because of the unknown and variable absorption effect due to sodium hydroxide, the use of an internal standard proved advisable. Bromine as sodium bromide was chosen, and bromine $K\alpha$ (1.041 Å) and tungsten $L\gamma 1$ (1.099 Å) were selected as the analytical lines best suited for comparison.

Accordingly, solutions for the determination of tungsten were prepared to contain 2 mg of bromide ion per gram of solution. The bromide ion was added as sodium bromide weighed precisely enough not to reduce appreciably the overall precision of the tungsten determination. With an argon-filled Geiger detector, the intensity ratio of the two analytical lines was established by measuring the counting interval required to accumulate about 82,000 counts for each line. With a krypton-filled Geiger detector this number of counts was doubled. The intensity ratio was computed as a ratio (R_t) of counting intervals. Corrections for background proved unnecessary—a marked advantage. Time for a complete determination, sample preparation and computation included: 30 minutes.

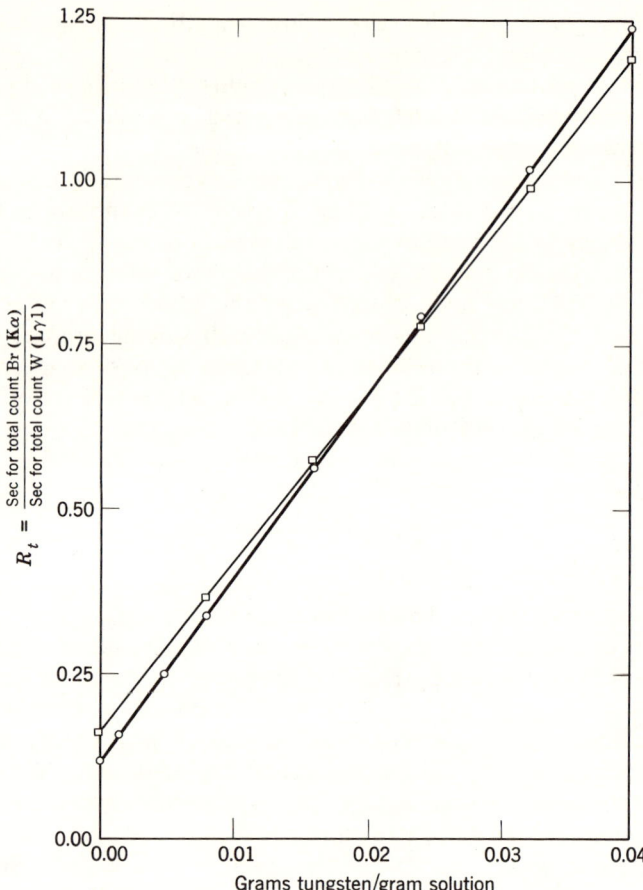

Fig. 9.18-2. Calibration curve for the determination of tungsten in solution with bromide as an internal standard, for two different proportional detectors. Squares = krypton-filled detector, total count 10(16,384); circles = argon-filled detector, total count 5(16,384). See J. E. Fagel, Jr., H. A. Liebhafsky, and P. D. Zemany, Ref. 36.

The following experimental details apply both to the determination of tungsten with bromine as internal standard, and to the experiments of Table 9.12-1. The solutions filled a 3-ml container made by sectioning a 10-ml beaker. To prevent evaporation and to maintain a fixed distance between x-ray tube window and sample surface, the beaker-section was covered with Mylar film, 0.0025 cm thick, placed in a plastic sample holder and pressed firmly against the sample drawer. The Mylar film attenuated the x-rays uniformly enough so as not to affect the precision of the results.

The calibration curves in Figure 9.18-2 were obtained on known samples prepared by weight from an aqueous standard solution of analyzed sodium tungstate dihydrate, and from another of analyzed sodium bromide. Each point is the average of three or more independent determinations, and the lines were fitted by the method of least squares.

To test the effectiveness of the internal standard in compensating absorption effects attributable to added sodium hydroxide, the experiments of Table 9.18–2 were carried out. The internal standard is clearly effective; R_t shows no trend attributable to an absorption effect.

Table 9.18–2. Determination of Tungsten in Sodium Hydroxide Solution[a]

Added NaOH, Normality	Number of Determinations	R_t, Ratio of Counting Times	S_A, Standard Deviation
None	15	0.7830	0.0042
0.8	11	0.7867	0.0023
4.0	10	0.7816	0.0050

[a] *Notes.* (1) Krypton-filled detector. (2) Tungsten content: 0.0240 g of tungsten per gram of solution. (3) Each determination complete and independent except that aliquots of the same solution were used for an entire series. (4) S_A is the standard deviation, as usually defined, for a single determination.

The intercepts in Figure 9.18–2 measure the ratios of background intensities. The uncertainties in this ratio will inevitably limit the usefulness of the method in detecting or determining traces of tungsten. The excellent linearity in the figure probably results from the fortuitous balancing of several factors. Although weight units are logical in x-ray work, one cannot predict the linearity observed in Figure 9.18–2; after all, there is no logical reason for omitting the background correction. As regards choice of units and treatment of background, it seems generally logical to proceed so as to obtain a simple calibration curve.

Finally, chemical determinations were carried out on three samples to give results for comparison with x-ray data (in parentheses): 0.026 (0.0283); 0.027 (0.0274); and 0.026 (0.0266), all in grams of tungsten per gram of solution. The estimated reliability of the chemical results is ± 0.001 g of tungsten per gram of solution. Even in the first case, where the difference between results by the two methods is unusually large, the difference could have been due entirely to combined experimental errors.

We turn next to *monitoring* and to techniques that resemble it. Obviously, an intersample comparison may suffer if variations occur in the intensity of the beam used for excitation. As obviously, this difficulty can be remedied by normalizing this intensity, which can be done by using a *compensatory*

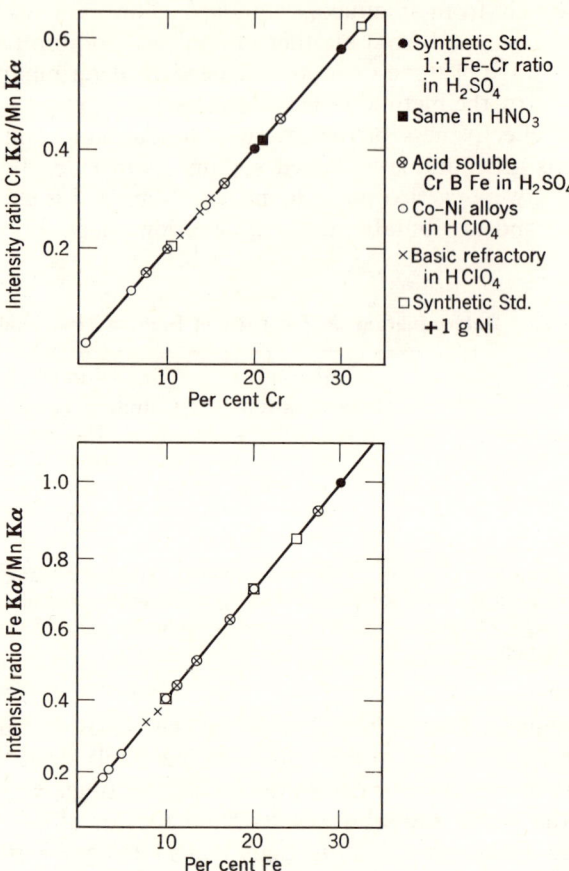

Fig. 9.18–3. The use of manganese as compensatory standard (control element) in the determination of iron and of chromium by intersample comparison under widely different conditions. The determinations were made on a multichannel spectrograph (Applied Research Laboratories' XIQ) programmed for the simultaneous determination of Fe, Cr, Mn, Ni, and Cu. Note the excellent linearity obtained and compare with Figure 9.10–3. See B. J. Mitchell and H. J. O'Hear, Ref. 63, p. 622, Fig. 4.

standard in a spectrograph so that the intensity of the beam entering a detector from this standard can be divided into I^s. The compensatory standard can be anything convenient. Preferably monitoring will be continuous, in which case the spectrograph must have at least two channels. Intermittent monitoring resembles commutation as practiced in Chapter 3: measurements of I will be alternated with measurements from the compensatory standard. Such monitoring will help only to remedy equipment difficulties.

Mitchell and O'Hear [63] made an important contribution to the routine

SAMPLE PREPARATION. REMEDIES AND COMPENSATIONS

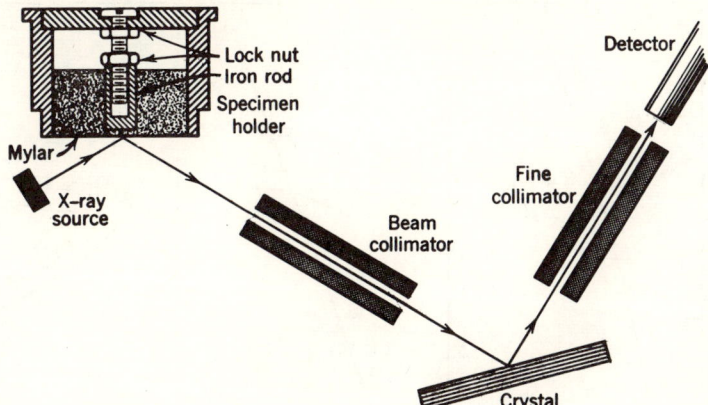

Fig. 9.18–4. Use of an iron rod as compensatory standard for equipment difficulties and variation of a (mass absorption coefficient and density variations) in the determination of manganese in different hydrocarbon base-stocks. See (7.4–2). Note that the density variation does not enter into absorptiometric measurements based on the mass absorption coefficient. See references to Chapter 3 in text. See also R. A. Jones, Ref. 64, p. 1342, Fig. 1.

determination of iron, chromium, and manganese in samples that can be dissolved and run on a multichannel spectrograph. Their contribution is the use of a dissolved "control element" as compensatory standard in the proportion of 0.5 g control element to 1.0 g sample in 100 ml. The element is chosen, not because it necessarily satisfies the criteria for an internal standard, but primarily because it is absent from the sample; thus, manganese is satisfactory as control element for the determination of iron and chromium in refractory alloys. Compensation of equipment difficulties, density or temperature changes in the sample, and interelement effects is surprisingly good as Figure 9.18–3 shows. Agreement of x-ray with chemical results was generally excellent. The method deserves consideration for other applications.

Jones [64] and Gunn [65] used solid metals in liquid samples as compensatory standards in an interesting way made clear by Figure 9.18–4, which is a preview of the Philips inverted-sample three-position spectrograph. In both cases, the need for such a standard arose owing to changes in the mass absorption coefficient and in density of the sample caused by variations in the gasoline base stock (see Sections 3.3 and 3.16). The choice of compensatory standard is made to suit the element being determined: Jones chose iron for a manganese determination (Fe $K\alpha$, 1.937 Å; Mn $K\alpha$, 2.103 Å); Gunn selected platinum for lead and used the $L\alpha$ of each. In Figure 9.18–4, the distance between the Mylar window and the end of the iron standard was adjustable: compensation was considered achieved at the position of the iron rod that gave substantially identical intensity ratios

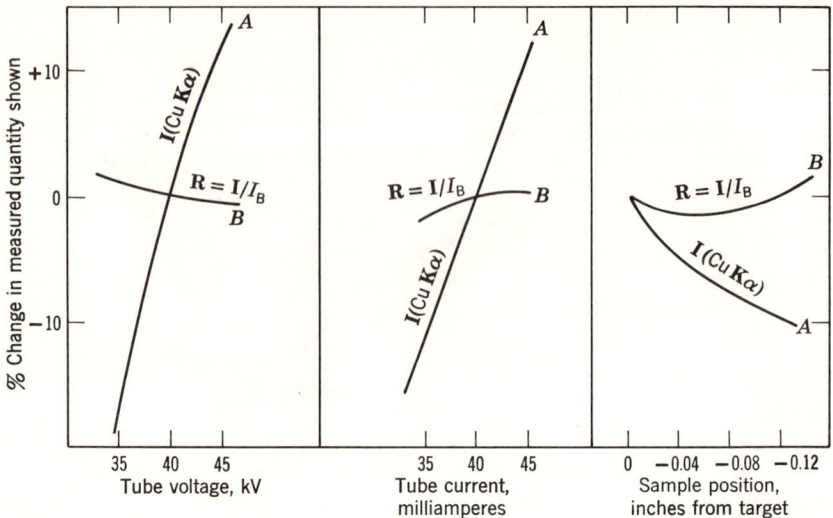

Fig. 9.18–5. Use of scattered background (near 1.4 Å; intensity I_B) as a compensatory standard for equipment difficulties. The sample was presumably a copper ore (Cu Kα = 1.5 Å). Note that the voltage and current ranges in the figure far exceed those to be expected during actual determinations. Note the sensitivity of the Cu Kα intensity to sample position; presumably the geometrical factors that determine the intensity distribution of the incident beam over the sample surface are responsible; see Chapter 5. See G. Andermann and J. W. Kemp, Ref. 66, p. 1306, Fig. 1.

Mn Kα/Fe Kα for two samples of identical manganese content in widely different base stocks. Satisfactory results were obtained.

The most generally useful compensatory standard is the scattered background, which was suggested for this purpose by Andermann and Kemp [66]. The scattered wavelength to be used had best be chosen empirically: a scattered target line, a portion of the scattered continuum, the scattered background near or at the wavelength of the analytical line (either Rayleigh or Compton scattering) might serve; see Figure 9.10–2.

Figure 9.18–5 shows that use of the scattered background in an intensity ratio $\mathbf{R} = \mathbf{I}/I_B$ is a more than adequate, though short of perfect, remedy for the common equipment difficulties. Figure 9.18–6 shows a gratifying effectiveness for the same ratio in compensating for differences in particle size (presumably mainly physical heterogeneity).

Andermann and Kemp show that, owing to the nature of x-ray absorption and of scattering, the ratio $\mathbf{I}/I_{scat}$ for a solid sample should be less sensitive to changes in composition than $\mathbf{I}$ itself, which is of course subject to strong absorption effects; and they give supporting experimental evidence. But they agree that the extent to which the use of $\mathbf{I}/I_B$ will compensate the changes

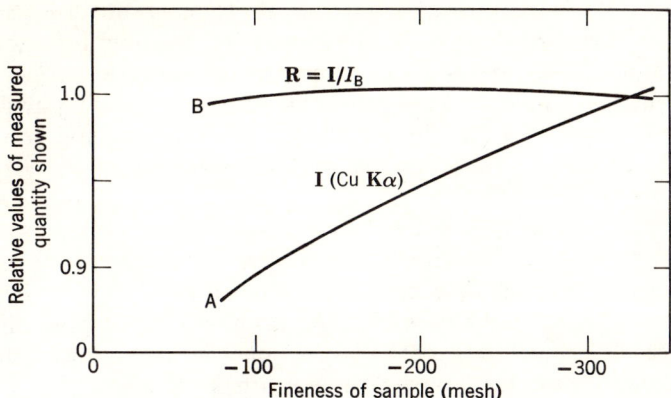

Fig. 9.18–6. Use of scattered background (near 1.4 Å; intensity I_B) as a compensatory standard for variation in particle size. The sample was a powdered ore, presumably a copper ore in which heterogeneity was mainly physical. The figure is in accord with general experience, which is that particle size effects disappear as powders are more finely ground. See G. Andermann and J. W. Kemp, Ref. 66, p. 1306, Fig. 2.

in absorption effects that accompany changes in composition cannot be predicted. It therefore seems safest to take the empirical approach and to establish for each application the scattered wavelength for which the use of I/I_B gives the best results. Enhancement effects cannot be compensated in this way because such effects do not appear in the scattered intensity. Scattering in x-ray emission spectrography reappears below.

We return now to *intrasample* comparisons with discussion of an important method known as "spiking," which is conveniently introduced by reconsidering the work of Pfeiffer and Zemany (Figure 7.12–1). Suppose these experiments had been done with an unknown amount x of initially present zinc on a filter-paper spot, the background N_B for the constant counting interval Δt being known. Suppose that after the measurement of I^x, $n(2)$ μg had been successively added, where $n = 1, 2, 3 \ldots$ so as to make the abscissas x, $x + 1(2)$, $x + 2(2)$, $x + 3(2)\ldots$. Then, in general with Δm in place of 2 μg,

$$\frac{I^x}{I^{x + n(\Delta m)}} = \frac{x}{x + n(\Delta m)} \qquad (9.18\text{–}3)$$

and x could be found by making two intensity measurements, or more if needed for increased precision; graphically, x would be located as the abscissa corresponding to ordinate I^x of a line through the origin. See Figure 9.18–7. Points to be noted: (1) (9.18–3) is a *linear* relationship, which implies absence or constancy of interelement effects; this places a limit on the ratio

$n(\Delta m)/x$. (2) $\mathbf{I}^x$ is the contribution made by the element sought *in the presence of* $n(\Delta m)$; if the contribution by this element is measured for another "unspiked" amount (say, $\mathbf{I}^{x'}$ for x' μg), $\mathbf{I}^x$ must be calculated by proportionality. (3) The method implies a kind of sample (dissolved, fused, powdered—in addition to supported) to which additions can properly and conveniently be made; when, as with solutions, sample measurement is by *volume*, it will be desirable to calculate *amounts* for use in (9.18–3). (4) The method assumes that N_B everywhere has the value obtained on a blank; see (9.10–2) and replace $\mathbf{W}^S$ by micrograms. Naturally, the requirements just given can be relaxed in actual determinations, which is especially justifiable when the precision (as in trace determinations) is low. (5) In principle, all the data in Figure 7.12–1 could have been obtained on one sample; hence the designation intrasample comparison is justified.

Rose and Cuttitta [67] give an excellent demonstration of usefulness of the method. To determine zinc (NBS value, 0.1% by weight) in an opal glass, they eliminated the silica, processed the residue after addition of cellulose by spiking, mixing, drying, grinding, and briquetting; then obtained $\mathbf{I}$ (Zn $\mathbf{K}\alpha$) on the spectrograph. They treated their results as in Figure 9.18–7. Results: 100-mg sample, 665 ppm of Zn; 50-mg sample, 670 ppm; atomic absorption gave 672 ppm. Their experimental points lay so well on straight lines that there is no point in reproducing them here.

Rose and Cuttitta [67] also determined bromine in saline waters of specific gravities exceeding 1.2 by the use of spiking. Here the matrix produces a pronounced absorption effect that makes an intersample comparison necessary, the standard being a known sodium bromide solution (no other salts). Results were again excellent, and the interested reader will wish to consult their work for a good way of evaluating the background correction.

No attempt has been made above to discuss the different ways of using calibration (working) curves, or of calculating results. These matters are old hat in analytical chemistry, and to introduce them here would be to increase the risk of confusion that is inherent in the great versatility of the comparative method as applied to x-ray emission spectrography. No neat delineation of the method is possible, but the following summary may be helpful:

1. *Intensity ratios* give more reliable results than do absolute intensities. The increase in reliability usually more than compensates for the inevitable increase in standard counting error s_C (Chapter 8).

2. Intensity ratios are most easily and conveniently measured on multichannel spectrographs.

3. A comparison of intensity ratios for standard and unknown can compensate (remedy) all difficulties in x-ray emission spectrography if standard

SAMPLE PREPARATION. REMEDIES AND COMPENSATIONS

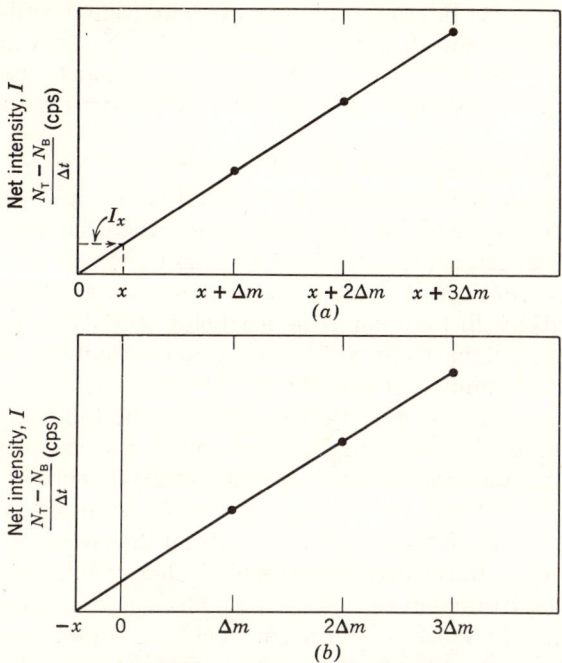

Fig. 9.18–7. Graphical illustrations of spiking technique. The two ways of finding the unknown x are obviously equivalent. (*a*) As based on Fig. 7.12–1. Abscissa is μg present. (*b*) As based on H. J. Rose, Jr., and F. Cuttitta, Ref. 67, p. 31, Fig. 8. Abscissa is μg of "spike."

and unknown are physically and chemically identical, a situation easiest to realize when the samples are homogeneous.

4. Equipment difficulties are the easiest, and difficulties due to heterogeneity the hardest, to compensate (remedy).

5. The different comparisons compensate interelement effects to varying degrees. Enhancement effects can be compensated only if they are produced so as to make comparison possible; a scattered x-ray used for comparison cannot do this. Absorption effects are much easier to compensate and are usually *at least ten times* as important as enhancement effects.

6. The seriousness of the various difficulties depends upon how the samples are prepared and handled. For best results, this should be done so as to minimize the difficulties the comparative method has to compensate.

9.19 The Choice of Internal Standards

In the days of Hevesy [5], the use of internal standards was the best way to get quantitative results. By now, the method has lost its preeminence, largely because proper internal standards are hard to find, and because it is often troublesome to add them to many samples so that they are uniformly distributed. Nevertheless, internal standards continue to be worth discussing, if only for the sharpened insight they give into x-ray emission spectrography.

Hevesy and his colleagues [5] found in their classical work on hafnium (Section 9.1) that intensity ratios for the original sample were often sensitive to composition, even with electron excitation, where interelement effects are small; they consequently used internal standards.

Glocker and Schreiber [43], using electron excitation, tried to determine vanadium in steel with titanium as internal standard. They found that the intensity ratio varied markedly with the tungsten content of the steel so that the apparent vanadium content decreased with the tungsten present. They considered that this deviation might have the following causes: (1) differential vaporization of V and Ti; (2) chemical reactions that converted either V or Ti, or both, into substances that suffer differential vaporization; (3) chemical reactions that destroy uniform distribution of V and Ti in the sample. These and similar processes are possible if, during electron bombardment, the sample is overheated; the possibility that such things will happen is one of the disadvantages of macroscopic electron excitation. Note that difficulties such as these are not interelement effects. These difficulties moved Glocker and Schreiber [43] to carry out the first extensive theoretical and experimental investigations of x-ray excitation, and of internal standards as a means of compensating interelement effects. This investigation may fairly be regarded as classic. Among several more recent investigations, that by Adler and Axelrod [68] deserves particular mention because their two-channel spectrograph permitted them to measure simultaneously the two intensities whose values determine the intensity ratio.

We have indicated above that an internal standard will not be completely effective if a change in composition influences the intensity ratio. Such influences could arise from differences in the extent to which the two analytical lines are absorbed or from differences in the extent to which the two lines are excited. Let us examine the way in which the addition of a third element A, the disturbing element, could influence the intensity ratio $I/I_{i\text{-}St}$ (9.18–2). In considering influences arising from differential absorption, we are concerned with the location of the absorption edge of A relative to the two analytical lines. In considering differential excitation, we are concerned with the location of the characteristic lines of A relative to the absorption edge of the element determined (E) and that of the internal standard ($i\text{-}St$). Four possible situations are shown in Figure 9.19–1, from which the qualitative influence on the intensity ratio is easy to deduce in each case. The following observations are in order:

SAMPLE PREPARATION. REMEDIES AND COMPENSATIONS

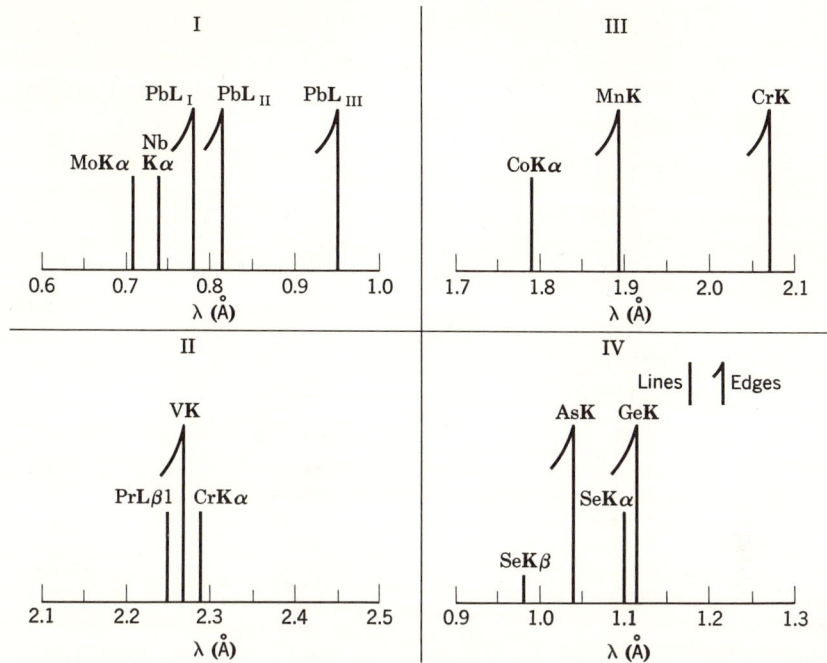

Fig. 9.19–1. The effect of absorption edges and strong emission lines on the relative intensities of the analytical lines of two elements. (I) An absorption edge on the long-wavelength side of the analytical lines. Lead was shown to have a minor effect on the Mo-Nb ratio. (II) An absorption edge between the two analytical lines. The addition of vanadium decreased the Pr-Cr ratio. (III) A strong line on the low-wavelength side of the two analytical absorption edges. Addition of cobalt had little effect on the Mn-Cr ratio. (IV) A strong line between the two analytical absorption edges. The effect of added selenium on the arsenic and germanium lines is shown in Fig. 9.19–2. See I. Adler and J. M. Axelrod, Ref. 68.

1. Interchanging the wavelengths of the analytical lines in Figure 9.19–1 will reverse the effect of the disturbing influence.
2. Cases II and IV are more likely to be serious than the other two.
3. The concentration of A helps determine how seriously the effectiveness of the standard is reduced by the presence of A.
4. An effect of the kind in Case I is exerted by the sample less E and i-St. The effect is reduced as the difference in wavelength between the analytical lines decreases, because differential absorption of the two analytical lines is then reduced.
5. The actual effect may be a superposition of several of the effects in Figure 9.19–1. For example, A usually has more than one characteristic line, and this will lead to such a superposition. Also, more than one disturbing element may be present.

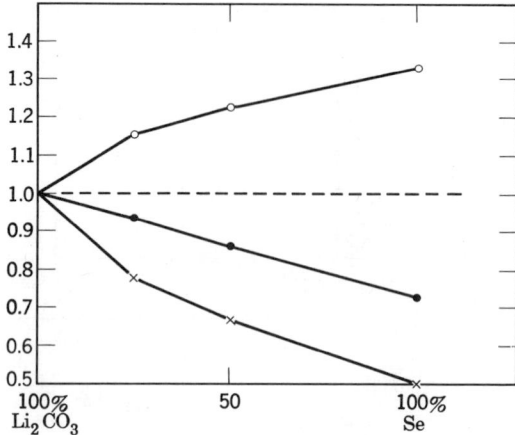

Fig. 9.19–2. Effect of added selenium on the analytical lines of arsenic and germanium. The ordinate in this figure is, for the upper curve, the normalized Ge-As intensity ratio and, for the lower curves, the normalized absolute intensity. The abscissa is the composition of the diluent added to the base material. The relation of analytical lines and absorption edges is shown in IV, Figure 9.19–1. Open circles = Ge Kα/As Kα; closed circles = Ge; crosses = As. Courtesy of I. Adler and J. M. Axelrod, Ref. 68.

6. Because the two analytical lines differ in wavelength, an internal standard can never compensate interelement effects completed. If Cases II and IV are avoided in selecting an internal standard, the use of such a standard will usually prove satisfactory. Special cases may require special calibration curves run with the disturbing elements present.

Data obtained by Adler and Axelrod [68] for an interesting example are shown in Figure 9.19–2, where the influence on the intensity ratio is a superposition of the following effects: (1) differential absorption of the analytical lines (germanium Kα and arsenic Kα) by the sample less arsenic and germanium, which varies in composition as shown on the abscissa (Case I); (2) differential excitation of these lines by selenium Kβ, small or negligible (Case III); (3) Excitation of germanium Kα by selenium Kα (Case IV).

This illustration shows that compensation by internal standards can be less than perfect. It is therefore not surprising that they have lost ground to less troublesome, though more approximate, methods of comparison.

9.20 More About Scattered X-Rays

Please see Sections 1.13–1.16 inclusive, 6.7, 9.10, and 9.16. Here we present additional information about scattered x-rays that has only specialized significance, for it applies only when light elements predominate in the sample. They may predominate in the matrix, binder included. The information was not previously presented to keep general discussions simple.

Attempts to calculate scattering coefficients have been made chiefly in connection with x-ray diffraction, where scattering is the principal concern. Such calculations are too complex (see Section 6.7) and of too little significance for analytical chemistry to warrant inclusion here. All the analytical chemist needs to know about scattering *in general* is that it decreases in importance relative to photoelectric absorption as Z increases and λ decreases. Inasmuch as photoelectric absorption increases as Z^4 and λ^3, this statement is not surprising. See (1.13–1) and (1.13–2). The position just taken is all the more justified because the analytical chemist concerned with intensities of scattering will usually have to measure them anyhow.

Compton scattering is not of first importance in analytical chemistry, yet it is receiving increased attention, not only as a cause of spectral interference but as an aid in establishing composition. Compton scattering, it will be remembered, increases relative to Rayleigh scattering with decreasing wavelength (Figure 1.14–3) and decreasing atomic number, and with increasing scattering angle ϕ (1.15–1). (The change with Z is attributed to an increasing ratio of *free* to *bound* electrons.) The ratio of Compton to Rayleigh scattering can thus be very large (short λ, small Z) or it can approach zero (long λ, large Z).

The significance of the Compton effect in x-ray emission spectrography was shown in part by Johnson and Stout [69], who were interested in trace determinations of the heavier metals in matrices of biological origin— that is, in the kind of work where the scattering being discussed is most likely to be troublesome. They recorded the spectrum of x-ray-tube lines scattered by each of four substances (paraffin, H_3BO_3, $CaSO_4$, and $BaSO_4$) for each of three different targets (Cu, Mo, W). This kind of experiment was mentioned earlier in the book as a way of testing the spectral purity of an x-ray source. Data for a tungsten target tube appear in Figure 9.20–1.

The results of Johnson and Stout are in accord with what is known of x-ray scattering. They obtained $\Delta\lambda_C = 0.0246 \pm 0.0033$ Å as the (approximate) Compton wavelength of the electron, the more precise value of which can be found from (1.15) as follows:

$$\Delta\lambda_C = \frac{h}{m_0 c} = 0.0242621 \pm 6(10^{-7})\,\text{Å} \qquad (9.20\text{--}1)$$

This equation applies for $\phi = 90°$. For the x-ray emission spectrograph of Johnson and Stout, as for most others, $\theta_1 + \theta_2 \simeq 90°$, these being the incident and the take-off angle at the sample. Johnson and Stout measured $\Delta\lambda_C$ directly because $\phi = 180 - (\theta_1 + \theta_2)$ degrees.

Also as expected, the Compton peaks they found are broader than the

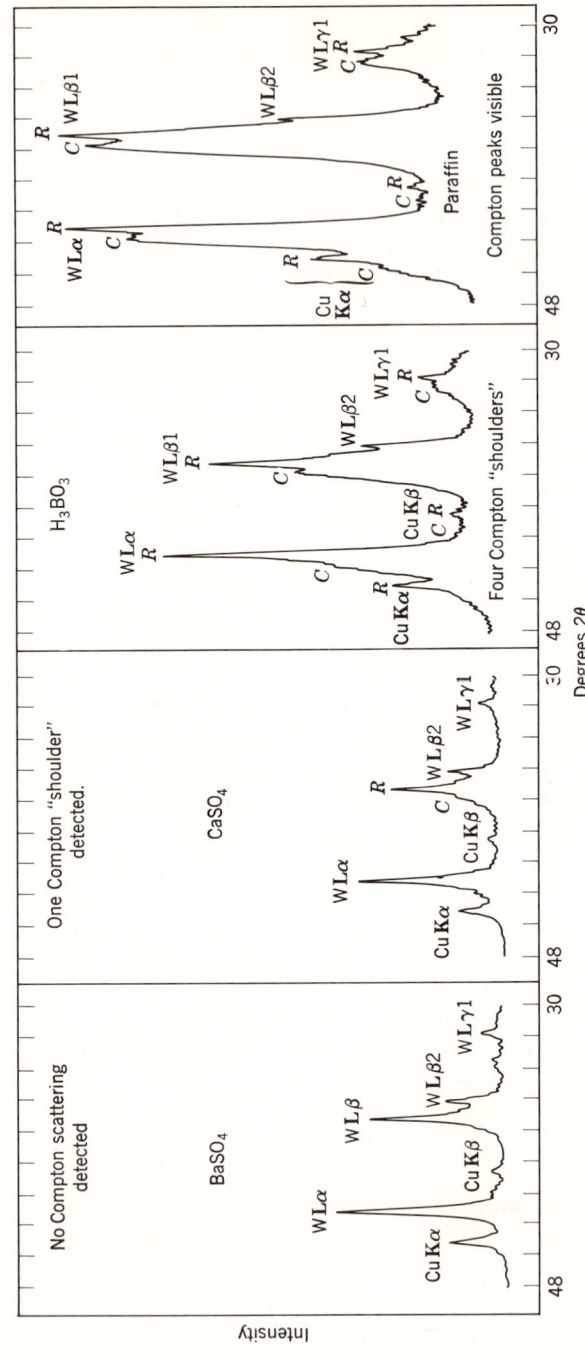

Fig. 9.20-1. Spectrum scattered by each of four substances exposed to x-ray beam from a tungsten-target Coolidge tube operated at 52.5 kV and 45 mA. The peaks represent characteristic lines from the tube in which copper was present as impurity. Rayleigh scattering (R) and Compton scattering (C) are shown. Intensities not given, but full deflection on rate-meter scale was 1000 cps. See C. M. Johnson and P. R. Stout, Ref. 69, p. 1922, Fig. 2.

corresponding Rayleigh peaks. Although the results show that spectral interference by these peaks is a hazard when traces must be determined in light-element matrices, it must not be forgotten that the abscissas in Figure 9.20–1 are compressed, and that peak intensities are low. Compare Figures 9.20–1 and 9.10–2. When interference does occur, changing the tube voltage, filtering the x-ray beam from the tube, and changing to a different target are possible remedies.

The ratio of Rayleigh to Compton scattering, namely

$$\mathbf{R}_{scat} = \frac{I_R}{I_C} \qquad (9.20-2)$$

can be used to establish the composition of a *binary* compound because $W_1 + W_2 = 1$ can be used in the calculation of results. The similarity in principle to the analysis of brass by Eddy and Laby is evident: see (9.18–1) and consider the Rayleigh and Compton peaks as *analogues* of the $K\alpha$ lines emitted by brass.

Dwiggins [70] was the first to take advantage of this situation. His experimental conditions appear in Table 9.20–1 and his results for pure hydrocarbons in Table 9.20–2.

Table 9.20–1. Experimental Data for Determination of C/H Ratio by X-Ray Scattering [70]

Bragg crystal, NaCl. Scattered line, W $L\alpha1$ (1.476Å); see Figure 9.20–1.
Liquid samples only. Norelco spectrograph. Scintillation detector.
Measured: Δt to 51,200 counts for Rayleigh and Compton peaks, and for background.
Counting times: 20 to 60 sec. Total time, determination and calculation, 20 min.
2θ values: Rayleigh peak, 30.60°; Compton peak, 31.21°; background, 29.75°.
Background cps subtracted from each peak cps.
Practical $\Delta\lambda$ exceeds theoretical from (9.20–1), no doubt because peaks are broadened in spectrograph.
Results: $\mathbf{R}_{scat}$ increases linearly with % C by weight above 80% C. Calibration data acceptable for at least 3 weeks.

Petroleum of course contains elements other than carbon and hydrogen, and the presence of additional elements (e.g., N and S) can complicate the analytical method. So long as nitrogen was below 1%, it could be considered as so much carbon, and the calibration curve of Table 9.20–1 could be regarded as giving the *sum* C + N in percentage by weight. Subtraction of the independently determined nitrogen content then gave the carbon. Sulfur is another story. The (larger) effect of its presence had to be found empirically by using standards of known sulfur contents. The result was

Table 9.20–2. **Typical Calibration Data and Results for Method of Table 9.20–1 Applied to Pure Hydrocarbons**

	Calibration data			
Hydrocarbon	R_{scat}, Measured	Carbon, %	R_{scat}, Calculated	Percentage Deviation in R_{scat}
n-Heptane	0.8359	83.90	0.8335	−0.29
Cyclohexane	0.9181	85.62	0.9122	−0.64
cis-Decahydronaphthalene	0.9640	86.87	0.9694	0.56
50% cis-Decahydronaphthalene, 50% toluene	1.0647	89.055	1.0694	0.44
Benzene	1.2149	92.25	1.2157	0.06

Calibration curve: % C (by weight) = $21.845 R_{scat} + 65.693$

Note: Curve based on extensive data including those above. Deviations will occur at hydrogen contents above 20% by weight. Measured values are averages of quadruplicate determinations.

	Determination of Percentage Carbon in Hydrocarbons		
	R_{scat}, Measured	Carbon, %	
		Detd.	Theor.
Methylcyclohexane	0.9112	85.60	85.63
cis-Decahydronaphthalene, after 3 weeks	0.9736	86.96	86.87
Xylene	1.1395	90.58	90.50
Toluene	1.1745	91.35	91.24

simple: a subtractive term (− 2.725% S) had to be included in the calibration equation of Table 9.20–1.

The method does all that one could reasonably ask. Accuracy and precision are satisfactory—several tenths per cent. Precision is near the standard counting error. Though probably not quite as precise as the best macrocombustion method, the x-ray method is so much less costly as to be generally preferable. Time for a determination could be reduced below 20 minutes with a sequential or multichannel spectrograph that can handle more than one sample [71].

Dwiggins' method is to be compared with the absorptiometric methods of Chapter 3. It might profitably be combined with absorptiometry and/or x-ray emission spectrography and extended to include the determination of other light elements.

For the use of $\mathbf{R}_{scat}$ and of I_C to compensate absorption effects, see Refs. 72 and 73. As I_C is the intensity of *modified, incoherent* scattering, it is preferable to I_R (*unmodified, coherent*) for this purpose when diffraction peaks (also *unmodified, coherent*) from macrocrystalline solids interfere. The relative usefulness of scattered-intensity ratios and of simple scattered intensities [66] to compensate absorption effects needs further study.

DIFFICULTIES AND KIND OF SAMPLE

9.21 Introduction

The principal objective here is to guide the analytical chemist in choosing which kind of samples (Section 9.17) had best be placed in the spectrograph to give the information needed. No distinction between *original* and *prepared* samples need be made here. It is of course understood that preparation involves change and perhaps destruction of the original: the statement that x-ray emission spectrography is ordinarily nondestructive applies only to what happens in the spectrograph.

To make the guide complete, sample preparation would have to be discussed in detail because its costs and the risks attending it are important factors. This is not possible here because such discussion must be brief and restricted to cases of special interest. Recourse to the original literature via the last appendix is recommended. So is the consulting of other books, especially of Refs. 13, 14, and 16.

9.22 Diluted Samples

Dilution can remedy difficulties due to heterogeneity and can mitigate interelement effects. The degree of dilution and the nature of the diluent determine the success of the remedy. Loss of intensity is a price that must always be paid.

Dilution is usually done with a substance relatively transparent to x-rays; for example, water [74], other solvent [75], cellulose [67], starch, starch and lithium carbonate [76], alumina, aluminum powder [77], or borate glass [78, 79]. Dilution can be brought about by dissolution to give liquid solutions (water or other solvents) or solid solutions by fusion: obviously this remedies all troubles due to heterogeneity. Dilution with a solid and relatively transparent binder (e.g., starch) will not always eliminate these troubles. Dilution offers opportunities for easy addition and uniform distribution of internal standards, and for chemical treatment (e.g., removal of objectionable substances) to make the x-ray determination easier or more reliable.

The extent of dilution required to reduce interelement effects to a tolerable level can be estimated as follows. First, enhancement effects, being usually far less important than absorption effects, will often be at a tolerable level when absorption effects have arrived there. (Assumption: no diluent produces an enhancement effect.) The extent of dilution required to make absorption effects tolerable can be estimated from the values of **a** and a before and after dilution; see Section 9.12, especially (9.12–2), and the example of aqueous sodium tungstate solution. The example shows the variation in critical depth with dilution.

Dilution ordinarily stabilizes a by reducing it to a value approaching a for the diluent. Claisse [78] originated the addition of compounds of heavy elements such as barium to stabilize a by *increasing* it. This procedure is useful in special cases where the loss in intensity of the analytical line can (or must) be tolerated.

Section 9.20 and earlier discussions of scattering show that a relative increase in scattered x-rays (hence in background) is one consequence of using a relatively transparent diluent, which must perforce contain only light elements. This disadvantage is mitigated somewhat because scattering approaches constancy with increasing dilution and increasing homogeneity.

Dilution can eliminate all difficulties deriving from variation in sample position or in x-ray intensity incident upon different portions of sample surface, such as results when one portion is nearer the target of the tube than another. These difficulties, which are more serious with heterogeneous than with homogeneous samples, were introduced in Section 7.12 and illustrated by Figure 7.12–2; it was pointed out that they can be alleviated by the spot-test method, which gives a supported sample. With dilution, it is possible to achieve homogeneity and to make certain that identical area will be exposed at identical height in all determinations. Whether dilution or making a thin film of the sample is the better remedy depends upon the problem.

It was mentioned in the discussion of Table 9.11–3 that dilution offers a way out of the kind of difficulty illustrated by the table. Dilution obviously can eliminate also difficulties (see below) associated with the surfaces of solid samples.

So much for a general discussion of the difficulties that can ideally be eliminated by dilution. We turn now to the individual characteristics of the various methods of producing diluted samples.

DILUTION WITH LIQUIDS. If the diluted sample is to be a liquid solution, there must be a cell to contain it. When evaporation from an open cell does not appreciably lower the liquid level (see Chapter 8) at which the x-ray

DIFFICULTIES AND KIND OF SAMPLE

beam is incident, it may be possible to dispense with a cell window provided no significant concentration change occurs. If a window is needed, there will be filtering and attenuation of x-rays. See Section 2.2, especially Table 2.2–1. Gas bubbles within the cell cannot be tolerated in the x-ray path. The settling of a precipitate is usually undesirable and can be serious. Consider, for example, that the action of x-rays precipitates a compound of lead that settles out of a solution in which that element is being determined. If the x-ray determination is made at the upper level of the liquid, the result will be too low; if at the bottom, as in an inverted-sample spectrograph, it will be too high. Heating of a solution in a cell by x-ray absorption could lead to trouble, but beam intensities high enough to cause expansion or boiling of a liquid sample in a reasonable time are not needed in the usual run of determinations. Nevertheless, solid samples are less likely to be affected by x-ray beams in ways that change cps values: even if chemical change occurs, the reaction products (if solid) will remain in place.

Many articles deal with liquid samples. Two, one by Bertin [80] and another by Campbell, Leon and Thatcher [81], give good overviews. Campbell [82] also describes an apparatus for continuous determinations on solutions. Cells have been made in many ways and of many materials; the requirements they and their windows must meet are fairly obvious. Disposable cells and windows suitable for most work can be bought. Mylar and polypropylene are the most common window materials.

Among cells of special interest, that by Bertin and Longobucco [83] is leakproof, practical, and generally useful in other than inverted-sample spectrographs; see Figure 9.22–1. When a vacuum path is mandatory in the determination of light elements, freezing of a liquid solution by use of liquid nitrogen, necessarily troublesome, may be desirable. Chan [84] gives a suitable cell. Such determinations can also be carried out in a cell with a window provided the position of the window is adequately reproducible and bubbles of gas do not interfere. In an ingenious cell of General Electric Company design, these conditions are met by providing a lower chamber that can remain empty when the vapor pressure of the diluent is low, or can be partially filled with a liquid of suitable vapor pressure.

The Philips Inverted-Sample Three-Position Spectrograph (see Chapter 10) is perhaps the best instrument for determinations on run-of-the-mill liquid samples.

DILUTION BY FUSION. Fusion has long been recognized as an indispensable way of "opening up" minerals for the determination of important constituents. Claisse [78] realized that fusion to give a glass disk could be used at once to give a diluted sample suitable for the spectrograph, to eliminate troubles caused by heterogeneity, and to cope with interelement effects by stabilizing a in (9.12–2). Accordingly, he developed the fusion of

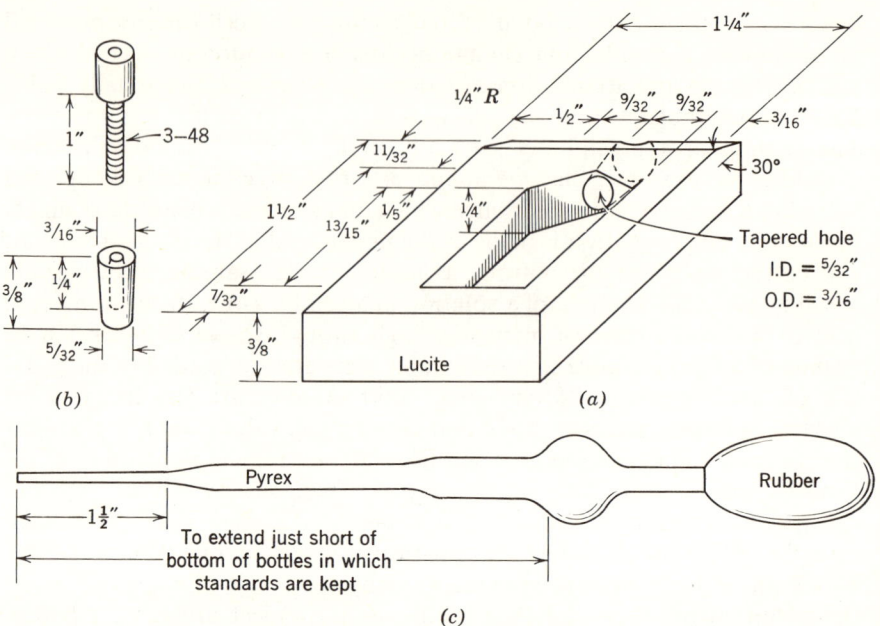

Fig. 9.22–1. Cells for liquid samples. Replaceable Mylar window (not shown) carefully stretched and cemented on with Pliobond rubber cement under pressure. Tapered plug inserted and withdrawn by use of Allen cap screw shown above it in figure. Pipet can fill or empty cell (volume, about 2 cc) in single operation. Cells have been used successfully up to 50 times under severe conditions. Simpler designs feasible. From E. P. Bertin and R. J. Longobucco, Ref. 83, p. 450, Fig. 5.

minerals (other than sulfide minerals) with fluxes based on borax (100 mg sample to 10 g of borax); when graphite or sulfides are present, oxidation must precede the borax fusion. Internal standards (when needed) or materials intended to stabilize a by increasing it may be added before the fusion. Claisse tested his methods on known minerals and plotted counting rate for the analytical line against content of element sought as established by chemical methods. Usually, the points lay close to a straight line, points from which established the "x-ray results" for comparison with the chemical. Excellent agreement between the two is shown in Table 9.22–1, which contains results for 11 sulfide ores. Experience shows that many determinations can be done satisfactorily without either internal standards or additives to increase a.

Drummond's fusion procedure [85] resembles that of Claisse and like it gives a sample capable of being inserted directly into the spectrograph. Original sample (3 g) and fused lithium tetraborate (9 g) are fused in a 97 Pt–3 Au crucible at 1250° long enough (10 minutes) to give a glass button

DIFFICULTIES AND KIND OF SAMPLE

Table 9.22–1. Comparison of Results for Sulfide Ores

Iron, %		Zinc, %		Lead, %		Copper, %	
Chemical	X-Ray	Chemical	X-Ray	Chemical	X-Ray	Chemical	X-Ray
11.8	11.6	(x)	(x)	0.89	0.80	0.26	0.22
11.8	12.0	11.7	11.7	49.1	49.0	0.07	0.27
17.6	17.6	13.0	13.3	35.6	35.8	0.07	0.08
24.8	24.8	14.0	13.9	17.4	18.2	0.08	0.24
27.7	27.7	5.5	5.7	26.1	26.0	0.12	0.24
29.8	29.8	8.8	8.7	5.74	5.75	0.45	0.45
34.2	34.7	2.6	2.6	1.05	1.05	0.16	0.00
36.5	36.5	3.3	3.9	1.30	1.60	(x)	(x)
36.9	36.8	6.9	6.9	3.44	3.55	(x)	(x)
40.2	40.2	4.7	4.7	1.40	1.35	1.40	1.40
38.6	38.9	4.5	4.3	0.78	0.90	0.91	0.95

on cooling. When the crucible is inverted at room temperature, the button drops out, whereupon it is ground and reheated in the crucible for 5 minutes at the same temperature. It is next annealed at 650° C for 30 minutes and cooled with the furnace. The flat side is polished with fine silicon carbide before being spectrographed. Lanthanum oxide can replace part of the tetraborate if a is to be increased. Some procedures do not mention polishing; others call for metal retaining rings during solidification of the sample; masking to expose always the same area in the spectrograph is often desirable.

Rose, Adler, and Flanagan [86] developed a fusion procedure applicable not only to silicate rocks but also to more complex minerals such as niobates and tantalites.

To stabilize a without unduly reducing intensity (hence sensitivity), 0.125 g La_2O_3 is fused at 1100° C for about 10 minutes with 0.125 g mineral and 1.000 g $Li_2B_4O_7$ after a preliminary heating in the graphite crucible at 750° C as needed to ensure safe removal of CO_2 and H_2O. After being cooled, the bead is removed from the crucible and enough boric acid is added to it as a normalization measure to give a combined weight of 1.300 g, prior to grinding at high speed in a special mill with balls and end plates of tungsten carbide. The ground powder is pressed into pellets or made into briquets; final pressure about 50,000 psi.

Here, an interpolation. Methods of sample preparation overlap, and several older methods may be combined to make a new. Nomenclature is inevitably confused and confusing. As an example, pellets (little balls) and briquets (little bricks) probably have issued from the same press with the same dies, likely as not cylindrical. Rather than attempt accurate description,

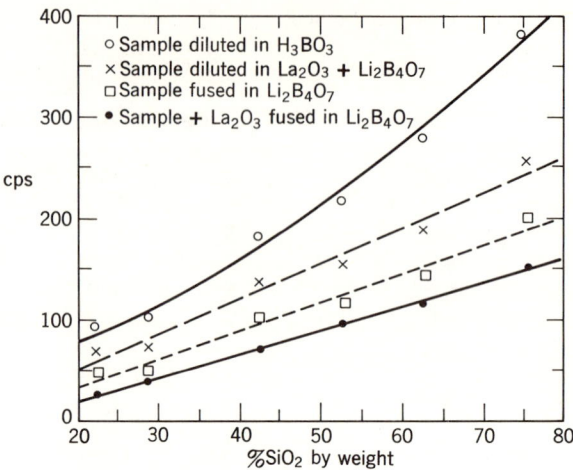

Fig. 9.22-2. Results of various ways of stabilizing a to mitigate difficulties due to absorption effects in the determination of silica. See I. Adler, Ref. 14, p. 122, Fig. 6, or H. J. Rose, I. Adler, and F. J. Flanagan, Ref. 86.

we repeat our injunction to consult the original literature. We arbitrarily speak of pelleting when nothing is added, and of briquetting when pressure is applied in other cases (see below).

Adler [14, p. 125] mentions that double-layered briquets are useful in connection with the fusion procedure under discussion. These are made by handpressing boric acid in a mold, spreading the ground fused sample (see above) evenly thereupon, and then completing the briquetting.

DILUTION WITH SOLIDS. As a first example, we cite the use of boric acid to make briquets in the determination of silica [14, pp. 121 and 122; and 86]. Figure 9.22–2 is a valuable comparison of the various methods of stabilizing a: (1) by use of a transparent solid diluent (H_3BO_3); (2) by fusion with a transparent flux ($Li_2B_4O_7$); (3) by use of an additive (La_2O_3) to increase a in each of the other cases. The results are not unexpected and speak for themselves.

A second example is the work of Rose and Cuttitta [67] already cited in Section 9.18. Here the elements to be determined were concentrated, and the concentrated solution absorbed by cellulose (made from chromatography paper) that acted as solid diluent. Briquetting followed.

An early example of briquetting with an *organic radical* and a *co-precipitate* as solid diluent is the work of Fagel, Balis, and Bronk [87]. For an investigation of the corrosion of stainless steels and Monel metals by molten carbonates, samples of alkali-metal carbonates containing 10^{-3} to $10^{-2}\%$ each of iron, manganese, nickel, copper, and chromium had

DIFFICULTIES AND KIND OF SAMPLE

to be analyzed. Direct x-ray emission spectrography was unsatisfactory, mainly because of the large negative absorption effects. The samples were dissolved and the heavy metals were precipitated with 8-hydroxyquinoline, co-precipitated aluminum acting as carrier. The dried precipitates were ground, briquetted at 6000 psi, and compared in the spectrograph with standards similarly precipitated from stock solutions. Satisfactory semiquantitative results were obtained without difficulty. The co-precipitant obviously serves also to *collect* the metals being determined.

Ion-exchange papers can also act as solid diluents. Ion-exchange x-ray emission techniques, begun by Grubb and Zemany (Section 7.18), developed among others by Luke [88], now rely largely upon paper loaded with about 50% ion-exchange resin or even upon cellulose modified to have ion-exchange properties. Campbell, Green, and Law [89] review the subject in an excellent article that describes a novel application in which a radio-active isotope ($_{79}$Au^{195}) monitors the extent to which the element sought (Au) is collected from solution, a device at once convenient and time-saving. The ion-exchange-paper disk, when ready for the gold determinations, is first mounted in front of the scintillation detector of the x-ray spectrograph. When the γ-ray count has been made, the x-ray count for the analytical line of gold is taken in the usual way. Comparison between standard and unknown is possible because the same quantity of $_{79}$Au^{195} is added to both, on which basis

$$\text{Au}_{ore}\,(\text{ppm}) = \frac{\text{Au}_{St}(\mu g)}{\text{Wt ore (g)}} \times \frac{(\text{cps, }\gamma\text{-ray})_{St}}{(\text{cps, }\gamma\text{-ray})_{U}} \times \frac{(\text{cps, x-ray})_{U}}{(\text{cps, x-ray})_{St}} \qquad (9.22\text{--}1)$$

The symbols are those of this book. The multiplication may be regarded as involving x-ray counts per γ-ray count for standard, and for unknown. About 10% of the gold was collected from 300 cm^3 of solution by the disks (diameter, 1.3 cm) in 16 hours. The results (in conventional units) are gratifying as Table 9.22-2 shows.

Table 9.22-2. **Results of Gold Assays (oz/ton) by Three Methods [89]**

Sample	X-Ray Emission (see text)	Atomic Absorption	Fire Assay
USBM ref. ore	0.215	0.206	0.206
USGS gold quartz st.	0.0776	0.0774	0.0761

The methods are not compared in other respects, but the x-ray procedure will doubtless be the best in some cases, especially when more than one metal must be determined.

9.23 Bulk Solids

The kind of sample considered here is illustrated by the two most common examples: an *object* or a solid *aliquot*. The former might be a manufactured object, a part thereof, a rock, or an artifact; the latter, part of an ingot, a coupon, a button cast from a melt, or a specimen of plastic. Objects generally go into the spectrograph as original samples mounted in plastic if necessary and rotated if needed; the experience of Bertin's laboratory [83, 90] (see also Section 7.14) is an excellent guide. As regards the needs for homogeneity, proper shape, and smooth surface, the aliquots have much in common with the fused samples of the previous section.

Much of this chapter and of Chapter 7 is so obviously applicable to determinations on bulk solids as to make repetition unnecessary. Among the important questions for such samples are the following. Does the surface have the bulk composition? If not, which composition is of interest? Both? Is there surface heterogeneity, or bulk heterogeneity, or are both heterogeneous? Even if both are homogeneous, does x-ray emission spectrography, which can be carried only down to the critical thickness, give the composition of the bulk? Should a bulk sample be run "as is," or is surface preparation advisable? Are localized determinations needed? If so, is masking useful? Should the sample be rotated? Is it wise, if feasible, to dissolve or fuse the sample and make determinations as in Section 9.22?

Obviously, internal standards or diluents cannot be added to bulk samples. Obviously, surface preparation must not contaminate the sample. Obviously also, surface difficulties will tend to be more serious when light elements are present. (Such difficulties will be less serious when the analytical lines are **K**, not **L**—an argument for 100-kV excitation of heavier elements.) It is less obvious that there can be surface difficulties induced by the careful surface preparation of "homogeneous" samples.

It is as embarrassing to define a "homogeneous" sample as it is to define a "monochromatic" x-ray beam. Even the purest bulk metals have grain boundaries; alloys naturally offer greater chances for complications. The work of Michaelis and Kilday [91] shows what can happen and what needs to be done. For example, if surface preparation is by grinding, the kind and size of the abrasive are important; so is whether the grinding is wet or dry. Figure 9.23–1 shows that grinding and the orientation of grinding marks can affect analytical-line intensity.

When there are segregated constituents in what might be considered a "homogeneous" alloy (e.g., Pb in hard steels), the softer constituent may be smeared over the sample surface by grinding, and the intensity of its analytical line unduly enhanced. Electrical (electrolytic or spark) machining seems promising for such cases. Brech [92] recommends that a servomechanism be used to keep constant the electrolyte spacing between the sample

DIFFICULTIES AND KIND OF SAMPLE

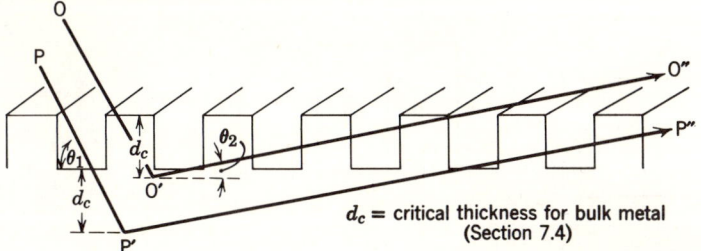

Fig. 9.23–1. Effect of surface finish on analytical-line intensity, oversimplified. Monochromatic x-ray beams OO′ and PP′, of equal initial intensity, penetrate to critical thickness for *smooth* bulk metal. The intensity of OO′ on emergence will be greater than that of PP′. The effect will depend upon the values of θ_1 and θ_2, upon the orientation of the grinding marks relative to the beam, and upon other factors. The concept of critical thickness becomes blurred. Rotation of the sample will stabilize the effect. Grinding with finer abrasive will reduce or eliminate it.

and a rotating disk. With electrical conductivity thus stabilized, the machining produces an "unsmeared," undistorted, smooth sample surface.

Masking of bulk solids is often useful, especially for standards, because it enables one to present always the same area to the x-ray beam. Localized determinations, those with the electron microprobe being the extreme example, are useful to establish the degree of heterogeneity of the sample surface.

9.24 Powdered Samples

Powders may be dry, or they may be suspended in liquids to form slurries, in which form they can become samples for x-ray emision spectrography with mineral assay or process control as the objective. When, as is usual, slurries flow while the x-ray determinations are made, an averaging over time can occur.

With powdered samples, heterogeneity is at its worst (see Section 9.17). Figure 9.24–1 is intended as a qualitative guide. The six kinds of samples in the figure, each of which contains copper and aluminum in equal volume, may be regarded as samples that fit into a spectrograph, to which considerations developed earlier in this chapter are applicable. Only examples based on absorption effects will be given. If the critical depth for each element is exceeded throughout, then 100% Al should be the spectrographic result for samples 1 and 2, with a result very near 0% for sample 3. With *vertical* incidence of the polychromatic beam and *vertical* take-off of the analytical line, samples 4 and 5 should give 50% Al by *volume*. If incidence (from the left) is at 60° and take-off (from the right) is at 30°, sample 5 should show a higher % Al than sample 4, and sample 6 could be intermediate. In each case pure Al is the standard for the determination.

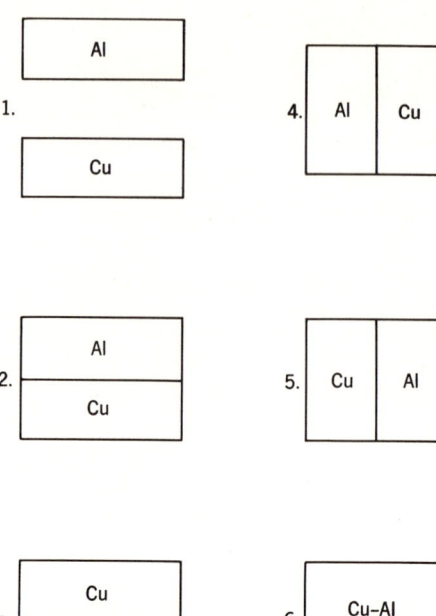

Fig. 9.24–1. Samples for the visualization of difficulties attending the x-ray emission spectrography of powders. A seventh sample composed of the two metals ground so fine as to make it indistinguishable spectrographically from sample 6 might have been included. But there are no sure guidelines to how fine such grinding must be or to whether all components can be satisfactorily ground. For unknown samples, empiricism is the only reliable guide—grinding to $(-270)\,(+325)$ mesh would seem adequate in most cases.

Now, imagine these to be not massive samples, but grains in a sample that has been ground or otherwise mechanically reduced. The same kinds of considerations apply, but one must now ask questions such as: are all grains the same size? the same shape? Uniformly packed, or are there voids, or segregations? It seems clear that the differences for massive samples can be modified by grinding and that sample 6 represents a limit. But, see caption of Figure 9.24–1.

If one wishes to approach reality closer, he must also ask: are enhancement effects present? If there are more than two components do the others contain an element being determined? What are the mass absorption coefficients of the other components? Can the sample be ground fine enough to approach sample 6 in uniformity of composition? Also, a closer approach to reality would have to include highly heterogeneous substances such as minerals.

The difficulties have not been consciously enlarged. It seems remarkable that theoretical treatments of the analytical-line intensities to be expected from powders have even been attempted, and more remarkable that they have met with success for synthetic mixtures. The first such theoretical treatment, by Claisse and Samson [93], related analytical-line intensity to particle size in void-free binary powdered mixtures. More recently, Berry,

DIFFICULTIES AND KIND OF SAMPLE

Furuta, and Rhodes [94], who summarize work which followed that just cited, have carried out a successful theoretical treatment that considers not only particle size but also packing fraction (i.e., presence of voids), and predicts variations not only in analytical-line intensity, but also in the back-scattered and transmitted x-ray intensities as well; their equations seem applicable also to slurries. The understanding thus obtained will help in the formulation of rules for increasingly reliable determinations in the field by portable x-ray spectrographs with radioisotopes as energy sources for the excitation of analytical lines.

Only a few of many existing references that deal with powdered samples will be mentioned. Gunn [95] showed for iron particles suspended in hydrocarbons that Fe $K\alpha$ intensity equivalent to that for dissolved iron was not reached until the particle diameter fell below 1 μ; at 8 μ, only 65% of the equivalent intensity was observed. An early study of the determination of calcium in wolframite, by Campbell and Thatcher [96], is interesting for these reasons. (1) It deals with samples in which the a of (9.12–2) is stabilized by the presence of a heavy element (tungsten) so that interelement effects are the same at all calcium contents; see Section 9.22. (2) The Ca $K\alpha$ intensity in the $(-270)(+325)$ fraction of both scheelite [$CaWO_4$] and wolframite [$(Fe, Mn)WO_4$] ores decreases with time of grinding; but (as was expected) the decrease was much more marked in the second case. (3) Briquetting of powder with methyl cellulose as diluent gave good results. (4) A final comparison leaves the impressions that x-ray emission spectrography on three kinds of samples (briquetted, fused, aqueous-solution) gave results reliable to $\pm 5\%$ W when the element was present to at least 0.5% by weight, and that these results are more reliable than those from a commercial analyst, by optical spectrography, or by wet methods. Some of the outstanding x-ray emission determinations on rocks will be presented in Chapter 10; as regards kind of sample, Baird, MacColl, and McIntyre [97] concluded:

> Preliminary studies using handpacked, ground rock material (mean particle size 2 microns, determined optically) yielded low precision values, particularly for Si and Al.... Our experience indicates that rocks with a diversity of minerals (micas, amphiboles, feldspars, and quartz) cannot be ground successfully to a uniformly fine powder. Mica, even after 25 hr grinding, sufficient to alter its diffraction pattern, remains flaky. For multicomponent material fusion seems to be required, with the major disadvantage of high dilutions in light-element work.

Madlem's careful, more recent work at Pomona confirms these conclusions [97].

Poole and Holloway [98] made an important contribution to the x-ray emission spectrography of powders when they included provision for the effects of heterogeneity in *regression equations*, a procedure identical in principle with that taken as regards equipment difficulties by Lucas-Tooth and Pyne [48]; see (9.13–4). Rock powders were made by grinding in an agate ball mill, capacity about 3 g; an electron microscope was used to show that a syenite (granite) sample milled for 30 minutes contained only very few grains of area exceeding 1 μ^2. These powders were pressed (7 tons) into disks ("pelletized") in plastic retaining cups. The data punched onto cards for computer handling included results of x-ray determinations on 31 standards (duplicate samples for each). Traces were determined on standards to which 200 ppm selenium or calcium as internal standard had been carefully added. Use of the correction factors based on the regression equations markedly improved the reliability of the results. The method deserves wider application.

Powders are so important in x-ray emission spectrography that the analytical chemist new to dealing with them may welcome the following brief and necessarily imperfect guide:

1. For highest precision, fuse or otherwise dissolve.
2. For usually acceptable precision, pellet or briquet.
3. If loose powders must be run, pack them reproducibly into the cell and smooth the surfaces. Standard grinding procedures, empirically established, are desirable. Inverted-sample spectrographs (Section 9.22) are often useful, especially for free-flowing powders.
4. The best possible comparison of unknowns with standards is always advisable. When warranted, regression methods should be used.

A final word about the packing fraction as a variable. Simple compacting by a brake-shoe-like device is adequate in the automated control (via x-ray emission) of cement manufacture [99], which indicates that the packing fraction should be easy to control even in field determinations if the sample is properly ground. In flowing slurries, such control should be achievable by time-averaged determinations. Of course, high precision will not come easily.

9.25 The Remaining Kinds of Samples

Supported samples were introduced in Chapter 7 in their simplest form: a plate of one metal upon another as substrate. Other forms have also been discussed; for example, residues from solutions evaporated on filter paper or on Mylar; and samples concentrated by ion-exchange. Sometimes it suffices to place powders on Scotch tape. Glass fibers and capillaries may prove useful. Porous materials satisfactory as (mechanical) filters can serve at once to collect the sample and to act as substrate; collectable impurities in gases can be determined in this way. It is occasionally desirable to cover the sample with a thin, protective, polymeric film. It is not possible to anticipate or list all the ingenious ways that might be devised of handling special samples. In general, the thinner the sample, the less uniformly it need be distributed over the substrate. When thickness is great enough so that absorption and enhancement effects begin to appear, rotation of the sample in the plane of the substrate may be useful; but, see Section 9.17.

Converted samples are those subjected to chemical change in order to improve spectrographic determinations. Already mentioned: precipitating or co-precipitating an ion by use of suitable reagent (usually organic), and combining stoichiometrically a light element with a heavier one easier to determine. Ashing a sample is another obvious way of converting it; the resulting powder often makes a satisfactory supported sample.

Radioactive samples, though uncommon in the usual analytical laboratory, are nevertheless very much with us. The main problems, as Mueller, Scotti, and Little [100] have pointed out, are to make sure that the samples are handled safely, and that radioactivity does not lead to an incorrectly increased analytical-line intensity. Proper encapsulation of the sample, effective use of lead shielding in and around the spectrograph, and pulse-height discrimination against γ-rays entering the detector are remedial measures usually required.

9.26 The "Light-Element Problem"

Despite notable progress since 1960, elements lighter than fluorine are still beyond reach of a *standard* spectrograph with *x-ray excitation*. There is no need to recapitulate difficulties already mentioned: these begin at the x-ray source and continue through sample, optical path, Bragg reflector, on into the x-ray detection system, electronic components included. We consequently recognize the "light-element" problem as the most serious that x-ray emission spectrography must face.

Fundamentally, the "light-element" problem is a problem common to *all* elements for analytical lines of long wavelength. But the light elements are unfortunately unique in having only such lines to offer; for example, the **K** spectrum of $_9$F lies near 18 Å—that is, in the ultrasoft region. The problem arises out of the way atoms are built—out of the energy relationships underlying the periodic table. See Chapters 1 and 4.

According to Chapter 4, the energy of a characteristic line is the *difference* in energy between *two* atomic states. Up to now, we have taken this energy to be (virtually) independent of chemical or physical state. This useful—hence justified—generalization rests on the assumption that x-ray spectra are concerned with *core* (inner-shell), and not with *valence* electrons.

The generalization must fail as atomic number decreases. The periodic table begins with $_1$H and $_2$He, neither of which can emit characteristic x-ray lines. But, consider now $_3$Li, the **K** spectrum of which appears when the *valence* electron fills a hole in the *core*; here, the **K** shell. For an atom built as is $_3$Li, it must follow that: (1) the **K** spectrum will be relatively long in wavelength because the **L** electron is attracted far less strongly to the nucleus than in an atom with greater nuclear charge. (2) The **K** spectrum of Li will be diffuse (it is actually a band) because the energy differences between the levels (orbitals) accessible to its **L** (valence) electron are appreciable relative to the energy of the **K** state. (3) Any change in the binding energy—hence chemical state—of the Li atom will change at least the energy of the initial atomic state concerned in the generation of the **K** spectrum. Such complications naturally exist in greater degree for the outer electrons of heavier atoms that afford greater opportunities for changes in chemical partners, coordinated species included; and in oxidation number. The transition metals, in which chemical changes can involve the electrons of inner shells, are noteworthy examples. There must naturally be concomitant changes in absorption edges and in characteristic lines.

Sulfur illustrates the situation. Of the various chemical influences on x-ray absorption, we mention only the wavelength shift of the **K** absorption edge. The wavelength of this edge is shorter the higher the oxidation state of the atom [101], for example, 5.0220 Å for sulfide sulfur in Cr_2S_3, and 4.9976 Å for sulfate sulfur in magnesium sulfate. Superimposed upon the influence of the state of oxidation is a smaller influence traceable to elements chemically combined with the sulfur—the edge in ZnS is found at 5.0156 Å. The shifts in wavelengths are very small. Comparable shifts were observed by Faessler and Goehring [102] in the wavelengths of the S **K** lines. See Figure 9.26–1. In this simple case, wavelength shift and line intensity are both of interest. Theory predicts and experiment establishes 2/1 as the intensity ratio of Kα1/Kα2 for sulfur. Where 3 lines appear in a spectrum in Figure 9.26–1, the presumption is that more than one kind of sulfur atom is present, and that an intensity ratio exceeding 2 results because certain lines coincide. The observed intensities in Figure 9.26–1 are those to be expected on the basis of this presumption, 2 being the known value of Kα1/Kα2 for every kind of sulfur atom.

Except in determinations of the lightest elements, these complexities are not important in x-ray emission spectrography as usually done. Analytical lines virtually unaffected by such complexities can generally be found. Any effect of wavelength shifts encountered can be reduced by taking N_T or $\overline{cps}$ at the empirically established goniometer setting of the peak intensity.

Insight into chemical binding, as distinct from analytical chemistry, is a different story. In principle, such insight can be provided by x-ray absorptiometry (position and fine structure of absorption edges); by photoelectron spectrography (binding energies of ejected electrons); and by highly refined x-ray emission spectrography (change in wavelengths and intensities of characteristic lines). The last method suffers because it measures energy

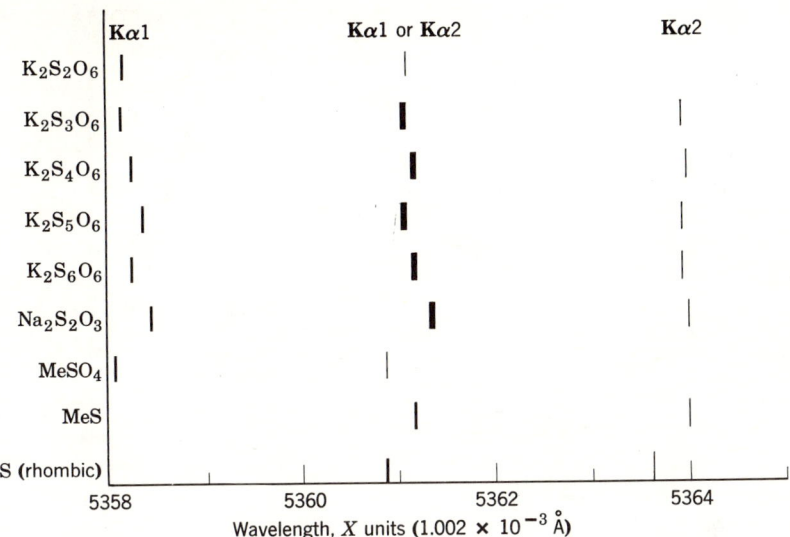

Fig. 9.26-1. The $K\alpha$ lines from sulfur in different chemical states. In the polythionates, the triplet results from the superposition of the $\alpha 1$, $\alpha 2$ doublets generated by sulfur of two different oxidation numbers. The maximum wavelength shift is about 0.004 Å. See A. Faessler and M. Goehring, Ref. 102.

differences between *two* atomic states while the others can give directly the energy with which an ejected electron was held. The *difference* ought to be less sensitive to changes in chemical binding than a *directly measured* value. The modern photoelectron spectrograph, magnetic or electrostatic, seems certain to dominate this field. See Chapter 10.

Qualitatively, then, the situation is this. Certain difficulties and complications in x-ray emission spectrography increase with the wavelength of the analytical line so that for any element the usefulness of characteristic lines as analytical lines often decreases in the order **K, L, M** ..., when all spectra can be conveniently excited. As the **K** lines of light elements are already long in wavelength, these elements are in the worst position. Consideration of absorption effects and of surface effects further darkens the prospect. The lighter the element sought, the greater the chance that the sample will contain others that have higher mass absorption coefficients. The addition of water to sodium tungstate solutions has a marked *positive* absorption effect on the **L** line of tungsten, but a marked *negative* effect on the **K** line of fluorine. The need for careful attention to sample surface is evident from the following experiment by the authors. A freshly polished sample of transformer steel containing 3.3% Si gave 263 cps for Si $K\alpha$; oxidation of the sample reduced the rate to 53, and a brief etching with nitric

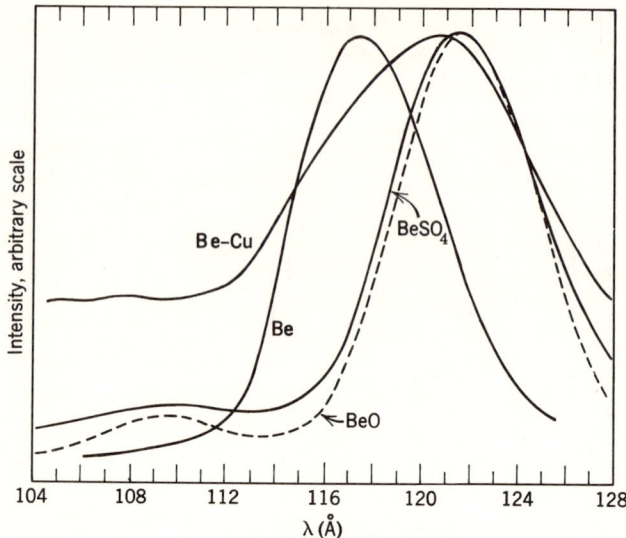

Fig. 9.26-2. Be K spectra from samples of four kinds indicated in the figure obtained in General Electric ultrasoft x-ray emission spectrograph with 3-keV electron excitation. Curved lignocerate Langmuir-Blodgett grating; Johann arrangement. X-ray excitation gave virtually identical spectra for Be and BeO. The alloy (2% Be by weight) gave cps values near 4000. See R. C. Ehlert and R. A. Mattson, Ref. 103, pp. 466 and 467, Figs. 14 and 15.

acid raised it to 725. The lighter the element, the more serious such problems.

In light of all the difficulties and complications, it is not surprising that no standard spectrograph with x-ray excitation can yet cope with the lightest elements. But remarkable progress has been made, of which the Henke x-ray tube (Section 9.4) is only one example. Baird [104] gives an excellent summary of the situation, and a useful bibliography, in a volume containing several other pertinent articles.

Fortunately, there is hope for the light elements in the greater efficiency of *electron excitation* as work with the electron microprobe continues to demonstrate—of course (see Figure 9.26-2), the "light-element problem" never disappears. X-ray excitation of these elements is inefficient because (1) intensity for excitation is liable to be low at wavelengths just shorter than those of an absorption edge, (2) x-ray scattering is relatively high at low atomic numbers, (3) negative absorption effects are usually serious, and (4) the characteristic-quantum yield is low. Clearly, electron excitation ought to be preferable.

If electron excitation over large (1 cm^2) sample areas, as in the Telsec Probe (Chapter 10), fulfills its promise, we shall have returned full circle to x-ray emission spectrography as practiced by Moseley and Hevesy. One

DIFFICULTIES AND KIND OF SAMPLE

often hears that electron excitation changes the sample to a greater extent than does x-ray excitation (Chapter 1). But, there are several things to be said. Trouble resulting from alteration of the sample must be judged for the *same analytical-line intensity*. Alteration of the sample may not cause trouble if new substances produced in this way remain in the sample. Good analytical results have been achieved with electron excitation. With both modes of excitation available to him, the analytical chemist has a choice. He may some day determine combined oxygen routinely and directly!

9.27 Localized Determinations

Though the two have features in common, it is well to distinguish between *localized determinations* on a large sample and the x-ray emission spectrography of *small samples*. Localized determinations are essential when one needs to establish variations in composition. Such variations can result, for example, from gradients in composition within a phase, from the existence of inclusions, or from the presence of grain boundaries. Qualitative or semiquantitative information may suffice. Scanning of a large sample is always desirable and sometimes mandatory. Really small samples can scarcely be scanned by the equipment discussed here. A list of possible applications for localized determinations would be very long.

Figure 7.12–2 exemplifies for small samples two important characteristics that carry over to localized determinations: obtaining the analytical line at sufficient intensity, and obtaining intensities independent of location. In that figure, $\overline{cps}$ values for Fe $K\alpha$ from the element ranged from 62 to 135, and the readings varied in three dimensions. These variations did not originate in the sample, which was a small iron foil. They resulted from changes in the x-ray intensity incident upon the sample at its different positions, and from variations with position in the probability that an analytical-line quantum reached the detector. Obviously, if one were to carry out localized determinations across a sample of iron foil, the $\overline{cps}$ values should be identical so far as the standard counting error permits. We shall not discuss either kind of variation. Localized determinations on standards can sometimes provide empirical correction factors helpful in connection with localized determinations on an unknown.

The crudest localized determinations are made by masking the sample in a conventional flat-crystal spectrograph. Analytical-line intensity will obviously be low if the exposed sample area is small, and exploration of the entire sample surface will be time-consuming. Soller-slit collimators had best be removed from the spectrograph because they may intercept the narrow beams.

In more sophisticated equipment, too diverse for description here [105–113], beam-limiting apertures are placed either between x-ray source and sample (in which case excitation is localized), or between sample and Bragg reflector (in which case the only analytical-line quanta entering the detector are those from the projected, localized sample area). Apertures may be tunnels, pinholes, or slits. The Bragg reflector is usually curved. Provision is often made for a manual or mechanical traverse of the sample so as to accomplish scanning.

The more sophisticated instruments for localized determinations with x-ray excitation are usually called *x-ray probes*, and they are offered by several manufacturers as auxiliaries for standard emission spectrographs. They have been called *milli-* and *macro* (sic!) probes to distinguish them from *x-ray emission electron microprobes*—electron microprobes for short—which are the ultimate means for localized x-ray emission spectrography, and which (though costly) dominate this field.

A LOOK AT THE ELECTRON MICROPROBE

9.28 General

In 1951, Castaing [114] demonstrated that an electron microscope could be converted into a useful x-ray emission spectrograph for determinations on a micron scale. The conversion consisted mainly in adding a second electrostatic lens to produce a narrower electron beam for the excitation of analytical lines, and in adding also an external spectrometer together with the auxiliaries needed to complete the spectrograph. In this way, determinations were made on somewhere near a $1\text{-}\mu$ cube touching the surface of a sample.

The usefulness of this remarkable tool is greatly increased when the capability to scan, introduced by Cosslett and Duncumb [115], is incorporated. In this way, areas for subsequent localized determinations can be selected, one might say, by television.

The electron microprobe is truly the prima donna among x-ray spectrographs—it may be the most remarkable *research* tool of the last several decades. Its importance in analytical chemistry is growing. It has its own Society (over 400 members) and its own literature, which—though extensive and specialized—says little about the *complete* cost of typical determinations. In an *analytical* laboratory, high cost per determination is easily justifiable only when the information found is important and unique. When there is enough work for the microprobe to do, this cost can be reduced by using a computer, not only to process output data, but to operate the probe as well. The prima donna then joins the chorus. An "on line scanning electron microscope/pseudo electron microprobe system" developed at IBM is interesting not only as a sophisticated example of this move to the chorus but also as a partial reversion of the microprobe to its origin [116].

A LOOK AT THE ELECTRON MICROPROBE

9.29 Sample Preparation and Standards

The smallness of the sphere of influence ($\sim$ a 1-μ cube) of the microprobe electron beam has mesmerized some into regarding this sphere as the sample being "analyzed." We consider the sample to be what is prepared and placed in the probe, and that the probe makes localized quantitative determinations on this sample, or scans it to give qualitative and semiquantitative information. The kinds of samples thus do not differ greatly from those already described for the standard spectrograph, but (1) samples for the probe must be stable in vacuum, (2) owing to the small area ($\sim$ 1 square micron) and the low depth of penetration of the electron beam, sample preparation (particularly sample-surface preparation) is more critical and more difficult for the probe than for standard spectrographs, and (3) certain thermal and electrical requirements must be met.

The strength of the microprobe is its ability to make localized determinations. Its use is ordinarily justified only when such determinations are needed. Many determinations are required to give *average composition* except for samples known to be pure. The microprobe and the standard spectrograph really complement each other. To illustrate: Is it enough to know that a trace of element A is in the sample? Use the standard spectrograph. Is it important to know whether A is localized; that is, exists in a few regions at relatively high weight-fractions? Use the microprobe.

In the microprobe, the small cross-section of the electron beam permits generation of an intense x-ray beam without the undue overheating (and consequent change) of the sample that plagued Hevesy [5] in his use of electron excitation. Cosslett [117] estimates that many materials can tolerate the generation of heat at rates near 10^4 kW/cm^2 in a spot 1 μ across because the heat is dissipated rapidly from such a small source, especially if the sample is a good thermal conductor. The sample surface must also be a good electrical conductor, for the electron beam may otherwise, owing to surface charge, excite at the wrong location. Samples that are insulators must be provided with a conducting surface layer.

The small scale of operations in the microprobe also makes difficult the preparation of satisfactory standards, these being after all only samples of a special kind. To the extent that microprobe determinations are *absolute*, pure elements (if compatible with the operating conditions) can serve as standards, and these are usually homogeneous enough to be satisfactory. But, a method is not truly absolute unless it is sufficiently reliable (accuracy and precision!) for a wide variety of samples, multicomponent samples included. The work of Hevesy [5] makes it unlikely that microprobe determinations will generally meet this test. Comparative methods (unknowns vs closely identical standards) would probably be welcome in microprobe work could they be easily applied. The scale of operations naturally makes this difficult. As has been mentioned, internal standards are poor prospects for the same reason. Homogeneity does not come easily on a micron scale!

NBS Special Publication 260 (July 1970) offers (p. 33) three microprobe standards: SRM 480 (a wafer), consisting of a core of W-20% Mo wire embedded in pure Mo onto which pure W has been deposited; and two sets (Au-Ag and Au-Cu) of color-coded wires covering the composition range in steps of 20% by weight. They are described as being highly homogeneous "at about the micrometer of spatial resolution." They have been extensively tested.

9.30 Calculations

With a pure element as standard, $\overline{cps}$ may reach $5(10^4)$ in an electron microprobe. When the element gives considerably lower $\overline{cps}$ values in an unknown, a dead-time correction is advisable when a proportional detector is used [118]. The problem is identical in principle with that discussed in Chapter 2.

Eddy and Laby [8] showed in 1930 that highly precise results could be obtained with electron excitation in a favorable case that required no calculations beyond (9.18-1). It was clear, however, that calculations would in general be required if pure elements were to serve as standards. Castaing [114] laid the foundation for these calculations by using (9.7-1) as a point of departure—but, with one important difference. He took the intensities to be the analytical-line intensities *as generated in the sample by electrons*—not as the intensities in the emergent beams. His procedure was logical for electron excitation because no x-rays then enter the sample. The procedure for x-ray excitation of large samples had a more empirical flavor as it dealt with emergent beams and relied heavily on the comparative method.

To compare calculations for the two cases, it is expedient to consider separately the calculations for the incident beams and for the emergent analytical line. The intensity of this line will be identical in the two cases if the analytical line is generated at identical intensity at identical depths in the sample. The calculations for the two incident beams are of course different.

The calculations for the incident electron beam in the microprobe may be based upon *average* electron scattering [21]; see Section 9.3. This generation of the analytical line by these scattered electrons is a function of atomic number; the variation is called the *atomic number effect*. The effect results from two factors: the *backscattering* factor and the *penetration* factor. These compensate to the extent that a "high scattering power in heavy elements is associated with low stopping power" [21]. This compensation increases the chance that (9.7-1) will hold. The calculations necessary when it does not hold are complex and can be done in different ways.

The x-rays generated in the sample encounter absorption and enhance-

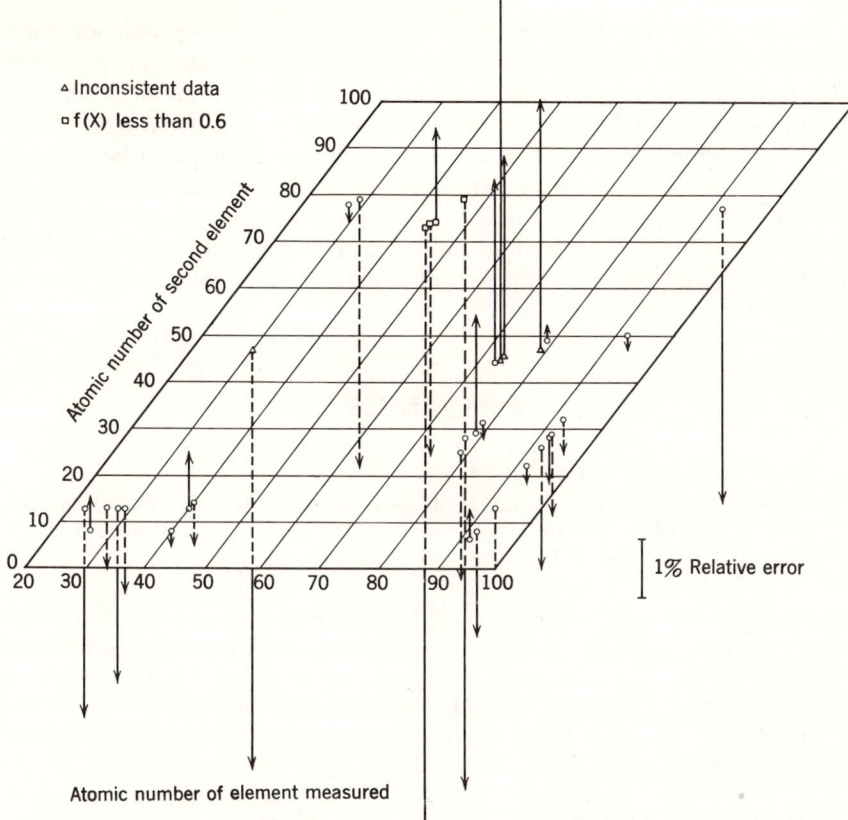

Fig. 9.30–1. Average relative errors, as evaluated by Heinrich, in results of microprobe determinations on binary systems of elements with atomic number as indicated. The errors do not vary systematically with atomic number. Relative errors can be estimated from the lengths of *solid* lines by use of the scale given. In determinations with f(X) below 0.6, there is risk of serious errors originating from x-ray absorption. See K. F. J. Heinrich, Ref. 119, p. 53.

ment effects like those already discussed, which were of course calculated for the standard spectrograph before the microprobe was invented. In addition, the excitation of the analytical line by the x-ray continuum generated in the sample must be calculated for the microprobe. This calculation is complex and somewhat uncertain. In all these calculations, the computer serves the microprobe well.

NBS Special Publication 298, issued October 1968, summarizes the situation as of June 1967. Probably the success of the calculations is best indicated by the results of Heinrich's reexamination [119] of earlier error diagrams for binary compounds. See Figure 9.30–1.

The success of the comparative method in standard x-ray emission spectrography with x-ray excitation makes it seem likely that an empirical approach will be at least a useful supplement to the absolute method (pure elements as standards) for the electron microprobe. The general reliability (accuracy and precision) of this absolute method cannot as yet be assessed because quantitative results are not now available for a sufficient diversity of samples and because the fundamental data (e.g., mass absorption coefficients) needed in the calculations are not always precisely known. Ziebold and Ogilvie [120] have two important papers on this subject.

More on the microprobe in Chapter 10.

REFERENCES

1. H. A. Liebhafsky, *Anal. Chem.*, **34**, 23A (1962).
2. (a) W. E. Burke and D. L. Wood, in *Advances in X-Ray Analysis*, Vol. 11, Plenum, New York, 1968, p. 204; (b) J. F. Cosgrove, D. W. Oblas, R. M. Walters, and D. J. Bracco, *Electrochemical Technology*, **6**, 137 (1968).
3. H. G. J. Moseley, *Phil. Mag.* [6], **26**, 1024 (1913).
4. D. Coster and G. von Hevesy, *Nature*, **111**, 79 (1923).
5. G. von Hevesy, *Chemical Analysis by X-Rays and Its Applications*, McGraw-Hill, New York, 1932.
6. F. Szabadváry, *J. Radioanal. Chem.*, **1**, 97 (1965). A memorial to von Hevesy in a new journal devoted to another kind of analytical chemistry in which he was also a pioneer.
7. A. Hadding, *Z. anorg. Chem.*, **122**, 195 (1922).
8. C. E. Eddy and T. H. Laby, *Proc. Roy. Soc. (London)*, **127A**, 20 (1930).
9. H. Friedman and L. S. Birks, *Rev. Sci. Instr.*, **19**, 323 (1948).
10. H. Friedman, L. S. Birks, and E. J. Brooks, *Am. Soc. Testing Materials Spec. Tech. Publ.*, No. 157, 3, 1954.
11. The most recent fundamental reviews are (a) W. J. Campbell and J. D. Brown, *Analytical Chemistry*, **36**, 312R (1964); (b) W. J. Campbell, J. D. Brown, and J. W. Thatcher, *Analytical Chemistry Annual Reviews*, **38**, 416R (1966); (c) W. J. Campbell and J. D. Brown, *ibid.*, **40**, 346R (1968); (d) W. J. Campbell and J. Gilfrich, *ibid.*, **42**, 248R (1970). The authors are grateful to Dr. Campbell for an advance copy of the 1972 review.
12. H. A. Liebhafsky, D. H. Wilkins, and F. Bernstein, *XIII Colloquium Spectroscopicum Internationale*, Adam Hilger, London, 1968, p. 58.
13. R. Jenkins and J. L. de Vries, *Practical X-Ray Spectrometry*, Philips Technical Library, Springer-Verlag, New York, 1967.
14. I. Adler, *X-Ray Emission Spectrography in Geology*, Elsevier Publishing Co., New York, 1966.
15. L. S. Birks, *X-Ray Spectrochemical Analysis*, Interscience, New York. (a) 1st ed., 1959; (b) 2nd ed., 1969.
16. E. P. Bertin, *Principles and Practice of X-Ray Spectrometric Analysis*, Plenum, New York, 1970.
17. (a) L. S. Birks, R. E. Seebold, A. P. Batt, and J. S. Grosso, *J. Appl. Phys.*, **35**, 2578 (1964); (b) L. S. Birks, R. E. Seebold, B. K. Grant, and J. S. Grosso, *ibid.*, **36**, 699 (1965).

REFERENCES

18. R. Glocker and H. Schreiber, *Ann. Phys.*, **85**, 1089 (1928).
19. (a) O. Eisenhut and E. Kaupp, *Z. Phys.*, **54**, 427 (1929); (b) G. R. Fonda and G. B. Collins, *J. Am. Chem. Soc.*, **53**, 113 (1931); **54**, 115 (1932); **55**, 123 (1933); (c) W. P. Jesse, *Rev. Sci. Instr.*, **6**, 47 (1935).
20. K. B. Stoddard, *Phys. Rev.*, **46**, 837 (1934).
21. P. Duncumb and P. K. Shields, *Brit. J. Appl. Phys.*, **14**, 617 (1963).
22. G. O. Archard and T. Mulvey, *Brit. J. Appl. Phys.*, **14**, 626 (1963).
23. I. Adler, Ref. 14, pp. 35 *et seq.*
24. J. V. Gilfrich and L. S. Birks, *Anal. Chem.*, **40**, 1077 (1968).
25. H. Kulenkampff, *Ann. Phys.*, **69**, 548 (1922).
26. B. L. Henke, in *Advances in X-Ray Analysis,* Vol. 5, Plenum, New York, 1962, p. 285.
27. P. D. Zemany, *ASTM Special Technical Publication*, No. 349, American Society for Testing Materials, Philadelphia, Pa., 1964, p. 1.
28. J. R. Rhodes, *The Analyst,* **91**, 683 (1966).
29. C. E. Crouthamel, Ed., *Applied Gamma-Ray Spectrometry* Pergamon, New York, 1969; one of a series of monographs on analytical chemistry and an excellent reference work.
30. A. Robert and P. Martinelli, in *Proceedings of the Symposium on Radiochemical Methods of Analysis,* Vol. 2, I.A.E.A., Vienna, 1965, p. 401.
31. R. L. Watson, personal communication to H.A.L., May 1, 1970.
32. T. F. Anater, *U.S. At. Energy Comm. Rep.*, WAPD-321, 1968.
33. W. D. Ashby and E. M. Proctor, *Picker Analyzer*, Picker X-ray Corporation, Cleveland, Ohio, No. 1, **4**, (1965).
34. K. W. Yee and R. D. Deslattes, *Rev. Sci. Instr.*, **38**, 637 (1967).
35. W. B. Eastman, in *Advances in X-Ray Analysis,* Vol. 6, Plenum, New York, 1963, p. 313.
36. J. E. Fagel, Jr., H. A. Liebhafsky, and P. D. Zemany, *Anal. Chem.*, **30**, 1918 (1958).
37. H. G. J. Moseley, *Phil. Mag.* [6], **26**, 1024 (1913); especially p. 1026.
38. K. P. Champion and R. N. Whitten, *Nature*, **199**, 1082 (1963).
39. H. A. Liebhafsky, H. G. Pfeiffer, E. H. Winslow, and P. D. Zemany, *X-Ray Absorption and Emission in Analytical Chemistry,* Wiley, New York, 1960, p. 172.
40. J. W. Criss and L. S. Birks, *Anal. Chem.*, **40**, 1080 (1968).
41. J. Sherman, *Spectrochim. Acta,* **7**, 283 (1955).
42. R. Glocker, *Phys. Z.*, **19**, 249 (1918).
43. R. Glocker and H. Schreiber, *Ann. Phys.*, **85**, 1089 (1928).
44. L. S. Birks and D. L. Harris, *Anal. Chem.*, **34**, 943 (1962).
45. Betty J. Mitchell and J. E. Kellam, *Appl. Spectr.*, **22**, 742 (1968).
46. H. A. Liebhafsky, H. G. Pfeiffer, E. H. Winslow, and P. D. Zemany, Ref. 39, pp. 168–170.
47. J. Sherman, *Am. Soc. Testing Materials Spec. Tech. Publ.*, No. 157, 27, 1954.
48. J. Lucas-Tooth and C. Pyne, in *Advances in X-Ray Analysis*, Vol. 7, Plenum, New York, 1964, p. 523.
49. H. J. Lucas-Tooth and B. J. Price, *Metallurgia,* **64**, 149 (1961).
50. H. J. Beattie and R. M. Brissey, *Anal. Chem.*, **26**, 980 (1954).
51. B. J. Alley and R. H. Myers, *Anal. Chem.*, **37**, 1685 (1965).
52. Betty J. Mitchell and F. N. Hopper, *Appl. Spectr.*, **20**, 172 (1966).
53. T. Shiraiwa and N. Fujino, *Jap. J. Appl. Phys.*, **5**, 886 (1966).
54. W. J. Campbell and H. F. Carl, *Anal. Chem.*, **28**, 960 (1956).

55. R. M. Brissey, *Anal. Chem.,* **24**, 1034 (1952).
56. H. F. Carl and W. J. Campbell, *Am. Soc. Testing Materials Spec. Tech. Publ.,* No. 157, 63, 1954.
57. L. S. Birks and E. J. Brooks, *Anal. Chem.,* **22**, 1017 (1950).
58. P. D. Zemany, *Spectrochim. Acta,* **16**, 736 (1960).
59. R. M. Dolby, *Proc. Phys. Soc. (London),* **73**, 81 (1959).
60. L. S. Birks, R. J. Labrie, and J. W. Criss, *Anal. Chem.,* **38**, 701 (1966).
61. E. P. Bertin and R. J. Longobucco, *Norelco Reporter,* **9**, 31 (1962).
62. E. P. Bertin, *Anal. Chem.,* **36**, 826 (1964).
63. B. J. Mitchell and H. J. O'Hear, *Anal. Chem.,* **34**, 1620 (1962).
64. R. A. Jones, *Anal. Chem.,* **31**, 1341 (1959).
65. E. L. Gunn, *Appl. Spectr.,* **19**, 99 (1965).
66. G. Andermann and J. W. Kemp, *Anal. Chem.,* **30**, 1306 (1958).
67. H. J. Rose, Jr., and F. Cuttitta, in *Advances in X-Ray Analysis,* Vol. 11, Plenum, New York, 1968, p. 23.
68. I. Adler and J. M. Axelrod, *Spectrochim. Acta,* **7**, 91 (1955).
69. C. M. Johnson and P. R. Stout, *Anal. Chem.,* **30**, 1921 (1958).
70. C. W. Dwiggins, Jr., *Anal. Chem.,* **33**, 67 (1961).
71. C. W. Dwiggins, Jr., *Anal. Chem.,* **36**, 1577 (1964).
72. R. C. Reynolds, *Amer. Mineral.,* **48**, 1133 (1963).
73. C. J. Carman, *Develop. Appl. Spectr.,* **5**, 45 (1966).
74. I. Adler and J. M. Axelrod, *Spectrochim. Acta,* **7**, 91 (1955).
75. G. Pish and A. A. Huffman, *Anal. Chem.,* **27**, 1875 (1955).
76. E. L. Gunn, *Anal. Chem.,* **29**, 184 (1957).
77. I. Adler and J. M. Axelrod, *Anal. Chem.,* **26**, 931 (1954).
78. F. Claisse, *Quebec Dept. Mines Prelim. Rept.,* No. 327, 1956; *Norelco Reptr.,***4**, 3 (1957).
79. *Spectrographer's News Letter,* No. 3, **7**, (1954).
80. E. P. Bertin, in *Advances in X-Ray Analysis,* Vol. 11, Plenum, New York, 1968, p. 1.
81. W. J. Campbell, M. Leon, and J. W. Thatcher, *U.S. Bureau of Mines Report of Investigations* 5497, 1959, 24 pp.
82. W. J. Campbell, *Appl. Spectr.,* **14**, 26 (1960).
83. E. P. Bertin and R. J. Longobucco, in *Advances in X-Ray Analysis,* Vol. 5, Plenum, New York, 1962, p. 447.
84. F. L. Chan, *Develop. Appl. Spectr.,* **5**, 59 (1966).
85. C. H. Drummond, *Appl. Spectr.,* **20**, 252 (1966).
86. H. J. Rose, I. Adler, and F. J. Flanagan, *Appl. Spectr.,* **17**, 81 (1963).
87. J. E. Fagel, Jr., E. W. Balis, and L. B. Bronk, *Anal. Chem.,* **29**, 1287 (1957).
88. C. L. Luke, *Anal. Chem.,* **36**, 318 (1964).
89. W. J. Campbell, T. E. Green, and S. L. Law, *American Laboratory,* American Laboratory Inc., Greens Farms, Conn., June 1970, p. 28.
90. E. P. Bertin and R. J. Longobucco, *Norelco Reptr.,* **9**, 31 (1962).
91. R. E. Michaelis and B. A. Kilday, Ref. 83, p. 405.
92. F. Brech, comment at the end of Ref. 91.
93. F. Claisse and C. Samson, in *Advances in X-Ray Analysis,* Vol. 5, Plenum, New York, 1962, p. 335.

94. P. F. Berry, T. Furuta, and J. R. Rhodes, in *Advances in X-Ray Analysis*, Vol. 12, Plenum, New York, 1969, p. 612.
95. E. L. Gunn, in *Advances in X-Ray Analysis*, Vol. 11, Plenum, New York, 1968, p. 164.
96. W. J. Campbell and J. W. Thatcher, in *Advances in X-Ray Analysis*, Vol. 2, Plenum, New York, 1960, p. 313.
97. (a) A. K. Baird, R. S. MacColl, and D. B. McIntyre, in *Advances in X-Ray Analysis*, Vol. 5, Plenum, New York, 1962, p. 412; (b) K. W. Madlem, *ibid.*, Vol. 9, 1966, p. 441.
98. A. B. Poole and S. M. Holloway, in *Advances in X-Ray Analysis*, Vol. 12, Plenum, New York, 1969, p. 534.
99. H. A. Liebhafsky, D. H. Wilkins, and F. Bernstein, *XIII Coll. Spectroscopicum Internationale*, Adam Hilger, London, 1968, p. 58.
100. J. I. Mueller, V. G. Scotti, and J. J. Little, in *Advances in X-Ray Analysis*, Vol. 2, Plenum, New York, 1959, p. 157.
101. (a) A. E. Lindh, "Röntgenspektroskopie," *Handbuch der Experimentalphysik (Wien u. Harms)*, Vol. XXIV:2, Akademische Verlagsgesellschaft, Leipzig, 1930, p. 291; (b) B. Lindström, "Roentgen Absorption Spectrophotometry in Quantitative Cytochemistry," *Acta Radiol. Suppl.*, **125**, 34 (1955).
102. (a) A. Faessler and M. Goehring, *Naturwissenschaften*, **39**, 169 (1952); (b) B. Lindström, "Roentgen Absorption Spectrophotometry in Quantitative Cytochemistry," *Acta Radiol. Suppl.*, **125**, 25 (1955).
103. R. C. Ehlert and R. A. Mattson, in *Advances in X-Ray Analysis*, Vol. 9, Plenum, New York, 1966, p. 456.
104. A. K. Baird, in *Advances in X-Ray Analysis*, Vol. 13, Plenum, New York, 1970, p. 26.
105. L. S. Birks and E. J. Brooks, *Anal. Chem.*, **27**, 437 (1955).
106. I. Adler and J. M. Axelrod, (a) *Norelco Reptr.*, **3**, 65 (1956); (b) *Am. Mineralogist*, **41**, 524 (1956); (c) *Econ. Geol.*, **52** (6), 694 (1957).
107. K. F. J. Heinrich, in *Advances in X-Ray Analysis*, Vol. 5, Plenum, New York, 1962, p. 516.
108. T. C. Loomis and K. H. Storks, *Bell Lab. Rec.*, **45**, 2 (1967).
109. T. C. Loomis, *Ann. N. Y. Acad. Sci.*, **137**, 284 (1966).
110. (a) E. P. Bertin, in *Advances in X-Ray Analysis*, Vol. 8, Plenum, New York, 1965, p. 231; (b) *ibid.*, Vol. 10, 1967, p. 462; (c) *Anal. Chem.*, **36**, 441 (1964).
111. K. Togel, *Siemens-Z.*, **36**, 497 (1962).
112. H. J. Rose, R. P. Christian, J. R. Lindsay, and R. R. Larson, *U.S. Geol. Surv. Prof. Paper* **650-B**, 128 (1969).
113. J. T. Campbell, F. W. J. Garton, and J. D. Wilson, *Talanta*, **15**, 1205 (1968).
114. (a) R. Castaing, *Recherche Aeronaut. (Paris)*, No. 23, 41 (1951); (b) R. Castaing and A. Guinier, *Anal. Chem.*, **25**, 724 (1953); (c) R. Castaing and J. Descamps. *J. Phys. Radium*, **16**, 304 (1955); (d) R. Castaing, *Adv. Electron. Electron. Phys.*, **13**, 317 (1960).
115. V. E. Cosslett and P. Duncumb, in *Proc. Intern. Conf. Electron Microscopy, Stockholm*, Academic, New York, 1957, pp. 12–14.
116. R. Pyle, R. B. Rogan, T. C. Hartmann, and M. A. Shulman, presented at the 161st ACS National Meeting, Los Angeles, March 29–April 2, 1971.
117. V. E. Cosslett, *Proc. Phys. Soc. (London)*, **65B**, 782 (1952).
118. K. F. J. Heinrich, D. Vieth, and H. Yakowitz, in *Advances in X-Ray Analysis*, Vol. 9, Plenum, New York, 1966, p. 208.
119. K. F. J. Heinrich, in *Advances in X-Ray Analysis*, Vol. 11, Plenum, New York, 1968, p. 40.
120. (a) T. O. Ziebold and R. E. Ogilvie, *Anal. Chem.*, **35**, 621 (1963); (b) *ibid.*, **36**, 322 (1964).

Chapter 10

Equipment and Selected Applications

L'embarras des richesses. Abbé D'Allainval

10.1 Introduction

In both the fields of this chapter there is such an "embarrassment of riches" that coverage must be restricted to include only what the analytical chemist might need to know to judge for himself the value of modern x-ray emission spectrography and related methods. Much of the material in our first book [1]—notably applications of x-ray methods in the life sciences—is not repeated. The reader interested in further information about equipment should, as we did, obtain from the principal manufacturers their excellent, generally short-lived, product literature. Illustrative applications have appeared in earlier chapters; many more are to be found in the guide to literature in Appendix VI.

The increase in the number of applications during the last decade is not surprising, but the increase in the complexity and the number of spectrographs offered by manufacturers here and abroad may come as a revelation. Events or items of particular interest after publication of the earlier book are as follows.

1. Great success of the electron microprobe.
2. Changes wrought by use of computers: for example, in control of equipment and in calculation and storage of results.

EQUIPMENT AND ILLUSTRATIVE APPLICATIONS

3. Introduction of solid-state circuitry.
4. Return to electron excitation.
5. Increasing number of useful Bragg reflectors.
6. Increasing attention to light elements, hence to vacuum optical paths and detector windows.
7. Enormous emphasis on speed and convenience, as through provisions for easy interchangeability of Bragg reflectors, collimators, and detectors; multiple sample loading; and modular construction that permits adding components to a simple system.
8. Growth of photoelectron spectrography.

Of course, greater sophistication costs money and sometimes leads to troubles—notably electronic troubles—difficult to locate; the switchover to color television comes to mind as a mild example. But the benefits on balance have made the changes worthwhile; for example, owing to the use of solid-state circuitry wherever possible in modern spectrographs, the minimum warm-up time is now set by the x-ray tube, and operation under good conditions is more stable; there is less heat to be rejected.

EQUIPMENT AND ILLUSTRATIVE APPLICATIONS

10.2 Evolution of a Flat-Crystal Spectrograph System

For relevant basic information needed here, see earlier chapters, especially 1, 2, 5, and 9.

Figure 5.13–1 shows the elements of an x-ray emission spectrograph and the interrelation of various x-ray instruments. As x-ray diffraction was well established when modern x-ray emission spectrography began, it followed naturally that many early spectrographs were converted diffractometers. Diffractometer-spectrographs (Figure 10.2–1) followed. Instruments such as this are valuable in teaching. They can be adapted to absorptiometry, polychromatic or monochromatic. The changeover from diffractometer to spectrograph requires only about 20 minutes if the operator is experienced. The principal items in the conversion are substitution of a high-intensity x-ray tube for one with line focus; insertion of Bragg crystal, Soller slit, sample holder, and sample; and changing (usually) and relocating the detector.

The next evolutionary step was a simple spectrograph *system*, such as is delineated in Figure 10.2–2. The reader will appreciate that this does not illustrate an important alternative method of collecting information, to be discussed later, in which the detector charges a capacitor.

Recent evolutionary steps have culminated in a spectrograph system that can determine up to 28 elements (down to and including fluorine) in each

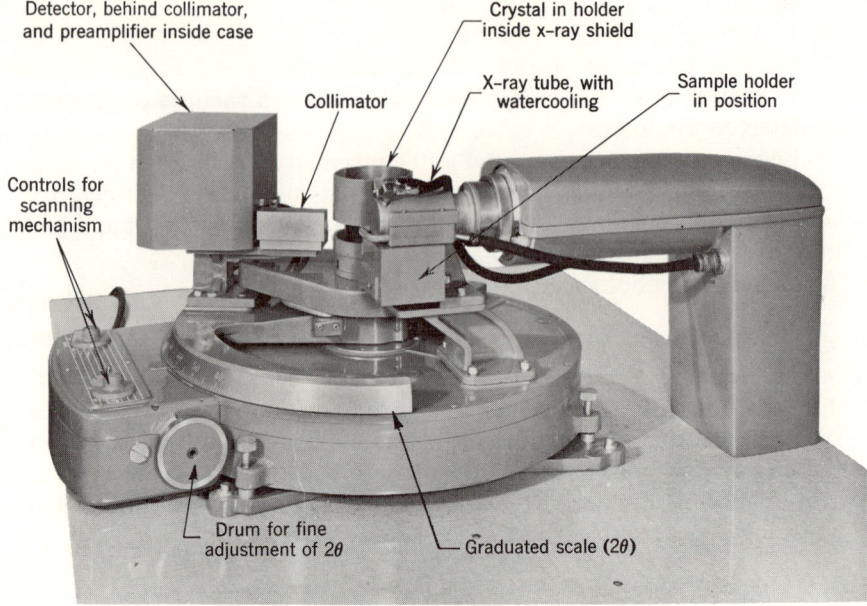

Fig. 10.2-1. The General Electric XRD-5 D/S spectrogoniometer arranged for x-ray emission work. Since the axis of rotation is vertical, heavy auxiliary equipment may be attached without disturbing the apparatus or increasing mechanical wear.

of 10 samples without further attention after the operator has completed loading and placed the system under control of the computer. The spectrograph resembles that of Figure 10.2–3, but is of 10-sample capacity.

The (larger) production system (XRD-410) shows what the computer (here called the digital controller) can do. After the proper instructions (the "program") have been inserted in the digital controller, determinations on the first sample begin when the proper vacuum has been reached. According to the program, the controller directs the goniometer to move sequentially to the 2θ positions of the required analytical lines, calculates by the comparative method the amount present of each element sought, and types out (or otherwise records) the results. Again, according to the program, the required set of determinations is repeated for each of the samples (up to 9) remaining, the samples being moved into position as the controller directs.

In semiautomatic operation of the system, the operator issues instructions step by step to the controller and observes how the system responds; "trouble-shooting" is done in this way. In manual operation, which is also

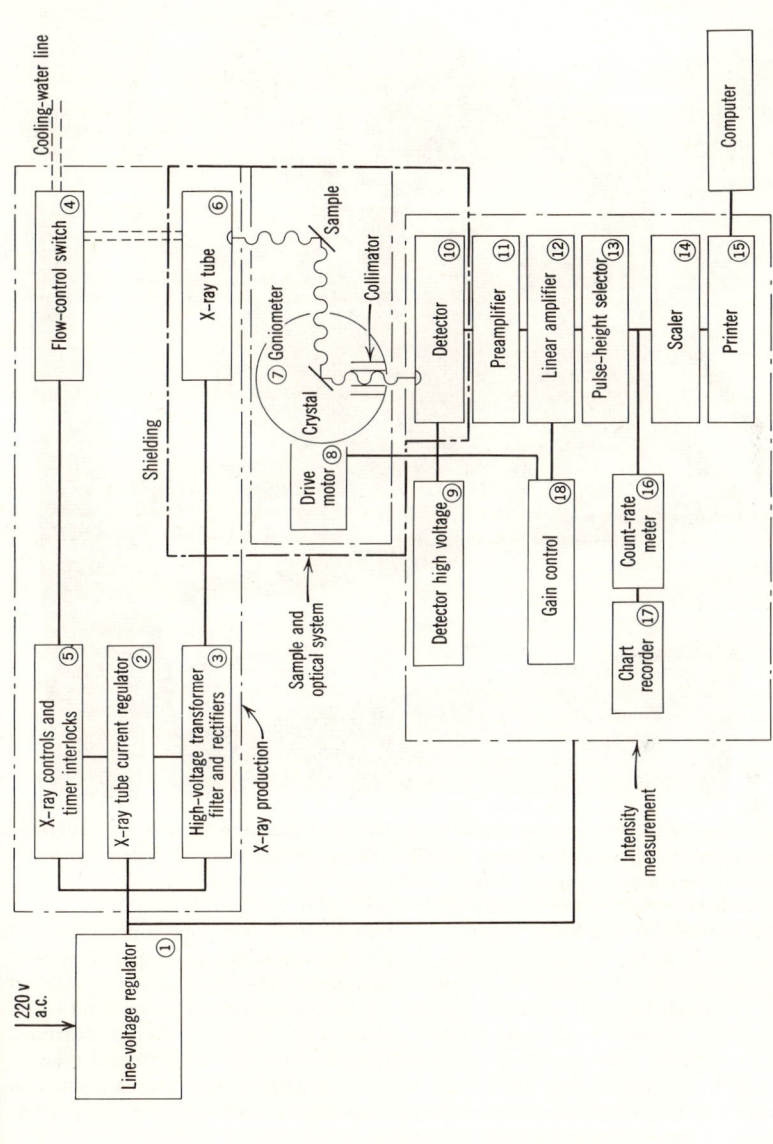

Fig. 10.2-2. A block diagram of a modern x-ray spectrograph system intended for the determination of one element at a time in an analytical laboratory not primarily concerned with routine work.

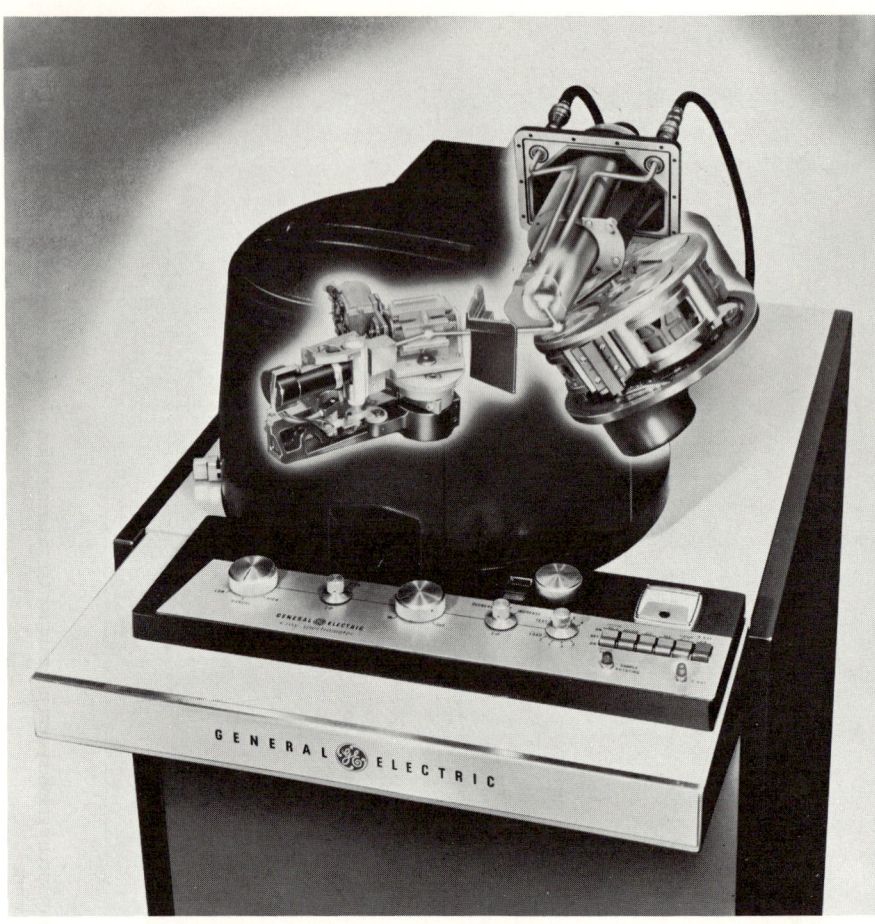

Fig. 10.2-3 The XRD-6VS, a recent General Electric x-ray emission vacuum spectrograph. Components visible in upper half of figure are all in vacuum when spectrograph is operative. (1) X-ray beam for excitation originates at Cr or W target of dual-target Coolidge tube (first "impact burst" between and in front of electrical leads). (2) Polychromatic beam strikes one of four samples in carousel that holds indexed, tilted sample very near tube window (second "impact burst"). (3) Analytical line and other x-rays pass through coarse, short Soller-slit collimator and strike (third "impact burst") one of four Bragg reflectors that can be selected and precisely positioned without breaking vacuum. (4) Analytical line and background x-rays pass fine Soller-slit collimator and enter detector system, which contains a flow proportional detector (white vertical tube) that abuts a scintillation detector (black horizontal tube) in tandem arrangement. At will, either detector can be made inoperative by having zero potential difference across it. The scintillation detector may be replaced by a xenon-filled sealed proportional detector. Courtesy General Electric Company.

EQUIPMENT AND ILLUSTRATIVE APPLICATIONS

possible, the controller serves only for counting and timing. The system can be "interfaced" with (joined to) a larger computer or with production-logging equipment.

Two ways of changing 2θ enter into computerized operation. The first, *scanning*, which is done at low angular velocities, needs no comment. The second, *slewing*, which is done at angular velocities of (say) 300°/minute moves the goniometer rapidly to the position of an analytical line to be counted and minimizes time between determinations on a sample. Rotation of the sample in its plane is provided for in the production equipment.

The evolutionary process sketched here has run a similar course in other firms (e.g., Philips) that manufacture x-ray equipment. The combination of spectrograph (Figure 10.2–3) and diffractometer operating on a common power source makes for versatility at reduced capital cost and can be used also for absorptiometric measurements.

10.3 Evaluation of a Flat-Crystal X-Ray Emission Spectrograph System

Siemens America has issued a comprehensive chart for the comparative evaluation of x-ray emission spectrograph systems. A modified tabular version of this chart follows as a convenient index of changes that have occurred in the last decade. The analytical chemist must decide for himself how many of these features he needs. (See Table 10.3–1.)

Table 10.3–1. Features of Modern X-Ray Emission Spectrograph System. Based on Features of the Siemens SRS

1. The X-ray generator
   Power: 4000 W
   Constant potential
   Maxima for x-ray tube: 60 kV; 80 mA
   Solid-state rectifier
   Closed-loop stabilization (internal reference)
   Tube current to 0.03 %
   Tube voltage to 0.03 %

2. The X-ray tube
   Asymmetrically positioned target
   Targets (one per tube) and power rating:
   Cr (2600 W)   Au (3000 W)
   Mo (3000 W)   W (3000 W)

3. The spectrometer and associated parts
   Maximum sample: 53.5 mm (dia), 50 mm thick
   Sample-to-target distance: 25 mm
   Sample positions: 2 (10 optional)
   Sample rotation provided
   Bragg reflectors: Up to 4
   Scanning ranges:
   Flow proportional detector 0–147°
   Scintillation detector 0–114°
   Scanning speeds (°/min):
   $\frac{1}{8}, \frac{1}{4}, \frac{1}{2}, 1, 2,$ and 4
   Slewing speed 300°/min
   Precision of goniometer setting $\pm 0.002°$

Table 10.3–1. (*Continued*).

Soller slits 0.4° (coarse) 0.15° (fine)
Slits interchangeable in vacuum
Likewise for the two detectors
Time for exhaust to 100 μ:
   Less than 6 sec (safety interlock)
Sine function preamplifier
Choice of order for Bragg reflection
Ambient temperature control for
   spectrometer (optional)
	When changing detectors
	When changing 2θ
	Ratemeter specifications:
	   12 counting ranges; log scale
	Background suppression
	Counting capabilities
	   Fixed time; fixed count; which-
	   ever is first
	Printer: 16-digit simultaneous
	   Prints counts, time, and event

4. The Detector System
   Ambient temperature control for
     scintillation detector
   Gas for flow proportional
     detector
   Pressure stabilization of gas for
     detector
   Window thickness for flow
     proportional detector: down
     to 0.4 μ
   Visual display of detector output
     and pulse-height analyzer
     position
   Automatic adjustment of pulse-
     height-analyzer window
	5. Automation
	   Compatible with spectrograph
	     system
	   Manual operation possible
	   128 analytical lines programmable
	   Calibration curve computed and
	     stored
	   Direct readout in percentage
	   Memory stores up to 256 multiple-
	     element programs.
	   Input/output either
	     Punched tape or teletype
	   Automatic control of 2θ
	     Accuracy and precision, each
	     0.002°

We repeat an important point made earlier: In most cases, sample preparation and not the spectrograph system, if functioning properly, will limit the reliability of results.

10.4 Multichannel X-Ray Emission Spectrograph Systems with Curved Crystals

In the main, we have been concerned so far with systems in which the intensity of one analytical line at a time is measured, wherefore only one Bragg reflector and one detector are used at once. For the determination of more than one element in a sample, operation must then be sequential; i.e., over a succession of 2θ values. The sequence may involve changing either the Bragg reflector or the detector, or both.

Determinations in the ultraviolet and visible ranges have long been carried out by the millions each year, especially in the aluminum industry, on automated Applied Research Laboratories (ARL) Quantometers.

EQUIPMENT AND ILLUSTRATIVE APPLICATIONS

These are *multichannel* instruments that make simultaneously all the determinations needed on a sample. One channel, complete with appropriate Bragg reflector and detector, is allocated to each analytical line. In addition, there is a standard or monitor channel, in which the sample might be a metal disk, that makes it possible to use the comparative method and express the results in a ratio system.

The original ratio system worked like this. The output of each detector in an analytical channel (Geiger, "semiproportional," or proportional—usually with gas flow) is accumulated in a capacitor. The output of the detector in the standard channel is recorded. When the charge in the capacitor for the standard channel reaches a predetermined value, the recorder disconnects the detectors in all the channels. Each capacitor is then appropriately connected in turn to the amplifier and recorder, and a value of the ratio for each channel (except the standard) may appear on a chart. Modern systems are more sophisticated.

The number of photons accumulated by the capacitor is of course proportional to the charge, hence the voltage, on the capacitor. This number of photons is equal to the intensity multiplied by the counting interval chosen, which is set for the standard channel. A short counting interval makes for speed, a great advantage of multichannel systems. To attain speed, this counting interval must be minimized, and the increased intensity resulting from the use of Bragg crystals curved according to Johansson (Section 5.11) is therefore welcome. Spectrographs thus equipped are being described as "fully focusing."

Among the multichannel x-ray spectrographs offered by ARL are MXQ, VXQ, VPXQ, and PCXQ. The letters have the following meanings: X, x-ray; Q, quantometer; M, modular; V, vacuum; P, production (i.e., rapid, large, for routine determinations); PC, process control. As the number and diversity of the features of these instruments are reminiscent of Table 10.3–1, we cannot do more than give the most important characteristics, not always with specific attribution.

Figures 10.4–1 and −2 show the geometry of the quantometers. Salient features not yet mentioned follow:

1. End-window x-ray tubes. Be windows as thin as 0.127 mil. Targets: W, Rh, Cr, Pt.
2. Wavelength range 0.36–24.9 Å (down to and including oxygen).
3. Pulse-height selection when needed.
4. Attenuation by filters used to increase composition range accessible.
5. All determinations on a sample simultaneously completed in seconds to a few minutes.
6. Computer can be used to maximum capability (control, reading of capacitors, calculations, readout).

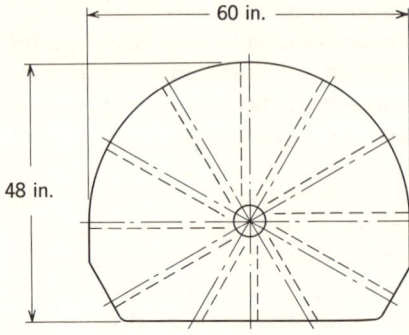

Fig. 10.4–1. Top-view schematic diagram of ARL-PXQ with maximum number (22) of channels indicated. The channels come in pairs, one (broken line) being below the other (see Figure 10.4–2). The standard (monitor) channel occupies the space allocated above to one pair. This figure and the next by courtesy of Davidson, Gilkerson, and Kemp, Pittsburgh Conference on Analytical Chemistry and Applied Spectroscopy, March 1958.

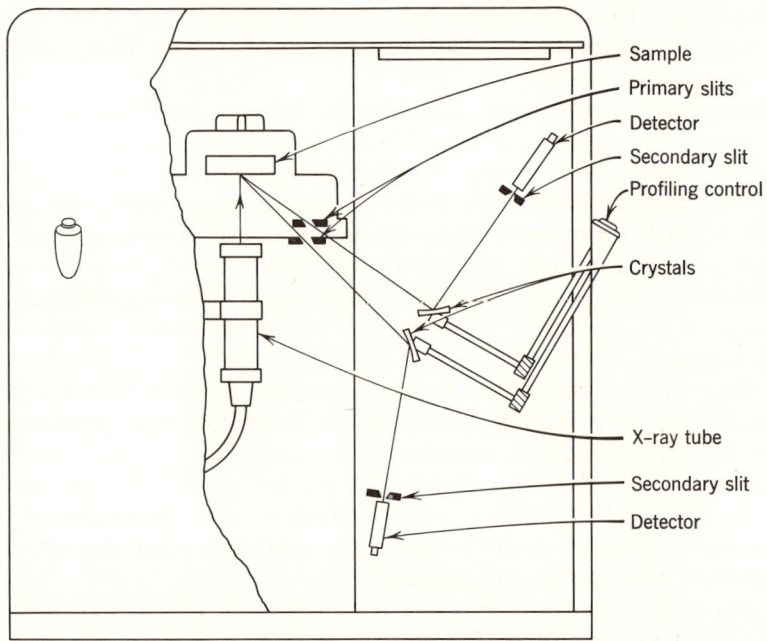

Fig. 10.4–2. Schematic diagram related to Figure 10.4–1. Note that an *end-window* x-ray tube is used to maximize the number of possible channels. The slits are adjustable to give the best balance between intensity and resolution for the determinations being made.

7. Reading of individual channels (voltages or ratios): (a) Sequentially, strip chart recorder, 1 channel in 2 sec. (b) Digital voltmeter-line printer, 1 channel/sec. (c) Analog output to computer, up to 50 channels/sec.

8. Equipment available for reading counts accumulated (N_T).

9. Solids, slurries, and liquids can all be handled.

10. Precision seems comparable with best achievable under given conditions.

EQUIPMENT AND ILLUSTRATIVE APPLICATIONS

11. Provision can be made for incorporating channels (up to 3) for sequential determinations of elements suited to channel thus incorporated. Each such channel replaces 2 fixed channels.

10.5 The Philips Inverted-Sample Three-Position Spectrograph

Although one of the earlier modern spectrographs produced specifically for x-ray emission, the Philips inverted-sample three-position spectrograph [2] is still attractive as meeting most x-ray emission needs of small laboratories, and as perhaps the most convenient way of handling liquid samples. It cannot of course deal with light elements. See Figure 10.5-1.

Three of the valuable features of this spectrograph deserve special mention as facilitating chemical analysis: (1) The provision for x-ray excitation of the sample from underneath. This feature, especially valuable for liquids, requires that the base of the cell be a good window for x-rays, and that it undergo no deformation that changes the x-ray path. Thin Mylar serves

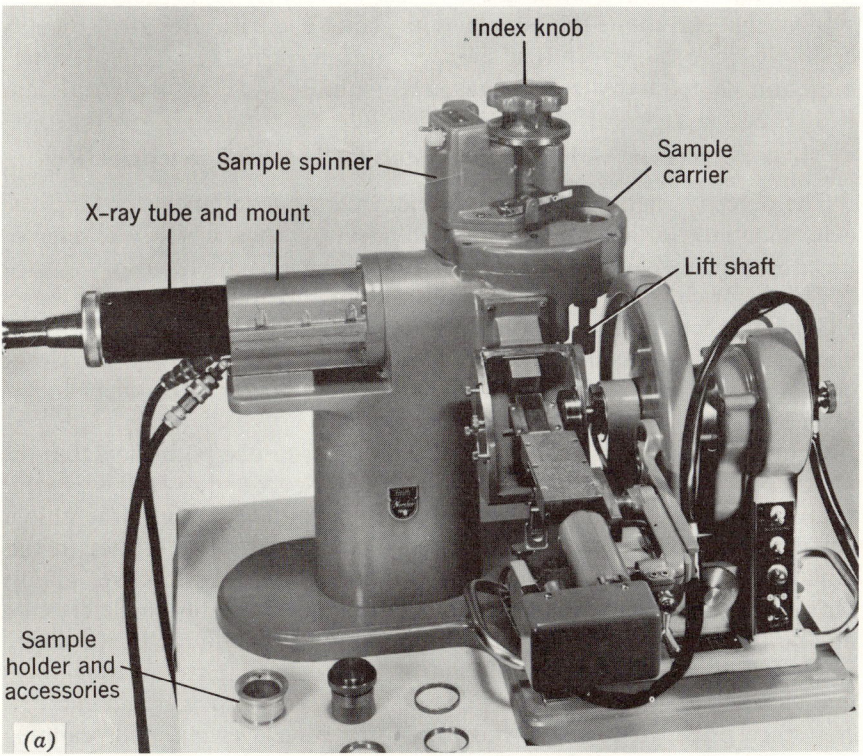

Fig. 10.5-1. (a) Photograph of the Philips Inverted-Sample Three-Position Spectrograph. (b) Schematic drawing of optical path. Courtesy of M. Tomaino and A. De Pietro, Ref. 2.

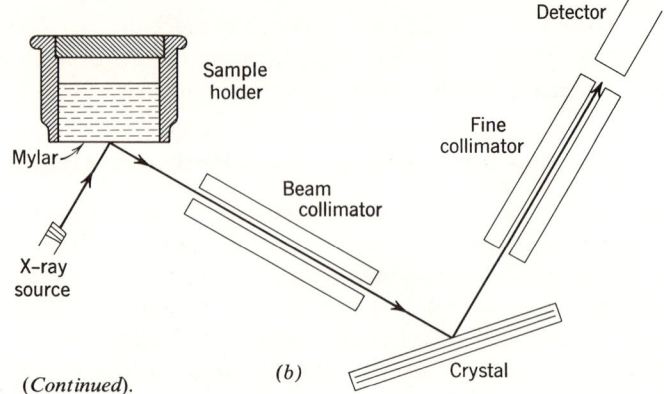

Fig. 10.5-1. *(Continued).*

well, but even this admirable material is subject to deterioration by x-rays. Note that a sample could reach an unprotected x-ray tube through a ruptured window! (2) The provision of more than one sample position so that a sample can be loaded while the analysis of another is in progress. (3) A spinner that can rotate the sample in its plane. These features are especially desirable for routine work. The spectrograph gave good results in the determination of tetraethyllead in gasoline with Pb $L\beta1$ as analytical line.

10.6 Equipment with Curved Crystals for Localized Determinations

The objective here is to describe briefly certain equipment for localized determinations, attachable to standard spectrographs, which was referred to in Section 9.27. The equipment serves for small samples also.

The Heinrich probe [3] is essentially an aperture in the sample drawer of a General Electric spectrograph that limits the area from which an analytical line generated in a completely irradiated sample can reach the flat Bragg crystal and, from there, the detector. The arrangement necessarily suffers from low intensity if the aperture is small.

Chapter 5 and discussions of the electron microprobe both show that the Johansson arrangement (crystal bent to radius R, ground to $R/2$) is capable of giving higher intensities (often an order of magnitude or more, other things equal) in localized determinations. Adler and Axelrod [4] exploited this advantage in building a small spectrograph for nondestructive, mainly semiquantitative, determinations on small samples, notably single crystals, and x-ray diffraction spindles (samples for x-ray powder diffraction patterns). See Ref. 1, p. 206. Philips now offers a similar spectrograph as an attachment to their standard x-ray emission spectrograph, with monocular and micrometer stage provided to make possible positioning, viewing, and scanning the sample. Conversion from flat- to curved-crystal operation takes about an hour. Operation in vacuum is possible.

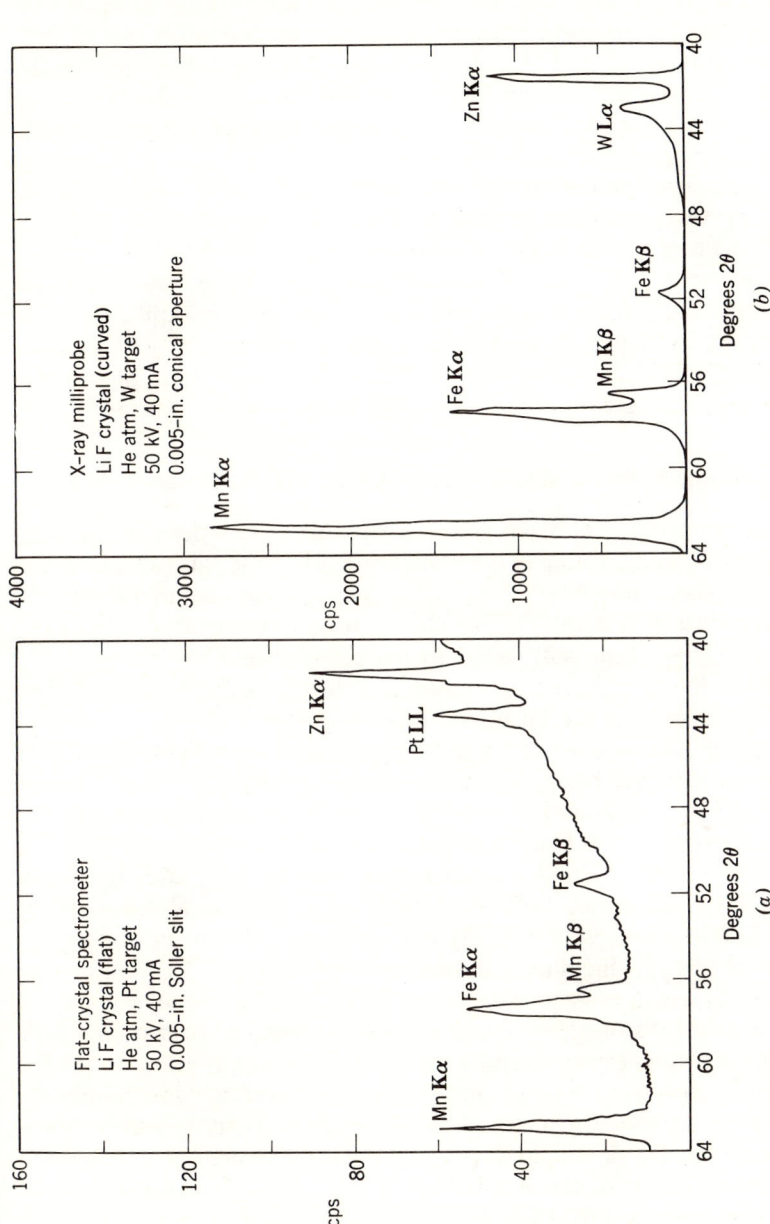

Fig. 10.6-1. Results of 2θ-scans for a diffraction spindle of helvite made with (*a*) a General Electric flat-crystal spectrograph and with (*b*) a curved-crystal attachment thereto. Curved crystal bent to 14-inch radius; ground to 7-inch radius. Note the difference in intensities. As was pointed out in Chapter 6, spectrographic information of this kind often facilitates the interpretation of diffraction patterns. See H. J. Rose, Jr., *et al.*, Ref. 5, p. *B* 133, Fig. 6.

A curved-crystal attachment was specially built for the Geological Survey by General Electric to make possible a similar conversion for one of their flat-crystal spectrographs [5]. Johansson-type LiF crystals cover the 2θ-range from 23 to 148.5°. Figure 10.6–1 shows how much better are the results obtained on a small sample with the spectrograph converted to the Johansson arrangement.

With equipment similar to that just described, Luke [6] improved the sensitivity of the ion-exchange-resin technique (Section 7.18) to where as little as 0.01 μg of a metal could be detected. After separation by chemical methods, the element to be determined was concentrated on an ion-exchange-resin disk 1/8 in. across. X-ray emission spectrography was then carried out on 28 metals exchanged as cations, and 15 exchanged as anions (e.g., MoO_4^{2-}, $PtCl_6^{2-}$, $Co(CN)_6^{3-}$). This is one example of a localized determination where a microprobe would seem to offer no advantages.

10.7 Portable X-Ray Emission Spectrographs with Radioactive Sources

Portable spectrographs make possible rapid determinations in the field. This saves time and trouble if the results are adequate, and it makes on-the-spot decisions possible. These spectrographs can be useful for the same reasons in a large plant. They can be useful adjuncts to larger laboratory instruments. They seem well suited to the classroom. They are generally simpler than the equipment in Figure 2.25–4 although these solid-state detector spectrographs are becoming more portable.

A portable, battery-operated spectrograph of *total* weight near 15 lb was developed on the basis of British work [7, 8] by the Texas Nuclear Corporation [9]. A radioactive source (Section 9.6) either excites the sample directly or generates the characteristic line of a "target element" for this purpose. The source, behind a shutter (diameter, 0.5 in.), is coaxial with the scintillation detector, both contained in a cylindrical housing ($2\frac{1}{2}$ by 8 in.). The source is so weak (say, 10^{-7} times the intensity of an x-ray tube) that no special shielding is needed and $\overline{cps}$ values of 10^3 to 10^5 from pure elements are attainable. Balanced Ross filters provide wavelength selection, and results of the sequential readings needed appear on a simple scaler. The instrument promises to have many applications even though the standard counting error is unavoidably large. A special version containing 6 filters has been used on board ship for determinations of the principal constituents (Mn, Ni, Fe, Co) in the manganese nodules present in enormous amounts on the floor of the Pacific and of the Indian Ocean [10].

Telsec Instruments offer an interesting 42 lb, dual-channel spectrograph that has other features in common with the instrument just described. In each channel, a sealed proportional detector receives x-rays that have passed from the sample through one of a pair of balanced Ross filters. Pulse-

EQUIPMENT AND ILLUSTRATIVE APPLICATIONS

height discrimination is provided. The difference in counts from the two detectors is accumulated for a preset counting interval and displayed on a meter. Operation is thus simultaneous and not sequential.

The Pitchford Manufacturing Corporation has produced an x-ray emission spectrograph with collimator, LiF crystal, goniometer, and Geiger detector that weighs less than 130 lb. The probe unit (23 lb; 15 × 17½ × 6 in.) can be held on any flat surface; or put on rod or tubing by use of self-positioning, adjustable shoes. Alternatively, unknown and standard may be placed in a two-position sample holder. Various readouts are possible. The equipment is intended mainly for plant use to identify materials.

10.8 The Electron Microprobe

Figures 10.8–1 and −2 pretty well explain themselves. The optical microscope needed for viewing the area to be explored and for locating the electron beam has been omitted from Figure 10.8–1. Baird and Zenger [11] have pointed out that the reflecting microscope usually provided to serve metallurgy is inadequate for geochemistry, and they have replaced it with one operating on transmitted light.

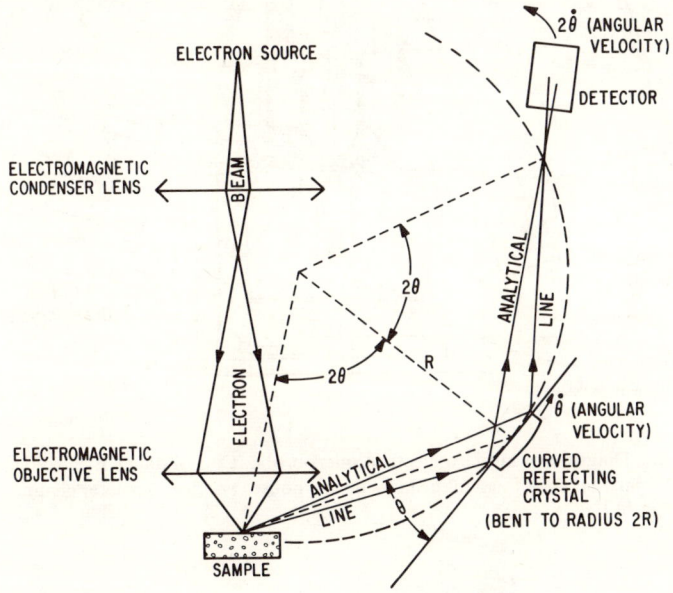

Fig. 10.8–1. Schematic diagram, not to scale, illustrating operation of an electron microprobe. Only one of four spectrometer channels is shown. Sample and Bragg crystal (Johansson arrangement) lie on focusing circle. Detector is near enough this circle to intercept focused analytical line at full intensity. This figure and the next by courtesy of Consolidated Electrodynamics Corporation.

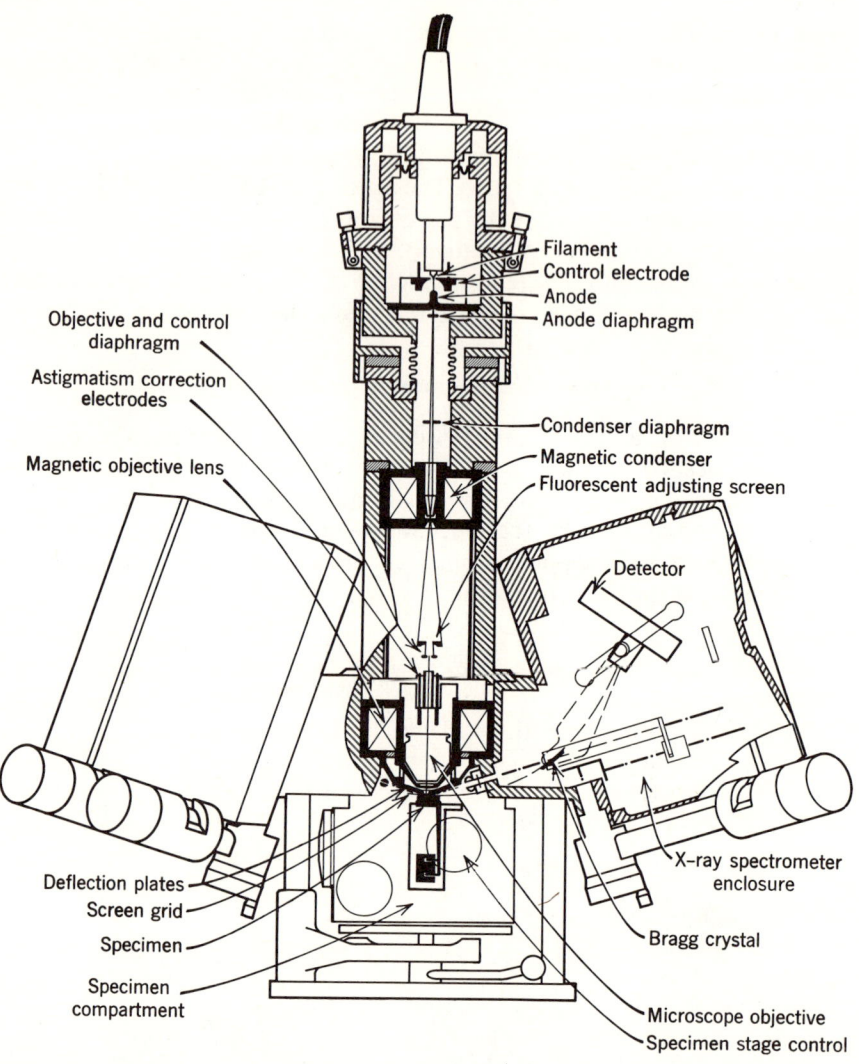

Fig. 10.8-2. Diagram of modern electron microprobe (Consolidated Electrodynamics Type 27-101) with parts labeled. Each of the large drums houses two spectrometer channels. Compare this with Figure 10.8-1.

EQUIPMENT AND ILLUSTRATIVE APPLICATIONS

That ability to scan the sample is vital to the microprobe (Section 9.28) is illustrated by Figure 10.8-3, which shows photographed cathode-ray oscilloscope images (700 ×) of a two-phase Al-Ag alloy in which the Al-rich phase contained 55 atomic % Al [12]. The three most important kinds of images are the characteristic x-ray image, the back-scattered electron image, and the specimen-current image, all of which are shown in the figure and understandable on the basis of Section 9.3. Images can also be created by secondary electrons from the specimen.

Figure 10.8-3 also contains *content maps* for the two elements of the alloy. These are produced by having a special pulser modulate the brightness of the image in direct proportion to the rate at which photons are being detected. In this way, it is possible to compensate a shortcoming of the human eye (logarithmic response to linear intensity variation) and make the contrast in the image sharper and more nearly representative of composition.

Figure 10.8-3 shows that scanning will generally reduce the need for quantitative determinations, and that it will give foreknowledge of where these had best be made when they are needed.

The electron microprobe and the computer seem made for each other. As the probe examines only about 1 μ^3 ($\sim$ 1 pg) of a sample at once, an enormous amount of data must be collected and processed if one wishes to establish localized composition with the greatest certainty of which the probe is capable [13]. The McCrone Microprobe Data Converter, compatible with all microprobes and most small computers, is an "interfacing system" developed to meet this need. It provides control for 81 microprobe parameters. With probe and computer thus "interfaced": (1) all elements from boron upward can be determined in 1-cubic-micron volume elements of interest; (2) data from the probe, corrected by computer, appear as analytical results in about 12 minutes; (3) overnight a square centimeter of sample can be scanned automatically for 3 different elements, about 10^{-14} g of each being detectable in each 1-μ^3 volume element (*not* in the sample!). It is no wonder [13] that computers with small memories are at a disadvantage.

Bayard [13] cites favorable experience with the Si(Li) detector and energy resolution to replace Bragg reflection for the microprobe (Section 2.25). All major elements above Mg can be determined at one location on the sample in 15 sec. Five thousand such locations within a 1-mm square can be thus examined in less than 24 hr and the result typed out automatically as a concentration map (upper limit, 6 elements). The replacement of sequential Bragg reflection by simultaneous energy resolution increases the number of routine samples done per day from 100 to 700. Through the combined efforts of Nuclear Equipment Corporation and Walter C. McCrone and Associates, a probe of this kind has been placed on the market.

The usefulness of the electron microprobe is easiest to demonstrate in

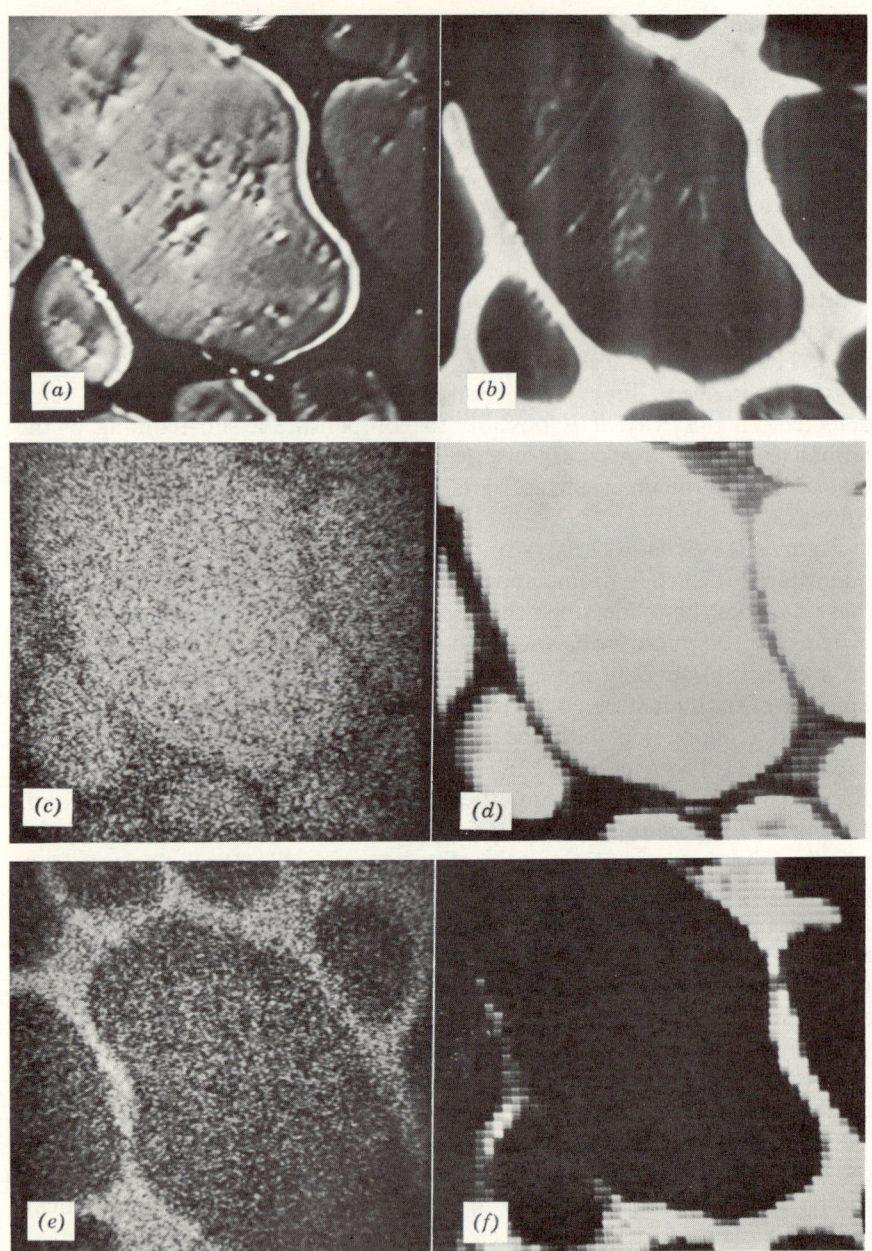

Fig. 10.8-3. Cathode-ray oscilloscope information display for a two-phase Al-Ag alloy (663 ×). (*a*) Back-scattered electron image; Ag-rich phase, light areas. (*b*) Specimen-current image, Ag-rich phase, dark areas. (*c*) Image created by Ag Kα. (*d*) Content map for Ag. (*e*) Image created by Al Kα. (*f*) Content map for Al. See J. S. Solomon and W. L. Baun, Ref. 12, p. 11, Fig. 1.

studies of metallic diffusion [14], but this has become commonplace. More noteworthy now is the *identification* and *analysis* of *single, pure phases* in minerals, an outstanding example of which is the case of sionite (Si-O-N; i.e., silicon oxynitride, Si_2N_2O) found as crystals up to near 200-μ long in a meteorite [15]. For this material, the microprobe determinations gave the following weight percentages to be compared with values in parentheses calculated from the formula: Si, 56.6 (56.1); N, 31.5 (27.9); O, 13.1 (15.9). As the last two elements, being light, are difficult to determine, these are excellent results. More of the same kind of work (on more costly samples) will be found in the "Apollo 11" issue of *Science* [16] and in subsequent accounts of work on lunar samples. We remind the reader that complete analyses via the microprobe are easiest to make on single, pure phases.

The electron microprobe appears to best advantage on important problems that cannot otherwise be solved. We give two examples.

In an important vacuum device, W. W. Welbon, General Electric Company, was able to trace leakiness in thin composite seals to the dissolving of a nickel plate by Cu-Au brazing alloy [17]. The determinations were made by traversing the seals at the rate of 6 μ/min across the stationary electron beam and recording sequentially the intensities of a characteristic line for each of the metals Ni, Au, and Mo, one spectrometer being used for each line. See Figure 10.8-4.

Few, if any, assignments in analytical chemistry are more important and more complex than investigations of what happens to nuclear-fuel elements during service [18]. Not only are highly localized determinations essential, but also the sample to be examined needs to be small because of its high radioactivity. Among the problems investigated are the redistribution of fuel, the distribution of fission products, the presence of substances resulting from interaction of fuel and the stainless-steel cladding, and the nature of inclusions. We have room only for the latter.

Metallographic photographs revealed particles of a gray nonmetallic phase. Electron microprobe examination of one such particle (area, about 500 μ^2) gave a bright specimen-current image. Such an image immediately shows that light elements predominate, for its brightness varies inversely with mean atomic number. An Al $K\alpha$ image showed the predominant light metal to be aluminum. In Figure 10.8-3, photographs *b* and *e* tell the same kind of story. X-ray images for Pu, U, Ba, Cs, Fe, and Te completed the analytical data that led to this conclusion: alumina had been introduced as impurity during the ball-milling of the fuel. Being insoluble in the solid UO_2–PuO_2 matrix, the Al_2O_3 had migrated to the cooler end of a long columnar fuel grain, 0.55 mm from the stainless-steel wall. Iron from this wall had been transported inward and formed inclusions around which the alumina had gathered. The alumina "gettered" BaO, Cs_2O, and Te. A more convincing example of the value of scanning would be hard to find:

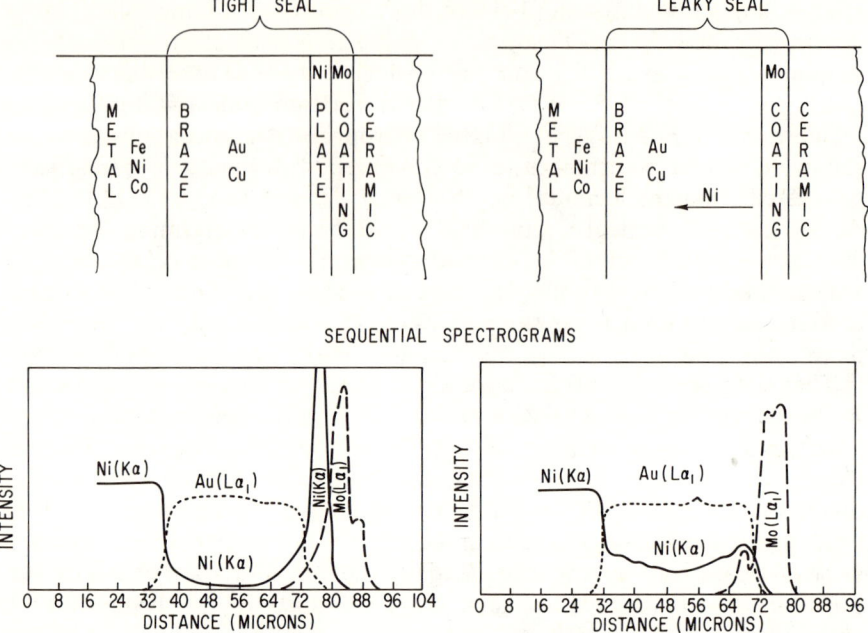

Fig. 10.8-4. Intensities of characteristic Ni, Au, and Mo lines measured sequentially via an electron microprobe in which ceramic-metal seals were traversed. Note how the intensity of Ni(Kα), after beginning in the Fernico alloy (reader's left) at a value common to both seals, diverges sharply in the region that contains the braze and the molybdenum coating. Clearly the nickel plate in the leaky seal has been dissolved by the braze. See H. A. Liebhafsky, et al., Ref. 17, p. 71, Fig. 4.

quantitative results were not needed. The usefulness of the microprobe in other aspects of the nuclear-fuel work was equally impressive.

10.9 Electron Excitation of Larger Sample Areas

The success of the electron microprobe has stimulated a reappraisal of electron excitation over large (~ 1 cm^2) sample areas and provided guide lines for the development of the needed equipment. This return to electron excitation could well be the most significant recent change in x-ray emission spectrography. We shall describe the Telsec Betaprobe, as one spectrograph of this type is called, and show results also for the JPX-3 Primary X-ray Analyzer, a similar instrument that differs in several important respects. As usual, the names tend to confuse.

Enough will be said about the Telsec Probe to make possible a meaningful comparison with the electron microprobe (see above). The Telsec Probe

EQUIPMENT AND ILLUSTRATIVE APPLICATIONS   479

has two contiguous vacuum systems, each with its own backing and diffusion pump, joined by an aperture for the electron beam, the aperture being so small that the electron-gun chamber is maintained at the necessary 10^{-5} torr, even though the pressure in the sample-(flat-crystal) spectrometer chamber is a hundred times as great. Results: pump-down time from atmospheric pressure is only 30 sec for a metal sample—longer (up to 3 min) for clays not previously outgassed to reduce pump-down time. Contamination of the electron gun is prevented—even so, the filament therein has a limited operating life (say, 24 hr; but many determinations!) and planned replacements had best be made.

A simple grid focuses the electron beam through the aperture. This beam would normally form on the sample an enlarged, *but unstable*, image of the tip of the filament. A simple electromagnetic lens is consequently added to focus the electrons to a spot about 1 mm in diameter on the sample. Scanning coils cause the beam usually to traverse a sample area of 10×10 mm; the side of this square can be changed from 1 to 16 mm to meet special requirements. Beam currents of 0.1, 0.2, 0.5, and 1.0 mA are available at up to 15 kV. Beam stability is 0.1%. Nonconducting powdered samples must be made conducting as by admixing graphite; they are subsequently put in a lead ring or recessed lead disk and pressed. Lower beam currents are often desirable to reduce heating. A determination usually takes about 10 sec once excitation has begun.

Three versions of the Telsec Probe are available. The B.100 is for nonroutine work and performs sequential determinations over a 2θ-range up to 150° with the aid of six Bragg crystals mounted in vacuum and manually selected from outside. The B.200 is a multichannel spectrograph with up to 12 channels complete to include the detector; the ancillary electronics (pulse-height selection, counting) is in a single unit that is automatically switched to each detector in turn so that the intensity can be printed out for each element. The B.300 is a fully automatic multichannel instrument that gives simultaneous teletype print-out of all intensities: 30 sec for sample insertion and pump-down with percentage results for 12 elements about 15 sec later. For element coverage and related data, see Table 10.9-1.

The Telsec Probe has already found important and diverse application in Britain. Applications Report 18 by the manufacturer is of special interest as it compares the Probe with a conventional x-ray emission spectrograph for determinations on clays. The Corporate Laboratories of the British Steel Corporation have a Probe with 9 fixed channels arranged to determine F, Mg, Al, Si, P, S, Ca, Mn, and Fe. Telsec Instruments Limited has literature describing these applications and others.

An examination of this literature has led to these conclusions. (1) The results obtained are generally in accord with information about x-rays, equipment, and samples given in this book. (2) The excellent linearity of

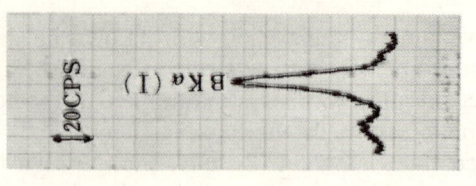

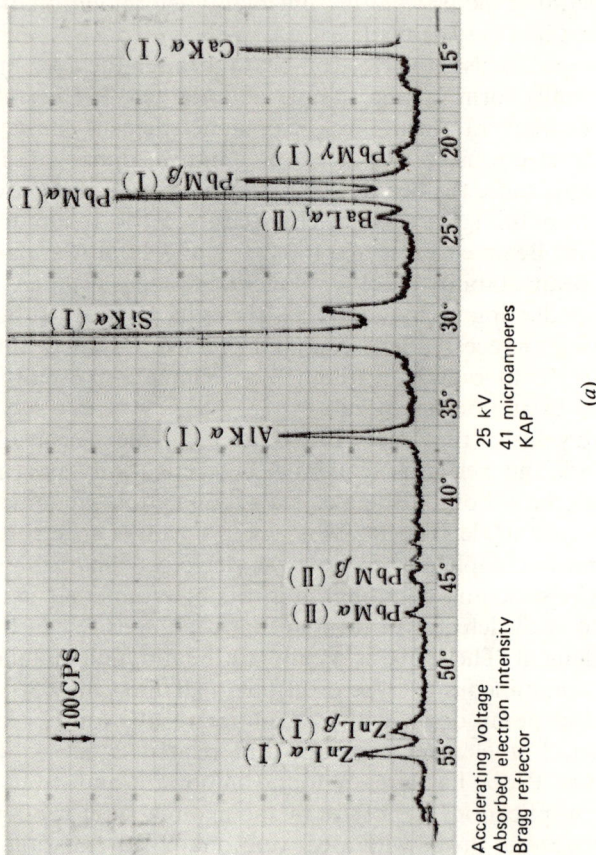

(a)

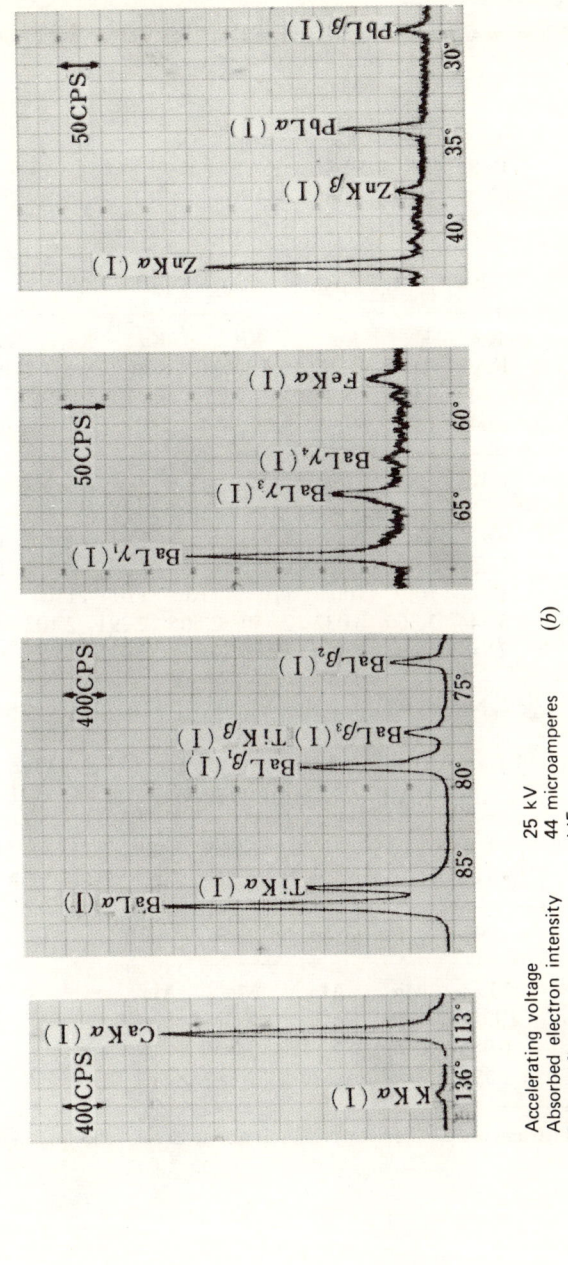

Accelerating voltage	25 kV
Absorbed electron intensity	44 microamperes
Bragg reflector	LiF

(b)

Fig. 10.9-1. The qualitative analysis of a borosilicate glass on the JPX-3 emission spectrograph with electron excitation up to 25 kV. Note that K, L, and M lines are all present at acceptable intensities. Courtesy of Japan Electron Optics Laboratory Co.

Table 10.9–1. Element Coverage and Related Data for Telsec Betaprobes[a]

Element	$_5$B	$_6$C	$_7$N	$_8$O	$_9$F			
Analytical line	Kα	Kα	Kα	Kα	Kα			
Bragg reflector	PbSt	OHM	PbSt	KAP	KAP			
Wavelength, Å	67	44.4	31.603	23.707	18.307			
Energy, keV	0.19	0.28	0.39	0.52	0.68			

Element	$_{11}$Na	$_{12}$Mg	$_{13}$Al	$_{14}$Si	$_{15}$P	$_{16}$S	$_{17}$Cl	
Analytical line	Kα	Kα	Kα	Kα	Kα	Kα	Kα	
Bragg reflector	KAP	ADP	EDDT	EDDT	KBr	NaCl	NaCl	
Wavelength, Å	11.909	9.889	8.339	7.126	6.155	5.373	4.729	
Energy, keV	1.04	1.25	1.49	1.74	2.01	2.31	2.62	

Element	$_{19}$K	$_{20}$Ca	$_{21}$Sc	$_{22}$Ti	$_{23}$V	$_{24}$Cr	$_{25}$Mn	$_{26}$Fe	$_{27}$Co
Analytical line	Kα	Kα	Kα	Kα	Kα	Kα	Kα	Kα	Kα
Bragg reflector	LiF 100	LiF 100	LiF 100	LiF 110	LiF 110	LiF 110	LiF 110	LiF 110	LiF 110
Wavelength, Å	3.744	3.360	3.032	2.750	2.505	2.291	2.103	1.937	1.791
Energy, keV	2.98	3.69	4.09	4.51	4.95	5.41	5.87	6.41	6.93

Element	$_{28}$Ni	$_{29}$Cu	$_{30}$Zn	$_{38}$Sr	$_{42}$Mo	$_{50}$Sn	$_{56}$Ba
Analytical line	Lα	Lα	Lα	Lα	Lα	Lα	Lα
Bragg reflector	KAP	KAP	KAP	EDDT	NaCl	LiF 100	LiF 100
Wavelength, Å	14.595	13.357	12.282	6.865	5.409	3.603	2.780
Energy, keV	0.85	0.93	1.01	1.87	2.29	3.44	4.46

Element	$_{74}$W	$_{78}$Pt	$_{79}$Au	$_{80}$Hg	$_{82}$Pb
Analytical line	Mα	Mα	Mα	Mα	Mα
Bragg reflector	EDDT	KBr	KBr	KBr	NaCl
Wavelength, Å	6.985	6.050	5.845	5.666	5.290
Energy, keV	1.77	2.05	2.12	2.19	2.34

[a] Courtesy Telsec Instruments Limited. Information about Bragg reflectors in Table 5.7–1.

EQUIPMENT AND ILLUSTRATIVE APPLICATIONS

calibration curves is unexpected and gratifying; many examples are given. (3) The Probe is clearly superior to the conventional spectrograph for the determination of light elements; for example, C in white cast iron. For the Probe, these elements are not a class apart. (4) As atomic number increases, the conventional spectrograph probably gains the upper hand; for it can use **K** and **L** spectra where the Probe uses **L** and **M**. (5) As regards interelement effects, the Probe has the expected advantage, interelement effects being much reduced. This was convincingly demonstrated by using the Probe to determine chromium on the 60 steels for which Lucas-Tooth and Pyne carried out a regression treatment of results obtained by x-ray excitation (Section 9.13). The same working curve could be drawn through the new results and the old, with the important difference that the Probe results lay so close to the working curve as to make a regression treatment unnecessary, or nearly so. (6) The great usefulness of standards (comparative method) is clearly demonstrated: although the calculation of corrections, mandatory in much electron microprobe work, could similarly be done here, it is unlikely to be often needed.

Determinations of light elements seem as easy to carry out as those of

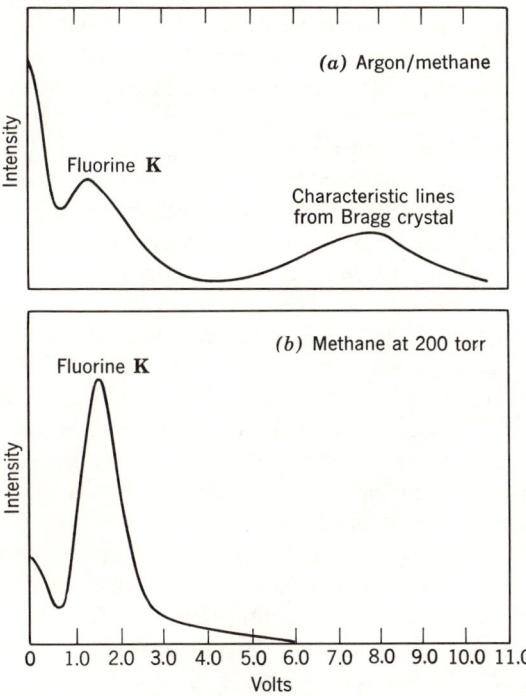

Fig. 10.9–2. Improved detector conditions for the determination of fluorine. The abscissa is the voltage setting for pulse-height discrimination. Courtesy Telsec Instruments Limited.

the heavier; Figure 10.9–1 shows representative spectra. The determination of fluorine is a useful illustrative example because it marks the lower limit of several conventional vacuum spectrographs. The Bragg crystal is KAP. The flow detector has a 2-μ polycarbonate window, and methane at 200 torr as filler gas. Figure 10.9–2 shows, in accord with Chapter 2 and elsewhere, the great improvement that results. Relevant data: intensity, adequate for 80 cps; counting interval, 40 sec; background equivalent to 0.3% F.

10.10 Modern Photoelectron Spectrography

It should be emphasized that the outer electron locations [in atoms] do not necessarily apply to molecular compounds, or to the atoms in a liquid or solid state, i.e., on the target of an x-ray tube. A. H. Compton and S. K. Allison [19]

Robinson was identified with magnetic photoelectron spectrography approximately from the first World War to the second. (See Sections 1.12 and 4.5). Since then, K. Siegbahn and his colleagues have made notable progress in the field [20] and increased Robinson's great precision still further so that it is now possible to measure binding energies of electrons to a fraction of an electron volt. Even more recently, the magnetic field for controlling the electron path has been replaced by an electrostatic, with the advantages that size and cost of equipment are reduced. In electrostatic photoelectron spectrography, the necessary magnetic shielding is accomplished by using mu-metal, or something similar. Equipment of this kind is now offered by Varian and by others [21].

As concerns us, photoelectron spectrography (or *E*lectron *S*pectroscopy for *C*hemical *A*nalysis, ESCA, acronym indicated) is the substitution of a characteristic (photo) *electron* line, related to binding energy of the electron as discussed in Section 1.12, for a characteristic *x-ray* line emitted because this photoelectron was ejected. The advantages and disadvantages of this substitution are implicit in the quotation given above and discussed in Section 9.26.

Figure 10.10–1 is to be compared with Figure 9.26–2. It is probable that the former spectra are more precise, easier to measure, better for establishing electron binding energies, and adequate for *qualitative* identification. Although characteristic-electron-line *energies* are being measured and tabulated in abundance, it is at least doubtful whether they will replace tables of characteristic-line *wavelengths* in *quantitative* determinations. As a contribution to the current furor about polywater, ESCA methods have been used to show that it contains Na^+, K^+, SO_4^{2-}, CO_3^{2-}, Cl^-, NO_3^-, BO_3^{3-}, SiO_4^{4-}, and organic carbon residues—"all the news that's fit to print?" [22]. As a counterbalance to this incursion of ESCA into qualitative analysis, the reader may wish to examine the use of soft x-ray spectra to establish binding energies [23].

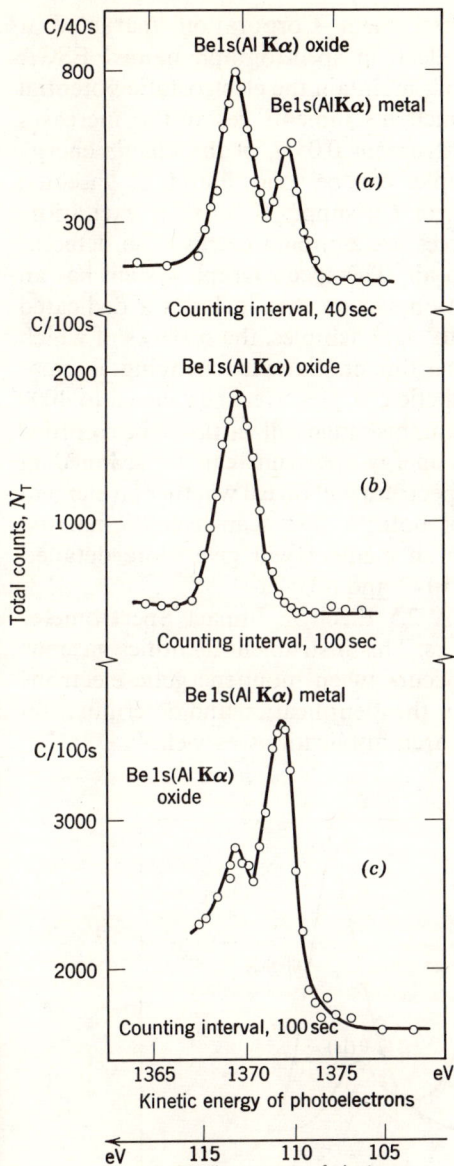

Fig. 10.10–1. Photoelectronspectra from (a) Be evaporated in vacuum (b) this Be heated in air and (c) this Be reduced with Zr, all excited by Al Kα. For energy relationships, see Sec. 1.12. The 1s electron is a **K** electron, and the Be **K** spectrum is emitted when the 2s (or **L**) electron fills the hole left by photoemission of the 1s (or **K**) electron. The *qualitative* change in the Be/BeO ratio is clearly indicated. After Siegbahn et al., Ref. 20, p. 77, Fig. V.1.

During 1971, the McPherson Instrument Corporation marketed a versatile, precise electrostatic photoelectron spectrograph named ESCA 36 because the hemispherical shells that maintain the electrostatic potential have a mean radius of 36 cm. In such instruments resolution increases with this diameter, being for this instrument 0.05% of the *kinetic* energy of the photoelectrons collected. Samples can be solid, liquid, or gaseous; cooled or heated. Targets of Mg, Al, or Cu supply **K** lines for excitation. Monoenergetic electron or UV sources are optional extras. The detector is an electron multiplier exposable to air. The spectrograph system has an automated control and data acquisition system that includes a dedicated computer. A sample wheel holds up to eight samples, the surfaces of which can be cleaned by argon-ion bombardment. Sample changing is easy.

The ESCA 36 measures electron kinetic energies over the range 4 to 4000 eV. Auger electrons (Section 7.16) within this range will naturally be recorded along with the photoelectrons. The energy spectrum can be scanned at variable rates: a rapid scan over this spectrum will reveal whether an element is present, be it free, or combined, or both; a slow scan over the narrow spectral region thus marked out for an element will give more detailed information about it. See Figures 10.10–2 and –3.

McPherson also markets an ESCA 2.5 Electron Impact Spectrometer for determinations on gases and vapors. This instrument identifies gaseous molecules via the energy loss that occurs when monoenergetic electrons strike them. It promises to be useful for the identification and determination of pollutants in air, and to have research applications as well.

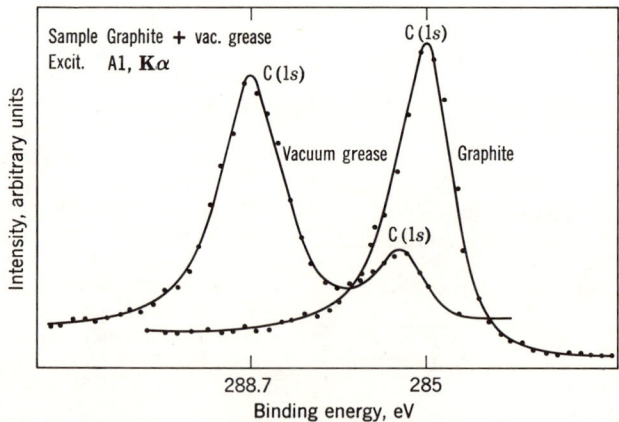

Fig. 10.10–2. Different binding energies of carbon demonstrated by modern photoelectron spectrography. A carbon C (1s) peak appears for pure graphite at 285 eV. Vacuum grease has C (1s) peaks at 288.7 and 286 eV. Carbon is thus bound differently in vaccum grease than in graphite. Courtesy of McPherson Instrument Corporation, a subsidiary of GCA Corporation.

SELECTED APPLICATIONS OF BROAD SIGNIFICANCE

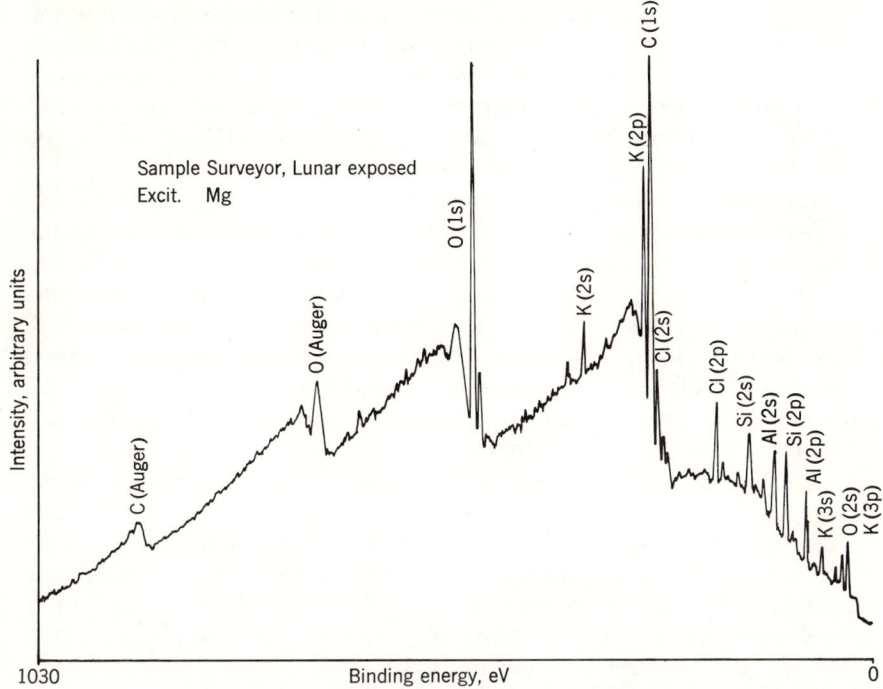

Fig. 10.10-3. Wide-range scan on a modern photoelectron spectrograph of a portion of the Surveyor Moon Probe exposed to lunar radiation for two years. Note that carbon and oxygen show both Auger- and photoelectron peaks. Courtesy of the McPherson Instrument Corporation, a subsidiary of GCA Corporation.

SELECTED APPLICATIONS OF BROAD SIGNIFICANCE

Applications presented above showed the capabilities of equipment. Applications that follow are chosen to illustrate the value of x-ray emission spectrography in an analytical laboratory and in the plant. Nonmetallic samples will be emphasized because metals and alloys have already had their share of attention and because they usually present fewer difficulties.

10.11 Qualitative and Semiquantitative Determinations

Qualitative determinations come so naturally that they need neither definition nor description. Semiquantitative determinations cover much ground, but may be defined operationally as those in which the reliability is the best possible in short time with minimum effort.

The Fluo-X-Spec Laboratory of Denver has long been prominent in this field. As M. L. Salmon has recently described its methods [24], we summarize below only the principal points of the usual procedures of this laboratory. Not all samples need the full treatment; special samples may require something more. Reference 24 is the source of the highlights that follow.

1. *Elements covered*: $_{22}$Ti and heavier.
2. *Samples*: liquids; slurries; residues from evaporation or ashing; homogeneous solids, directly or as fine powders; heterogeneous solids, ground fine enough so that an aliquot is of average sample composition.
3. *Sample containers*: plastic cups (diameter, 3/8 to 1 in.) sealed with Mylar films by means of plastic retaining rings. Adapter rings present cups reproducibly to the x-ray beam. Sealed cups are filed for possible future use. Adequacy of sample preparation is established operationally by x-ray spectrographic examination.
4. *Prepared standard samples*: No one method for all cases. For powdered unknowns, prepare powdered standards (1) by grinding to 325 or 400 mesh for constituents of like densities; (2) by evaporating solutions of heavy constituents onto light constituents in powder form if density differences are large; (3) by fluxing first, then grinding for cases where (1) and (2) fail.
5. *X-ray equipment*: Philips. Inverted geometry. For scanning from 0.15 to 2.85 Å, use W-target tube up to 100 kV. When needed, use Mo-target tube for determination of individual elements with analytical lines near 1 Å.
6. *Bragg crystal*: LiF as best single compromise.
7. *Collimation*: Soller slits, 4 by 0.005 in. between sample and Bragg crystal; $1\frac{1}{2}$ by 0.023 in. between crystal and detector.
8. *Detector*: scintillation, at 900 V; selected as giving adequate response for Ba Lα, the analytical line of longest wavelength.
9. *Pulse-height selection* with automatic adjustment of voltage window to changing 2θ.
10. *2θ-scanning:* at variable rates up to 32°/minute. Time constants of associated electronic circuitry variable from 0.02 to 16 sec. Lower time constants desirable for rapid scanning.
11. *Six standard scanning procedures* suffice for most samples. They are made on the same paper, usually with different colored inks. Not all six are always needed. Details in Ref. 24. Principal characteristics of each procedure follow. Number 1 gives complete qualitative information; pulse-height selection is used. Semiquantitative data only for elements free of spectral interference by characteristic lines of W target. Number 2 includes higher-order Bragg reflection; pulse-height selection omitted. Supplementary intensity information also. Number 3 eliminates interference by characteristic lines from W target by action of composite filters (2 mils Al, 1 mil brass, 0.5 mil Fe) over tube window. Facilitates determina-

tion of W, Ge, Hg, Au, Se, Pt, Tl, Ga, Re, and Ta. Number 4 reduces spectral interference by nearest-neighbor elements through insertion of filter with appropriate absorption edge to act on beam from sample. Number 5 gives improved spectra for elements with atomic numbers between 38 and 74 by use of 100-kV, 16.5-mA excitation. In other scanning procedures, 50 kV, 33 mA. Number 6 increases intensity at wavelengths beyond 1 Å by use of He path. Air path in other scanning procedures.

12. *Qualitative results*: obtained visually by placing chart over transparent template on transparent rotary drum with interior illumination—an ingenious development of the Fluo-X-Spec Laboratory. On the template are typed for a 2θ position (horizontal axis) the element, x-ray line, and order of Bragg reflection for all possible peaks the chart could show near that position; sometimes over 50 possibilities exist. For any peak, single or composite, on the chart, all possibilities are visible. Crosschecks at other 2θ positions permit elimination of possibilities not applicable to sample and confirmation of the analytical lines actually generated.

13. *Semiquantitative results*: obtained from chart in "chart-units" with any changes in chart scale allowed for. Background is estimated and subtracted from peak height; see Figure 9.10-2. Background is also used to normalize sample and instrumental conditions by using a ratio analogous to $(N_T - N_B)/N_B$ in our symbols, which is like using a scattered x-ray line as a reference standard. Special attention must be paid to the background near an absorption edge.

What has been described above may well be the best existing approach of its kind for complex and diverse samples. Results for uranium are impressive. To show this, we give for each of six standards the % U by weight chemically established followed by the *range*, in parentheses, of 10 replicate x-ray determinations: 0.049 (0.045–0.055); 0.108 (0.10–0.12); 0.179 (0.17–0.18); 0.248 (0.23–0.25); 0.458 (0.45–0.48); 0.665 (0.65–0.67). For most of many actual ores, the agreement is almost as good. Method development in the Fluo-X-Spec Laboratory is continuing.

Anater [25] has developed a less comprehensive scheme of qualitative and semiquantitative analysis that is interesting because it shows what can be done in x-ray emission spectrography with an almost irreducible minimum of effort. Compromise standard operating conditions are chosen that depend upon the equipment available. Normalization factors (NF's) are next measured on known standards for each element but one within the range of the equipment, the one element (in his case Fe) being used as reference. The following equations describe the procedure:

$$\frac{(NACD)_{Fe}}{W_{Fe}} = Q_{Fe} \qquad \frac{(NACD)_E}{W_E} = Q_E \qquad (10.11\text{-}1)$$

$$\frac{Q_{Fe}}{Q_E} = (NF)_E \qquad (10.11\text{-}2)$$

where the W's are percentages by weight and NACD's, the "net adjusted chart divisions," measure peak heights on the same scale for Fe and for element E, with the heights properly corrected for background. Unknowns of all kinds are run as originally received if the sample holder can accommodate them. For any element E in an unknown, W_E can be calculated from these equations if the needed NF's are known and if all elements in the sample give analytical lines within range of the equipment. Clearly, the unknown need not contain Fe, for

$$\frac{Q_E}{Q_{E'}} = \frac{(NF)_E}{(NF)_{E'}} \qquad (10.11\text{-}3)$$

When not all elements in the sample can be determined, the assumption

$$W_E + W_{E'} + W_{E''} \ldots + W_n = 100 \qquad (10.11\text{-}4)$$

is made for the n elements determined, and the results are reported on this basis. They will obviously be too high.

Anater applied this scheme to 27 elements above and including $_{13}$Al. He claims reliability within a factor of two with most results precise to $\pm 20\%$ (relative) of the amount present. Representative NF's vary from 80 for Al (least sensitive to method) to 0.74 for Ca (most sensitive).

10.12 Selected Determinations on Complex Materials

Minerals are the prime examples under this heading. Reference 1, pp. 199–209, gives a general discussion, and the book by Adler [26] should be consulted. Here we mention only that Carl and Campbell [27] have divided minerals conveniently into four groups from the viewpoint of x-ray emission spectrography.

THUCOLITE. The classical work of Faessler in Hevesy's laboratory [28] serves to set the stage. Thucolite is a remarkable mineral that contains an appreciable proportion of a volatile carbon compound, not a hydrocarbon; thucolite was investigated at Freiburg mainly in connection with the age of the earth. The mineral fumed and did not change shape when ignited in air. The residual skeleton, 37% the original weight, collapsed on touch to form a fine powder in which 23 elements were found by x-ray emission spectrography with electron excitation and with the photographic plate as detector. The characteristic lines for this powder are listed in Table 2.4–1, and their wavelengths attest to a resolution probably unsurpassed in work

SELECTED APPLICATIONS OF BROAD SIGNIFICANCE

of this kind. Quantitative results (with emphasis on Th, U, and Pb) were obtained by the use of internal standards and by ingenious reasoning. The account of this investigation will repay close study today.

WORK AT POMONA COLLEGE. In 1961, Baird, MacColl, and McIntyre [29], using an x-ray emission vacuum spectrograph with a tungsten-target tube—all virtually as supplied by Philips—presented the results of a careful investigation of the precision and sources of error in the determination of the major light elements in granitic rocks. They concluded that the method offers the geologist his first opportunity to obtain quickly and relatively cheaply high-precision silicate analyses in large number. Borax being used as flux, there was complete insensitivity to sodium and (usually) poor sensitivity to magnesium. Several years later, Baird and Henke [30] had used one of Henke's new x-ray sources [31] to generate Cu L lines for the excitation of analytical lines in 17 rock samples, powdered and compressed into smooth pellets, but not fused. Their results, obtained by comparison with quartz, appear in Table 10.12–1.

Table 10.12–1. Precisions of Silicate Analyses by Several Methods [30]

Element	Approx. Composition, w/o	Intralaboratory Analytical Precision, Std. Dev., s, w/o				
		X-Ray	Wet Chemistry	Emission	Flame	Neutron Activation
O	48.	0.4	0.3	—	—	0.9
Na	2.6	.02	.15	0.11	0.07	—
Mg	0.5	.01	.15	.12	—	—
Al	9.3	.04	.19	.53	—	—
Si	30.0	.10	.14	1.1	—	—
K	3.7	.03	.21	0.15	0.07	—
Ca	2.5	.01	.07	.14	—	—
Ti	0.22	.003	—	—	—	—
Fe	3.1	.02	.21	.14	—	—
	99.9					

As these samples were unfused, it was possible to determine O, Na, and Mg. The results become even more remarkable when one remembers that fusion is normally required for high precision. There is no better evidence of the value of x-ray emission spectrography than this table. It is the only method by which all elements, *oxygen included*, can be directly determined. The standard deviations speak for themselves.

The work at Pomona College includes a study [32] of 949 samples of plutonic rock from Rattlesnake Mountain based on x-ray results obtained with a spectrograph that permitted determinations of Na, Mg, Al, Si, P, S, K, Ca, Ti, Mn, and Fe on a single preparation of a silicate rock. The work is noteworthy for, among other things, the extensive use of computers to automate the x-ray determinations and to expedite the extensive statistical treatment of the data. Reference 32 is a useful guide to related papers published from Pomona College.

RARE EARTHS. The ores of these elements have always been a thorn in the side of the classical analytical chemist. For this, the chemical similarity of the elements is of course responsible, and this similarity makes the isolation of pure compounds difficult. Moseley [33] was consequently pushed into making the first attempt at quantitative determinations by x-ray emission spectrography: he had to conclude from the x-ray spectra that what he thought was praseodymia contained rare-earth elements in these proportions: La, 50; Ce, 35; Pr, 15. *Caveat emptor* must have come to his mind!

Although x-ray emission spectrography has an overriding advantage here over classical analytical methods, reliable x-ray results presuppose careful work as Lytle, Botsford, and Heller point out [34]; see also Ref. 1, pp. 205–206. Kunzendorf and Wollenberg [35] have recently shown what can be done by exciting plane rock surfaces with collimated 59.5 keV γ-rays from a $_{95}Am^{241}$ source, and by using a Ge(Li) detector in conjunction with a 1024-channel pulse-height analyzer. By unfolding (see Section 9.16) the complex spectrum obtained from a known powdered mixture of rare-earth oxides, they obtained data by use of which it was possible to reverse the process and calculate a spectrum for a sample that contained La, Ce, Pr, and Nd. The calculated and measured spectra agreed within 3%. From this result and other work, they concluded that their method could detect rare-earth elements when present near 10 ppm, but that conventional x-ray emission spectrography carried out on powdered samples would give results of higher reliability.

PORTLAND CEMENT. C. W. Moore [36] has submitted to ASTM a suggested method for the determination of Si, Al, Fe, Ca, Mg, K, Ti and Sr in portland cement by x-ray emission spectrography. The cement is fused after having been mixed with lithium tetraborate in an amount that gives a flux/cement ratio of 3:1 in the resulting bead, which is ground to a fine powder subsequently briquetted to give a flat surface. The comparative method with a suitable permanent reference standard is used. Results are obtained by use of calibration curves, or by calculations that use intensity ratios and allow for loss of weight on ignition. The method is directly applicable also to determinations on many of the raw materials from which cement is made; appropriate standards are needed, and corrections for interelement effects must be modified. A flow proportional detector is used for all elements with analytical lines longer than 2 Å; a scintillation detector is used for Fe and Sr. The analytical lines are $K\alpha$'s, except that $K\beta$ (3.090 Å) is employed for Ca. The minimum number of counts taken for an element in the reference standard ranges from $2.0(10^3)$ for Sr to $4.8(10^5)$ for Fe. Table 10.12–2 shows the precision attained.

Table 10.12–2. Precision Attained in Suggested Method for Portland Cement[a]

Oxide	Average Composition, w/o	Single-Operator-Day Standard Deviation, s, w/o	Single-Operator Multiday Standard Deviation, s, w/o	Single-Operator Multiday Relative Standard Deviation, s, %
SiO_2	22.05	0.05[c]	0.05[d]	0.23
Al_2O_3	3.79	0.07	0.05	1.32
Fe_2O_3[b]	2.65	0.007	0.006	0.23
CaO	66.71	0.09	0.14	0.21
MgO	1.26	0.04	0.05	3.97
K_2O	0.40	0.005	0.005	1.25
TiO_2	0.17	0.005	0.005	2.94

[a] Data supplied by C. E. Carter, Research Chemist, Lone Star Cement Corporation.
[b] Fe_2O_3 data corrected for calcium interelement effect.
[c] Numbers in this column are standard deviations based on single determinations made during a single day on each of 10 portions of a single cement. One operator throughout.
[d] Numbers in this column are standard deviations based on single determinations made on single portions prepared on each of six different days from the same cement with determinations closely following the preparation. One operator throughout.

LARGE-SCALE APPLICATIONS

Please read Section 3.17. The two applications described below use servomechanism systems (control loops). The first loop was not closed in 1968, but may be closed by now.

Process control and computers are being so rapidly and happily joined that they will never be rent asunder. Analog error signals (i.e., continuous signals such as might result by comparing two rate meters) must pass through an analog-digital converter before a computer can use them; the computer output must undergo the reverse transformation before control can be exercised in a closed loop. A dedicated (small) computer is often used in this way. Access to a central (large) computer may be needed for extensive storage of information or for further calculations. Computerized control is more flexible, more versatile, and easier to expand than pure analog control. described in Section 10.4.

10.13 Process Control of Flowing Mineral-Slurry Streams

The Lake Dufault Mine, Noranda, Quebec, produces copper and zinc. Determinations of these elements (and of Fe if needed) are made sequentially and automatically on seven slurry streams [37] by an ARL Production Control X-Ray Quantometer (PCXQ), one of the instruments in the line described in Section 10.4.

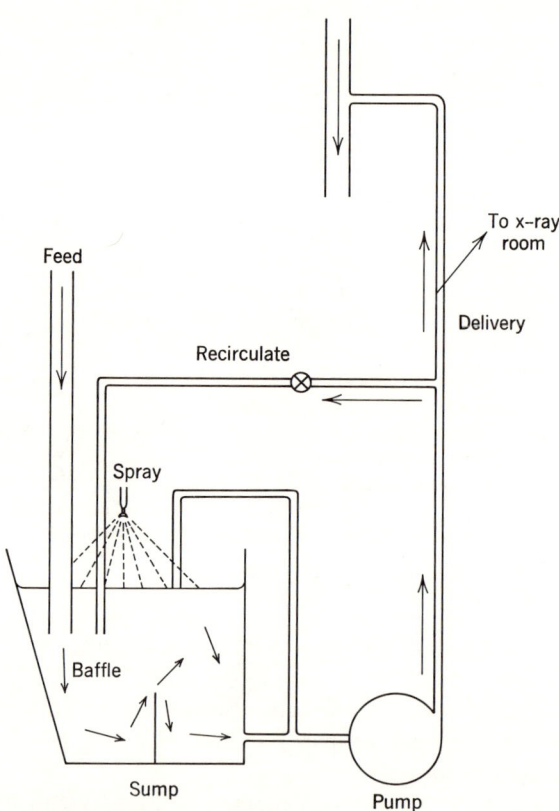

Fig. 10.13-1. System for treatment and delivery of mineral slurries to PCXQ. 20 to 30 gallons of slurry per minute are taken from the main stream. As needed, stationary screens remove undesired material such as wood fibers. Sand in the mill-head slurry is ground to reduce it from -10 to -28 mesh. The fine water-spray breaks up froth and releases entrapped air. Pipes are polyethylene. A 3/8-in. tube inclined at 45° (not shown) leads to manifold and valve system in x-ray room; slurry volume here is minimal. See C. L. Lewis, R. A. Hall, J. W. Anderson, and W. H. A. Timm, Ref. 37, p. 4, Fig. 6.

LARGE-SCALE APPLICATIONS

The ore is treated by crushing, grinding, and selective flotation to make copper and zinc concentrates. A less promising kind of sample for rapid, precise determinations is hard to imagine, but it does have the advantage that a rapid flow of slurry makes for automatic averaging (see Figure 10.13-1). As Table 10.13-1 shows, and several years of satisfactory operation prove, the results are highly reliable, in large part because the slurries can be reproducibly handled, and because scattered x-rays can be used empirically to measure adequately the solids content of the slurries. The following ranges of 19 chart readings for each element on the same circulating slurry are narrow indeed: Cu, 10.0–10.1; Zn, 51.6–52.1; Fe, 64.5–65.6. ARL also offers an automatic briquetting system for the PCXQ.

Table 10.3-1. X-Ray and Chemical Results for Seven Slurries [37]

Sample	X-Ray		Chemical	
	Cu, w/o	Zn, w/o	Cu, w/o	Zn, w/o
Mill heads	3.60	8.70	3.97	8.89
	3.70	8.00	3.86	8.02
	4.24	10.01	4.25	10.10
Unit cell concentrate	20.70	4.20	21.17	4.50
	24.90	2.80	24.84	3.24
	19.40	4.25	19.73	4.39
Cyclone overflow	1.27	8.35	1.32	8.78
	0.92	11.00	0.93	10.39
	2.20	13.25	2.38	13.10
Copper concentrate	19.80	5.05	20.06	5.44
	21.70	4.05	21.75	3.66
	20.55	4.09	20.00	4.30
Copper tailing	0.31	9.10	0.32	9.30
	0.29	10.25	0.30	10.52
	0.24	10.80	0.23	10.32
Zinc concentrate	0.47	52.36	0.50	52.75
	0.43	50.05	0.58	50.28
	0.54	51.85	0.40	51.90
Final tailing	0.19	0.71	0.22	0.86
	0.14	0.48	0.15	0.50
	0.14	0.49	0.12	0.50

10.14 Automated Cement Manufacture

At Tijeras, N.Mex., the Ideal Cement Company has a closed control loop for making its product. In normal operation, an x-ray emission system (General Electric X-Ray Emission Gage, XEG) continuously determines four elements (Ca, Al, Si, and Fe) in the raw cement mixture as it issues from the ball mill at rates up to about 2000 tons/day. The computer in the system checks the analytical information every 6 to 10 minutes and accumulates it for 1 hour. At the end of the hour, the computer uses this information properly weighted to calculate whether the composition of the raw cement mixture corresponds to certain "holding points" for the finished product.

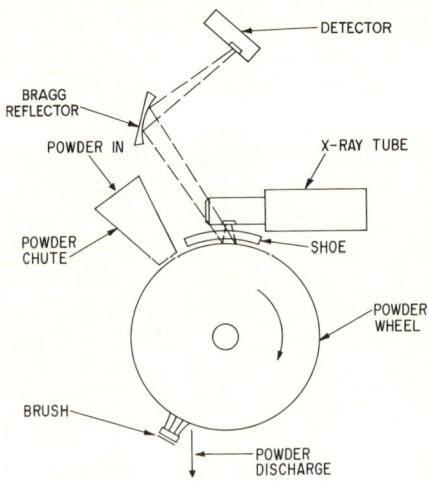

Fig. 10.14-1. Sampling in closed-loop system for cement manufacture. The rim of the powder presenter, a rotating wheel, is suitably indented to retain the powder that is the sample. Before the characteristic lines are excited, the powder is compacted by the brakeshoe-like device (center) to make the sample smooth and uniform. The x-ray beams pass through a 1-mil beryllium window in the shoe. See Ref. 17, p. 73, Fig. 6.

Any significant deviations of the calculated results from these holding points are then corrected by appropriate automatic adjustments of the weighers for the raw-material feeds. On-line experience with this x-ray emission control system has been favorable: sampling has been improved and the cost of analyses lowered; manufacturing equipment now lasts longer because process control is better; and more uniform cement is being made at lower cost.

Figure 10.14-1 shows the more interesting components of the x-ray control system (see caption). The reliability of the control procedure is shown in Table 10.14-1.

Table 10.14–1. Reliability of X-Ray Emission Control of Cement Manufacture

Constituent	Percentage by Weight	Accuracy[a]	Precision (Standard Deviation)
CaO	45	± 0.5%	± 0.15% (1 part in 300)
SiO_2	15	± 0.5%	± 0.15% (1 part in 100)
Al_2O_3	3	± 0.5%	± 0.15% (1 part in 20)
Fe_2O_3	2	± 0.1%	± 0.025% (1 part in 80)

[a] When errors affecting precision are negligible, the percentage by weight as measured should not differ from the true value by more than the limit shown; e.g., if 45% CaO is measured under these conditions, the true value will lie between 44.5 and 45.5% CaO. These are the limits set by experience with actual mixes.

TODAY AND TOMORROW

What part do x-rays aided by electrons play in analytical chemistry today? What part tomorrow? There are no firm answers.

No one knows how many x-ray determinations (absorptiometric, diffractometric, spectrographic) are done in the United States each day. Two straws in the wind: during the period 1956–59, the Fluo-X-Spec Laboratory processed some 30,000 samples. In the middle sixties, the Wisconsin Centrifugal Foundry was doing determinations around the clock at the rate of 25,000/month. These thousands will multiply. Mountain ranges and the ocean floor can provide quite a few samples, and those from the moon command a lot of work!

What part ought the methods in this book to play? The answer will largely determine their future role, and it will depend on their relative merits. To say more is to invite controversy. An attempt was made to assay for this book the relative merits of the principal methods with which those treated herein must compete—an attempt that would have meant broadening the comparison between x-ray and optical emission made in Ref. 1, pp. 237–239. The attempt was abandoned as futile. Analytical chemists have been caught in a flash flood of new methods and new equipment. While as yet they have not drowned, neither have they had time to know their uncounted new blessings well enough to reach a consensus about relative merits. There is a present risk that the spirit of "My big brother can lick your big brother" (or, as regards traces and localized determinations, of "My little brother is littler than your little brother") will obtrude upon any discussion looking toward such a consensus.

A consensus may appear if relative calm ensues when the competitive struggles now under way are settled. Examples of such struggles follow.

X-RAYS VERSUS ELECTRONS. These now approach interchangeability in the information they can give when they reach a detector. Will either predominate in the future? Will ESCA become important in analytical chemistry as contrasted to the determination of binding energies?

WAVELENGTH VERSUS ENERGY RESOLUTION. Will Bragg reflection have to yield to solid-state detection coupled with pulse-height selection for all elements heavier than, say, sodium? Can Bragg reflection continue dominant in x-ray emission methods for process control? Germanium, first made by R. N. Hall with an impurity content near 1 in 10^{12}, is now offered by General Electric as a detector that needs cooling only during use, and that matches the best Si(Li) in performance [38].

X-RAY VERSUS ELECTRON EXCITATION. Can x-rays compete with electrons in the excitation of large areas when light elements must be determined? Will a Telsec Probe (or something similar) for light elements and a spectrograph with x-ray excitation and energy resolution be the best possible future combination for a laboratory that must do nonroutine determinations of elements across the widest achievable range of atomic numbers?

X-RAY VERSUS OPTICAL METHODS. Can x-ray emission continue its inroads on optical? Will the aluminum industry switch to x-ray emission spectrography? What about atomic absorption? emission?

X-RAY VERSUS γ-RAYS AND ACTIVATION ANALYSIS. Government subsidies excluded, what place will each make for itself?

ELECTRON MICROPROBE VERSUS ION MICROPROBE. In the latter, a beam of ions comparable in diameter at the sample surface with the beam of an electron microprobe sputters ions from the sample for detection in a mass spectrometer. Can the ion microprobe become the best instrument for localized determinations? Bayard [39] seems to think so.

ANALYTICAL CHEMIST AND (NOT VERSUS!) COMPUTER. Will computerization and automation give the analytical chemist too much free time? "Free time" here includes "enforced free time" traceable to equipment troubles.

CENTRALIZED VERSUS DISPERSED ANALYTICAL LABORATORIES. Will many small, dispersed laboratories continue viable for nonroutine work? Or, will the increasing need for a variety of costly and sophisticated equipment lead to a relatively few large, centralized laboratories in which such equipment (e.g., computers) can be used to capacity?

COST VERSUS RESULTS. Will the complete cost of isolated, sophisticated determinations be justified in most cases by the value of the results as a basis for action? Will the criterion of need-to-know come to be applied more rigorously to such determinations?

CLASSICAL ANALYTICAL CHEMISTRY VERSUS MODERN INSTRUMENTAL METHODS. Will future analytical chemists know what happens when concentrated ammonium hydroxide is added to copper sulfate solution? Will they need to?

There is no need to extol the methods of this book. If, despite their manifest advantages, they cannot withstand the heat of competition, then—in the words of President Truman—"they'd better get out of the kitchen!"

REFERENCES

1. H. A. Liebhafsky, H. G. Pfeiffer, E. H. Winslow, and P. D. Zemany, *X-Ray Absorption and Emission in Analytical Chemistry*, Wiley, New York, 1960.
2. M. Tomaino and A. De Pietro, *Norelco Reptr.*, **3**, 57 (1956).
3. K. F. J. Heinrich, in *Advances in X-Ray Analysis*, Vol. 5, Plenum, New York, 1962, p. 516.
4. (a) I. Adler and J. M. Axelrod, *Norelco Reptr.*, **3**, 65 (1956); (b) *Am. Mineralogist*, **41**, 524 (1956); (c) *Econ. Geol.*, **52** (6), 694 (1957).
5. H. J. Rose, Jr., R. P. Christian, J. R. Lindsay, and R. R. Larson, *U.S. Geol. Survey Professional Paper 650-B*, p. B128 (1969).
6. C. L. Luke, *Anal. Chem.*, **36**, 318 (1964).
7. S. H. U. Bowie, A. G. Darnley, and J. R. Rhodes, *Trans. Inst. Min. Met.*, **74**, 361 (1964–1965).
8. J. R. Rhodes, *Analyst*, **91**, 683 (1966).
9. J. R. Rhodes and T. Furuta, in *Advances in X-Ray Analysis*, Vol. 11, Plenum, New York, 1968, p. 249.
10. J. R. Rhodes, personal communication to H. A. L., March 1968.
11. A. K. Baird and D. H. Zenger, in *Advances in X-Ray Analysis*, Vol. 9, Plenum, New York, 1966, p. 487.
12. J. S. Solomon and W. L. Baun, *American Laboratory*, December 1970, p. 10.
13. M. Bayard, *American Laboratory*, September 1970, p. 10.
14. R. Castaing and A. Guinier, *Anal. Chem.*, **25**, 724 (1953).
15. C. A. Andersen, K. Keil, and B. Mason, *Science*, **146**, No. 3641, 256 (1964).
16. Various authors, in *Science*, **167**, No. 3918 (1970).
17. H. A. Liebhafsky, D. H. Wilkins, and F. Bernstein, in *XIII Colloquium Spectroscopicum Internationale*, Adam Hilger, London, 1968, p. 58.
18. (a) C. E. Crouthamel, C. E. Johnson, N. R. Stalica, C. A. Seils, and K. E. Anderson in *Annual Report, Argonne National Laboratories*, 1969, Chapter IV, p. 5; (b) C. E. Johnson and C. E. Crouthamel, *J. Nucl. Mat.*, **34**, 101 (1970); (c) C. E. Crouthamel and C. E. Johnson, paper given at 160th National Meeting, American Chemical Society, Chicago, September 14–18, 1970.
19. A. H. Compton and S. K. Allison, *X-Rays in Theory and Experiment*, Van Nostrand, 1935, p. 797. Quotation added to as shown.
20. K. Siegbahn, *et al.*, *ESCA, Atomic, Molecular and Solid State Structure Studied by Means of Electron Spectroscopy*, Almquist and Wiksells, Uppsala, 1967.

21. *C&EN*, March 22, 1971, p. 60.
22. *New York Times*, September 27, 1970, p. 56.
23. D. W. Fischer, *Appl. Spectr.* **25**, 263 (1971).
24. M. L. Salmon, in *Handbook of X-Rays,* E. F. Kaeble, Ed., McGraw-Hill, New York, 1967, p. 35–1.
25. T. F. Anater, *U.S. Atomic Energy Commission Report WAPD-321* (1968).
26. L. Adler, *X-Ray Emission Spectrography in Geology,* Elsevier, New York, 1966.
27. H. F. Carl and W. J. Campbell, *Am. Soc. Testing Materials Spec. Tech. Publ.,* No. 157, 63, 1954.
28. Georg (von) Hevesy, *Chemical Analysis by X-Rays and Its Applications,* McGraw-Hill, New York, 1932, p. 96.
29. A. K. Baird, R. S. MacColl, and D. B. McIntyre, in *Advances in X-Ray Analysis,* Vol. 5, Plenum, New York, 1962, p. 412.
30. A. K. Baird and B. L. Henke, *Anal. Chem.,* **37**, 727 (1965).
31. B. L. Henke, in *Advances in X-Ray Analysis,* Vol. 5, Plenum, New York, 1962, p. 285.
32. A. K. Baird, D. B. McIntyre, and E. E. Welday, *Geol. Soc. Am. Bull.,* **78**, 191 (1967).
33. H. G. J. Moseley, *Phil. Mag.* [6], **27**, 703 (1914); also p. 710.
34. F. W. Lytle, J. I. Botsford, and H. A. Heller, *U.S. Bur. Mines Rept. Invest.,* No. 5378, December 1957.
35. H. Kunzendorf and H. A. Wollenberg, *Nucl. Inst. and Meth.,* **87**, 197 (1970).
36. C. W. Moore, Chief, Quality Control Systems, Lone Star Cement Corporation, private communication dated February 26, 1971, to H. A. L.
37. C. L. Lewis, R. A. Hall, J. W. Anderson, and W. H. A. Timm, *The Canadian Mining and Metallurgical Bulletin,* April 1968.
38. (a) R. N. Hall and T. J. Soltys, *IEEE Trans. Nucl. Sci.* **NS–18**, 160 (1971); (b) R. D. Baertsch, *ibid.,* p. 166.
39. M. Bayard, *Am. Lab.,* **3,** No. 4 (April 1971), p. 15.

Appendix I

X-Ray Safety

The manufacturers of x-ray equipment produce equipment that has been very carefully engineered and safety checked. However, it is impossible to anticipate all configurations and uses. Each system should on installation and on any modification be carefully monitored by a qualified safety engineer. Periodic inspections and continuous personnel monitoring are essential.

X-ray damage to tissues is insidious and the effects are cumulative. The damage may be done before any ill effects are felt. Some early workers in the field suffered severe injury and even death [1] because they were not aware of the injury that can result from overexposure to radiation. The subject has now been studied and recommended procedures have been established [2]. These have been summarized, and federal [3] (and state) rules and regulations for safe operation have been described.

Present-day x-ray equipment manufacturers are aware of the hazards, and provide safe equipment and recommend safe procedures insofar as it lies within their control. Nevertheless, certain conditions of operation, or homemade equipment, may present hazards. Most recent harmful exposures to radiation [4] were caused by failure to use the safety devices provided, such as shielding or interlocks, or failure to follow the recommendations for safe operation of the equipment.

The unit of biological radiation exposure is the Radiation Absorbed Dose (Rad). A Rad corresponds to the absorption of 100 erg of energy per gram of body tissue. The biological damage is measured in the unit Roentgen Equivalent Man (Rem) since different kinds of radiation (x-rays, alpha particles, neutrons) cause different degrees of damage. One Rad of x-rays is equivalent to one Rem. The maximum permissible limit of radiation exposure [3] is 3 Rem/quarter, or 5 Rem/year. These limits apply to industrial exposure and do not include medical diagnostic radiation or the dose absorbed by an individual due to natural radiation sources, such as radioactive material normally present in the air or soil; or cosmic rays, which subject everyone to an exposure of a fraction of a Rem each year.

Radiation dosage is measured by portable survey meters, which are often calibrated in M Rad per hour (1000 M Rad = 1 Rad). They consist of a battery-operated Geiger or proportional detector tube which has a rate meter output. Radiation dosimeters utilizing photographic film, or thermoluminescent dosimeters (TLD) mounted in finger rings or clip-on badges, integrate the dosage over long periods (weeks or longer), while simple ionization detectors and TLD are used for shorter (one day) periods of monitoring. Film dosimeter services are commericially available that provide the dosimeter and periodically measure the accumulated dose by measuring the density after development, or integrating the light emitted.

Every laboratory using x-ray analytical equipment should have available a portable survey meter to monitor the x-ray equipment.

In addition to the x-ray tube itself the high-voltage rectifiers can become a source of x-rays when they become gassy; thus it is necessary to survey in the vicinity of the rectifiers periodically to detect this potential radiation hazard.

Radioactive sources such as are used with energy resolution, thickness gauges, and such, emit lower levels of radiation than an x-ray tube, yet they can constitute a radiation hazard. This hazard should be determined before any use is made of the source. They should be handled carefully. The actual radioactive material is often covered with a thin and fragile foil. If the foil is broken, dangerous material may be released.

The dangers attending high voltages in the x-ray generator and electronics are well known. Safety interlocks should not be bypassed. Adequate grounding should be installed. The high voltages in the x-ray tube power supply are particularly dangerous.

REFERENCES

1. R. D. Carman and A. Miller, *Radiology,* **3**, 408 (1924); V. P. Blair, J. B. Brown, and W. G. Hamm, *Radiology,* **19**, 337 (1932).
2. H. Blatz, *Introduction to Radiological Health,* McGraw Hill, New York, 1964.
   W. D. Claus, *Radiation Biology and Medicine,* Addison Wesley Pub. Co., Reading, Mass., 1958.

REFERENCES

3. *Rules and Regulations: Conditions and Limitations on the General License Provisions of 10 CFR 150.20*, Division of Licenses and Regulations, U.S. Atomic Energy Commission, Washington, D.C., 1964; U.S. Department of Commerce, *National Bureau of Standards Handbooks 60*, 1955, and *62*, 1957, Washington, D.C.; *USAEC Manual*, "Radiation Standards" Chap. 0525, 1963.
4. TID 5360, *A Summary of Industrial Accidents in USAEC Facilities, Supplement 3 revised*, 1961.

Appendix II

The X-Ray Absorption Edges of the Elements

The data in Appendices II, III, and IV are based on the following.

1. J. A. Bearden, "X-Ray Wavelengths," *U.S. At. Energy Comm. Rep.* NYO–10586, 1964, 533 pp.
2. R. D. Dewey, R. S. Mapes, and T. W. Reynolds, "Computed X-Ray Wavelengths with Wavelength Tables," Reynolds Metals Co., Metallurgical Research Div., Richmond, Va., 1966, 198 pp.
3. R. D. Dewey, R. S. Mapes, and T. W. Reynolds, "A Study of X-Ray Mass Absorption Coefficients with Tables of Coefficients," Reynolds Metals Co., Metallurgical Research Div., Richmond, Va., 1967, 51 pp.
4. S. Fine and C. F. Hendee, "Table of X-Rays K and L Emission and Critical Absorption Energies for All the Elements," *Nucleonics,* **13** (3), 36–37 (1955); *Norelco Rep.,* **3**, 113–115 (1956).
5. K. F. J. Heinrich, "X-Ray Absorption Uncertainty" [Including Table of Mass Absorption Coefficients], in *The Electron Microprobe,* E. D. McKinley, K. F. J. Heinrich, and D. B. Wittry, Eds., Wiley, New York, 1966. pp. 296–377.
6. B. L. Henke, R. L. Elgin, R. E. Lent, and R. B. Ledingham, "X-Ray Absorption in the 2-to-200 Å Region," U.S.A.F. Rep. AFOSR 67–1254, 1967.

For each edge, the energy (E) is given in keV and the initial absorption wavelength (λ) in Å.

K and L Absorption Edges

Z	Element	K		L$_I$		L$_{II}$		L$_{III}$	
		E	λ	E	λ	E	λ	E	λ
1	H								
2	He								
3	Li		226.500						
4	Be	0.115	111.000						
5	B	0.188	66.200^a						
6	C	0.282	43.680						
7	N	0.397	30.990						
8	O	0.533	23.320						
9	F	0.692	*17.900*						
10	Ne	0.874	14.300						
11	Na	1.080	11.570						
12	Mg	1.309	9.512	0.062	197.300	0.050	249.300	0.056	250.700
13	Al	1.562	7.948	0.087	142.500	0.076	*169.000*	0.075	*177.000*
14	Si	1.840	6.738	0.118	*105.000*	0.101	*126.000*	0.100	*130.000*
15	P	2.143	5.784	0.153	*83.200*	0.130	*97.300*	0.129	*99.900*
16	S	2.471	5.019	0.193	*65.500*	0.164	*77.000*	0.163	*78.800*
17	Cl	2.824	4.397	0.237	*53.700*	0.204	*62.400*	0.202	*64.100*
18	A	3.203	3.871	0.286	*44.400*	0.247	*51.400*	0.245	*52.400*
19	K	3.607	3.437	0.340	*37.200*	0.296	*43.400*	0.293	*44.600*
20	Ca	4.034	3.070	0.403	*32.100*	0.346	35.130	0.342	35.490
21	Sc	4.486	2.762	0.462	*27.800*	0.400	*31.800*	0.396	*32.800*
22	Ti	4.965	2.497	0.529	24.300	0.460	27.900	0.454	28.500
23	V	5.463	2.269	0.626	21.200	0.519	24.400	0.511	25.100
24	Cr	5.987	2.070	0.694	19.000	0.582	21.900	0.572	22.400
25	Mn	6.537	1.896	0.768	16.000	0.649	19.200	0.638	19.800
26	Fe	7.112	1.743	0.846	15.100	0.721	17.200	0.708	17.900
27	Co	7.712	1.608	0.929	13.700	0.797	15.620	0.782	15.920
28	Ni	8.339	1.488	1.016	12.400	0.878	14.240	0.861	14.530
29	Cu	8.993	1.381	1.109	*11.300*	0.965	13.014	0.945	13.290
30	Zn	9.673	1.283	1.208	10.400	1.057	11.860	1.034	12.130
31	Ga	10.386	1.1958	1.316	9.517	1.155	10.828	1.134	11.100
32	Ge	11.115	1.1166	1.426	8.773	1.259	9.924	1.228	10.187
33	As	11.877	1.0450	1.536	8.107	1.368	9.125	1.333	9.367
34	Se	12.666	0.9797	1.662	7.503	1.485	8.407	1.444	8.646
35	Br	13.483	0.9203	1.791	6.959	1.605	7.753	1.559	7.984
36	Kr	14.330	0.8655	1.923	6.470	1.732	7.168	1.680	7.392
37	Rb	15.202	0.8155	2.067	6.008	1.866	6.644	1.806	6.862
38	Sr	16.106	0.7697	2.217	5.592	2.008	6.173	1.940	6.387

^{a} Italic values are by interpolation of Moseley plot.

K and L Absorption Edges

Z	Element	K		L$_I$		L$_{II}$		L$_{III}$	
		E	λ	E	λ	E	λ	E	λ
39	Y	17.037	0.7277	2.372	5.217	2.155	5.756	2.079	5.962
40	Zr	17.997	0.6888	2.535	4.879	2.305	5.378	2.227	5.579
41	Nb	18.985	0.6530	2.698	4.575	2.464	5.031	2.370	5.230
42	Mo	20.002	0.6198	2.867	4.304	2.628	4.719	2.523	4.913
43	Tc	21.048	0.5891	3.047	4.058	2.797	4.436	2.681	4.630
44	Ru	22.123	0.5605	3.230	3.835	2.973	4.180	2.844	4.369
45	Rh	23.229	0.5340	3.421	3.629	3.156	3.943	3.013	4.130
46	Pd	24.365	0.5092	3.619	3.437	3.344	3.723	3.187	3.907
47	Ag	25.531	0.4859	3.822	3.256	3.540	3.516	3.368	3.700
48	Cd	26.727	0.4641	4.034	3.085	3.742	3.326	3.554	3.505
49	In	27.953	0.4437	4.250	2.926	3.951	3.147	3.744	3.324
50	Sn	29.211	0.4247	4.681	2.777	4.167	2.982	3.939	3.156
51	Sb	30.499	0.4067	4.706	2.639	4.389	2.829	4.140	3.000
52	Te	31.817	0.3897	4.942	2.510	4.616	2.688	4.345	2.856
53	I	33.168	0.3738	5.186	2.388	4.851	2.554	4.556	2.720
54	Xe	34.551	0.3584	5.442	2.274	5.092	2.429	4.772	2.593
55	Cs	35.966	0.3445	5.700	2.167	5.341	2.314	4.993	2.474
56	Ba	37.414	0.3310	5.964	2.068	5.547	2.205	5.220	2.363
57	La	38.894	0.3184	6.235	1.978	5.860	2.105	5.452	2.261
58	Ce	40.410	0.3065	6.516	1.893	6.181	2.012	5.690	2.166
59	Pr	41.958	0.2952	6.802	1.814	6.408	1.926	5.932	2.079
60	Nd	43.538	0.2845	7.095	1.739	6.691	1.844	6.177	1.997
61	Pm	45.152	0.2743	7.398	1.6674	6.981	1.7676	6.427	1.9191
62	Sm	46.801	0.2646	7.707	1.6002	7.278	1.6953	6.683	1.8457
63	Eu	48.486	0.2555	8.024	1.5381	7.584	1.6271	6.944	1.7761
64	Gd	50.207	0.2468	8.343	1.4784	7.898	1.5632	7.211	1.7117
65	Tb	51.965	0.2384	8.679	1.4223	8.221	1.5023	7.484	1.6497
66	Dy	53.761	0.2305	9.013	1.3692	8.553	1.4445	7.762	1.5916
67	Ho	55.593	0.2229	9.365	1.3190	8.874	1.3905	8.048	1.5368
68	Er	57.484	0.2157	9.725	1.2706	9.243	1.3386	8.336	1.4835
69	Tm	59.374	0.2088	10.097	1.2250	9.601	1.2892	8.632	1.4334
70	Yb	61.322	0.2022	10.479	1.1818	9.968	1.2428	8.933	1.3862
71	Lu	63.311	0.1959	10.869	1.1402	10.346	1.1985	9.241	1.3405
72	Hf	65.345	0.1898	11.262	1.0997	10.734	1.1548	9.555	1.2972
73	Ta	67.405	0.1839	11.672	1.0613	11.126	1.1137	9.872	1.2553
74	W	69.517	0.1784	12.092	1.0247	11.535	1.0745	10.199	1.2155
75	Re	71.670	0.1730	12.522	0.9894	11.952	1.0371	10.530	1.1773
76	Os	73.869	0.1679	12.968	0.9558	12.382	1.0041	10.868	1.1408
77	Ir	76.111	0.1629	13.416	0.9236	12.824	0.9671	11.215	1.1058

K and L Absorption Edges

Z	Element	K		L$_I$		L$_{II}$		L$_{III}$	
		E	λ	E	λ	E	λ	E	λ
78	Pt	78.400	0.1582	13.880	0.8931	13.277	0.9341	11.568	1.0723
79	Au	80.729	0.1536	14.353	0.8638	13.739	0.9026	11.925	1.0400
80	Hg	83.109	0.1492	14.835	0.8353	14.215	0.8722	12.290	1.0091
81	Tl	85.532	0.1450	15.344	0.8081	14.700	0.8434	12.660	0.9793
82	Pb	88.008	0.1409	15.863	0.7820	15.204	0.8154	13.039	0.9507
83	Bi	90.540	0.1369	16.391	0.7571	15.725	0.7887	13.422	0.9234
84	Po	93.113	*0.1332*	16.440	*0.7321*	16.250	*0.7632*	13.812	*0.8973*
85	At	95.730	*0.1295*	17.495	*0.7089*	16.787	*0.7388*	14.207	*0.8722*
86	Rn	98.407	*0.1260*	18.047	*0.6866*	17.337	*0.7151*	14.609	*0.8550*
87	Fr	101.131	*0.1226*	18.630	*0.6652*	17.900	*0.6925*	15.017	*0.8250*
88	Ra	103.909	*0.1193*	19.222	*0.6445*	18.475	*0.6707*	15.433	*0.8028*
89	Ac	106.738	*0.1161*	19.823	*0.6248*	19.063	*0.6499*	15.854	*0.7815*
90	Th	109.641	0.1131	20.449	0.6059	19.689	0.6299	16.283	0.7607
91	Pa	112.599	*0.1101*	21.088	*0.5875*	20.312	*0.6104*	16.716	*0.7410*
92	U	115.597	0.1072	21.745	0.5695	20.932	0.5919	17.158	0.7223
93	Np	119.380	*0.1038*	22.417	*0.5527*	21.596	*0.5741*	17.614	*0.7039*
94	Pu	121.600	*0.1019*	23.097	*0.5365*	22.262	*0.5571*	18.066	*0.6867*
95	Am	124.790	*0.0993*	23.793	*0.5211*	22.944	*0.5403*	18.525	*0.6693*

M Absorption Edges

Z	Element	M_I		M_{II}		M_{III}		M_{IV}		M_V	
		E	λ	E	λ	E	λ	E	λ	E	λ
17	Cl	0.020	591.0	—	—	—	—	—	—	—	—
18	A	0.026	476.0	—	—	—	—	—	—	—	—
19	K	0.033	373.0	—	—	—	—	—	—	—	—
20	Ca	0.040	309.0	—	—	—	—	—	—	—	—
21	Sc	0.046	264.0	—	—	—	—	—	—	—	—
22	Ti	0.054	228.0	—	—	—	—	—	—	—	—
23	V	0.061	201.0	—	—	—	—	—	—	—	—
24	Cr	0.072	171.0	—	—	—	—	—	—	—	—
25	Mn	0.082	150.0	—	—	—	—	—	—	—	—
26	Fe	0.093	132.0	—	—	—	—	—	—	—	—
27	Co	0.104	118.0	—	—	—	—	—	—	—	—
28	Ni	0.120	103.0	—	—	—	—	—	—	—	—
29	Cu	0.135	91.6	0.090	137.0	—	—	0.015	793.4	—	—
30	Zn	0.151	81.8	0.106	116.0	—	—	0.022	546.2	—	—
31	Ga	0.169	73.1	0.125	98.6	0.115	107.5	0.030	405.0	—	—
32	Ge	0.190	65.1	0.137	89.8	0.132	93.4	0.041	299.4	—	—
33	As	0.211	58.5	0.156	79.1	0.150	82.2	0.052	238.3	—	—
34	Se	0.234	52.8	0.177	69.9	0.170	72.5	0.066	187.6	—	—
35	Br	0.265	46.6	0.198	62.3	0.191	64.7	0.082	150.2	—	—
36	Kr	0.294	42.0	0.225	54.8	0.217	57.0	0.095	129.7	—	—
37	Rb	0.328	37.6	0.250	49.5	0.240	51.5	0.114	108.4	0.112	109.9
38	Sr	0.358	34.5	0.280	44.1	0.270	45.8	0.136	90.9	0.134	92.2
39	Y	0.394	31.4	0.312	39.7	0.300	41.3	0.159	77.8	0.156	78.9
40	Zr	0.435	28.4	0.348	35.5	0.335	36.9	0.187	66.1	0.184	67.1
41	Nb	0.468	26.4	0.379	32.7	0.362	34.1	0.207	59.8	0.204	60.6
42	Mo	0.507	24.4	0.412	30.0	0.394	31.4	0.232	53.2	0.228	54.2
43	Tc	0.551	22.5	0.449	27.5	0.429	28.8	0.260	47.5	0.257	48.1
44	Ru	0.591	20.9	0.486	25.4	0.467	26.5	0.290	42.6	0.288	43.0
45	Rh	0.637	19.4	0.531	23.3	0.506	24.4	0.321	38.5	0.315	39.2
46	Pd	0.684	18.1	0.573	21.6	0.546	22.6	0.354	34.9	0.349	35.4
47	Ag	0.734	16.8	0.619	20.0	0.588	21.0	0.389	31.7	0.383	32.2
48	Cd	0.781	15.8	0.666	18.6	0.632	19.6	0.423	29.2	0.420	29.5
49	In	0.839	14.7	0.716	17.3	0.678	18.2	0.464	26.7	0.456	27.1
50	Sn	0.894	13.8	0.772	16.0	0.720	17.2	0.506	24.5	0.497	24.9
51	Sb	0.952	13.0	0.822	15.0	0.774	16.0	0.546	22.6	0.536	23.1
52	Te	1.010	12.2	0.873	14.1	0.822	15.0	0.586	21.1	0.575	21.5
53	I	1.071	11.5	0.929	13.3	0.873	14.1	0.630	19.6	0.618	20.0
54	Xe	1.147	10.8	0.989	12.5	0.926	13.3	0.677	18.3	0.662	18.7
55	Cs	1.199	10.3	1.048	11.8	0.981	12.6	0.722	17.1	0.704	17.6
56	Ba	1.266	9.78	1.111	11.1	1.036	11.9	0.770	16.0	0.750	16.5

M Absorption Edges

Z	Element	M_I E	λ	M_{II} E	λ	M_{III} E	λ	M_{IV} E	λ	M_V E	λ
57	La	1.330	9.31	1.173	10.50	1.092	11.30	0.823	15.00	0.801	15.40
58	Ce	1.401	8.84	1.240	9.99	1.152	10.70	0.870	14.20	0.851	14.50
59	Pr	1.476	8.39	1.305	9.49	1.210	10.20	0.923	13.40	0.898	13.70
60	Nd	1.544	8.02	1.372	9.03	1.266	9.78	0.969	12.70	0.946	13.00
61	Pm	1.642	7.55	1.439	8.61	1.327	9.34	1.019	12.10	0.994	12.40
62	Sm	1.689	7.33	1.512	8.19	1.388	8.92	1.073	11.50	1.048	11.80
63	Eu	1.767	7.01	1.584	7.82	1.450	8.54	1.129	10.90	1.101	11.20
64	Gd	1.849	6.70	1.653	7.49	1.511	8.20	1.185	10.40	1.153	10.70
65	Tb	1.937	6.39	1.737	7.13	1.583	7.82	1.245	9.95	1.211	10.20
66	Dy	2.019	6.13	1.805	6.86	1.642	7.54	1.304	9.50	1.266	9.79
67	Ho	2.104	5.89	1.886	6.57	1.715	7.22	1.365	9.07	1.327	9.34
68	Er	2.184	5.67	1.973	6.28	1.783	6.95	1.430	8.66	1.385	8.95
69	Tm	2.291	5.41	2.071	5.98	1.861	6.66	1.498	8.27	1.451	8.54
70	Yb	2.387	5.19	2.165	5.72	1.948	6.36	1.566	7.91	1.518	8.16
71	Lu	2.488	4.98	2.262	5.48	2.025	6.11	1.637	7.57	1.586	7.81
72	Hf	2.601	4.76	2.366	5.23	2.109	5.87	1.718	7.21	1.664	7.45
73	Ta	2.698	4.59	2.459	5.04	2.184	5.67	1.783	6.95	1.725	7.18
74	W	2.812	4.40	2.566	4.83	2.273	5.45	1.864	6.64	1.803	6.87
75	Re	2.926	4.23	2.676	4.63	2.361	5.25	1.946	6.37	1.879	6.59
76	Os	3.047	4.06	2.792	4.44	2.453	5.05	2.033	6.09	1.963	6.31
77	Ir	3.171	3.90	2.908	4.26	2.551	4.85	2.119	5.84	2.040	6.07
78	Pt	3.296	3.76	3.036	4.08	2.649	4.67	2.204	5.62	2.129	5.82
79	Au	3.379	3.66	3.149	3.93	2.744	4.51	2.307	5.37	2.220	5.58
80	Hg	3.566	3.47	3.287	3.77	2.848	4.35	2.392	5.18	2.291	5.40
81	Tl	3.702	3.34	3.418	3.62	2.957	4.19	2.483	4.99	2.389	5.18
82	Pb	3.853	3.21	3.558	3.48	3.072	4.03	2.586	4.79	2.484	4.98
83	Bi	4.003	3.09	3.709	3.34	3.186	3.89	2.694	4.60	2.586	4.79
84	Po	4.147	2.98	3.863	3.20	3.312	3.74	2.798	4.43	2.681	4.62
85	At	4.350	2.85	4.008	3.09	3.428	3.61	2.905	4.26	2.780	4.45
86	Rn	4.524	2.74	4.156	2.98	3.536	3.50	3.014	4.11	2.882	4.30
87	Fr	4.678	2.65	4.324	2.86	3.654	3.39	3.125	3.96	2.986	4.15
88	Ra	4.811	2.57	4.477	2.76	3.779	3.28	3.237	3.82	3.093	4.00
89	Ac	5.019	2.47	4.637	2.67	3.892	3.18	3.352	3.69	3.202	3.87
90	Th	5.176	2.39	4.810	2.57	4.030	3.07	3.474	3.56	3.313	3.74
91	Pa	5.355	2.31	4.993	2.48	4.164	2.97	3.597	3.44	3.416	3.62
92	U	5.532	2.24	5.177	2.39	4.293	2.88	3.712	3.33	3.533	3.50

N Absorption Edges

Z	Element	N_I λ	N_{II} λ	N_{III} λ	N_{IV} λ	N_V E	N_{VI} E	N_{VII} E	E	E
43	Tc	162.7	226.7	263.9	635.9	707.0	0.076	0.054	0.046	0.019	0.017
44	Ru	145.5	206.4	239.4	570.6	633.2	0.085	0.060	0.051	0.021	0.019
45	Rh	136.8	188.2	217.5	512.5	568.6	0.090	0.065	0.056	0.024	0.021
46	Pd	122.5	172.0	198.1	461.4	511.7	0.101	0.072	0.062	0.026	0.024
47	Ag	110.5	157.5	180.8	416.3	461.6	0.112	0.078	0.068	0.029	0.026
48	Cd	101.0	144.5	165.3	376.4	417.3	0.122	0.085	0.074	0.032	0.029
49	In	91.1	132.8	151.5	341.1	378.0	0.136	0.093	0.081	0.036	0.032
50	Sn	85.8	122.2	139.0	309.7	343.1	0.144	0.101	0.089	0.040	0.036
51	Sb	76.5	112.7	127.8	281.7	312.0	0.161	0.109	0.097	0.044	0.039
52	Te	72.0	104.1	117.6	256.7	284.3	0.172	0.119	0.105	0.048	0.043
53	I	66.8	96.3	108.5	234.4	257.5	0.185	0.128	0.114	0.052	0.047
54	Xe	62.0	89.2	100.2	214.3	237.2	0.199	0.138	0.123	0.057	0.052
55	Cs	58.2	82.7	92.7	196.3	217.3	0.212	0.149	0.133	0.063	0.057
56	Ba	54.6	76.8	85.8	180.6	199.2	0.226	0.161	0.144	0.068	0.062
57	La	51.3	71.4	79.6	166.5	183.0	0.241	0.173	0.155	0.074	0.067
58	Ce	48.2	66.5	73.9	153.7	168.4	0.257	0.186	0.167	0.080	0.073
59	Pr	45.4	62.0	68.7	142.1	155.1	0.272	0.199	0.180	0.087	0.079
60	Nd	43.5	57.9	64.0	131.6	143.2	0.284	0.213	0.193	0.094	0.086
61	Pm	41.4	54.3	60.3	122.0	132.3	0.298	0.227	0.205	0.101	0.093
62	Sm	39.6	51.1	57.0	113.4	122.5	0.312	0.242	0.217	0.109	0.101
63	Eu	37.5	48.1	53.8	105.5	113.6	0.330	0.257	0.230	0.117	0.109
64	Gd	35.5	45.3	50.9	98.3	105.4	0.348	0.273	0.243	0.126	0.117

Z	Element							λ	λ					E	E
65	Tb	33.6	42.7	48.2	91.7	98.0				0.367	0.289	0.256	0.135	0.126	
66	Dy	31.9	40.3	45.7	85.7	91.2				0.388	0.307	0.270	0.144	0.135	
67	Ho	30.3	38.1	43.4	80.1	85.1				0.408	0.324	0.265	0.154	0.145	
68	Er	28.7	36.1	41.2	75.1	79.4				0.431	0.343	0.300	0.165	0.156	
69	Tm	27.2	34.2	39.2	70.4	74.2				0.454	0.362	0.315	0.175	0.167	
70	Yb	25.9	32.4	37.3	66.2	69.4	1063.8			0.478	0.382	0.331	0.187	0.178	0.011
71	Lu	24.6	30.5	35.2	61.3	64.4	839.7			0.503	0.405	0.351	0.202	0.192	0.014
72	Hf	23.4	28.7	33.2	56.9	59.8	665.1			0.529	0.431	0.372	0.217	0.207	0.018
73	Ta	22.2	27.0	31.4	52.9	55.5	528.4			0.557	0.458	0.394	0.234	0.223	0.023
74	W	21.0	25.4	29.6	49.2	51.6	421.2	446.3		0.587	0.486	0.417	0.251	0.240	0.029
75	Re	20.0	24.0	28.1	45.8	48.1	336.7	360.4		0.619	0.514	0.440	0.270	0.257	0.036
76	Os	18.9	22.7	26.6	42.7	44.8	270.0	291.8		0.652	0.545	0.465	0.290	0.276	0.045
77	Ir	18.5	21.4	25.2	39.8	41.7	217.1	237.0		0.688	0.577	0.491	0.310	0.296	0.057
78	Pt	17.0	20.3	23.9	37.1	39.0	175.1	193.0		0.725	0.608	0.517	0.333	0.317	0.070
79	Au	16.2	19.2	22.6	34.7	36.4	141.6	157.5		0.765	0.645	0.547	0.356	0.340	0.087
80	Hg	15.3	18.1	21.4	32.4	34.0	114.8	128.9		0.806	0.682	0.577	0.381	0.363	0.107
81	Tl	14.5	17.2	20.3	30.4	31.9	100.1	108.9		0.849	0.720	0.608	0.407	0.388	0.123
82	Pb	13.4	16.2	19.3	28.4	29.9	85.7	89.0		0.894	0.760	0.641	0.435	0.414	0.144
83	Bi	13.1	15.4	18.3	26.7	28.0	77.0	75.5		0.941	0.801	0.674	0.464	0.441	0.160
84	Po	12.5	14.6	17.4	25.0	26.3	69.2	68.5		0.991	0.845	0.710	0.494	0.470	0.178
85	At	11.7	14.2	16.5	23.5	24.8	62.3	62.2		1.040	0.891	0.747	0.525	0.499	0.198
86	Rn	11.3	13.2	15.7	22.1	23.3	56.2	56.6		1.092	0.938	0.785	0.560	0.530	0.220
87	Fr	10.4	12.9	15.0	20.8	22.0	50.7	51.5		1.144	0.988	0.825	0.593	0.561	0.244
88	Ra	10.3	11.9	14.2	19.7	20.8	45.4	46.9		1.198	1.040	0.870	0.627	0.594	0.270
89	Ac	9.8	11.3	13.7	18.7	19.7	41.5	42.8		1.254	1.093	0.901	0.661	0.627	0.298
90	Th	9.4	10.7	13.0	17.7	18.7	37.6	39.1		1.316	1.149	0.951	0.696	0.661	0.329
91	Pa	9.0	10.2	12.4	16.9	17.8	35.7	37.1		1.367	1.207	0.993	0.731	0.694	0.346
92	U	8.5	9.7	11.9	16.1	17.0	33.0	34.2		1.433	1.268	1.037	0.767	0.728	0.375

O Absorption Edges

Z		λ O_I	λ O_{II}	λ O_{III}	λ O_{IV}	λ O_V	λ O_{VI}	E O_I	E O_{II}	E O_{III}	E O_{IV}	E O_V	E O_{VI}
56	Ba	1000.0						0.012					
57	La	891.4						0.013					
58	Ce	796.2						0.015					
59	Pr	712.5						0.017					
60	Nd	638.8						0.019					
61	Pm	573.8						0.021					
62	Sm	516.3						0.024					
63	Eu	465.3						0.026					
64	Gd	420.1						0.028					
65	Tb	379.8	1000.0	1287.3				0.032	0.012	0.009			
66	Dy	344.0	869.5	1119.9				0.036	0.014	0.011			
67	Ho	312.0	757.7	975.8				0.039	0.016	0.012			
68	Er	283.3	661.6	892.1				0.043	0.018	0.014			
69	Tm	257.7	578.8	745.5				0.048	0.021	0.016			
70	Yb	234.7	507.4	653.5				0.052	0.024	0.018			
71	Lu	214.1	445.5	573.9				0.057	0.027	0.021			
72	Hf	195.5	392.1	504.9				0.063	0.031	0.024			
73	Ta	178.7	345.6	445.1	1000.0	1049.5		0.069	0.035	0.027	0.012	0.011	
74	W	163.6	305.1	392.9	883.5	927.3		0.075	0.040	0.031	0.014	0.013	
75	Re	143.3	269.4	347.5	782.0	820.7		0.082	0.045	0.035	0.015	0.015	
76	Os	137.6	239.0	307.4	693.2	727.5		0.090	0.051	0.040	0.017	0.017	
77	Ir	126.4	212.0	273.1	615.5	666.0		0.098	0.058	0.045	0.020	0.019	
78	Pt	116.2	198.4	242.7	547.3	574.4		0.106	0.065	0.051	0.022	0.021	
79	Au	107.0	167.7	216.0	487.4	511.6		0.115	0.078	0.057	0.025	0.024	
80	Hg	98.6	149.4	182.8	434.7	456.3		0.125	0.082	0.064	0.028	0.027	
81	Tl	90.9	130.5	168.0	398.3	407.5		0.136	0.095	0.073	0.031	0.030	
82	Pb	84.0	114.1	146.9	347.3	364.5		0.147	0.108	0.084	0.035	0.034	
83	Bi	77.6	99.9	128.7	311.0	326.4		0.159	0.124	0.096	0.039	0.037	
84	Po	71.8	91.7	118.1	278.9	292.7	1000.0	0.172	0.135	0.104	0.044	0.042	0.017
85	At	66.5	84.3	108.6	250.4	262.8	789.5	0.186	0.147	0.114	0.049	0.047	0.018
86	Rn	61.6	77.5	99.8	225.2	236.3	625.0	0.201	0.159	0.124	0.055	0.052	0.019
87	Fr	57.2	71.4	91.9	202.7	212.9	496.2	0.216	0.173	0.134	0.061	0.058	0.024
88	Ra	53.1	65.8	84.7	182.7	191.7	396.9	0.233	0.188	0.146	0.067	0.066	0.031
89	Ac	49.3	61.1	79.4	164.8	173.0	315.1	0.251	0.200	0.156	0.075	0.071	0.039
90	Th	45.9	57.8	74.4	148.9	156.3	252.1	0.270	0.214	0.166	0.083	0.079	0.048
91	Pa	42.7	53.7	69.2	134.6	141.3	202.2	0.290	0.230	0.178	0.092	0.087	0.061
92	U	39.7	50.0	64.4	121.9	127.9	162.5	0.311	0.247	0.192	0.101	0.096	0.076

Appendix III

X-Ray Emission Spectra of the Elements

The values in italic in the following table have been calculated from a Moseley plot. The **M** and **N** lines included are for general information only. These lines tend to be more diffuse and vary greatly in relative intensity. Nondiagram lines of considerable intensity are also common. Transitions between **N** states fall into the same spectral region as some transitions to the **N** states.

As a general rule, the uncertainties in the line position are of the order of the ratio of the chemical binding energies to the x-ray line energy. The relative intensities of doublets and triplets (**K** $\alpha 1$, $\alpha 2$, $\beta 1$) tend to remain constant unless one of the transition states is near a valence level.

K Lines, Å

Z	Element	Kα Unresolved α1–α2	Kα1	Kα2	Kβ3	Kβ1	Kβ2	Kβ5	Kβ4
3	Li	228.0000							
4	Be	114.0000							
5	B	67.6000							
6	C	44.7000							
7	N	31.6000							
8	O	23.6000							
9	F	18.3000							
10	Ne	14.6100				14.4500			
11	Na	11.9100				11.5800			
12	Mg	9.8900				9.5210			
13	Al	8.3400	8.3390	8.3420		7.9605			
14	Si	7.1262	7.1254	7.1279		6.7530			
15	P	6.1578	6.1568	6.1598		5.7960			
16	S	5.3730	5.3722	5.3750		5.0316			
17	Cl	4.7288	4.7278	4.7307		4.4034			
18	A	4.1928	4.1918	4.1947		3.8860			
19	K	3.4424	3.7414	3.7445		3.4539			
20	Ca	3.3595	3.3584	3.3617		3.0897			
21	Sc	3.0320	3.0309	3.0342		2.7796		2.7634	
22	Ti	2.7497	2.7485	2.7522		2.5139		2.4985	
23	V	2.5050	2.5036	2.5074		2.2840		2.2695	
24	Cr	2.2910	2.2897	2.2936		2.0849		2.0709	
25	Mn	2.1030	2.1018	2.1058		1.9102		1.8971	
26	Fe	1.9373	1.9360	1.9400		1.7566		1.7442	
27	Co	1.7903	1.7890	1.7929		1.6208		1.6089	
28	Ni	1.6592	1.6579	1.6617		1.5001		1.4886	
29	Cu	1.5419	1.5406	1.5444		1.3922	1.3811	1.3816	
30	Zn	1.4365	1.4352	1.4390		1.2953	1.2837	1.2848	
31	Ga	1.3414	1.3401	1.3440	1.2084	1.2080	1.1960	1.1981	
32	Ge	1.2554	1.2541	1.2580	1.1294	1.1289	1.1169	1.1195	
33	As	1.1772	1.1759	1.1799	1.0578	1.0573	1.0450	1.0488	
34	Se	1.1061	1.1048	1.1088	0.9927	0.9922	0.9799	0.9843	
35	Br	1.0410	1.0397	1.0438	0.9333	0.9328	0.9205	0.9255	
36	Kr	0.9814	0.9801	0.9841	0.8790	0.8785	0.8661	0.8708	0.8653
37	Rb	0.9250	0.9236	0.9278	0.8275	0.8270	0.8148	0.8219	0.8164
38	Sr	0.8766	0.8753	0.8794	0.7835	0.7829	0.7708	0.7764	0.7699
39	Y	0.8302	0.8288	0.8331	0.7413	0.7407	0.7286	0.7345	0.7278
40	Zr	0.7873	0.7859	0.7902	0.7023	0.7017	0.6899	0.6959	0.6890
41	Nb	0.7476	0.7462	0.7504	0.6663	0.6658	0.6542	0.6608	0.6532
42	Mo	0.7107	0.7093	0.7136	0.6329	0.6323	0.6210	0.6270	0.6201

K Lines, Å

Z	Element	Kα Unresolved α1–α2	Kα1	Kα2	Kβ3	Kβ1	Kβ2	Kβ5	Kβ4
43	Tc	0.6764	0.6750	0.6793	0.6019	0.6013	0.5902	*0.5962*	*0.5994*
44	Ru	0.6445	0.6431	0.6474	0.5731	0.5725	0.5617	0.5679	0.5609
45	Rh	0.6147	0.6133	0.6176	0.5462	0.5456	0.5351	0.5411	0.5341
46	Pd	0.5868	0.5854	0.5898	0.5211	0.5205	0.5102	0.5167	0.5093
47	Ag	0.5608	0.5594	0.5638	0.4977	0.4971	0.4870	0.4931	0.4859
48	Cd	0.5364	0.5350	0.5394	0.4737	0.4751	0.4653	*0.4727*	*0.4649*
49	In	0.5135	0.5121	0.5165	0.4552	0.4545	0.4450	0.4509	0.4442
50	Sn	0.4921	0.4906	0.4951	0.4359	0.4352	0.4258	0.4318	0.4252
51	Sb	0.4719	0.4704	0.4748	0.4177	0.4171	0.4079	0.4138	0.4073
52	Te	0.4528	0.4513	0.4558	0.4007	0.4000	0.3911	*0.3973*	*0.3905*
53	I	0.4348	0.4333	0.4378	0.3846	0.3839	0.3752	*0.3814*	*0.3746*
54	Xe	0.4178	0.4163	0.4209	0.3694	0.3687	0.3603	*0.3665*	*0.3598*
55	Cs	0.4018	0.4003	0.4048	0.3551	0.3544	0.3461	*0.3523*	*0.3456*
56	Ba	0.3866	0.3851	0.3897	0.3415	0.3408	0.3328	0.3382	0.3323
57	La	0.3722	0.3707	0.3753	0.3287	0.3280	0.3201	0.3255	0.3195
58	Ce	0.3586	0.3571	0.3617	0.3165	0.3158	0.3082	0.3135	0.3077
59	Pr	0.3466	0.3441	0.3487	0.3050	0.3043	0.2968	*0.3021*	*0.2962*
60	Nd	0.3333	0.3318	0.3365	0.2940	0.2933	0.2862	*0.2915*	*0.2856*
61	Pm	0.3217	0.3202	0.3248	0.2836	0.2829	0.2760	*0.2813*	*0.2754*
62	Sm	0.3105	0.3090	0.3137	0.2738	0.2730	0.2663	0.2711	*0.2657*
63	Eu	0.2999	0.2984	0.3031	0.2643	0.2636	0.2572	*0.2622*	*0.2566*
64	Gd	0.2899	0.2884	0.2930	0.2554	0.2546	0.2482	0.2528	0.2476
65	Tb	0.2802	0.2787	0.2834	0.2468	0.2461	0.2399	*0.2449*	*0.2393*
66	Dy	0.2710	0.2695	0.2742	0.2366	0.2379	0.2319	0.2362	*0.2313*
67	Ho	0.2623	0.2608	0.2655	0.2308	0.2301	0.2243	0.2286	*0.2238*
68	Er	0.2539	0.2524	0.2571	0.2234	0.2227	0.2170	0.2212	*0.2164*
69	Tm	0.2459	0.2443	0.2491	0.2164	0.2156	0.2101	0.2140	*0.2095*
70	Yb	0.2383	0.2367	0.2414	0.2096	0.2088	0.2036	0.2073	*0.2030*
71	Lu	0.2309	0.2293	0.2341	0.2031	0.2023	0.1973	0.2008	*0.1967*
72	Hf	0.2238	0.2222	0.2270	0.1969	0.1961	0.1912	*0.1949*	*0.1906*
73	Ta	0.2171	0.2155	0.2203	0.1901	0.1852	0.1850	0.1889	0.1845
74	W	0.2106	0.2090	0.2138	0.1852	0.1844	0.1795	0.1831	0.1789
75	Re	0.2045	0.2029	0.2076	0.1797	0.1789	0.1742	0.1778	0.1736
76	Os	0.1984	0.1968	0.2016	0.1744	0.1736	0.1691	0.1726	0.1684
77	Ir	0.1926	0.1910	0.1959	0.1694	0.1685	0.1641	0.1675	0.1635
78	Pt	0.1920	0.1904	0.1855	0.1645	0.1637	0.1594	0.1627	0.1588
79	Au	0.1818	0.1802	0.1851	0.1548	0.1590	0.1548	0.1580	0.1542
80	Hg	0.1767	0.1751	0.1800	0.1549	0.1545	0.1504	0.1535	0.1498
81	Tl	0.1717	0.1701	0.1750	0.1510	0.1501	0.1461	0.1492	0.1455
82	Pb	0.1670	0.1654	0.1703	0.1468	0.1460	0.1421	0.1451	0.1416

K Lines, Å

Z	Element	Kα Unresolved α1 − α2	Kα1	Kα2	Kβ3	Kβ1	Kβ2	Kβ5	Kβ4
83	Bi	0.1624	0.1608	0.1657	0.1428	0.1419	0.1382	0.1411	0.1376
84	Po	0.1580	0.1564	0.1613	0.1389	0.1381	0.1344	*0.1372*	*0.1338*
85	At	0.1537	0.1521	0.1571	0.1352	0.1343	0.1307	*0.1334*	*0.1301*
86	Rn	0.1496	0.1480	0.1529	0.1316	0.1307	0.1272	*0.1298*	*0.1266*
87	Fr	0.1456	0.1440	0.1490	0.1281	0.1272	0.1238	*0.1264*	*0.1232*
88	Ra	0.1417	0.1401	0.1451	0.1247	0.1238	0.1205	*0.1230*	*0.1199*
89	Ac	0.1380	0.1364	0.1414	0.1214	0.1206	0.1173	*0.1197*	*0.1167*
90	Th	0.1344	0.1328	0.1378	0.1183	0.1174	0.1143	0.1167	0.1137
91	Pa	0.1309	0.1293	0.1343	0.1152	0.1143	0.1113	*0.1136*	*0.1107*
92	U	0.1275	0.1259	0.1310	0.1123	0.1114	0.1084	0.1107	0.1078
93	Np	0.1242	*0.1226*	*0.1278*	*0.1095*	*0.1086*	*0.1056*	*0.1079*	*0.1050*
94	Pu	0.1210	*0.1194*	*0.1246*	*0.1068*	*0.1059*	*0.1029*	*0.1052*	*0.1023*
95	Am	0.1179	*0.1163*	*0.1214*	*0.1042*	*0.1033*	*0.1004*	*0.1027*	*0.0998*
96	Cm	0.1149	*0.1133*	*0.1184*	*0.1017*	*0.1008*	*0.0979*	*0.1001*	*0.0973*

		L_{III}				L_{II}					L_I		
Z	Element	$L\alpha1$	$L\alpha2$	$L\beta2$	$L\beta6$	Ll	$L\beta1$	$L\eta$	$L\gamma1$	$L\beta3$	$L\beta4$	$L\gamma2$	$L\gamma3$
16	S							83.400					
17	Cl					68.280		67.680					
18	A					56.410		55.890					
19	K					47.550		47.010					
20	Ca	36.330				40.960	35.940	40.480					
21	Sc	31.350				35.590	31.020	35.130					
22	Ti	27.420				31.360	27.050	30.890					
23	V	24.250				27.770	23.880	27.340					
24	Cr	21.640				24.780	21.270	24.300		21.890			
25	Mn	19.450				22.290	19.110	21.850		19.430			
26	Fe	17.590				20.150	17.260	19.750		17.580			
27	Co	15.972				18.292	15.666	17.870		15.700			
28	Ni	14.561				16.693	14.271	16.270		14.270			
29	Cu	13.336				15.286	13.053	14.900		13.160			
30	Zn	12.254				14.020	11.983	13.680		12.095			
31	Ga	11.292				12.953	11.023	12.597		11.192			
32	Ge	10.436				11.945	10.175	11.609		10.365			
33	As	9.671				11.072	9.4141	10.734		9.581	9.640		
34	Se	8.990				10.294	8.7358	9.962		8.929			
35	Br	8.375				9.585	8.125	9.255		8.321			
36	Kr	7.817			7.510	8.947	7.576	8.627		7.767			
37	Rb	7.318	7.325		6.984	8.364	7.076	8.042		7.264	7.304		
38	Sr	6.863	6.870		6.519	7.836	6.624	7.517		6.788	6.821	6.046	
39	Y	6.449	6.456		6.094	7.356	6.212	7.041		6.367	6.403	5.645	
40	Zr	6.071	6.079	5.586	5.710	6.919	5.836	6.607	5.384	5.983	6.019	5.283	
41	Nb	5.724	5.732	5.238	5.361	6.518	5.492	6.211	5.036	5.633	5.668	4.954	
										5.310	5.346	4.654	

L Lines, Å

Z	Element	L_{III}						L_{II}			L_I			
		Lα1	Lα2	Lβ2	Lβ6	Ll	Lβ1	Lη	Lγ1	Lβ3	Lβ4	Lγ2	Lγ3
42	Mo	5.407	5.414	4.923	5.049	6.151	5.177	5.848	4.726	5.013	5.049		4.380
43	Tc	5.115	5.122	4.654	4.759	5.820	4.887	5.518	4.462	4.737	4.773		4.131
44	Ru	4.846	4.854	4.372	4.487	5.504	4.621	5.205	4.182	4.487	4.523		3.898
45	Rh	4.597	4.605	4.131	4.242	5.217	4.374	4.922	3.944	4.252	4.289		3.686
46	Pd	4.368	4.376	3.909	4.016	4.953	4.146	4.661	3.725	4.035	4.071		3.489
47	Ag	4.154	4.163	3.703	3.808	4.708	3.935	4.418	3.523	3.833	3.870		3.307
48	Cd	3.956	3.965	3.514	3.615	4.480	3.738	4.193	3.336	3.645	3.682		3.137
49	In	3.772	3.781	3.338	3.436	4.269	3.555	3.983	3.162	3.470	3.507		2.980
50	Sn	3.600	3.609	3.175	3.269	4.072	3.385	3.789	3.001	3.406	3.343		2.833
51	Sb	3.439	3.448	3.023	3.115	3.888	3.226	3.608	2.852	3.153	3.190		2.695
52	Te	3.289	3.298	2.882	2.971	3.717	3.077	3.438	2.712	3.009	3.047		2.567
53	I	3.149	3.158	2.750	2.837	3.558	2.937	3.280	2.582	2.874	2.912		2.447
54	Xe	3.017	3.025	2.626	2.715	3.421	2.804	3.144	2.462	2.746	2.785	2.338	2.331
55	Cs	2.892	2.902	2.512	2.593	3.267	2.684	2.993	2.348	2.629	2.667	2.237	2.233
56	Ba	2.776	2.786	2.404	2.483	3.136	2.568	2.863	2.242	2.516	2.555	2.139	2.134
57	La	2.666	2.675	2.303	2.379	3.006	2.459	2.740	2.142	2.211	2.449	2.046	2.041
58	Ce	2.562	2.571	2.209	2.282	2.892	2.356	2.620	2.049	2.311	2.350	1.960	1.955
59	Pr	2.463	2.473	2.119	2.191	2.784	2.259	2.512	1.961	2.217	2.255	1.879	1.874
60	Nd	2.370	2.381	2.036	2.104	2.676	2.167	2.409	1.878	2.127	2.167	1.801	1.796
61	Pm	2.282	2.293	1.956	2.024	2.578	2.080	2.312	1.799	2.042	2.082	1.728	1.724
62	Sm	2.200	2.210	1.882	1.947	2.482	1.998	2.219	1.727	1.962	2.001	1.659	1.655
63	Eu	2.121	2.132	1.812	1.874	2.395	1.920	2.132	1.657	1.887	1.926	1.596	1.590
64	Gd	2.047	2.058	1.746	1.805	2.312	1.847	2.049	1.592	1.815	1.854	1.533	1.530
65	Tb	1.977	1.986	1.683	1.742	2.235	1.777	1.973	1.530	1.747	1.786	1.476	1.472
66	Dy	1.909	1.920	1.624	1.682	2.159	1.711	1.897	1.473	1.682	1.721	1.423	1.416

67	Ho	1.845	1.856	1.567	1.624	2.086	1.648	1.826	1.417	1.620	1.660	1.370	1.364
68	Er	1.784	1.796	1.514	1.568	2.015	1.587	1.757	1.364	1.562	1.601	1.321	1.315
69	Tm	1.727	1.738	1.464	1.516	1.955	1.530	1.696	1.315	1.506	1.545	1.274	1.268
70	Yb	1.627	1.683	1.416	1.466	1.894	1.476	1.636	1.268	1.452	1.491	1.229	1.222
71	Lu	1.620	1.630	1.370	1.419	1.836	1.424	1.578	1.222	1.401	1.441	1.185	1.180
72	Hf	1.570	1.580	1.326	1.374	1.781	1.374	1.523	1.179	1.353	1.392	1.144	1.138
73	Ta	1.522	1.533	1.285	1.331	1.728	1.327	1.471	1.138	1.307	1.346	1.105	1.099
74	W	1.4764	1.4874	1.2446	1.2899	1.6782	1.2818	1.4211	1.0986	1.2627	1.3016	1.0681	1.0620
75	Re	1.4329	1.4440	1.2066	1.2510	1.6306	1.2386	1.3734	1.0610	1.2203	1.2592	1.0323	1.0261
76	Os	1.3912	1.4023	1.1698	1.2135	1.5850	1.1973	1.3279	1.0250	1.1796	1.2184	0.9981	0.9919
77	Ir	1.3513	1.3625	1.1353	1.1780	1.5409	1.1578	1.2845	0.9909	1.1409	1.1796	0.9655	0.9593
78	Pt	1.3130	1.3243	1.1020	1.1436	1.4995	1.1199	1.2429	0.9580	1.1039	1.1422	0.9343	0.9279
79	Au	1.2764	1.2877	1.0702	1.1109	1.4596	1.0835	1.2027	1.9265	1.0679	1.1065	0.9043	0.8978
80	Hg	1.2412	1.2526	1.0396	1.0798	1.4216	1.0487	1.1652	0.8965	1.0336	1.0722	0.8754	0.8692
81	Tl	1.2074	1.2188	1.0103	1.0496	1.3848	1.0151	1.1277	0.8675	1.0006	1.0392	0.8477	0.8413
82	Pb	1.1750	1.1865	0.9822	1.0210	1.3499	0.9829	1.0924	0.8397	0.9691	1.0075	0.8210	0.8147
83	Bi	1.1439	1.1554	0.9552	0.9933	1.3161	0.9520	1.0586	0.8131	0.9386	0.9769	0.7957	0.7892
84	Po	1.1139	1.1255	0.9294	0.9672	1.2829	0.9220	1.0252	0.7875	0.9091	0.9475	0.7724	0.7646
85	At	1.0850	1.0967	0.9046	0.9419	1.2526	0.8935	0.9934	0.7629	0.8814	0.9193	0.7477	0.7412
86	Rn	1.0572	1.0690	0.8836	0.9175	1.2231	0.8661	0.9632	0.7393	0.8544	0.8921	0.7248	0.7184
87	Fr	1.0305	1.0423	0.858	0.8939	1.1946	0.8394	0.9346	0.7165	0.8279	0.8659	0.7030	0.6966
88	Ra	1.0047	1.0166	0.8354	0.8709	1.1672	0.8138	0.9074	0.6946	0.8027	0.8407	0.6820	0.6754
89	Ac	0.9799	0.9918	0.8140	0.8489	1.1407	0.7890	0.8806	0.6706	0.7782	0.8161	0.6627	0.6550
90	Th	0.9560	0.9679	0.7935	0.8279	1.1151	0.7652	0.8545	0.6521	0.7548	0.7926	0.6422	0.6356
91	Pa	0.9328	0.9448	0.7737	0.8079	1.0908	0.7423	0.8295	0.6336	0.7323	0.7699	0.6239	0.6169
92	U	0.9106	0.9226	0.7547	0.7884	1.0671	0.7200	0.8051	0.6148	0.7103	0.7480	0.6052	0.5986
93	Np	0.8891	0.9010	0.7362	0.7691	1.0428	0.6985	0.7809	0.5965	0.6892	0.7267	0.5873	0.5810
94	Pu	0.8683	0.8803	0.7185	0.7515	1.0226	0.6777	0.7591	0.5789	0.6687	0.7062	0.5707	0.5640
95	Am	0.8481	0.8603	0.7014	0.7342	1.0012	0.6577	0.7398	0.5619	0.6489	0.6864	0.5544	0.5478
96	Cm	0.8289	0.8411	0.6850	0.7175	0.9808	0.6386	0.7208	0.5455	0.6298	0.6273	0.5386	0.5320

Selected M and N Lines, Å

Z	Element	Mα1	Mα2	Mβ	Mγ	M_{II}-M_{IV}*	N series
40	Zr	—	—	—	38.390	37.000	
41	Nb	—	—	—	34.900	33.100	
42	Mo	—	—	—	32.700	31.400	
43	Tc	—	—	—	30.100	28.800	
44	Ru	—	—	—	26.900	25.500	
45	Rh	—	—	—	25.010	24.450	
46	Pd	—	—	—	23.300	22.100	
47	Ag	—	—	—	21.820	20.660	
48	Cd	—	—	—	20.470	19.400	
49	In	—	—	—	19.210	18.240	
50	Sn	—	—	—	17.940	16.930	
51	Sb	—	—	—	16.920	15.980	
52	Te	—	—	—	15.930	15.020	
53	I	—	—	—	15.010	14.150	
54	Xe	—	—	—	14.180	13.310	
55	Cs	—	—	—	13.420	12.580	189
56	Ba	—	—	—	12.750	11.890	162
57	La	14.880		14.510	12.080	11.280	
58	Ce	14.040		13.750	11.530	10.690	
59	Pr	13.343		13.060	10.998	10.180	
60	Nd	12.680		12.440	10.505	9.700	
61	Pm	—		—	10.050	9.260	
62	Sm	11.470		11.270	9.600	8.840	
63	Eu	10.960		10.750	9.211	8.450	
64	Gd	10.460		10.254	8.844	8.120	
65	Tb	10.000		9.792	8.486	7.740	
66	Dy	9.590		9.357	8.144	7.460	
67	Ho	9.200		8.965	7.865	7.160	
68	Er	8.820		8.592	7.546	6.860	
69	Tm	8.480		8.249	7.318	6.540	
70	Yb	8.149		7.909	7.024	6.270	
71	Lu	7.840		7.601	6.768	6.020	
72	Hf	7.539		7.303	6.544	5.770	
73	Ta	7.252		7.023	6.312	5.570	
74	W	6.983	6.992	6.757	6.092	5.357	56
75	Re	6.729		6.504	5.885	5.150	
76	Os	6.490		6.267	5.682	4.955	
77	Ir	6.262	6.275	6.038	5.500	4.780	
78	Pt	6.047	6.058	5.828	5.319	4.601	
79	Au	5.840	5.854	5.624	5.145	4.432	
80	Hg	5.648	5.677	5.432	4.984	4.266	
81	Tl	5.460	5.472	5.249	4.823	4.116	
82	Pb	5.286	5.299	5.076	4.674	3.968	

* The two levels denote the electronic transition producing the line.

Selected M and N Lines, Å

Z	Element	Mα1	Mα2	Mβ	Mγ	M_{II}–M_{IV}	N series
83	Bi	5.118	5.130	4.909	4.532	3.834	
84	Po	4.955	4.958	4.736	4.361	3.680	
85	At	4.802	4.802	4.581	4.234	3.559	
86	Rn	4.655	4.657	4.436	4.124	3.448	
87	Fr	4.515	4.521	4.303	4.008	3.322	
88	Ra	4.383	4.392	4.178	3.892	3.220	
89	Ac	4.256	4.270	4.060	3.798	3.118	
90	Th	4.138	4.151	3.941	3.679	3.011	35
91	Pa	4.022	4.035	3.827	3.577	2.910	
92	U	3.910	3.924	3.716	3.479	2.817	33

N Spectrum of U (Å)

N_I–P_{II}	8.81	N_{II}–P_{IV}	10.40	N_V–$N_{VI,VII}$	34.8	N_{VII}–O_V	50.0
N_I–P_{III}	8.76	N_{III}–O_V	12.90	N_{VI}–O_{IV}	43.3		
N_I–$P_{IV,V}$	8.60	N_{IV}–N_{VI}	31.8	N_{VI}–O_V	42.1		

X-Ray Emission Energies, keV

Z	Element	K lines				L lines				
		Kβ2	Kβ1	Kα1	Kα2	Lγ1	Lβ2	Lβ1	Lα1	Lα2
1	H									
2	He									
3	Li			0.052						
4	Be			0.110						
5	B			0.185						
6	C			0.282						
7	N			0.392						
8	O			0.523						
9	F			0.677						
10	Ne			0.851						
11	Na		1.067	1.041						
12	Mg		1.297	1.254						
13	Al		1.553	1.487	1.486					
14	Si		1.832	1.740	1.739					
15	P		2.136	2.015	2.014					
16	S		2.464	2.308	2.306					
17	Cl		2.815	2.622	2.621					

X-Ray Emission Energies, keV

Z	Element	K lines				L lines				
		Kβ2	Kβ1	Kα1	Kα2	Lγ1	Lβ2	Lβ1	Lα1	Lα2
18	A		3.192	2.957	2.955					
19	K		3.589	3.313	3.310					
20	Ca		4.012	3.691	3.688				0.344	0.341
21	Sc		4.460	4.090	4.085				0.399	0.395
22	Ti		4.931	4.510	4.504				0.458	0.452
23	V		5.427	4.952	4.944				0.519	0.510
24	Cr		5.946	5.414	5.405				0.581	0.571
25	Mn		6.490	5.898	5.887				0.647	0.636
26	Fe		7.057	6.403	6.390				0.717	0.704
27	Co		7.649	6.930	6.915				0.790	0.775
28	Ni	8.328	8.264	7.477	7.460				0.866	0.849
29	Cu	8.976	8.904	8.047	8.027				0.948	0.928
30	Zn	9.657	9.571	8.638	8.615				1.032	1.009
31	Ga	10.365	10.263	9.251	9.234				1.122	1.096
32	Ge	11.100	10.981	9.885	9.854				1.216	1.186
33	As	11.863	11.725	10.543	10.507				1.317	1.282
34	Se	12.651	12.495	11.221	11.181				1.419	1.379
35	Br	13.465	13.290	11.923	11.877				1.526	1.480
36	Kr	14.313	14.112	12.648	12.597				1.638	1.587
37	Rb	15.184	14.960	13.394	13.335			1.752	1.694	1.692
38	Sr	16.083	15.834	14.164	14.097			1.872	1.806	1.805
39	Y	17.011	16.736	14.957	14.882			1.996	1.922	1.920
40	Zr	17.969	17.666	15.774	15.690	2.302	2.219	2.124	2.042	2.040
41	Nb	18.951	18.621	16.614	16.520	2.462	2.367	2.257	2.166	2.163
42	Mo	19.964	19.607	17.478	17.373	2.623	2.518	2.395	2.293	2.290
43	Tc	21.012	20.585	18.410	18.328	2.792	2.674	2.538	2.424	2.420
44	Ru	22.072	21.655	19.278	19.149	2.964	2.836	2.683	2.558	2.554
45	Rh	23.169	22.721	20.214	20.072	3.144	3.001	2.834	2.696	2.692
46	Pd	24.297	23.816	21.175	21.018	3.328	3.172	2.990	2.838	2.833
47	Ag	25.454	24.942	22.162	21.988	3.519	3.348	3.151	2.984	2.978
48	Cd	26.641	26.093	23.172	22.982	3.716	3.528	3.316	3.133	3.127
49	In	27.859	27.274	24.207	24.000	3.920	3.713	3.487	3.287	3.279
50	Sn	29.106	28.483	25.270	25.042	4.131	3.904	3.662	3.444	3.435
51	Sb	30.387	29.723	26.357	26.109	4.347	4.100	3.843	3.605	3.595
52	Te	31.698	30.993	27.471	27.200	4.570	4.301	4.029	3.769	3.758
53	I	33.016	32.292	28.610	28.315	4.800	4.507	4.220	3.937	3.926
54	Xe	34.446	33.644	29.802	29.485	5.036	4.720	4.422	4.111	4.098
55	Cs	35.819	34.984	30.970	30.623	5.280	4.936	4.620	4.286	4.272
56	Ba	37.255	36.376	32.191	31.815	5.531	5.156	4.828	4.467	4.451
57	La	38.728	27.799	33.440	33.033	5.789	5.384	5.043	4.651	4.635
58	Ce	40.231	39.255	34.717	34.276	6.052	5.613	5.262	4.840	4.823

X-Ray Emission Energies, keV

Z	Element	K lines				L lines				
		Kβ2	Kβ1	Kα1	Kα2	Lγ1	Lβ2	Lβ1	Lα1	Lα2
59	Pr	41.772	40.746	36.023	35.548	6.322	5.850	5.489	5.034	5.014
60	Nd	43.298	42.269	37.359	36.845	6.602	6.090	5.722	5.230	5.208
61	Pm	44.955	43.945	38.649	38.160	6.891	6.336	5.956	5.431	5.408
62	Sm	46.553	45.400	40.124	39.523	7.180	6.587	6.206	5.636	5.609
63	Eu	48.241	47.027	41.529	40.877	7.478	6.842	6.456	5.846	5.816
64	Gd	49.961	48.718	42.983	42.280	7.788	7.102	6.714	6.059	6.027
65	Tb	51.737	50.391	44.470	43.737	8.104	7.368	6.979	6.275	6.241
66	Dy	53.491	52.178	45.985	45.193	8.418	7.638	7.249	6.495	6.457
67	Ho	55.292	53.934	47.528	46.686	8.748	7.912	7.528	6.720	6.680
68	Er	57.088	55.690	49.099	48.205	9.089	8.188	7.810	6.948	6.904
69	Tm	58.969	57.576	50.730	49.762	9.424	8.472	8.103	7.181	7.135
70	Yb	60.959	59.352	52.360	51.326	9.779	8.758	8.401	7.414	7.367
71	Lu	62.946	61.282	54.063	52.959	10.142	9.048	8.708	7.654	7.604
72	Hf	64.936	63.209	55.757	54.579	10.514	9.346	9.021	7.898	7.843
73	Ta	66.999	65.210	57.524	56.270	10.892	9.649	9.341	8.145	8.087
74	W	69.090	67.233	59.310	57.973	11.283	9.959	9.670	8.396	8.333
75	Re	71.220	69.298	61.131	59.707	11.684	10.273	10.008	8.651	8.584
76	Os	73.393	71.404	62.991	61.477	12.094	10.596	10.354	8.910	8.840
77	Ir	75.605	73.549	64.886	63.278	12.509	10.918	10.706	9.173	9.098
78	Pt	77.866	75.736	66.820	65.111	12.939	11.249	11.069	9.441	9.360
79	Au	80.165	77.968	68.794	66.980	13.379	11.582	11.439	9.711	9.625
80	Hg	82.526	80.258	70.821	68.894	13.828	11.923	11.823	9.987	9.896
81	Tl	84.904	82.558	72.860	70.820	14.288	12.268	12.210	10.266	10.170
82	Pb	87.343	84.922	74.957	72.794	14.762	12.620	12.611	10.549	10.448
83	Bi	89.833	87.335	77.097	74.805	15.244	12.977	13.021	10.836	10.729
84	Po	92.386	89.809	79.296	76.868	15.740	13.338	13.441	11.128	11.014
85	At	94.976	92.319	81.525	78.956	16.248	13.705	13.873	11.424	11.304
86	Rn	97.616	94.877	83.800	81.080	16.768	14.077	14.316	11.724	11.597
87	Fr	100.305	97.483	86.119	83.243	17.301	14.459	14.770	12.029	11.894
88	Ra	103.048	100.136	88.485	85.446	17.845	14.839	15.233	12.338	12.194
89	Ac	105.838	102.846	90.894	87.681	18.405	15.227	15.712	12.650	12.499
90	Th	108.671	105.592	93.334	89.942	18.977	15.620	16.200	12.966	12.808
91	Pa	111.575	108.408	95.851	92.271	19.559	16.022	16.700	13.291	13.120
92	U	114.549	111.289	98.428	94.648	20.163	16.425	17.218	13.613	13.438
93	Np	117.533	114.181	101.005	97.023	20.774	16.837	17.740	13.945	13.758
94	Pu	120.592	117.146	103.653	99.457	21.401	17.254	18.278	14.279	14.082
95	Am	123.706	120.163	106.351	101.932	22.042	17.677	18.829	14.618	14.411
96	Cm	126.875	123.235	109.098	104.448	22.699	18.106	19.393	14.961	14.743
97	Bk	130.101	126.362	111.896	107.023	23.370	18.540	19.971	15.309	15.079
98	Cf	133.383	129.544	114.745	109.603	24.056	18.980	20.562	15.661	15.420
99	Es	136.724	132.781	117.646	112.244	24.758	19.426	21.166	16.018	15.764
100	Fm	140.122	136.075	120.598	114.926	25.475	19.879	21.785	16.379	16.113

Relative Intensities of X-Ray Lines

K series
$\alpha 1 = 100$
$\alpha 2 = 50$
$\beta 1 + \beta 3$ varies from 20 for $Z = 35$
to 42 for $Z = 92$
$\beta 2$ varies from < 1 at $Z = 40$
to 12 at $Z = 92$

L series
$\alpha 1 = 100$
$\alpha 2 = 11$
$\beta 1 = 55$
$\beta 2 = 28$
$\beta 3 = 3$
$\beta 4 = 1.5$
$\beta 6 = 1.6$
$\beta 15 = 3$
$l = 3.4$
$\eta = 1.8$
$\gamma 1 = 15$

M series
$\alpha 1 + \alpha 2 = 100$
$\beta = 65$

No conclusions should be drawn as to intensity relationships between different series of lines.

Appendix IV

Mass Absorption Coefficients of the Elements and of Selected Films

Satisfactory values of mass absorption coefficients are not available at all wavelengths for all atomic numbers. The following table is entirely new and represents our best judgment as of 1971. Extrapolations and interpolations were made on the basis of wavelength raised to the 2.8 power. For the transuranic elements, all values are extrapolated. No values are given above 30 Å for the heavier elements.

Numerical values of the mass absorption coefficients of the elements are given in columns headed by corresponding wavelength in the tables immediately following.

Mass absorption

Z	Element	0.1	0.15	0.20	0.25	0.3	0.4	0.5	0.6	0.7	0.8	0.9	1.0	Wavelength Å 1.5
1	H	0.29	0.32	0.34	0.35	0.37	0.38	0.40	0.42	0.43	0.44	0.45	0.45	0.49
2	He	0.11	0.12	0.12	0.13	0.14	0.14	0.17	0.86	0.20	0.22	0.23	0.25	0.35
3	Li	0.12	0.13	0.13	0.14	0.15	0.15	0.18	0.22	0.25	0.30	0.36	0.42	1.02
4	Be	0.13	0.13	0.14	0.14	0.15	0.16	0.19	0.23	0.28	0.34	0.43	0.53	1.54
5	B	0.13	0.14	0.14	0.15	0.16	0.19	0.24	0.31	0.40	0.54	0.70	0.92	2.87
6	C	0.14	0.14	0.15	0.16	0.17	0.23	0.31	0.42	0.59	0.83	1.14	1.54	4.79
7	N	0.14	0.15	0.16	0.17	0.20	0.28	0.40	0.59	0.88	1.26	1.76	2.37	7.38
8	O	0.14	0.15	0.16	0.18	0.23	0.34	0.53	0.83	1.27	1.84	2.56	3.45	10.74
9	F	0.14	0.15	0.18	0.21	0.26	0.43	0.70	1.14	1.77	2.57	3.57	4.81	14.96
10	Ne	0.14	0.16	0.19	0.24	0.31	0.54	0.93	1.54	2.38	3.45	4.80	6.47	20.11
11	Na	0.15	0.17	0.21	0.27	0.36	0.66	1.21	2.02	3.11	4.51	6.28	8.45	26.29
12	Mg	0.15	0.18	0.23	0.31	0.43	0.83	1.54	2.57	3.97	5.76	8.02	10.79	33.56
13	Al	0.15	0.19	0.26	0.36	0.50	1.03	1.93	3.22	4.96	7.21	10.03	13.50	41.97
14	Si	0.15	0.20	0.29	0.42	0.59	1.27	2.38	3.97	6.11	8.88	12.35	16.63	51.69
15	P	0.16	0.22	0.31	0.47	0.70	1.54	2.88	4.81	7.41	10.77	14.98	20.14	62.68
16	S	0.17	0.23	0.35	0.54	0.84	1.85	3.46	5.77	8.88	12.90	17.95	24.16	75.10
17	Cl	0.17	0.25	0.39	0.61	0.98	2.19	4.10	5.89	10.52	15.29	21.27	28.63	88.98
18	A	0.18	0.27	0.45	0.70	1.15	2.57	4.81	8.02	12.34	17.93	24.95	33.58	104.38
19	K	0.19	0.29	0.50	0.81	1.34	2.99	5.59	9.32	14.36	20.86	29.01	39.05	121.37
20	Ca	0.20	0.32	0.55	0.92	1.54	3.45	6.44	10.74	16.54	24.04	33.43	44.99	139.85
21	Sc	0.21	0.36	0.61	1.06	1.77	3.95	7.37	12.29	18.93	27.50	38.25	51.49	160.03
22	Ti	0.22	0.38	0.69	1.20	2.01	4.49	8.38	13.98	21.53	31.27	43.50	58.55	181.98
23	V	0.23	0.40	0.76	1.36	2.27	5.07	9.47	15.79	24.31	35.31	49.12	66.11	205.47
24	Cr	0.24	0.43	0.83	1.53	2.55	5.70	10.63	17.73	27.30	39.66	55.16	74.25	230.78
25	Mn	0.25	0.46	0.92	1.71	2.85	6.37	11.89	19.82	30.52	44.33	61.66	83.00	257.97
26	Fe	0.26	0.50	1.02	1.90	3.17	7.09	13.23	22.06	33.96	49.34	68.62	92.36	287.08
27	Co	0.28	0.56	1.13	2.11	3.52	7.86	14.66	24.45	37.64	54.68	76.06	102.37	318.19
28	Ni	0.30	0.61	1.25	2.33	3.88	8.68	16.19	27.00	41.57	60.38	83.99	113.04	43.84
29	Cu	0.32	0.68	1.38	2.56	4.27	9.55	17.82	29.71	45.74	66.45	92.42	120.48	48.58
30	Zn	0.35	0.72	1.51	2.81	4.69	10.47	19.54	32.58	50.17	72.88	101.37	136.44	53.65
31	Ga	0.37	0.76	1.73	3.08	5.13	11.46	21.39	35.66	54.91	79.76	110.94	149.32	59.31
32	Ge	0.40	0.82	1.81	3.35	5.59	12.49	23.31	38.87	59.84	86.93	120.91	162.75	65.10
33	As	0.42	0.88	1.96	3.65	6.08	13.59	25.36	42.28	65.09	94.56	131.52	177.03	71.02
34	Se	0.47	0.96	2.13	3.96	6.60	14.75	27.51	45.87	70.63	102.61	142.71	25.02	77.85
35	Br	0.50	1.04	2.31	4.29	7.15	15.97	29.78	49.66	76.46	111.07	154.49	27.29	84.93
36	Kr	0.53	1.12	2.49	4.63	7.72	17.25	32.18	53.65	82.60	120.00	22.05	29.69	92.27
37	Rb	0.57	1.21	2.69	4.99	8.32	18.59	34.68	57.82	89.03	129.33	23.99	32.29	100.38
38	Sr	0.61	1.30	2.89	5.37	8.96	20.00	37.32	62.22	95.80	18.71	26.03	35.03	108.89
39	Y	0.65	1.40	3.11	5.77	9.62	21.48	40.08	66.82	102.88	20.24	28.16	37.90	117.80
40	Zr	0.69	1.50	3.33	6.19	10.31	23.03	42.96	71.63	15.06	21.88	30.43	40.96	127.33
41	Nb	0.74	1.60	3.57	6.62	11.03	24.65	45.98	76.66	16.19	23.52	32.72	44.04	136.88
42	Mo	0.79	1.71	3.81	7.07	11.79	26.33	49.12	81.90	17.38	25.25	35.12	47.28	146.95
43	Tc	0.84	1.83	4.07	7.55	12.58	28.09	52.41	12.12	18.66	27.11	37.70	50.75	157.73
44	Ru	0.89	1.95	4.33	8.04	13.40	29.93	55.83	12.97	19.97	29.02	40.36	54.33	168.86
45	Rh	0.94	2.07	4.61	8.55	14.26	31.84	59.39	13.87	21.36	31.03	43.16	58.10	180.58
46	Pd	1.00	2.20	4.90	9.09	15.14	33.82	63.18	14.81	22.80	33.12	46.07	62.01	192.75
47	Ag	1.05	2.34	5.19	9.64	16.07	35.89	9.4	15.7	24.3	35.3	49.1	66.1	205.
48	Cd	1.10	2.48	5.51	10.22	17.03	38.04	10.0	16.8	25.8	37.6	52.2	70.3	218.
49	In	1.15	2.62	5.83	10.82	18.03	40.27	10.7	17.8	27.5	39.9	55.5	74.7	232.
50	Sn	1.20	2.77	6.16	11.44	19.06	42.58	11.9	19.9	30.7	44.7	62.1	79.5	260.
51	Sb	1.25	2.93	6.51	12.08	20.14	44.97	12.0	20.1	30.9	45.0	62.5	84.2	261.
52	Te	1.30	3.09	6.87	12.75	21.25	6.85	12.7	21.3	32.7	47.6	66.2	89.1	277.

K

coefficients

6.20	4.96	4.13	3.10	2.48	2.07	1.77	1.55	1.38	1.24	.827	.620	.496	.413	
2.0	2.5	3.0	4.0	5.0	6.0	7.0	8.0	9.0	10.0	15.0	20.0	25.0	30.0	
0.52	0.62	0.75	1.25	2.12	3.28	4.85	7.1	10.0	13.7	32.	69.	127.	208.	
0.71	1.04	1.48	3.55	6.90	11.60	18.1	26.6	37.7	51.	107.	268.	540.	970.	
2.18	3.98	6.60	15.2	28.80	48.80	76	113.0	157.	213.	402.	970.	1900.	3270.	
3.45	6.44	10.74	24.04	44.91	74.84	115.2	167.4	232.	312.	973.	2178.	4068	6778.	
6.43	12.01	20.03	44.80	83.69	139.4	214.7	312.0	434.	583.	1814.	4059.	7583.	12633.	
10.72	20.01	33.35	74.61	139.3	232.2	357.5	519.6	722.	970.	3020.	6760	12627	21037.	
16.51	30.83	51.38	114.9	214.6	357.7	550.7	800.4	1113.	1495.	4652.	10413.	19450.	32407.	K
24.02	44.85	74.75	167.1	312.3	520.4	801.2	1164.5	1619.	2175.	6768.	15149.	1005.	1675.	
33.45	62.45	104.0	232.8	434.8	724.7	1115.6	1621.4	2254.	3029.	9425.	932.	1741.	2900.	
44.98	83.98	139.9	313.0	584.8	974.5	1500.3	2180.5	3032.	4073.	659.	1475.	2756.	4593.	
58.79	109.75	182.9	409.1	764.2	1273.5	1960.6	2849.5	3962.	5323.	951.	2130.	3979.	6629.	
75.05	140.11	233.5	522.3	975.6	1625.9	2503.0	3637.8	5059.	467.	1453.	3252.	6075.	10122.	
93.86	175.22	292.0	653.2	1220.1	2033.3	3130.3	336.1	467.	628.	1954.	4373.	8168.	13608.	
115.58	215.78	359.6	804.4	1502.5	2503.9	305.3	443.8	617.	829.	2579.	5773.	10784.	17966.	
140.16	261.66	436.1	975.4	1822.0	251.1	386.6	561.9	781.	1049.	3266.	7310.	13655.	22750	
167.92	313.48	522.4	1168.6	2182.9	310.7	478.4	695.3	966.	1298.	4041.	9044.	16894.	28146.	
198.97	371.45	619.0	1384.7	225.1	375.2	577.6	839.5	1167.	1568.	4880.	10921.	20400.	33987	
233.39	435.70	726.1	156.2	291.9	486.5	748.9	1092	1510	2033.	6327.	14160.	26450.	44066	
271.38	506.63	844.4	190.8	356.4	594.0	914.5	1329.	1848.	2483.	7725.	17289.	32295.	53804.	
312.71	583.78	972.9	232.7	434.7	724.4	1115.	1620.	2254.	3028.	9422.	21086.	39387.	65624.	L_I
357.84	668.03	121.9	272.6	509.3	848.8	1306.	1899.	2641.	3547.	11039.	24705.	46146.	65378.	L_{II}
406.91	85.71	142.7	319.3	596.6	994.2	1530.	2224.	3093.	4155.	12930.	28939.	45874.	8845.	L_{III}
459.45	104.07	173.4	387.9	724.7	1207.	1859.	2702.	3757.	5048.	15707.	28265.	5960.	9930.	
516.03	117.54	195.9	438.1	818.5	1364.	2099.	3051.	4244.	5701.	17740.	32301.	6897.	11490.	
70.76	132.11	220.1	492.5	919.9	1533.	2360.	3430.	4770.	6408.	19938.	4155.	7761.	12930.	
79.21	147.88	246.4	551.3	1029.	1716.	2641.	3839.	5339.	7173.	18597.	4656.	8698.	14490.	
88.41	165.04	275.0	615.2	1149.	1915.	2948.	4285.	5959.	8005.	20828.	5184.	9684.	16134.	
98.04	183.02	305.0	682.3	1274.	2123.	3269.	4752.	6608.	8877.	2623.	5872.	10968.	18273.	
108.64	202.82	338.0	756.1	1412.	2353.	3623.	5266.	7323.	9838.	2918.	6531.	12199.	20325.	
119.96	223.95	373.2	834.8	1559.	2598.	4000.	5814.	8086.	10862.	3234.	7237.	13519.	22523.	
132.61	247.57	412.6	922.9	1723.	2872.	4422.	6428.	8939.	10315.	3584.	8021.	14982.	24961.	
145.58	271.78	452.9	1013.	1892.	3153.	4855.	7056.	8535.	11079.	3984.	8916.	16654.	27746.	
158.81	296.47	494.1	1105.	2064.	3440.	5296.	7697.	9389.	1409.	4386.	9816.	18336.	30549.	
174.08	324.98	541.6	1211.	2263.	3771.	5805.	7397.	10287.	1548.	4818.	10783.	20141.	33556.	
189.90	354.52	590.8	1321.	2468.	4113.	5574.	994.	1383.	1858.	5781.	12938.	24166.	40262.	
206.33	385.19	641.9	1435.	2682.	4469.	6090.	1122.	1561.	2088.	6525.	14604.	27279.	45447.	
224.45	419.02	698.3	1562.	2917.	4315.	877.	1275.	1774.	2383.	7416.	16598.	31004.	51652.	
243.49	454.57	757.6	1694.	3165.	4699.	971.	1412.	1963.	2638.	8208.	18371.	34315.	57169.	
263.41	491.75	819.5	1833.	3424.	704.	1084.	1575.	2191.	2943.	9159.	20498.	38289.	62686.	M_I
284.71	531.50	885.8	1981.	3313.	790.	1217.	1769.	2460.	3305.	10286.	23020.	42998.	55280.	
306.07	571.37	952.3	2130.	3579.	860.	1324.	1925.	2678.	3597.	11193.	25050.	46791.	60946.	
328.59	613.43	1022.	2286.	568.	946.	1456.	2117.	2944.	3955.	12307.	27543.	39976.	66612.	$M_{II}\ M_{III}$
352.68	658.40	1097.	2454.	624.	1040.	1602.	2328.	3238.	4350.	13535.	30293.	44632.	39444.	
378.03	704.85	1174.	2385.	678.	1131.	1741.	2531.	3520.	4728.	14713.	32929.	48987.	44717.	
403.77	753.78	1256.	2423.	739.	1233.	1898.	2758.	3836.	5154.	16036.	29040.	30192.	50304.	
431.00	804.61	1341.	430.	804.	1340.	2063.	2999.	4170.	5603.	17433.	31764.	33872.	56432.	
459.	857.	1429.	467.	873.	1455.	2240.	3255.	4527.	6082.	18924.	34727.	37812.	62996.	M_{IV}
489.	913.	1522.	502.	937.	1562.	2406.	3497.	4863.	6533.	20327.	22293.	41641.	18612.	
519.	970.	1485.	546.	1020.	1700.	2618.	3805.	5291.	7108.	19536.	24793.	46311.	20439.	M_V
581.	1086.	1480.	588.	1098.	1831.	2819.	4097.	5736.	7654.	20067.	27429.	12951.	21578.	
585.	1093.	282.	632.	1181.	1969.	3031.	4405.	6126.	8230.	21593.	29983.	14367.	23935.	
619.	1157.	302.	677.	1265.	2109.	3247.	4719.	6563.	8816.	21580.	32605.	15196.	25317.	

$L_I\quad L_{II}\ L_{III}$ 527 $M_I\ M_{II}\quad M_{III}\quad\quad M_{IV}\ M_V$

Mass absorption

Z	Element	0.1	0.15	0.20	0.25	0.3	0.4	0.5	0.6	0.7	0.8	0.9	1.0	1.5
53	I	1.36	3.26	7.24	13.44	22.40	7.24	13.5	22.5	34.6	50.4	70.1	94.3	293.
54	Xe	1.42	3.43	7.63	14.16	23.59	7.66	14.2	23.8	36.6	53.3	74.1	99.7	310.
55	Cs	1.48	3.61	8.03	14.90	24.82	8.08	15.0	25.1	38.7	56.2	78.2	105.3	327.
56	Ba	1.53	3.80	8.44	15.66	26.10	8.52	15.9	26.5	40.8	59.3	82.5	111.0	345.
57	La	1.60	3.99	8.86	16.45	27.42	8.98	16.7	27.9	42.9	62.4	86.8	116.9	363.
58	Ce	1.66	4.19	9.30	17.27	28.78	9.45	17.6	29.4	45.2	65.7	91.4	123.1	382.
59	Pr	1.72	4.39	9.76	18.11	4.45	9.94	18.5	30.9	47.5	69.1	96.1	129.4	402.
60	Nd	1.80	4.60	10.23	18.98	4.67	10.4	19.4	32.4	49.9	72.6	101.0	135.9	422.
61	Pm	1.86	4.82	10.71	19.88	4.90	10.9	20.4	34.0	52.4	76.2	106.0	142.7	443.
62	Sm	1.93	5.04	11.21	20.80	5.14	11.4	21.4	35.7	55.0	79.9	111.2	149.7	465.
63	Eu	2.02	5.28	11.73	21.76	5.39	12.0	22.4	37.4	57.7	83.8	116.5	156.9	487.
64	Gd	2.09	5.52	12.26	3.38	5.64	12.6	23.5	39.2	60.3	87.6	122.0	164.2	478.
65	Tb	2.18	5.76	12.80	3.54	5.91	13.2	24.6	41.0	63.2	91.8	127.7	171.7	501.
66	Dy	2.26	6.02	13.37	3.70	6.18	13.8	25.7	42.9	66.0	95.9	133.5	179.6	469.
67	Ho	2.33	6.28	13.95	3.87	6.46	14.4	26.9	44.8	69.0	100.3	139.5	187.8	489.
68	Er	2.42	6.55	14.55	4.05	6.75	15.0	28.1	46.8	72.1	104.8	145.8	196.3	107.
69	Tm	2.50	6.82	15.16	4.23	7.05	15.7	29.3	48.9	75.4	109.5	152.3	205.1	113.
70	Yb	2.58	7.11	15.90	4.42	7.36	16.4	30.6	51.1	78.7	114.4	159.1	214.1	118.
71	Lu	2.66	7.40	2.48	4.61	7.68	17.1	32.0	53.3	82.1	119.3	166.0	223.5	124.
72	Hf	2.75	7.71	2.59	4.80	8.01	17.8	33.3	55.6	85.6	124.4	173.0	232.9	131.
73	Ta	2.82	8.02	2.70	5.01	8.35	18.6	34.7	57.9	89.3	129.7	180.4	242.8	136.
74	W	2.90	8.34	2.81	5.22	8.70	19.4	36.2	60.4	93.0	135.1	188.0	253.0	143.
75	Re	2.96	8.67	2.93	5.44	9.06	20.2	37.7	62.9	96.9	140.8	195.8	249.4	150.
76	Os	3.03	9.00	3.05	5.66	9.44	21.0	39.3	65.5	100.9	146.6	204.0	260.1	157.
77	Ir	3.10	9.35	3.17	5.89	9.82	21.9	40.9	68.2	105.0	152.5	212.2	231.5	165.
78	Pt	3.17	9.71	3.30	6.13	10.2	22.8	42.5	70.9	109.2	158.7	210.1	240.0	172.
79	Au	3.23	10.08	3.43	6.37	10.6	23.7	44.2	73.8	113.6	165.1	218.2	249.0	177.
80	Hg	3.30	1.60	3.57	6.62	11.0	24.6	46.0	76.7	118.1	171.5	191.6	257.9	189.
81	Tl	3.36	1.67	3.71	6.89	11.4	25.6	47.8	79.7	122.8	178.4	198.3	63.5	197.
82	Pb	3.41	1.73	3.86	7.16	11.9	26.6	49.7	82.9	127.7	176.5	205.3	66.6	207.
83	Bi	3.45	1.80	4.01	7.44	12.4	27.7	51.6	86.1	132.6	152.7	212.3	69.6	216.
84	Po	3.52	1.87	4.17	7.73	12.8	28.7	53.7	89.5	137.8	157.8	54.1	72.7	225.
85	At	3.56	1.94	4.33	8.03	13.3	29.8	55.7	92.9	143.1	163.1	57.1	76.8	238.
86	Rn	3.61	2.02	4.49	8.33	13.8	31.0	57.8	96.4	141.6	168.5	59.7	80.4	250.
87	Fr	3.66	2.09	4.65	8.64	14.4	32.1	60.0	100.0	119.8	174.0	62.1	83.6	260.
88	Ra	3.70	2.17	4.83	8.96	14.9	33.3	62.2	103.7	123.6	179.6	64.2	86.4	268.
89	Ac	3.75	2.25	5.00	9.29	15.4	34.5	64.5	107.5	127.6	48.5	67.4	90.8	282.
90	Th	3.81	2.33	5.19	9.63	16.0	35.8	66.8	111.4	131.6	50.2	69.9	94.1	292.
91	Pa	3.86	2.42	5.39	9.98	16.6	37.1	69.3	110.5	135.7	52.3	72.7	97.9	304.
92	U	3.91	2.51	5.58	10.35	17.4	38.5	71.8	90.8	139.9	54.3	75.5	101.5	316.
93	Np	3.95	2.58	5.66	10.7	18	40	83	93	143.0	57	77	104.	328.
94	Pu	4.00	2.66	5.74	11.0	19	41	85	96	39.1	59	79	107	342.
95	Am	4.05	2.74	5.82	11.4	20	43	87	99	41	61	81	110	354.
96	Cm	2.50	2.82	5.90	11.8	21	44	89	101	43	63	84	113	367.
97	Bk	2.56	2.90	5.97	12.3	21	45	91	104	46	65	86	117	382.
98	Cf	2.62	2.96	6.03	12.8	22	47	89	106	48	67	89	121	395.
99	Es	2.68	3.03	6.09	13.3	22	50	71	110	50	69	92	125	412.
100	Fm	2.74	3.10	6.15	13.7	23	51	73	35	52	71	95	128	427.

Convert $\lambda \to$ KeV : Energy (KeV) = $12.398 \times \dfrac{1}{\lambda(\text{Å})}$

coefficients

KeV

6.199	4.96	4.13	3.10	2.48	2.07	1.77	1.55	1.38	1.24	0.827	0.620	0.496	0.413
2.0	2.5	3.0	4.0	5.0	6.0	7.0	8.0	9.0	10.0	15.0	20.0	25.0	30.0
655.	1132.	324.	725.	1355.	2258.	3477.	5053.	7027.	9440.	15831.	34719.	16259.	27088.
693.	1110.	351.	786.	1469.	2448.	3769.	5478.	7619.	10235.	17221.	9323.	17414.	29012.
731.	222.	370.	827.	1546.	2576.	3966.	5764.	8016.	10769.	18564.	9873.	18442.	30725.
771.	236.	394.	882.	1648.	2746.	4228.	6144.	8545.	9855.	20012.	10461.	19540.	32554.
755.	250.	417.	934.	1745.	2908.	4478.	6508.	9051.	10506.	21612.	11061.	20661.	34422.
765.	266.	443.	992.	1854.	3090.	4758.	6915.	8338.	10195.	5564.	12454.	23262.	38755.
766.	283.	471.	1055.	1971.	3284.	5056.	7349.	8853.	10887.	5960.	13339.	24916.	41511.
805.	298.	497.	1111.	2076.	3460.	5327.	7743.	9382.	8404.	6263.	14016.	26181.	43619.
171.	320.	533.	1193.	2230.	3716.	5721.	7134.	9150.	8913.	6631.	14841.	27721.	46184.
177.	331.	551.	1234.	2305.	3842.	5915.	7546.	7048.	9466.	7000.	15667.	29264.	48754.
186.	349.	581.	1301.	2430.	4050.	6236.	7194.	7478.	10044.	7460.	16695.	31184.	51954.
197.	367.	613.	1371.	2561.	4269.	5769.	7606.	7910.	10627.	7943.	17778.	33207.	55323.
208.	388.	647.	1448.	2704.	4507.	6112.	6022.	8374.	10887.	8453.	18918.	35336.	58871.
218.	407.	679.	1519.	2838.	4730.	5723.	6357.	8840.	2890.	8994.	20130.	37600.	62642.
229.	427.	712.	1594.	2977.	4369.	6022.	6707.	9327.	3073.	9561.	21399.	39971.	66597.
239.	446.	744.	1665.	3110.	4603.	4873.	7082.	2433.	3268.	10169.	22759.	42512.	54286.
252.	472.	787.	1760.	3288.	4300.	5141.	7472.	2585.	3472.	10804.	24179.	45164.	57823.
265.	495.	825.	1847.	3450.	4537.	5417.	7632.	2745.	3687.	11474.	25679.	47966.	61517.
278.	520.	866.	1938.	3240.	4747.	5701.	2095.	2914.	3914.	12180.	27258.	39587.	65211.
293.	547.	912.	2041.	3415.	3917.	6032.	2224.	3093.	4155.	12929.	28936.	42537.	59733.
306.	571.	952.	2130.	3571.	4092.	6124.	2361.	3284.	4411.	13726.	30721.	45622.	63626.
321.	599.	999.	2236.	3258.	4309.	1727.	2511.	3492.	4691.	14597.	32670.	48933.	37825.
336.	628.	1046.	2341.	3406.	4530.	1837.	2670.	3713.	4988.	15520.	34735.	43618.	41124.
352.	658.	1097.	2455.	3562.	4767.	1951.	2836.	3944.	5299.	16488.	29811.	46504.	44644.
369.	689.	1149.	2324.	3004.	4786.	2077.	3019.	4199.	5640.	17551.	31973.	49560.	48379.
386.	721.	1202.	2444.	3144.	1433.	2207.	3208.	4461.	5993.	18647.	34038.	31491.	52468.
397.	742.	1237.	2172.	3314.	1525.	2348.	3413.	4747.	6377.	19843.	29965.	34034.	56784.
423.	791.	1318.	2268.	3458.	1623.	2498.	3631.	5050.	6784.	21110.	31981.	36864.	61416.
442.	826.	1377.	2370.	3455.	1723.	2652.	3855.	5362.	7110.	18501.	33997.	39791.	66048.
463.	865.	1442.	2027.	1098.	1829.	2817.	4094.	5693.	7648.	19714.	22997.	42956.	21595.
484.	905.	1508.	2126.	1166.	1943.	2992.	4348.	6047.	8124.	20961.	24795.	46314.	23810.
505.	943.	1447.	2222.	1238.	2063.	3176.	4616.	6420.	8625.	17953.	26718.	49906.	26221.
534.	997.	1510.	2322.	1310.	2184.	3362.	4887.	6796.	9130.	19300.	28660.	17312.	28844.
559.	1043.	1305.	2424.	1386.	2310.	3557.	5170.	7190.	9659.	20448.	30902.	19023.	31694.
581.	1085.	1356.	2328.	1464.	2440.	3757.	5460.	7593.	10201.	21674.	33017.	20880.	34787.
600.	1121.	1410.	2497.	1544.	2573.	3961.	5758.	8007.	10757.	15740.	33026.	24647.	41063.
631.	1005.	1459.	872.	1630.	2716.	4181.	6077.	8452.	9673.	16748.	14827.	27696.	46142.
654.	1121.	1520.	923.	1724.	2873.	4424.	6429.	8942.	10250.	17797.	16640.	31082.	51783.
680.	947.	1331.	964.	1801.	3001.	4620.	6715.	9339.	10852.	18837.	17647.	32963.	54918.
707.	981.	1381.	1019.	1903.	3172.	4883.	7098.	8406.	9091.	19924.	19373.	36188.	60289.
729.	1000.	1435.	1070.	2020.	3360.	5170.	7450.	N_I	N_{II}	N_{III}	N_{IV} N_V		
755.	1021.	1470.	1120.	2140.	3750.	5460.							
785.	1042.	1400.	1170.	2260.	3960.	5750.							
530.	1063.	1450.	1220.	2370.	4180.	6080.							
420.	910.	510.	1270.	2490.	4420.	6430.							
435.	930.	525.	1320.	2620.	4620.								
452.	950.	540.	1370.	2750.	4880.								
470.	980.	555.	1420.	2880.	5100.								
M_{II}	M_{III}	$M_{IV} M_V$											

Additional Mass Absorption Coefficients for Light Elements

		Energy (keV)						
		0.309	0.248	0.206	0.177	0.155	0.137	0.124
	Element			Wavelength (Å)				
		40	50	60	70	80	90	100
2	He	2360	4670	8000	12,600	18,500	25,800	34,500
3	Li	7500	14,000	23,000	34,400	48,400	65,000	83,000
4	Be	16,500	29,100	45,500	65,000	88,000	114,000	142,000
5	B	28,300	47,900	22,000	3650	5100	6700	8500
6	C	44,100	3140	4890	7000	9500	12,400	15,700
7	N	2960	5100	7900	11,400	15,600	20,500	26,000
8	O	4560	7900	12,400	17,900	24,400	32,000	40,600
9	F	6600	11,500	17,900	25,800	34,900	45,100	56,000
10	Ne	10,300	17,800	27,300	38,400	51,000	64,000	78,000
11	Na	14,100	23,900	35,800	49,200	64,000	78,000	93,000

Mass Absorption Coefficients for Selected Films and Gases

Wavelength Å	Formvar $(C_5H_7O_2)_x$	Mylar $(C_{10}H_8O_4)_x$	Air O_2 21%, N_2 78%, A 1%	P_{10} (CH_4) 10%, A 90%
2.0	14.	14.	21.	230.
4.0	113.	116.	148.	162.
6.0	372.	384.	481.	467.
8.0	850.	870.	1090.	1020.
10.0	1580.	1630.	2020.	1850.
12.0	2600.	2680	3310.	3010.
14.0	3920.	4040.	4980.	4500.
16.0	5600.	5800.	7100.	6400.
18.0	7500.	7800.	9500.	8400.
20.0	9900.	10200.	12400.	10900.
22.0	12500.	12900.	15700.	13500.
24.0	8200.	8500.	14100.	16400.
26.0	10100.	10400.	17100.	19600.
28.0	12000.	12400.	20400.	22800.
30.0	14300.	14700.	24000.	26300.
32.0	16700.	17200.	2290.	29700.
34.0	19300.	19900.	2650.	33300.
36.0	22100.	22800.	3040.	36900.
38.0	25000.	25800.	3460.	40500.
40.0	28200.	29100.	3810.	37600.

Mass Absorption Coefficients for Selected Films and Gases

Wavelength Å	Formvar $(C_5H_7O_2)_x$	Mylar $(C_{10}H_8O_4)_x$	Air O21%, N78%, A1%	P_{10} (CH_4) 10%, A90%
42.0	31500.	32500.	4270.	40900.
44.0	3250.	3350.	4780.	42600.
46.0	3640.	3760.	5300.	45600.
48.0	4050.	4170.	5900.	48900.
50.0	4450.	4590.	6400.	52000.
52.0	4910.	5100.	6300.	
54.0	5400.	5600.	7000.	
56.0	5900.	6100.	7600.	
58.0	6400.	6600.	8200.	
60.0	7000.	7200.	8900.	
62.0	7500.	7800.	9700.	
64.0	8100.	8400.	10400.	
66.0	8700.	9000.	11200.	
68.0	9400.	9700.	12100.	
70.0	10000.	10300.	12900.	
72.0	10700.	11100.	13800.	
74.0	11400.	11800.	14700.	
76.0	12200.	12500.	15600.	
78.0	12900.	13300.	16600.	
80.0	13600.	14100.	17600.	
82.0	14500.	14900.	18600.	
84.0	15300.	15800.	19700.	
86.0	16100.	16600.	20800.	
88.0	17000.	17500.	21900.	
90.0	17800.	18400.	23100.	
92.0	18800.	19400.	24200.	
94.0	19700.	20300.	25400.	
96.0	20600.	21300.	26700.	
98.0	21600.	22300.	28000.	
100.0	22600.	23300.	29200.	

Appendix V

a. Characteristic Photon Yields of the Elements*
b. Absorption Jump Ratios

a. Characteristic Photon Yields

Z	K	L	M	Z	K	L	M
1				18	0.097		
2				19	0.118		
3				20	0.142	0.001	
4				21	0.168	0.001	
5				22	0.197	0.001	
6	0.001			23	0.227	0.002	
7	0.002			24	0.258	0.002	
8	0.003			25	0.291	0.003	
9	0.005			26	0.324	0.003	
10	0.008			27	0.358	0.004	
11	0.013			28	0.392	0.005	
12	0.019			29	0.425	0.006	
13	0.026			30	0.458	0.007	
14	0.036			31	0.489	0.009	
15	0.047			32	0.520	0.010	
16	0.061			33	0.549	0.012	
17	0.078			34	0.577	0.014	

* J. W. Colby, *Advances in X-Ray Analysis,* Vol. 11, Plenum, New York, 1968, pp. 297–306.

a. Characteristic Photon Yields

Z	K	L	M	Z	K	L	M
35	0.604	0.016		68	0.932	0.240	0.011
36	0.629	0.019		69	0.934	0.251	0.012
37	0.653	0.021	0.001	70	0.937	0.262	0.013
38	0.675	0.024	0.001	71	0.939	0.272	0.014
39	0.695	0.027	0.001	72	0.941	0.283	0.015
40	0.715	0.031	0.001	73	0.942	0.293	0.016
41	0.732	0.035	0.001	74	0.944	0.304	0.018
42	0.749	0.039	0.001	75	0.945	0.314	0.019
43	0.765	0.043	0.001	76	0.947	0.325	0.020
44	0.779	0.047	0.001	77	0.948	0.335	0.022
45	0.792	0.052	0.001	78	0.949	0.345	0.024
46	0.805	0.058	0.001	79	0.951	0.356	0.026
47	0.816	0.063	0.002	80	0.952	0.366	0.028
48	0.827	0.069	0.002	81	0.953	0.376	0.030
49	0.836	0.075	0.002	82	0.954	0.386	0.032
50	0.845	0.081	0.002	83	0.954	0.396	0.034
51	0.854	0.088	0.002	84	0.955	0.405	0.037
52	0.862	0.095	0.003	85	0.956	0.415	0.040
53	0.869	0.102	0.003	86	0.957	0.425	0.043
54	0.876	0.110	0.003	87	0.957	0.434	0.046
55	0.882	0.118	0.004	88	0.958	0.443	0.049
56	0.888	0.126	0.004	89	0.958	0.452	0.052
57	0.893	0.135	0.004	90	0.959	0.461	0.056
58	0.898	0.143	0.005	91	0.959	0.469	0.060
59	0.902	0.152	0.005	92	0.960	0.478	0.064
60	0.907	0.161	0.006	93	0.960	0.486	0.068
61	0.911	0.171	0.006	94	0.960	0.494	0.073
62	0.915	0.180	0.007	95	0.960	0.502	0.077
63	0.918	0.190	0.007	96	0.961	0.510	0.083
64	0.921	0.200	0.008	97	0.961	0.517	0.088
65	0.924	0.210	0.009	98	0.961	0.524	0.093
66	0.927	0.220	0.009	99	0.961	0.531	0.099
67	0.930	0.231	0.010	100	0.961	0.538	0.106

b. Absorption Jump Ratios

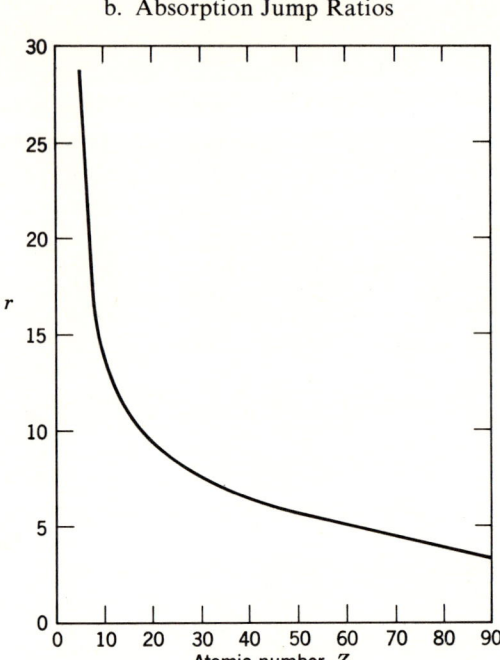

Fig. A-1. Absorption jump ratios, $r = \mu_m/\mu_{m'}$. μ_m and $\mu_{m'}$ are the absorption coefficients on the short and long wavelength sides of the **K** edge respectively.

Appendix VI

Determination of Elements by X-Ray Emission Spectrography. A Guide to the Recent Literature (1964–1970)

"The end crowns all", Shakespeare, *Troilus and Cressida, Act* IV Sc. 5, l. 223

The following table and corresponding references of x-ray spectrographic analyses of various elements have been compiled from reviews concerning "X-Ray Absorption and Emission" which were published in 1966, 1968, and 1970 in the biennial fundamental reviews of *Analytical Chemistry*. The references for these three reviews are as follows:

W. J. Campbell, J. D. Brown, and J. W. Thatcher, *Anal. Chem.,* **38**, 416R (April 1966).
W. J. Campbell and J. D. Brown, *Anal. Chem.,* **40**, 346R (April 1968).
W. J. Campbell and J. V. Gilfrich, *Anal. Chem.,* **42**, 248R (April 1970).

Earlier reviews of this field are as follows:

H. A. Liebhafsky and E. H. Winslow, *Anal. Chem.,* **28**, 583 (April 1956).
Ibid., **30**, 580 (April 1958).
H. A. Liebhafsky, E. H. Winslow, and H. G. Pfeiffer, *Anal. Chem.,* **32**, 240R (April 1960).
Ibid., **34**, 282R (April 1962).
W. J. Campbell and J. D. Brown, *Anal. Chem.,* **36**, 312R (April 1964).

For an earlier version of Appendix VI, see

H. A. Liebhafsky, H. G. Pfeiffer, E. H. Winslow, and P. D. Zemany, *X-Ray Absorption and Emission in Analytical Chemistry,* Wiley, New York, 1960, p. 328.

References to X-Ray Spectrographic Determination of Various Elements

Aluminum
    In alloys and metals (67, 69, 105, 111, 112, 123, 155, 196, 234, 341)
    In minerals and ores (22, 68, 87, 88, 161, 167, 170, 178, 186, 198, 253, 283, 287, 288, 289, 321, 323, 328, 338, 339, 349, 351, 355)
    In other materials (81, 148, 190, 210, 265, 266, 269, 270, 279, 328, 346)
Antimony (43, 106, 162, 196, 219, 250, 356)
Argon (158)
Arsenic (43, 44, 106, 153, 224, 226, 290, 356)
Barium (46, 137, 149, 180, 192, 211, 245, 261, 336, 352)
Beryllium (97, 215)
Bismuth (226, 244)
Boron (97, 109, 110, 151, 152, 173, 174, 216)
Bromine (77, 89, 163, 195, 278, 295, 309, 343)
Cadmium (61, 144, 162)
Calcium (130, 246, 269, 345)
    In cement materials and slags (20, 34, 81, 171, 210, 212, 265, 266, 279, 299, 355)
    In minerals and ores (13, 22, 68, 87, 88, 96, 137, 146, 147, 161, 170, 186, 189, 198, 249, 253, 283, 287, 288, 293, 303, 321, 323, 328, 338, 339, 351, 355)
    In organics (5, 19, 46, 58, 149, 211, 256)
    In other materials (16, 50, 126, 178, 190, 263, 270, 278, 307, 341)
Carbon (90, 109, 113, 151, 152, 173, 174, 179, 326, 327)
Cerium (71, 247)
Chlorine (5, 84, 120, 148, 168, 186, 195, 211, 223, 257, 278, 285, 286, 292, 343, 360)
Chromium
    In alloys and metals (37, 74, 95, 145, 156, 180, 205, 300, 308, 315, 329)
    In other materials (19, 28, 50, 236, 246)

REFERENCES TO DETERMINATIONS OF VARIOUS ELEMENTS 537

Cobalt (221, 233, 236, 246, 264)
  In alloys and metals (70, 205, 300, 306, 307)
  In organics (15, 19)
  In other materials (79, 80, 258, 259, 276, 359)
Copper
  In alloys and metals (6, 11, 51, 67, 78, 98, 107, 134, 180, 183, 205, 237, 239, 241, 268, 278, 300, 306, 307, 362)
  In other materials (50, 79, 80, 85, 168, 233, 236, 258, 264, 298, 314, 335)
Fluorine (26, 151, 152, 214, 229)
Germanium (356)
Gold (40, 47, 65, 78, 98, 124, 231, 235, 237, 312, 362)
Hafnium (1, 24, 114, 122, 213, 251, 267, 340)
Hydrogen (90, 327)
Iodine (159, 163, 217, 322)
Iridium (312)
Iron (130, 148, 221, 246, 264, 269)
  In alloys and metals (37, 66, 70, 111, 134, 145, 180, 205, 234, 273, 300, 306, 329, 341)
  In cement materials (20, 34, 81, 210, 265, 266, 279)
  In minerals and ores (13, 22, 85, 88, 91, 92, 94, 96, 108, 118, 147, 161, 170, 186, 198, 249, 253, 271, 283, 284, 287, 288, 293, 323, 328, 332, 338, 339, 349, 351, 355)
  In organics (15, 19, 99, 131, 133)
  In other materials (27, 50, 79, 80, 86, 167, 178, 182, 263, 270, 278, 302)
Lead (125, 192, 219)
  In alloys and metals (78, 225, 234, 241, 306)
  In minerals and ores (29, 347, 361)
  In organics (46, 136, 309)
  In other materials (50, 55, 226, 244, 263, 276, 343, 352)
Lithium (215)
Magnesium (22, 87, 170, 287)
  In minerals and ores, slags, and cement materials (81, 210, 279, 288)
  In other materials (152, 161, 167, 180, 190, 218, 245, 306, 328, 338, 339, 351)
Manganese (17, 37, 95, 96, 180, 198, 205, 234, 236, 246, 264, 306, 323, 355)
  In alloys and metals (74, 75, 143, 156, 185, 300, 307)
  In minerals and ores (13, 146, 161, 283, 328, 338, 339)
  In slags (299)
Mercury (79, 209, 226)
Molybdenum
  In alloys and metals (74, 75, 95, 138, 315, 329, 334)
  In other materials (17, 33, 50, 85, 118, 200, 208, 220, 254, 258, 313, 335, 344)

Nickel
　In alloys and metals (66, 70, 74, 112, 134, 145, 164, 180, 222, 234, 306, 307, 315)
　In other materials (17, 18, 27, 50, 73, 79, 80, 88, 233, 236, 246, 258, 263, 264, 269, 278, 344, 360)
Niobium (132, 221, 259, 291, 358)
　In alloys (64, 196, 320)
　In minerals and ores (172, 201, 202, 281)
Nitrogen (109, 113, 151, 152, 174)
Osmium (319)
Oxygen (12, 103, 151, 152, 174, 229)
Palladium (98, 237)
Phosphorus (5, 115, 137, 161, 196, 206, 211, 215, 257, 270, 302, 355)
Platinum (98, 237, 312)
Plutonium (102, 119, 255)
Potassium (5, 13, 25, 88, 161, 186, 198, 204, 249, 256, 269, 270, 283, 287, 303, 323, 338, 339, 351)
Rare-earth Elements (4, 9, 23, 36, 116, 141, 240, 269, 282, 311, 358)
　In metals (67, 184, 185, 191, 207, 273)
　In rare-earth mixtures (3, 35, 135, 193, 272, 275, 330, 331)
　In other materials (185)
Rhenium (305)
Rhodium (124)
Rubidium (30, 137, 192, 197, 317)
Ruthenium (319)
Scandium (140)
Selenium (2, 59, 130, 224, 226, 232, 356)
Silicon (269)
　In alloys and metals (95, 155, 180, 196, 324, 329, 341, 345)
　In cement materials (81, 210, 265, 266, 279)
　In minerals and ores (13, 22, 68, 72, 87, 88, 96, 146, 147, 161, 178, 186, 198, 253, 287, 288, 289, 328, 338, 339, 351, 355)
　In other materials (50, 60, 62, 167, 190, 248, 270, 302, 328, 346)
Silver (42, 47, 48, 76, 78, 98, 117, 162, 195, 237, 268, 277, 362)
Sodium (151, 152, 190, 216, 218, 228, 243, 302, 324, 338, 339, 351)
Strontium (57, 63, 126, 128, 129, 130, 137, 161, 176, 192, 252, 317, 333, 337, 350, 352, 357)
Sulfur (5, 21, 22, 41, 109, 127, 175, 181, 186, 194, 206, 211, 214, 230, 242, 257, 292, 294, 296, 309, 353, 354, 360)
Tantalum (139, 157, 172, 201, 202, 221, 259, 260, 280, 281, 301, 320, 325)
Technetium (220)
Tellurium (45, 76, 232)
Thallium (130, 226)

REFERENCES TO DETERMINATIONS OF VARIOUS ELEMENTS

Thorium (29, 31, 32, 38, 101, 104, 188, 358)
Thulium (165)
Tin (43, 53, 56, 64, 106, 121, 125, 168, 196, 219, 234, 241, 250, 268, 276, 304, 316, 334, 356)
Titanium (130, 148, 221, 269, 291)
  In metals (74, 196, 205, 300)
  In minerals and ores (13, 108, 118, 137, 161, 170, 186, 249, 253, 262, 283, 288, 321, 323, 338, 339, 349, 355)
  In other materials (19, 50, 203, 247, 258, 259, 270, 358)
Tungsten (10, 54, 75)
Uranium (27, 150, 180, 188, 358)
  In minerals and ores (29, 38, 177, 201, 202)
  In other materials (100, 101, 142, 169, 238, 254, 255, 274, 310, 352)
Vanadium (18, 75, 83, 196, 199, 236, 258, 269, 291)
Yttrium (7, 8, 135, 187, 297, 330, 331, 358)
Zinc (77, 130, 236, 264, 268)
  In metals and ores (6, 11, 14, 37, 52, 91, 98, 107, 166, 180, 183, 227, 234, 237, 306, 361)
  In organics (15, 46, 131, 149, 154, 211)
  In other materials (27, 82, 270, 342)
Zirconium
  In alloys and metals (39, 76, 122, 138, 196, 213, 285, 318, 334)
  In minerals and ores (1, 93, 108, 137, 160, 321, 348)
  In other materials (16, 24, 49, 233, 267, 276, 291)

Literature References Cited

1. A. M. Abdel Gawad, *Am. Mineral.,* **51**, 464 (1966).
2. C. H. Albright, K. E. Burke, and M. M. Yanak, *Talanta,* **16**, 309 (1969).
3. E. Aleksiev, *Izv. Geol. Inst. Bulgar. Akad. Nauk,* **10**, 5 (1962).
4. E. Aleksiev and R. Boyadjieva, *Geochim. Acta,* **30**, 511 (1966).
5. G. V. Alexander, *Anal. Chem.,* **37**, 1671 (1965).
6. R. Alvarez and R. Flitsch, *Nat. Bur. Std. (U.S.), Misc. Publ.,* **260–5** (1965).
7. A. V. Antonev, V. K. Renev, P. K. Spitsyn, and N. V. Troneva, *Zavodsk. Lab.,* **35**, 672 (1969).
8. E. Asada and S. Matsuda, *Bunseki Kagaku,* **15**, 1104 (1966).
9. *Ibid.,* **16**, 790 (1967).
10. D. Babusci, *Anal. Chim. Acta.,* **32**, 175 (1965).
11. L. Backerud, *Appl. Spectrosc.,* **21**, 315 (1967).
12. A. K. Baird and B. L. Henke, *Anal. Chem.,* **37**, 727 (1965).
13. D. F. Ball, *Analyst,* **90**, 258 (1965).
14. T. K. Ball and R. H. Filby, *Geochim. Cosmochim. Acta,* **29**, 737 (1965).
15. S. A. Bartkiewicz and E. A. Hammatt, *Anal. Chem.,* **36**, 833 (1964).
16. A. H. Beavers, J. B. Fehrenbacher, P. R. Johnson, and R. L. Jones, *Soil Sci. Soc. Am. Proc.,* **27**, 408 (1963).
17. J. A. Belk and B. Clayton, *X-Ray Optics and Microanalysis,* Hermann, Paris, 1966, p. 409.
18. J. G. Bergmann, C. H. Ehrhardt, L. Granatelli, and J. L. Janik, *Anal. Chem.,* **39**, 1258 (1967).
19. *Ibid.,* p. 1331.
20. I. D. Berkhoer, *Zavodsk. Lab.,* **29**, 1172 (1963) (Eng. Ed.).
21. M. Berman and S. Ergun, *Fuel,* **47**, 285 (1968).
22. M. Berman and S. Ergun, *U.S. Bur. Mines, Rept. Invest. 7124* (1968).
23. S. S. Berman, P. Semeniuk, and D. S. Russell, *Can. Spectrosc.,* **14**, 68 (1969).
24. P. J. Bermudez, *Anales Real Soc. Espan. Fis. Quim.,* **60**, 297 (1964).
25. F. Bernstein, in *Advances in X-Ray Analysis,* Vol. 7, Plenum, New York, 1964, p. 555.
26. F. Bernstein and R. A. Mattson, in *Advances in X-Ray Analysis,* Vol. 10, Plenum, New York, 1967, p. 494.
27. H. J. Berthold and D. Ankner, *Z. Anal. Chem.,* **226**, 13 (1967).
28. K. Beyermann, H. J. Rose, Jr., and R. P. Christian, *Anal. Chim. Acta,* **45**, 51 (1969).
29. S. B. Bhattacherjee and M. N. Kumar, *Anal. Chem.,* **36**, 1400 (1964).
30. H. E. Bishop, *Brit. J. Appl. Phys. (J. Phys. D),* Ser. 2, **1**, 673 (1968).
31. G. R. Blank and W. H. Dingeldein, *Rept. NLCO–955,* 1965.
32. G. R. Blank and W. H. Dingeldein, *Rept. NLCO–963,* 1965.
33. V. I. Bochenin, *Zavodsk. Lab.,* **33**, 1371 (1967) (Eng. Ed.).
34. R. Boirat, *Bull. Soc. Franc. Ceram.,* **78**, 41 (1968).
35. G. V., Bondarenko and M. A. Blokhin, *Zavodsk. Lab.,* **30**, 1663 (1964) (Eng. Ed.).
36. *Ibid.,* **33**, 628 (1967) (Eng. Ed.).
37. G. Bonissoni and M. Paganelli, *Met. Ital.,* **8**, 268 (1966).

LITERATURE REFERENCES CITED

38. W. Borchert and E. Donderer, *Neues Jahrb. Mineral., Abh.*, **110**, 142 (1969).
39. M. Brill, *Z. Anal. Chem.*, **244**, 36 (1969).
40. J. M. Brinkerhoff and R. Forsyth, *Rept. AEC NYO 3160 1*, December 1965.
41. C. Brown and R. Kanaris-Sotiriou, *Analyst*, **94**, 782 (1969).
42. G. Brunner, *Isotopenpraxis*, **4**, 99 (1968).
43. De P. Bruyne, *Fermentation*, **1**, 28 (1964).
44. K. E. Burke and M. M. Yanak, *Anal. Chem.*, **41**, 963 (1969).
45. K. E. Burke, M. M. Yanak, and C. H. Albright, *Anal. Chem.*, **39**, 14 (1967).
46. W. E. Burke, L. S. Hinds, G. E. Deodato, E. D. Sager, Jr., and R. E. Borup, *Anal. Chem.*, **36**, 2404 (1964).
47. P. G. Burkhalter, *Rept, IAEA STI/Pub/198,365*, 1969.
48. P. G. Burkhalter, *Int. J. Appl. Radiat. Isotopes*, **20**, 353 (1969).
49. A. J. Busch and C. G. Goldbeck, *Rept. NBL-195*, 1963, p. 39.
50. A. J. Busch and C. G. Goldbeck, *Rept. NBL-210*, 1964, p. 41.
51. A. Carnevale and A. J. Lincoln, *Developments in Applied Spectroscopy*, Vol. 5, Plenum, New York, 1966, p. 31.
52. K. G. Carr-Brion, *Analyst*, **89**, 346 (1964).
53. *Ibid.*, **90**, 9 (1965).
54. K. G. Carr-Brion and K. W. Payne, *Analyst*, **93**, 441 (1968).
55. G. Cecchetti, F. C. Ramusino, and R. Intonti, *Met. Ital.*, **8**, 333 (1964).
56. B. R. Chamberlain and R. J. Leech, *Talanta*, **14**, 597 (1967).
57. K. P. Champion, J. C. Taylor, and R. N. Whittem, *Anal. Chem.*, **38**, 109 (1966).
58. K. P. Champion and R. N. Whittem, *Analyst*, **92**, 112 (1967).
59. F. L. Chan, in *Advances in X-Ray Analysis*, Vol. 7, Plenum, New York, 1964, p. 542.
60. *Ibid.*, Vol. 9, 1966, p. 515.
61. F. L. Chan, *Developments in Applied Spectroscopy*, Vol. 7A, Plenum, New York, 1969, p. 3.
62. F. L. Chan, *Proc. SAC Conference, Nottingham*, W. Heffer and Sons Ltd., Cambridge, England, 1965, p. 89.
63. B. W. Chappell, W. Compston, P. A. Arriens, and M. J. Vernon, *Geochim. Cosmochim. Acta*, **33**, 1002 (1969).
64. K. L. Cheng and E. P. Bertin, *RCA Rev.*, **25**, 379 (1964).
65. A. Chow and F. E. Beamish, *Talanta*, **13**, 539 (1966).
66. E. T.-K. Chow and E. P. Cocozza, *Appl. Spectr.*, **21**, 290 (1967).
67. L. J. Christensen, J. M. Khan, and W. F. Brunner, *Rev. Sci. Instr.*, **38**, 20 (1967).
68. D. H. J. Christie and S. Bergstol, *Acta Chem. Scand.*, **22**, 421 (1968).
69. J. E. Cline and S. Schwartz, *J. Electrochem. Soc.*, **114**, 605 (1967).
70. E. P. Cocozza and A. Ferguson, *Appl. Spectr.*, **21**, 286 (1967).
71. S. Cohen and F. R. Bryan, *Appl. Spectr.*, **22**, 342 (1968).
72. U. M. Cowgill, *Rept. ORNL-IIC-10*, 1967, p. 613.
73. W. O. Crawford and W. D. McHenry, *Rept. AD620141*, June 1965.
74. F. Creton and B. Moschin, *Met. Ital.*, **8**, 425 (1963).
75. F. Creton and B. Moschin, *Rev. Met. Mem. Sci.*, **61**, 379 (1964).
76. T. J. Cullen, *Developments in Applied Spectroscopy*, Vol. 3, Plenum, New York, 1964, p. 97.

77. F. Cuttitta and H. J. Rose, *Appl. Spectr.*, **22**, 321 (1968).
78. H. A. Das and J. Zonderhuis, *Rec. Trav. Chim.*, **85**, 837 (August 1966).
79. K. E. Daugherty, Ph. D. Thesis, Univ. of Washington, Seattle, Wash., 1964.
80. K. E. Daugherty, R. J. Robinson, and J. I. Mueller, *Anal. Chem.* **36**, 1869 (1964).
81. Z. DeBeer and F. Creton, *Met. Ital.*, **8**, 308 (1966).
82. C. Decroly, M. Ghodsi, and R. Winand, *Mem. Sci. Rev. Met.*, **62**, 163 (1965).
83. I. G. Dem'yanikov and I. I. Chernitsyna, *Zavodsk. Lab.*, **32**, 1319 (1966).
84. R. D. Deslattes and R. E. LaVilla, *Appl. Optics*, **6**, 39 (January 1967).
85. L. S. Diaz and R. J. Anderson, *Mining Congr. J.*, July 1965, p. 94.
86. C. G. Dodd and D. J. Kaup, *Clay Min. Bull.*, **5**, 290 (1963).
87. J. C. Dumesnil and G. Perrault, *Can. Spectr.*, **12**, 3 (1967).
88. J. A. Dunne and N. L. Nickle, *Rept. ORNL-IIC-10*, 1967, p. 336.
89. P. J. Dunton, *Appl. Spectr.*, **22**, 99 (1968).
90. C. W. Dwiggins, Jr., *Anal. Chem.*, **36**, 1577 (1964).
91. B. Dziunikowski, *Nukleonika*, **9**, 829 (1964).
92. B. Dziunikowski, *Trans. Inst. Min. Metal.*, **76B**, 202 (1967).
93. B. Dziunikowski, B. Holynska, and J. Stachurski, *Chem. Anal.*, (Warsaw), **12**, 1107 (1967).
94. B. Dziunikowski and Z. Skrzeszewski, *Nukleonika*, **12**, 85 (1967).
95. S. Eckhard and R. Marotz, *Z. Anal. Chem.*, **215**, 23 (1966).
96. W. D. Egan and F. A. Achey, in *Advances in X-Ray Analysis,* Vol. 11, Plenum, New York, 1968, p. 150.
97. R. C. Ehlert and R. A. Mattson, in *Advances in X-Ray Analysis,* Vol. 9, Plenum, New York, 1966, p. 456.
98. J. D. Eick, H. J. Caul, D. L. Smith, and S. D. Rasberry, *Appl. Spectrosc.*, **21**, 324 (1967).
99. R. Emmericz, J. Gilewicz Wolter, and B. Holynska, *Zavodsk. Lab.*, **32**, 193 (1966) (Eng. Ed.).
100. S. Enomoto, *Rept. CEA-R-3369*, 1968.
101. D. Ertel and W. Wettstein, *Rept. EUR-FNR-365*, 1967.
102. D. Ertel and W. Wettstein, *U.S. At. Energy Comm., Rept. KFK-747*, 1968.
103. B. P. Fabbi and A. Volborth, *Norelco Reptr.*, **15**, 93 (1968).
104. M. C. Farquhar and M. M. English, in *Advances in X-Ray Analysis,* Vol. 7, Plenum, New York, 1964, p. 584.
105. L. Fergason, *Rev. Sci. Instr.*, **37**, 964 (1966).
106. C. L. Fillmore, A. C. Eckert, and J. V. Scholle, *Appl. Spectrosc.*, **23**, 502 (1969).
107. D. W. Fischer, *J. Appl. Phys.*, **36**, 2048 (1965).
108. D. W. Fischer, *Wright Patterson Air Force Base, Ohio, Rept., AFML-TR-69-143*, 1969.
109. D. W. Fischer and W. L. Baun, in *Advances in X-Ray Analysis,* Vol. 9, Plenum, New York, 1966, p. 329.
110. D. W. Fischer and W. L. Baun, *J. Appl. Phys.*, **37**, 768 (1966).
111. *Ibid.*, **38**, 229 (1967).
112. D. W. Fischer and W. L. Baun, *Phys. Rev.*, **145**, 555 (1966).
113. D. W. Fischer and W. L. Baun, *Wright Patterson Air Force Base, Ohio, Tech. Rept. AFML-TR-65-255*, 1965.

LITERATURE REFERENCES CITED

114. W. Fischer, K. Biesenberger, W. Bohmer, and K. Reinhardt, *Z. Anal. Chem.,* **216**, 61 (1966).
115. J. Flechon and J. P. Fleck, *Bull. Soc. Chim. France,* **8**, 2189 (1965).
116. W. Funasaka, T. Ando, and Y. Tomida, *Bunseki Kagaku,* **17**, 1133 (1968).
117. J. Furuta and E. Hiraoka, *Ann. Rept. Radiation Center Osaka Prefect.,* **5**, 78 (1965).
118. M. J. Gallagher, *Trans. Inst. Mining Met.,* **76**, B155 (1967).
119. M. Ganivet and T. Arnal, *Anal. Chim. Acta,* **39**, 73 (1967).
120. K. J. Garska, *Anal. Chem.,* **40**, 809 (1968).
121. M. S. Garson and J. H. Bateson, *Trans. Inst. Mining Met.,* **76**, B165 (1967).
122. O. R. Gates and E. J. Brooks, *Rept. NRL-6427,* 1966.
123. J. V. Gilfrich and D. C. Sullivan, *Norelco Reptr.,* **10**, 127 (1963).
124. G. H. Glade, in *Advances in X-Ray Analysis,* Vol. 11, Plenum, New York, 1968, p. 185.
125. G. H. Glade and H. R. Post, *Appl. Spectr.,* **22**, 123 (1968).
126. E. F. Gloyna, S. K. Bhagat, and W. A. Felsing, Jr., *J. Water Pollution Control Federation,* **35**, 893 (1963).
127. S. C. Goadby and J. F. Stephens, *Fuel,* **46**, 19 (1967).
128. M. Goldman and R. P. Anderson, *Anal. Chem.,* **37**, 718 (1965).
129. M. Goldman, R. P. Anderson, and W. Gee, *U.S. At. Energy Comm., Rept. UCD-108,* 1963, p. 75.
130. M. Goldman and E. D. Beckman, *Developments in Applied Spectroscopy,* Vol. 6, Plenum, New York, 1968, p. 13.
131. M. Goldman, C. K. Hui, and R. P. Anderson, *Rept. UCD-472-112,* 1966.
132. R. Gonzales, E. Muratori, P. Frere, and R. Durand, *Mem. Sci. Rev. Met.,* **64**, 403 (1967).
133. H. Goto, K. Hirokawa, and F. Maeda, *Japan Analyst,* **13**, 402 (1964).
134. H. Goto and A. A. Saito, *Sci. Res. Inst. Tohoku Univ., Ser. A,* **20**, 59 (1968).
135. K. H. Grothe and W. Fischer, *Z. Anal. Chem.,* **204**, 161 (1964).
136. E. L. Gunn, *Appl. Spectr.,* **19**, 99 (1965).
137. P. Hahn-Weinheimer and H. Ackermann, *Z. Anal. Chem.,* **194**, 81 (1963).
138. E. A. Hakkila, R. G. Hurley, and G. R. Waterbury, *Anal. Chem.,* **36**, 2094 (1964).
139. *Ibid.,* **40**, 818 (1968).
140. *Ibid.,* **41**, 665 (1969).
141. E. A. Hakkila, R. G. Hurley, and G. R. Waterbury, *Appl. Spectr.,* **22**, 5 (1968).
142. E. A. Hakkila, R. G. Hurley, and G. R. Waterbury, *U.S. At. Energy Comm., Rept. LA-3159,* 1964.
143. *Ibid., Rept. LA-3160,* 1964.
144. *Ibid., Rept. LA-3305,* 1965.
145. E. A. Hakkila and G. R. Waterbury, *Anal. Chem.,* **37**, 1773 (1965).
146. J. Hanon and R. Winand, *Mem. Sci. Rev. Met.,* **62**, 45 (1965).
147. M. von Hautecler and M. Lesir, *Archiv Eisenhuttenw.,* **12**, 1165 (1964).
148. S. Hayashi, K. Sugahara, and K. Teranishi, *Japan Analyst,* **16**, 1328 (1967).
149. R. F. Haycock, *J. Inst. Petrol.,* **50**, 123 (1964).
150. J. A. Hayden, *Talanta,* **14**, 721 (1967).
151. B. L. Henke, in *Advances in X-Ray Analysis,* Vol. 7, Plenum, New York, 1964, p. 460.
152. *Ibid.,* Vol. 8, 1965, p. 269.

153. S. Hirano and Y. Ujihira, *Japan Analyst,* **12**, 747 (1963).
154. H. Hirata, A. Amemiya, and K. Date, *Japan Analyst,* **16**, 99 (1967).
155. K. Hirokawa and A. Saito, *Z. Anal. Chem.,* **237**, 419 (1968).
156. K. Hirokawa, T. Shimanuki, and H. Goto, *Sci. Repts. Res. Inst. Tohoku Univ., Ser. A15*, 1963, p. 124.
157. K. Hisano and K. Oyama, *Bunseki Kagaku,* **17**, 1373 (1968).
158. W. Hoffmeister, *Z. Anal. Chem.,* **245**, 244 (1969).
159. B. Holynska and J. Jankiewicz, *Chem. Anal.,* **14**, 219 (1969).
160. B. Holynska and L. Langer, *Anal. Chim. Acta,* **40**, 115 (1968).
161. P. R. Hooper, *Anal. Chem.,* **36**, 1271 (1964).
162. J. A. Hope and J. S. Watt, *Intern. J. Appl. Radiation Isotopes,* **16**, 9 (1965).
163. C. R. Hudgens and G. Pish, *Anal. Chem.,* **37**, 414 (1965).
164. E. C. R. Hunt, *Lab. Methods,* **77**, 135 (1968).
165. R. G. Hurley, E. A. Hakkila, and G. R. Waterbury, *Rept. LA-3549*, 1966.
166. A. Ichiryu and T. Sawada, *Japan Analyst,* **14**, 7 (1965).
167. Y. Ishii, *Japan Analyst,* **14**, 1120 (1965).
168. Y. Ishii, H. Kawamura, and S. Yagi, *Japan Analyst,* **17**, 1 (1968).
169. R. B. Jacob, *U.S. At. Energy Comm., Rept. HW-83474*, 1964.
170. J. Jegou-Vilnat and M. Dubois, *Bull. Soc. Fr. Ceram.,* **79**, 45 (1968).
171. V. F. Kahler, *Radex-Rundsch.,* **4**, 256 (1968).
172. M. M. Kakhana, *Zavodsk. Lab.,* **30**, 541 (1964) (Eng. Ed.).
173. H. Kamada, R. Inoue, M. Terasawa, Y. Gohshi, H. Kamei, and I. Fujii, *Anal. Chim. Acta,* **46**, 107 (1969).
174. H. Kamada, T. Ui, S. Kimoto, and M. Sato, *Japan Analyst,* **16**, 952 (1967).
175. R. Kanaris-Sotiriou and G. Brown, *Analyst,* **94**, 780 (1969).
176. V. N. Karev and L. I. Reshetova, *Zavodsk. Lab.,* **31**, 534 (1965) (Eng. Ed.).
177. J. O. Karttunen and W. R. Harmon, *Spectrochim. Acta,* **24B**, 301 (1969).
178. Y. Kawsaki and E. Asada, *Japan Analyst,* **12**, 501 (1963).
179. J. M. Khan, D. L. Potter, and R. D. Worley, *Rept. UCRL-7826*, 1964.
180. C. A. Kienberger and A. R. Flynn, *Union Carbide Corp., Oak Ridge, Tenn., Rept. K1638*, January 1966.
181. W. R. Kiley and J. A. Dunne, *Am. Soc. Testing Materials Spec. Tech. Publ.,* **No. 349**, 24 (1963).
182. Y. S. Kim, *Anal. Chem.,* **39**, 664 (1967).
183. J. Kinnunen, P. Rautavalta, and M. Koponen, *Metallurgia,* **75**, 189 (1967).
184. H. R. Kirchmayr, *Acta. Phys. Austr.,* **24**, 234 (1966).
185. H. R. Kirchmayr and D. Mach, *Z. Metallk.,* **55**, 247 (1964).
186. L. T. Kiss, *Anal. Chem.,* **38**, 1731 (1966).
187. J. F. Klecka, *Rept. UCRL-17144*, 1966.
188. *Ibid., Rept. UCRL-17166*, 1966.
189. D. R. Knoke and H. F. Waldron, *Advances in X-Ray Analysis,* Vol. 8, Plenum, New York, 1965, p. 448.
190. S. Konno, A. Nagashima, F. Abe, and E. Asada, *Japan Analyst,* **14**, 1093 (1965).
191. E. K. Korchemnaya and V. I. Naumova, *Zavodsk. Lab.,* **28**, 1370 (1962) (Eng. Ed.).

192. H. M. Koster, *Contrib. Mineral. Petrol.,* **12**, 168 (1966).
193. R. A. Kravchenko Berezhnoi and L. I. Polezhaeva, *Zavodsk. Lab.,* **31**, 530 (1965) (Eng. Ed.).
194. S. D. Kullbom, W. K. Pollard, and H. F. Smith, *Anal. Chem.,* **37**, 1031 (1965).
195. N. Kunimine, H. Ugazin, K. Yabe, and E. Asada. *Japan Analyst,* **13**, 679 (1964).
196. H. De Laffolie, *Arch. Eisenhuttenw.,* **7**, 535 (1967).
197. R. Laib, in *Advances in X-Ray Analysis,* Vol. 8, Plenum, New York, 1965, p. 443.
198. R. A. Laidley, *Appl. Spectr.,* **22**, 420 (1968).
199. M. F. Landi and A. Battaglia, *Met. Ital.,* **8**, 650 (1967).
200. A. P. Langheinrich and J. W. Forster, in *Advances in X-Ray Analysis,* Vol. 11, Plenum, New York, 1968, p. 275.
201. O. Latorre and P. J. Bermudez, *Anales Fis. Quim.,* **61B**, 667 (1965).
202. O. Latorre and P. J. Bermudez, *Rept. JEN 134-DQ/1-39,* 1964.
203. Y. G. Lavrentev, *Zavodsk. Lab.,* **30**, 217 (1964) (Eng. Ed.).
204. R. I. Lawson, *Geol. Sur. G. Brit.,* **25**, 85 (1966).
205. C. L. Lewis, W. L. Ott, and N. M. Sine, *The Analysis of Nickel,* Pergamon, 1966, p. 85.
206. R. A. Libby, *Anal. Chem.,* **40**, 1507 (1968).
207. F. Lihl, *Rept. AD-428773,* 1962.
208. A. L. Lingard and M. G. Willigman, *Proc. S. Dakota Acad. Sci.,* **62**, 170 (1963).
209. W. B. Link, K. S. Heine Jr., J. H. Jones, and P. Wattinglington, *J. Assoc. Offic. Agr. Chemists,* **47**, 391 (1964).
210. F. W. Locher and W. Richartz, *Zement-Kalk-Gips,* **15**, 10 (1962).
211. R. Louis, *Z. Anal. Chem.,* **201**, 336 (1964).
212. A. Lubecki and M. Wasilewska, *J. Radioanal. Chem.,* **1**, 25 (1968).
213. C. L. Luke, *Anal. Chim. Acta,* **41**, 453 (1968).
214. *Ibid.,* **43**, 245 (1968).
215. *Ibid.,* **45**, 365 (1969).
216. *Ibid.,* p. 377.
217. P. K. Lund and J. C. Mathies, *Am. J. Clin. Pathol.,* **40**, 132 (1963).
218. P. K. Lund, D. A. Morningstar, and J. C. Mathies, *Biochem. Biophys. Res. Commun.,* **14**, 177 (1964).
219. H. M. Luschow and H. V. Steil, *Z. Anal. Chem.,* **245**, 304 (1969).
220. F. Lux, F. Ammentorp-Schmidt, and W. Opavsky, *Z. Anorg. Allgem. Chem.,* **341**, 172 (1965).
221. R. Maeda and K. Hino, *Bunseki Kagaku,* **17**, 1239 (1968).
222. B. A. Malyukov, Y. M. Ukrainskii, and V. E. Korolev, *Zavodsk Lab.,* **33**, 1133 (1967) (Eng. Ed.).
223. *Ibid.,* p. 1723.
224. V. E. Mamaev and K. T. Protasov, *Sb. Nauch Trud. Vses. Nauchno-Issled. Gornmetallurg. Inst. Tsvet. Metall.,* **9**, 163 (1965).
225. V. J. Manners, J. V. Craig, and F. H. Scott, *J. Inst. Metals,* **95**, 173 (1967).
226. F. J. Marcie, *Environ. Sci. Technol.,* **1**, 164 (1967).
227. S. Margolinas, *Rept. ORNL-IIC-10,* 1967, p. 805.
228. N. Matano, K. Ono, and T. Fujii, *Japan Analyst,* **17**, 560 (1968).

229. R. A. Mattson and R. C. Ehlert, in *Advances in X-Ray Analysis,* Vol. 9, Plenum, New York, 1966, p. 471.
230. J. Merritt and E. J. Agazzi, *Anal. Chem.,* **38**, 1954 (1966).
231. I. W. Mitchell, N. M. Saum, and C. L. Hiltrop, *Norelco Reptr.,* **11**, 39 (1964).
232. T. Miura and K. Tsutsumi, *Japan Analyst,* **13**, 860 (1964).
233. B. Montford, *Can. Spectr.,* **13**, 4, (1968).
234. L. H. Moorhead, *et al., Rept. AD 652 733,* 1967.
235. A. F. Morhnheim, *Plating,* **50**, 725 (1963).
236. A. W. Morris, *Anal. Chim. Acta,* **42**, 397 (1968).
237. B. W. Mulligan, H. J. Caul, S. D. Rasberry, and B. F. Scribner, *J. Res. Nat. Bur. Std.,* **68A**, 5 (1964).
238. J. A. Murray and T. H. Bartlett, *Norelco Reptr.,* **11**, 132 (1964).
239. R. H. Myers, D. Womeldorph, and B. J. Alley, *Anal. Chem.,* **39**, 1031 (1967).
240. T. Nakajima, H. Kawaguchi, and Y. Ouchi, *Bunseki Kagaku,* **16**, 832 (1967).
241. H. Nakanishi, M. Yoshimura, K. Yoshisako, and K. Itsuki, *Japan Analyst,* **13**, 1131 (1964).
242. S. Natelson, *Rept. ORNL-IIC-10,* 1967, p. 282.
243. S. Natelson, *Trans. N.Y. Acad. Sci.,* **26**, 3 (1963).
244. S. Natelson and K. de Paritosh, *Microchem. J.,* **7**, 448 (1963).
245. S. Natelson, A. N. Vassilevsky, K. de Paritosh, and W. R. Whitford, *Microchem. J.,* **8**, 295 (1964).
246. W. Nichiporuk, A. Chodos, E. Helin, and H. Brown, *Geochim. Cosmochim. Acta,* **31**, 1911 (1967).
247. H. Nickel and H. J. Stocker, *Z. Anal. Chem.,* **206**, 95 (1964).
248. R. C. Nickols, Jr., *Norelco Reptr.,* **11**, 37 (1964).
249. J. Nicolas, M. Quintin, and P. Douillet, *Bull. Soc. Franc. Ceram.,* **74**, 11 (1967).
250. F. V. Nikitin, *Zavodsk. Lab.,* **31**, 966 (1965).
251. K. Ohno and N. Matano, *Bunseki Kagaku,* **18**, 213 (1969).
252. W. R. Oliver, *Lab. Pract.,* **17**, 690 (1968).
253. E. W. Orrell and P. J. Gidley, *Trans. Brit. Ceram. Soc.,* **63**, 19 (1964).
254. H. Parthey, *Z. Anal. Chem.,* **209**, 398 (1965).
255. P. A. Pella and A. V. Baechmann, *Anal. Chim. Acta,* **47**, 431 (1969).
256. G. Peter and T. Tuchscheerer, *Z. Anal. Chem.,* **220**, 351 (1966).
257. B. Piccolo, D. Mitcham, and R. T. O'Connor, *Appl. Spectr.,* **22**, 502 (1968).
258. E. D. Pierron and R. H. Munch, *Monsanto Techn. Rev.,* **9**, 16 (1964).
259. M. V. R. V. Pinheiro, *Revta Port. Quim.,* **7**, 193 (1965).
260. M. R. Portafaix, *Method. Phys. Anal.,* **4**, 170 (1968).
261. S. A. Prokopovich and E. R. McCartney, *Analyst,* **92**, 253 (1967).
262. S. M. Przhiyalgovskii, V. N. Smirnov, and A. L. Yakubovich, *Zavodsk. Lab.,* **33**, 186 (1967) (Eng. Ed.).
263. R. Puschel, *Mikrochim. Acta,* **4**, 770 (1965).
264. R. Puschel, *Talanta,* **16**, 351 (1969).
265. R. Rabot and R. Alegre, *Silicates Ind.,* **27**, 181 (1962).

266. *Ibid.*, p. 250.
267. E. Ramous, *Met. Ital.*, **10**, 473 (1964).
268. S. D. Rasberry, H. J. Carl, and A. Yezer, *Spectrochim. Acta*, **23B**, 345 (1968).
269. K. A. Rayburn, *Appl. Spectr.*, **22**, 726 (1968).
270. A. J. Regis, *ASME, Proc. National Incinerator Conf.* (1966), p. 195.
271. T. Reistad, *Jernkontorets. Ann.*, **151**, 216 (1967).
272. G. I. Rekhkolainen, *Zavodsk. Lab.*, **31**, 536 (1965) (Eng. Ed.).
273. *Ibid.*, **32**, 191 (1966) (Eng. Ed.).
274. J. Renaud and P. Desson, *Energie Nucl.*, **10**, 93 (1968).
275. V. K. Renev and V. I. Ganopolskii, *Zavodsk. Lab.*, **29**, 1076 (1963).
276. J. R. Rhodes, *Analyst*, **91**, 683 (1966).
277. J. R. Rhodes, *Rept. ORNL-IIC-10*, 1967, p. 442.
278. L. P. Rigdon, *Rept. UCRL-14879*, 1966.
279. J. M. M. Robinson and E. P. Gertiser, *Mater. Res. Std.*, **4**, 228 (1964).
280. P. A. Romans, W. J. Niebuhr, and J. R. Hauger, *U.S. Bur. Mines, Rept. Invest. 6483*, 1964.
281. H. J. Rose, Jr. and R. Brown, in *Advances in X-Ray Analysis*, Vol. 7, Plenum, New York, 1964, p. 598.
282. H. J. Rose, Jr. and F. Cuttitta, *Appl. Spectr.*, **22**, 426 (1968).
283. H. J. Rose, Jr., F. Cuttitta, and R. R. Larson, *U.S. Geol. Surv., Profess. Papers*, 525-B, B-155 (1965).
284. R. S. Rubinovich, *Zavodsk. Lab.*, **30**, 539 (1964) (Eng. Ed.).
285. J. S. Rudolph, O. H. Kriege, and R. J. Nadalin, *Developments in Applied Spectroscopy*, Vol. 4, Plenum, New York, 1965, p. 57.
286. J. S. Rudolph and R. J. Nadalin, *Anal. Chem.*, **36**, 1815 (1964).
287. J. Sahcres, *Bull. Cent. Rech. Pau*, **2**, 137 (1968).
288. B. S. Sanderson and J. A. Yeck, in *Advances in X-Ray Analysis*, Vo. 10. Plenum, New York, 1967, p. 474.
289. C. Savelli, *Met. Ital.*, **8**, 671 (1967).
290. M. Schlunz and A. Koster-Pflugmacher, *Z. Anal. Chem.*, **234**, 188 (1968).
291. H. Schneider and H. Schumann, *Z. Anal. Chem.*, **235**, 160 (1968).
292. E. Schnell, *Monatsch. Chem.*, **96**, 1302 (1965).
293. G. Seibel and J. Y. LeTraon, *Intern. J. Appl. Radiation Isotopes*, **14**, 259 (1963) (in French).
294. A. Servasier, *Rev. Inst. Francais Petrole*, **19**, 339 (1964).
295. C. Shenberg, J. Gilat, and H. L. Finston, *Anal. Chem.*, **39**, 730 (1967).
296. Y. Shibuya, E. Nishiyama, and K. Yanagase, *Japan Analyst*, **16**, 123 (1967).
297. J. Shiokawa, T. Shin-Ike, G. Adachi, and T. Ishino, *J. Chem. Soc. Japan*, **87**, 131 (1966).
298. T. Shono, M. Tanaka, and K. Shinra, *Japan Analyst*, **16**, 1209 (1967).
299. A. Silber and H. Blaas, *Berg-Huettenmaenn, Monatsh. Montan. Hochschule Leoben*, **109**, 237 (1964).
300. N. M. Sine and C. L. Lewis, *Talanta*, **12**, 389 (1965).

301. A. N. Smagunova, N. F. Losev, and V. I. Lipskaya, *Zavodsk. Lab.*, **31**, 201 (1965) (Eng. Ed.).
302. G. S. Smith, *Chem. Ind. London*, **22**, 907 (1963).
303. J. Smuts, *Norelco Reptr.*, **11**, 9 (1964).
304. J. Smuts, C. Plug, and J. Van Niekerk, *J. S. African Inst. Mining and Met.*, 462, (April 1967).
305. M. W. Solt, J. S. Wahlberg, and A. T. Myers, *Talanta*, **16**, 37 (1969).
306. E. F. Spano and T. E. Green, *Anal. Chem.*, **38**, 1341 (1966).
307. E. F. Spano, T. E. Green, and W. J. Campbell, *U.S. Bur. Mines, Rept. Invest. 6565* (1964).
308. J. Sparks and S. C. Britton, *Sheet Metal Ind.*, **41**, 447 (June 1964).
309. D. C. M. Squirrell, *Proc. SAC Conference, Nottingham*, W. Heffer and Sons, Ltd., Cambridge, England, 1965, p. 132.
310. W. C. Stoecker, *U.S. At. Energy Comm., Rept. MCW-1477*, 1963.
311. I. C. Stone, Jr. and K. A. Rayburn, *Anal. Chem.*, **39**, 356 (1967).
312. A. Strasheim and F. T. Wybenga, *Appl. Spectr.*, **18**, 16 (1964).
313. Y. A. Studennikov, R. A. Belova, and N. F. Loser, *Zavodsk. Lab.*, **33**, 1715 (1967) (Eng. Ed.).
314. B. A. Stulov, *Zavodsk. Lab.*, **33**, 1717 (1967) (Eng. Ed.).
315. M. Sugimoto, *Bunseki Kagaku*, **12**, 475 (1963), Eng. translation *RS 10–130* (February 1964).
316. T. R. Sweatman, Y. C. Wong, and K. S. Toong, *Trans. Inst. Mining Met.*, **76**, B149 (1967).
317. A. Taddeucci and M. Barbieri, *Met. Ital.*, **8**, 281 (1966).
318. B. L. Taylor, *Proc. SAC Conference, Nottingham*, W. Heffer and Sons Ltd., Cambridge, England, 1965, p. 81.
319. H. Taylor and F. E. Beamish, *Talanta*, **15**, 497 (1968).
320. R. W. Taylor, *Developments in Applied Spectroscopy*, Vol. 4, Plenum, New York, 1965, p. 65.
321. R. Tertian, C. Fagot, and M. Jamey, *Publ. Group Avan. Methodes Spectrogr.*, **4**, 267 (1963).
322. J. F. Tinney and J. L. Cate, *Rept. UCRL-50007-68-2*, 1968, p. 51.
323. R. O. Toubes and J. B. Polonio, *An. Real. Soc. Espan. Fis. Quim., Ser. B*, **64**, 311 (1968).
324. C. J. Toussaint and G. Vos, *Anal. Chem.*, **38**, 711 (1966).
325. C. J. Toussaint and G. Vos, *Anal. Chim. Acta*, **33**, 279 (1965).
326. C. J. Toussaint and G. Vos, *Appl. Spectr.*, **18**, 171 (1964).
327. C. J. Toussaint and G. Vos, *Rept. EUR-488F*, 1964.
328. J. Y. Traon and G. Seibel, *Intern. J. Appl. Radiation Isotopes*, **14**, 365 (1963).
329. A. Tsukamoto, I. Shimizu, and M. Ohata, *Nippon Kinzoku Gakkaishi*, **32**, 473 (1968).
330. K. Tsutsumi, *Japan Analyst*, **13**, 635 (1964).
331. *Ibid.*, p. 645.
332. V. P. Tsvetkov and V. K. Kalosha, *Zavodsk. Lab.*, **30**, 958 (1964) (Eng. Ed.).
333. T. Tuchscheerer, *Z. Anal. Chem.*, **207**, 1 (1965).

334. G. L. Vassilaros and J. P. McKaveney, *Talanta,* **16,** 195 (1969).
335. P. A Verkhovodov, *Zavodsk Lab.,* **33,** 1137 (1967) (Eng. Ed.).
336. A. Visapaeae, *Teknillisen Kem. Aikl.,* **20,** 563 (1963).
337. A. Visapaeae, *Valtion Tek. Tutkimuslaitos, Tiedotus, Sarja VI,* **53,** 3 (1963).
338. A. Volborth, *Appl. Spectr.,* **18,** 1 (1965).
339. A. Volborth, *Nevada Bur. Mines, Reptr.,* **6** (1963).
340. G. Vos, *Anal. Chim. Acta,* **47,** 243 (1969).
341. F. Wagner, *Z. Anal. Chem.,* **198,** 98 (1963).
342. J. C. Wagner, E. H. Bicknese, and F. R. Bryan, *Appl. Spectr.,* **21,** 176 (1967).
343. J. C. Wagner and F. R. Bryan, in *Advances in X-Ray Analysis,* Vol. 9, Plenum, New York, 1966, p. 528.
344. J. C. Wagner and E. J. Violante, *Appl. Spectr.,* **19,** 195 (1965).
345. V. F. Wagner, *Arch. Eisenhuettenw.,* **39,** 759 (1968).
346. G. E. Walden, A. D. Condrey, and K. A. Sells, *Rept. Y-DA-590,* 1964.
347. J. S. Watt, *Rept. ORNL-IIC-10,* 1967, p. 663.
348. G. R. Webber and J. D. Volbrath, *Can. Spectr.,* 12, 105 (1967).
349. P. Wecht and E. Schultz, *Tonind. Z. Keram. Rundschau,* **8,** 75 (1962).
350. S. B. Weed and R. A. Leonard, *Soil Sci. Soc. Proc.,* **27,** 474 (1963).
351. E. E. Welday, A. K. Baird, D. B. McIntyre, and K. W. Madlem, *Am. Mineralogist,* **49,** 889 (1964).
352. P. W. West, A. M. G. MacDonald, and T. S. West, Eds., "Analytical Chemistry, 1962," *Proc. Intern. Symp. Birmingham Univ.,* Elsevier, 1963.
353. D. W. Wilbur, *Rept. UCRL-14379,* 1965.
354. D. W. Wilbur and J. W. Gofman, in *Advances in X-Ray Analysis,* Vol. 9, Plenum, New York, 1966, p. 354.
355. A. Wittman, J. M. Bourdieu, and D. Jorre, *Rev. Met.,* **63,** 529 (1966). Eng. translation available as **BISI 5068.**
356. F. Wlotzka, *Z. Anal. Chem.,* **215,** 81 (1966).
357. G. J. Wonsidler and R. S. Sprague, *Anal. Chim. Acta,* **31,** 51 (1964).
358. L. W. Wray, *Rept AECL-2526,* 1965.
359. F. T. Wybenga, *Appl. Spectrosc.,* **19,** 193 (1965).
360. M. Yamashita and S. Watanabe, *Japan Analyst,* **18,** 143 (1969).
361. L. Zanaroli, *Met. Ital.,* **8,** 338 (1964).
362. S. J. Zanin and G. E. Hooser, *Appl. Spectr.,* **22,** 105 (1968).

Index

Absorptiometry, 127-170
  comparative, 156-158; see also Comparative absorptiometry
  concluding remarks, 168
  differential, 136-144
    across an absorption edge, 136-143
    applications at Oak Ridge, 141-143
    bromine determinations at Dow, 140, 141
  equation for, 131
  exploratory experiments with various gases, 153-155
  of liquids, 158-160, 163
  modified differential, 144
  monochromatic, equipment for, 132, 133
  with monochromatic beams, 131-144
  point-to-point exploration of treated carbon brushes, 155, 156
  polychromatic beams, 144-168
    of gases, 146
    of liquids, 146
    of solids, 146
  for process control, 160-162
  simple, 131
  of solids, 150-153
  of steel strip by the General Electric Raymike® thickness gauge, 164-168
  used for identification, 158, 159
Absorption, application of Beer's law to, 18, 144, 145, 148-150
  chemical influences in, 53
  in detectors, 60
  equation, 18, 23
  by gases in proportional detector, 91
  leading to scattering, 28, 29
  measurements by Barkla, 19-22
  photoelectric, 28, 29
  scattering, 28
  studies by Roentgen, 19
  by various substances, 60
Absorption coefficient, linear, 23
  units of, 23
  see also absorption coefficient
Absorption edge method, see Absorptiometry, differential
Absorption edges, 24
  discovery by Barkla and Sadler, 24
  discussion, 24-27
  and quantum theory, 26
    relationship between, 26, 27
  relationship with characteristic x-ray lines, 49-51
  wavelength shift with change of chemical state, 24-26
  x-ray, of the elements, tables, 504-512
Absorption effects, 387
  calculation of, 258-263
  example of, in diffraction, 261, 262
Absorption factor, in diffraction, 253, 258-261

552 INDEX

Absorption (interelement) effects, estimation of, 391-393
  negative, 261, 282, 386-391
  positive, 261, 282, 386-391
Absorption jump ratios, 139, 177
  graph of r versus Z, 534
Absorption of x-rays, by detector gases, 60
  discussion, 17-29
  by solid detectors, 60
Accuracy, 328-330
  of comparative x-ray methods, 329, 330
Aging of equipment, 376
Alnico, determination of, 383
Aluminum in Alnico, determination of, 383
Aluminum in TiO_2, determination, 111
Amplification, in detectors, 76, 78, 82
Amplifier, 62
Analysis, determination of elements by x-ray emission spectrography, bibliography, 535-549
  of variance, 338, 343-347
  see also Absorptiometry; X-ray emission spectrography, references
Analytical lines, excitation of, 358
  ultrasoft, x-ray tubes for excitation of, 368, 369
Analyzing crystals, 214, 215; see also Bragg reflectors
Anticoincidence, 65
  circuit, 108
Angstroms, (Å), conversion from kilo X units (1000 XU) to, 14
Applied Research Laboratories (ARL), multichannel x-ray spectrographs, 466-469
Applied Research Laboratories Production Control X-ray Quantometer (PCXQ), use in process control of Cu and Zn production, table, 494, 495
Applied Research Laboratories Production X-Ray Quantometer (ARL-PXQ), schematic diagram, 468
Assessment of x-ray diffraction methods, 264-276
  advantages and disadvantages, 264, 265
  automation and computerization, camera method, 270-273
  of the diffractometer, 273-275
  conclusion, 275, 276
  Dow ZRD Search-Match program, 268-270
  illustrative applications, 265-268
ASTM diffraction file, 239, 251-257
Atomic absorption coefficient, 23
Atomic number, significance proved by Moseley, 45-49
Atomic scattering factor, 253
Atomic structure, Bohr theory of, 47, 48
Attenuation, by detector window, 60
  of polychromatic beams by plastics, 151
  in spectrograph, 60
Auger effect, 51, 52
Auger electrons, 11, 28, 318
Auger-electron spectrography, 318, 319
  adapted to qualitative determinations on surfaces, 318
  for the determination of light elements, 318, 319
  recorded by ESCA, 36, 486, 487
Automation, of diffraction methods, 270-275
Avalanche detectors, 99
Avogadro's Number, 23, 29, 234
  and x-ray diffraction, 41-44

Background, difficulties caused by, 377
  in diffractometry, 252
  from electron and x-ray excitation, 45, 46
  treatment of, 379
Background correction, 339, 340
Balanced filter, 199, 200
Bank notes, genuine, x-ray emission spectrum from, 310
  histograms from x-ray spectrographic chart recordings of, 309-312
  x-ray emission spectrographic examination, 309-312
Barium, determination of, in vacuum tubes, 312, 313
  by differential absorptiometry, 137, 138
Beer's law, in absorptiometry, 145, 153
  application to x-ray absorption, 18, 23, 28, 132, 145, 153
  and plating thickness, 286, 287
Benzoic acid, determinations, 266
Beta filter, 199
Bibliography, of x-ray spectrographic determinations of various elements, 535-549
Binomial distribution, equation and discussion, 332-334

INDEX 553

Bohr atom, 47, 186-188
Boron, determination in thin films, 317
Borosilicate glass, qualitative analysis of, with use of the JPX-3 emission spectrograph, 480, 481, 483, 484
Bragg reflection, from crystals with cubic symmetry, 238
  from curved crystals, 218-223
    the Cauchois arrangement, transmission, 220-223
    the Johann and the Johansson arrangements, reflection, 221-223
  discussion, 37-39
  by a flat crystal, 201-218
  interference with, by background and by other lines, 109
  order of, 37, 38
  unwanted x-rays from, 218
Bragg reflectors, care of, 216
  choice of, 212-216
  digest of operating details for flat, 216-218
  Langmuir-Blodgett gratings best for the light elements, 213, 216, 450
  temperature effects on, 217
  unwanted wavelengths sometimes generated by, 115, 218
  useful range of, list, 214
  for x-ray emission spectrography (table), 214, 215
Bragg spectrometer, 136
Bragg's law, 37, 201, 206, 209, 210, 212, 213, 217, 219, 227, 233, 236, 241, 242, 279
  derivation, 38
Bremsstrahlung, 17
Bromine, by absorption edge method, 138, 140, 141
  in saline water, determination of, 426

Cadmium sulfide, activated, photoconducting x-ray detectors, 94
Carbon, determination in thin films, 315-317
Carbon brushes, treated, examination by x-ray absorptiometry, 155, 156
Carbon percentage in hydrocarbons, determined by x-ray scattering, 433, 434
Cathode rays, 3
Cauchois arrangement, 220-223

Čerenkov radiation, 59
Cesium film on silicon, determined by Auger-electron spectrography, 318
Characteristic line generator, 198
  use of, 412
Characteristic lines, and absorption edges, 49-51
  excitation of, 358
  origin of, 179-186
  wavelength, history of, 40
  see also X-ray emission lines, characteristic
Characteristic-photon (fluorescence) yield, 51, 52
Characteristic photon yields of the elements table, 532, 533
Characteristic x-ray emission spectra, differences between electron and x-ray excitation of, 45, 46
  discovery, 19-22
  from electron excitation, 45, 46
  origin of, 49-51, 179-186
  of platinum, measurement by Bragg, 40, 41, 50
  revealed by Barkla's absorption measurements, 19-22
  from x-ray excitation, 20, 21, 45, 46
  see also X-ray emission lines, characteristic
Characterization of plastics by x-ray absorptiometry, 150, 151
Chemical effects, on absorption and emission, 53
Chemical influences in x-ray absorption and emission, 53
Chemical state, influence in x-ray emission spectrography of light elements, 53
Chlorine, in chlorinated hydrocarbon polymers, determination by x-ray absorptiometry, 152, 153
Choking, 82
Chromite-olivene, determination of iron in, 389
Chromium, bromine interference, 116
Circuitry, electronic, functions of, 61-65, 68
  solid-state, 65
Circuits, electronic, in detector, 61
Clipping time, 62, 63, 82
Cobalt $K\alpha$ Lines, calculation of intensity, from monolayer of cobalt, 176-179, 225, 226, 295, 296

x-ray intensity losses in x-ray emission spectrography illustrated by, 225, 226, 295, 296
Coincidence loss, 82
  examples of, 119
  in Geiger detector measurement, 83
Collimation in the x-ray spectrograph, 201-205
Collimator, best choice of, 205
  transmittance, actual, 204
Color changes, caused by x-rays, 53
Comparative absorptiometry, determination of tetraethyl lead fluid in gasoline by, 159, 160
  identification of pure compounds by, 158, 159
  in the laboratory, discussion of, 156-158
Comparative method in absorptiometry, 132
Comparative method in x-ray emission spectrography, 417-427
  its great importance emphasized, 417
Comparison of electron and x-ray excitation, 370-371
Compensatory standard, 423
Compton effect, 11, 34-36, 252, 304, 424, 430-435
Compton scattering, *see* Compton effect
Computers, and analytical chemist, 498
  in controlled cement manufacture, 496
  and electron microprobe, 475
  use of in diffraction, 268-275
Conduction band, 95
Consolidated Electrodynamics Corporation, Type 27-101, electron microprobe, diagram, 474
Constant potential x-ray tube power supply, 8-11, 364
Consumer risk, 349, 351
Continuous x-ray spectrum, Bremsstrahlung, 17
  discussion, 13-17, 171-173
  intensity distribution with wavelength, 13-17, 171-173
  short-wavelength limit, 13-17, 171-173
Coolidge tubes, advantages, 6, 7
  characteristics, 6, 7
  lines due to impurities in spectrum, 175, 176
  targets available in, 6, 7

voltage circuits for, 6-10
  as x-ray source in chemical analysis, 12
Copper isotope $_{29}Cu^{64}$, 87
Copper-zinc determination via electron excitation, 362
Corrosion and x-ray absorptiometry, 128, 129
Counting, 64
Counting error, *see* Standard counting error
Counting rate, as function of, detector voltage, 112
  window width, 113
Critical absorption wavelength, 24-27
Critical thickness (depth), 290-294
Crystal, surface preparation of, 207
Crystal identification, card file for, 256, 257
Crystal imperfections, 207
Crystallite size by x-ray diffraction, 276
Crystallography, Bragg's law and, 233-239
  descriptive data for the six crystal symmetry systems, 236
  laws of crystal symmetry and of rational indices, 234
  Miller indices, 234-240, 242, 253, 258
Crystals, analyzing, choice of, 212-216; *see also* Bragg reflectors
Crystal symmetry, law of, 234
Cubic structure, 44
Cubic system, 236
  equation for, 235
Curved crystals, 218-223

Dead time, 82
  of Geiger detectors, 83
Debye-Scherrer camera, 244
Debye-Scherrer method, 241-247
  exposure time, 246
Decay time, of scintillators, 73
Detection of x-rays, 3, 4, 58-126
  by excitation of a phosphor, 58, 59
  importance of absorption in, 58
  by ionization in a gas, 58, 59
  by the photographic process, 4, 58, 59, 66-69
  separation of charge in certain crystals, 58, 59
  use of intensifying technique in the photographic process for, 20, 21, 69
Detection systems, *see* Systems for detection of x-rays

INDEX

Detector, accumulative, 59, 81
  instantaneous, 59, 81
Detector systems, 58ff
Detectors, best choice, 118
  flow proportional, 75-80, 90-93, 118-121
  gas-filled ionization, 75-82
    RC, time constant of the circuit, 81, 82
  Geiger, 75-80, 82-84, 118
  in 1969, 118-124
  lithium-drifted, Si(Li) and Ge(Li), 94-103
  phosphor-photoelectric, 72, 73
  photoconducting, activated cadmium sulfide, 94
  photoelectric, 70
  photographic plate (or film), 4, 20, 21, 58, 59, 66-69
  photomultiplier, 70-72
  proportional, 75-80, 84-90, 92, 93, 118
    efficiency, 90
  scintillation, 70, 73-75, 118-121
  solid ionization, 99-103, 122
  solid proportional, "avalanche," 99
  x-ray, 55ff
Determination of elements by x-ray emission spectrography, a guide to the recent literature (1964-1970), 535-549
Development, photographic, 68
Differential absorptiometry, *see* Absorptiometry, differential
Diffraction of x-rays, general, applications, 265-268
  comparison of methods for, 264, 265
  by crystals, 36, 37
  intensity of lines, 251
  limitations of, 264, 265
  mixtures, determination of, 259
  x-ray, 232-280
    absorption effects, 258-262
    Bragg's resolution of the Pt L spectrum, 40, 41
    in chemical analysis, 232-280
    by crystals, 37-41
    determination of crystallite size, 276, 277
    determinations on powdered samples, 239-253
      diffractometry, 247-251
      Hull-Debye-Scherrer method, 241-247
      intensity of Bragg reflection, 251-253
      introductory remarks, 239-241
    discussion, 227-229, 232, 233
    fingerprint method, 241, 266
    interpretation of the diffraction pattern, 253-258
    miscellaneous methods, 276-279
    parafocusing in, 227-229
    spectrometer, 39-41, 223-225
    studies of linear polymers and fibers, 277-279
    use of internal standards, 262, 263
    see also Assessment of x-ray diffraction methods
Diffraction line width, equation for, 277
Diffraction pattern, aluminum, 257
  interpretation, 253-258
  platinum, 240
  polymers, 278
  silica, 267
  tungsten, 243, 248
Diffraction peaks in x-ray emission, 415
Diffractometer, collimation in, 228
  parafocusing, 247
  x-ray, 223-225
    automation and computerization of, 273-275
    collimation scheme for, 227-229
    diagram of a parafocusing, 247
Diffractometer geometry, 247, 250
Diffractometer-spectrograph, 461, 462
Diffractometry, 247-251
  background in, 252
Dilution in sample preparation, 436
Dispersion, *see* Resolution
Divergence of x-ray beam
  reflected by crystal, 206-209
  in Soller slit, 201-205
Dosage, x-ray, 58
Double crystal spectrometer, 208
Dow *ZDR* Search-Match, 268-270
Drift, due to evaporation, 338
  electronic, 65
  in long-term data, 344, 345
Dual Bragg reflector, 213
Dual-target x-ray tube, 367, 368
Duane-Hunt rule, 14
Dynodes, 71

Edge crystal spectrograph, 229, 230
Edges, absorption, 24; *see also*

Absorption edges
Effective wavelength of a polychromatic x-ray beam, determination of, 16, 144, 145, 148-150
Efficiency, of x-ray excitation, 16, 17, 174, 176-179
Einstein equivalence law, 12, 26, 181, 318
Electron excitation of larger sample areas, Telsec Betaprobe and JPX-3 primary x-ray analyzer, 478-484
Electron excitation of x-rays, 12-17, 45, 46, 315, 360-363
  compared with x-ray excitation, 370, 371
  used in the HEED technique, 313-317
Electronic circuitry, functions of, 61-65
Electronic production of x-rays, 12
Electron microprobe, 473-478
  calculations, 454-456
  Consolidated Electrodynamics Corporation, Type 27-101, diagram, 474
  detector for, 124
  general discussion of, 452
  sample preparation and standards, 453, 454
  use, in detecting cause of leakiness in vacuum device, 477, 478
    in identification and analysis of single, pure phases in minerals, 477
    in nuclear-fuel element problem, 477, 478
    of its scanning ability, 475, 476
    when extreme localization is needed, 324
Electron penetration of x-ray target, 17
Electrons, characteristic lines excited by, 359, 360ff
  recoil, 35
  scattering of, 361
  x-ray energy levels, the periodic table, and, 192
Electrons and x-rays, 11
Electron volts, 16
Element determinations, by x-ray emission spectrography, bibliography, 535-549
Emission, x-ray, introduction to, 45
Emission energies, x-ray table of, 521-523
Emission lines, *see* Characteristic lines; characteristic x-ray emission spectra; and x-ray emission lines

Emission spectra, x-ray, of the elements, tables, 513, 521; *see also* Characteristic x-ray emission spectra; X-ray emission lines, characteristic
Energy-level diagram, x-ray, for uranium, 184, 185
Energy levels of atoms, x-ray, quantum numbers and electrons for, 191
Energy, radiant, 14
Energy resolution, 58ff, 100, 103-118
  pulse-height distribution, 103-105
  sharpness of, 105-107
  spectrograph systems, 121
  with wavelength resolution, 109-116
  window setting procedure, 111
Enhancement (interelement) effects, 386-391
Equipment, difficulties due to, 375
  and illustrative applications, 461-487
  and selected applications, introduction, 460, 461
ESCA (Electron Spectroscopy for Chemical Analysis), 484-487
Escape peaks, 74, 87, 88, 116, 117
Excitation of analytical lines, comparison of electron and x-ray excitation, 370, 371
  by electrons, 358, 359
  by ions, 359
  by protons, 359
  by x-rays, 358, 359
Excitation of x-rays, by electrons, 12; *see also* Electron excitation of x-rays; Excitation of analytical lines; Ion excitation of x-rays; Proton excitation of x-rays; Radioactive sources; Radioactive x-ray sources; and X-ray excitation of x-rays
Exposure time, in Debye-Scherrer method, 246

Fano factor, 105, 121
Faraday number, 41, 44
FET, 99
Field effect transistor, 99
Filler gas, for detector, 83
  for proportional detector, 84, 92
Film-measuring mechanism for diffraction patterns, 254
Film thickness, determination, 281-303

INDEX 557

by use of various radioactive sources, 301, 302
by x-ray emission spectrography, 290-298
of aluminum on semiconducting silicon, 299, 300
of cadmium sulfide and of other II-VI compounds, 302, 303
of films near the monolayer level, 313-317
of multiple plating, 300, 301
of permalloy, 303
of tin plated on steel, by attenuation of an unresolved beam from the substrate, 282-284
use of Norelco tin coating weight gauge, 283, 284
various examples, by attenuation of a characteristic line from the substrate, 285-290
of zinc or aluminum on steel, by attenuation of a characteristic line from the substrate, 285
Films, thin, and Auger-electron, spectrography, 318, 319
Filter, $K\beta$, for diffraction, 244
Filtering, 198, 199ff
Filtering of polychromatic x-ray beams, 19, 199-201
by method of Ross, 199, 200
Fine structure, 195, 196
Flat Bragg reflectors, 201-218
Flow proportional detectors, 75-80, 90-93, 118-121
Fluidized bed, x-ray absorptiometry of, 155
Fluorescence yield, see Characteristic-photon yield
Fluorine, determination in thin films, 317
Fluorite, as internal standard, 266
Fluoroscopy, definition, 127
Fluo-X-Spec Analytical Laboratory, methods and techniques, 487-489
F-test, 344, 346
Full wave rectifier, 8
Full wave power supply, 8
Fundamental parameters and interelement effects, 399
FWHM (*Full Width* of Gaussian curve at *Half Maximum*), collimator-limited, 24
definition, 103, 104
of diffraction pattern from different crystals, 211
on Gaussian curves, 103-105
intrinsic, 205
for line broadening processes, 210
used in rating of detectors, 106
of x-ray line, increased by Bragg reflection, 205

Gases, absorptiometric experiments with, 153-155
Gas-filled ionization detectors (chambers), 75-82
RC, time constant of the circuit, 81, 82
Gas (or ion) tubes, advantages, 5, 6
characteristics, 4-6
Gaussian distribution, equation and discussion, 329-336, 348, 349
of Rutherford-and-Geiger radioactivity data, 335, 336
of x-ray emission data on tungsten, 335, 336
Geiger detectors, 75-80, 82-84, 118
coincidence loss, in measurement with, 83
dead time, 83
Ge (Li) detectors, 94-103
General Electric Heinrich milliprobe, use of, in vacuum tube problems, 312, 313
General Electric Raymike® 2000 Thickness Gauge, 164-168
General Electric x-ray emission gauge, XEG, use in large-scale process control of cement production table, 496, 497
General Electric x-ray photometer, 162-164
General Electric XRD-3 spectrograph, use in vacuum tube problem, 312, 313
General Electric XRD-5 D/S spectrogoniometer, arranged for x-ray emission work, 461, 462
General Electric XRD-6VS, x-ray emission vacuum spectrograph, 464
General Electric XRD-410 production system, computerized, 462, 465
Germanium, 95
Glass, coloration of by x-rays, 53
opal, determination of, 426
Goniometer reset error, 341-343, 345-347
Graphite as Bragg reflector, 208

Grating, x-ray, ruled, reflecting, 42
Gratings, Langmuir-Blodgett, 212-216
  Rowland, 219
Guinier Camera, 270, 272
  parafocusing in, 227

Hafnium, discovery of, 356, 357
Hanawalt-Rinn-Frevel system, 254-257
"Hard" x-rays, 4, 54
HEED (High-Energy Electron Diffraction), use of, 313-317
Heisenberg uncertainty principle, 36, 205
Henke tube, 365-370, 450
Heterogeneous sample, treatment of, 415
Hexagonal system, 236
Hittorf-Crookes tube, 3
Hull-Debye-Scherrer method, 241-247, 253, 264, 265, 268, 270
Hydrogen, optical spectrum, 194, 195

Identification of pure compounds, 158, 159
Impedance matching, 62
Instability of equipment, 65, 375, 376
Instantaneous readout, 65
Integrated versus peak intensities, 252
Intensifying technique, for photographic detection of x-rays, 20, 21, 69
Intensity, comparative, denoted by intensity ratio, 13
  in diffraction, 251-253
  emitted from different regions of sample, 204
  integrated, 14
  lines, relative, 174
  measurement of x-ray, 13, 58-126
  relative, of lines in various series, 174
  of tube targets, 369
Intensity maximum of continuous spectrum, 13, 172, 173
Intensity of Bragg reflection, variation with order, 383
Intensity versus angle for Bragg reflectors, 207
Intercepts, of unit crystallographic plane, 234
Interelement-effect corrections, calculated, comparison of, 386, 405, 406
Interelement effects, calculated from fundamental parameters, 399-405
  comparative study on six stainless steels, of three methods for, 403-405
  mitigation of, by dilution, 435
  see also Absorption effects; enhancement effects
Interference, method of reducing, 115
  by target lines, 175
  by tungsten and molybdenum lines, 175
  spectral, traceable to Bragg crystal, 115
Internal standards, in x-ray diffraction, 258-263
  in x-ray emission, 419, 427
Interplanar spacing, 233-235
Intrinsic ("natural") line width, 205-207
Ion-exchange membranes, use in trace determinations, 322-324
Ion-exchange papers as solid diluents, 441
Ion excitation of x-rays, 314, 359
Ionization chambers, see Gas-filled ionization detectors
Ionization detectors, solid, problems of, 99, 100
  in x-ray photometer, 162-166
Ion pairs, 76-79, 81, 84, 85
Ion tubes, see Gas tubes
Iron isotope, $_{26}Fe^{55}$, for absorptiometry, 133-136
Iron oxides, determination of, 390

Japan Optical Company, see JPX-3 primary x-ray analyzer
Johann arrangement, 221-223, 450
Johansson arrangement, 221-223, 270, 271, 467, 470, 472, 473
JPX-3 Primary X-ray Analyzer, 478, 480, 481, 483, 484

K-capture, 87, 88
K spectra, history of, 21
K state, 49
Kossel photography, 303

Langmuir-Blodgett gratings, 212-216
L lines, excitation of different groups, 193
L spectra, of platinum, diagram, 50
L state, 49
Large-scale applications of x-ray emission spectrography, 493-497
Latent image, 58, 68
Lattice constant determination, 257
LEED (Low-Energy Electron Diffraction),

INDEX   559

useful in Auger-electron spectrography, 318, 319
"Light-element" problem, 447-451
Light elements, influence of chemical state in x-ray emission spectrography, 53, 448, 449
  use of Langmuir-Blodgett gratings in x-ray emission spectrography for, 213, 216
Limit, short-wavelength, 14
Linear absorption coefficient, 23
Linear circuits, 61
Line designation, 21, 22
Lines, see Characteristic x-ray emission spectra; X-ray emission lines, characteristic
Line shape, 205, 206
Line width, intrinsic, 205
Liquids, absorptiometry of, 158-160, 163
Lithium drifted detectors, Si(Li) and Ge(Li), 94-103
LL, LM, and LN states, 52
Logic circuits, 61
Lorentz factor, 253
Losses, in x-ray spectrograph, 225-227

McPherson Instument Corporation's ESCA 2.5 Electron Impact Spectrometer, 486
McPherson Instrument Corporation's ESCA 36, 486, 487
Magnetic photoelectron spectrograph, use of, 26, 27, 49, 484, 485
Manipulative errors and standard counting error, 345-347
Mass absorption coefficient, 23, 28
  additivity, 24
  discussion, 23-26, 28, 29
  equation, for mixture of elements, 24
  as function of wavelength, 24, 25
  independence of chemical or physical state, 23
  photoelectric equation, 29
  plotted against wavelength for aluminum, copper, and lead, 25
Mass absorption coefficients, of the elements, in 0.1Å-30 Å region, table, 525-530
  of Formvar and Mylar films, in 2 Å-100 Å region, table, 530, 531

of light elements, in 40 Å-100 Å region, table, 530
of selected gases, in 2 Å-100 Å region, table, 530, 531
Methane-argon as filler for proportional detector, 84
Microprobe, electron, see Electron microprobe
Miller indices, 234-239
Modified scattering, 28-36
Molybdenum, by absorption edge method, 142
  as target material, 6, 7
Molybdenum-tungsten ore, determination of, 417
Molybdenum-tungsten ratio, determination of, 382
Molybdenum x-ray spectrum, 46
Monochromatic x-ray beams, 198, 205-207
  discussion, 127-131
  four kinds of, 130
  from radioactive isotopes, 130, 131
  selection by use of diffraction, 37-39
  simple absorptiometry with, 131-136
Monochromator, Bragg crystal, 225
Monoclinic system, 236
Mosaic structure of crystal surface, effects of, 207
Moseley, H. G. J., x-ray researches, 45-49, 188
Moseley's equation, 47
Multichannel analyzer, 103
Multiple regression in interelement effects, 394
Multiplication, in detectors, 82
  in proportional detector, 84

Naphthalene, 73
Nickel filter, for suppression of copper K_β, 199
Nobel awards for x-ray research, 4, 20
Nomenclature, x-ray instruments, 223-225
Nondispersive spectrum analysis, 64
Norelco Tin Coating Weight Gauge, 283, 284
Normal error curve, see Gaussian distribution

Optical arrangements, of spectrometers compared, 222-225
Optical spectra and x-ray spectra, 194-197

Orders of Bragg reflection, 37, 38
Orthorhombic system, 236
Oscillator, damped, 205, 206
Overlapping lines, unfolding of, 406
Oxygen determination in thin films, 314-317

Parafocusing, 227-229, 247-249
  use of Guinier camera, 270-273
Particle size, determined by diffraction, 276, 277
Pass band, 199
Pauli exclusion principle, 188-190, 193
Peak and integrated intensities, 252
Peaks in x-ray emission spectrography, 378
Periodic table, 47-49, 188, 192
Permalloy films, determination, of composition and thickness, 303
Philips inverted-sample, three-position spectrograph, 469, 470
Phosphor-photoelectric detector, 72, 73, 146
Phosphorus determination in silica films, 300
Photocathode, 70
Photoconducting detector, 94
Photoelectric absorption, 28, 29
Photoelectric effect and quantum theory, 11, 12
Photoelectric mass absorption coefficient, equation for, 29
Photoelectron spectrography, modern, 484-487
Photographic plate or film as x-ray detector, use in x-ray detection, 4, 20, 21, 58, 59, 66-69
Photographic x-ray detection, 68, 69
Photometer, x-ray, commercial, applications, 162, 164
  definition, 225
  diagram, 148, 163
  modification of simple, 156-158
  simple, 146-148
Photon yields, characteristic, table, 532, 533
Pigments, determination of, 122, 123
Pitchford Manufacturing Corporation's x-ray emission spectrograph, 473
Planck's constant, determination of, 14, 15
Plastics, characterization of by x-ray absorptiometry, 150, 151
Plateau, voltage, 79
Platinum, diffraction pattern of, 240, 241
  L peaks, measured by Bragg, 40, 41
  L spectra, 41, 50
Poisson distribution, 332-334
Polarization factor, 253
Polarization of x-rays, and background intensity, 385, 386
  by scattering, 33
Polychromatic x-ray beams, 16, 129, 130
  effective wavelength, 16
  filtering of, 19
  generation, 12, 13
Polymer structure by x-ray diffraction, 227
Portland cement, composition, by-x-ray emission spectrography, table, 492, 493
Potassium chloride, effect of x-rays on, 53
Powder camera, 244
Powder diffraction, 241-276
Powder Diffraction File, 239, 254-257
Power, total, of x-ray beam, 16
Power supplies for x-ray tubes, 7-11
Preamplifier, FET, 99
Precision, 328
Pressure control, for proportional detectors, 93
Process control by x-ray absorptiometry, 160-168
Producer risk, 349
Proportional detectors, 75-80, 84-90, 92, 93, 118
  applications, 87
  irregularities in, 108
  pressure control for, 93
Proton excitation of x-rays, 314, 359
Pulse-height analyzers, definition, 63, 64
  multichannel, 100, 103, 117, 118, 123
Pulse-height discriminator, definition, 63, 64
Pulse-height distributions and the Gaussian, 103-105
Pulse-height selection, ABC's of, 107-109
  energy resolution, 63, 64
  to reduce interferences, 383
  use in reducing background, 384, 385
Pulse-height selector, definition, 64
Pure compounds, identification by comparative absorptiometry, 158, 159

INDEX 561

Qualitative trace determinations, statistical considerations, 349-351
Quantrol (Applied Research Laboratories), control of tin-plate manufacture by, 298
Quantum theory, 187-197
  in study, of absorption edges, 26, 27
  of electron and x-ray interplay, 11-16
Quartzite, determination by diffraction, 261-263

Radiant energy, definition, 14
Radioactive sources, for excitation, 371
Radioactive x-ray sources, for absorptiometry, 130
Radiography, definition of, 127
Raman effect, 21, 36
Range, angular, of spectrometer, 204
Rare earths, analysis, by x-ray emission spectrography, 492
Rational indices, law of, 234; see also Miller indices
Ray Mike®, thickness gauge, 81, 167-168
Reciprocity, 69
Reflection, Bragg, 37, 238, 262-266
  total, 42
Refraction of x-rays, 41-44
Regression treatments of interelement effects, 351-354, 393-399, 403-406
Rejection of pulses, 65
Relative intensities of x-ray lines, table, 524
Relative resolution (RR), detectors compared, 121
  as measure of energy resolution, 106, 107
Reliability, 328-331
  of x-ray emission spectrography, 328-354
Resolution, of adjacent lines, 406ff
  definition of, 209, 210
  by detectors, 106
  energy resolution, 58ff, 100
  equation for, 212, 213
  in proportional detectors, 85, 87
  relative, 106
  of x-rays, by filtering, 199-201
Rhombohedral system, 236
Risk, consumer, 349, 351
  producer, 349
Ritz combination principle and emitted x-rays, 186, 188
Rock analyses by x-ray emission spectrography table, 491
Rocking curves of crystals, 207-209
Roentgen, W. C., discovery of x-rays, 3, 4
  Nobel Prize awarded to, 4
  production of secondary x-rays discovered by, 20
  x-ray absorption studies by, 19
  x-ray papers of, 4
Ross filter method, 199-200
Rotation of samples, in absorptiometry, 143
  in diffraction measurements, 244
  in x-ray emission spectrography, 415
Rowland grating, 219
Rydberg constant, 186

Safety in use of x-ray equipment, 501-503
Sample preparation, for diffraction, 243, 248, 249
  in x-ray emission spectrography, 414-435
Satellite lines, 52
Scaling, 64
Scattered background, as standard, 424
Scattering, modified, 34
Scattering of x-rays, 28-36, 430-435
  for determining binary mixtures, 433
  differences between Compton, Rayleigh, and Thomson, 36
  percentage of carbon in hydrocarbons determined by, 433, 434
Scintillation detectors, 70, 73-75, 118-121
Screening constant, 187
Secondary x-ray emission, see X-ray excitation of x-rays
Selected applications, of x-ray emission spectrography, 487-493
Selection, pulse height, 63
Selection rules and x-ray emission spectra, 193
Selection of x-ray wavelengths, 198-231
Semiconductors, 95
Semiconductor detector used for diffraction measurement, 240
Servomechanism systems, 160-162
Shaping of pulses, 62, 82
Shift of wavelength, due to chemical effects, 53
Short-wavelength limit, of the continuous spectrum, 12-15
Siemens SRS x-ray emission spectrograph

system, 465, 466
Silicon and Germanium, 95
Si(Li) detector, 94-103
Silicosis, 265
Silver, in photographic film, determination of, 380
Soap films, as Bragg reflectors, 262-266
Sodium iodide in detectors, 73
"Soft" x-rays, 4
  tubes for, 368
Soils, determination of, 266
Solid ionization detector systems, unaided energy resolution using, 100-103
Solid ionization detectors, 94, 99-103, 121-124
Solid proportional detectors, "avalanche", 99
Solids, absorptiometry of, 150-153
Soller collimator, or slit system, 201-205
Spectra, x-ray, 171-197
  characteristic, excitation of, 45
  crude correlation with other kinds of, 55
  of Nb, Mo and Co, aided by energy resolution, 114
Spectral interference, composition of stainless steels by such unfolding, 411-414
  and unfolding of pulse-height distributions, 406-414
Spectrograph, photoelectron, electrostatic, 485
  magnetic, 26, 49, 484, 485
  x-ray emission, 39, 47, 223-225
    Applied Research Laboratories (ARL) multichannel, 466-469
    Applied Research Laboratories Production X-ray Quantometer (ARL-PXQ), schematic diagram, 468
    Birks edge-crystal, 229, 230
    comparison of spectrometers for, 223
    equipment with curved crystals for localized determinations, 470-472
    estimated losses in, 225-227
    evaluation of, a flat-crystal emission spectrograph system, 465, 466
    a flat-crystal spectrograph system, 461-465
    General Electric XRD-5 D/S spectrogoniometer, 461, 462
    General Electric XRD-6VS vacuum spectrograph, 464
    General Electric XRD-410 production system, computerized, 462, 465
    JPX-3 primary x-ray analyzer, 478, 480, 481, 483, 484
    multichannel systems with curved crystals, 466-469
    multichannel spectrograph systems with curved crystals, 466-469
    Philips inverted-sample, three-position spectrograph, 469, 470
    Pitchford Manufacturing Corporation equipment, 473
    portable, with radioactive source, 472, 473
    resolution (dispersion) in, 209-212
    Siemens SRS system, features, table, 465, 466
    Telsec Betaprobe, 478-484
    use of General Electric X-Ray Emission Gage, XEG, in control of cement production, table, 496, 497
  Spectrograph system, x-ray, block diagram, 463
Spectrometers, x-ray, 39-41, 223-225
  for spectrographs, comparison of, 223
Spectrophotometer, x-ray, of Barringer, 141, 142
  of Frevel and North, 140, 141
  of Hughes and Wilczewski, 131, 132
Spectrum, characteristic x-ray, from a target, 174
  by electrons and x-rays, 45
  from solid state detector, 100-102
  of trace elements, 110
  x-ray, continuous, 14-17, 171-173
  of x-ray tube, 368
  with impurities, 175, 176
Sphalerite, determination of, 268
Spiking method, illustration of, 427
Spinthariscope, 70
Stainless steels, composition of, by fundamental parameters, 403
  by "unfolding", 412
Standard counting error, comparable with manipulative errors, 345-347
  counting strategy, 347, 348
  definition and discussion, 331-336
  guaranteed reliability, 347-349
  high, 344, 345

INDEX 563

measurement of, in complex cases, 340, 341
  as an operating criterion, 336-338
  practical, in more complex cases, 338-340
  relative, 339
Standard deviation, compared with standard counting error, 336-338, 340, 341, 343-345
  definition and equation, 105, 330, 331
  equation for overall, 105, 331
  observed value on Gaussian curves, 103, 104
Standards, for electron microprobe, 360
  internal, 427
Steel strip, x-ray gauging, General Electric Thickness Gauge, 161, 162, 164-168
Straumanis film loading method, 245
Stripped atoms, spectra of, 196, 197
Strontium ion, aqueous, spectrum of, 381
Structure factor, 253
Sulfur, in petroleum products, $_{26}Fe^{55}$ method and other methods for, 134-136
Sulfur K absorption edge, chemical influences on, 448
Sulfur K from sulfur in different chemical states, 448, 449
Surface preparation of Bragg crystals, 208
Symmetry, systems of, 235, 236
Systems for detection of x-rays, absorption problems, 60
  amplification, 62
  discussion, 59
  "flip-flop" circuits, 64
  functions of electronic circuitry in, 59, 61-65, 68
  impedance matching, 62
  instability in electronic circuitry of, 65
  instantaneous readout, 65
  pulse rejection, 65
  use of solid-state devices in, 61, 65, 68

Tandem detectors, efficiency of, 119
Tantalum and niobium, determination of, 406ff
Target in Coolidge tube, contamination of, 6
Targets, available in Coolidge tubes, 6, 7
  choice of target in x-ray emission spectrography, 367-370
  dual-target x-ray tube, 367, 368
  interchangeable, in gas tubes, 6
  studies with thin metal foil, 171-173
Telsec Betaprobe, advantages of, 478-484
  element coverage and related data, table, 482
  three versions discussed, 478-484
Telsec Instruments, portable dual-channel spectrograph, 472, 473
Temperature effects on Bragg reflectors, 217, 239, 253
Tetraethyl lead fluid, in gasoline by comparative absorptiometry, 159, 160
Tetragonal system, 236
Texas Nuclear Corporation's portable spectrograph, 472
Thickness, critical, see Critical thickness
Thickness gauging of steel strip, 164-168
Thickness of plating, 282-298
Thick-target x-ray spectra, 12, 13, 172, 173
Thin target spectrum, 171-173
Thorium by absorption edge method, 142
Thucolite and x-ray emission spectrography, 66, 490, 491
Time constant, for diffractometers, 250
  RC of gas-filled ionization detectors, 81, 82
"Today and Tomorrow," 497-499
Total reflection in Soller slit, 202
Trace determinations, qualitative, 349-351
  minimum amount guaranteed detectable (MAGD), 351
  simple, 304-325
  of barium deposits in vacuum tubes, 312, 313
  discussion, use of spot-test and other techniques, 304-309
  examination of bank notes, 309-312
Transmission of x-rays, by air and He in spectrographs, 60
  as function of absorber thickness, 19
  by various spectrograph windows, 60
Transmitted beam, components of, 28
Triclinic System, 236
Tubes, x-ray, for excitation of ultrasoft analytical lines, 368, 369; see also Coolidge tubes; Gas (or ion) tubes
Tungstate solutions, estimation of absorption effects in, 391-393
Tungsten, in solution, determination of, 419

Unmodified scattering, 28-34
Uranium, x-ray energy-level diagram of, 184, 185
U-Nb-Zr Alloy by absorption edge method, 142, 143
Unfolding, by computation, 409
  of overlapping lines, 406
Unit plane, choice of, 234
Units, of concentration, 374

Vanadium, chromium, manganese, in steel, 408
Van de Graaf generator, 359
Variance, analysis of, 338, 343-347
  definition and equation, 330, 331
Voltage for excitation, 364

Wavelength, critical absorption, discusssion, 24-27
  resolution by Bragg reflection aided by energy resolution, 109-116
  selection, 198-231
    by filtering, 199-201
  units for, 14
Wavelength distribution in continuous spectrum, 172, 173
Wavelengths of characteristic lines, discussion, 40, 41
Windows, thin, for detectors, 90ff
  transmission of, 60, 91
  two, on detector, 77, 84

X-ray absorptiometry, *see* Absorptiometry
X-ray absorption, *see* Absorption
X-ray absorption edges of the elements, tables, 504-512
X-ray and optical spectra, 194, 195
X-ray beam power, equation for calculation, 16
X-ray detection, *see* Detection of x-rays
X-ray detection systems, *see* Systems for detection of x-rays
X-ray detectors, *see* Detectors
X-ray diffraction, *see* Diffraction, x-ray
X-ray diffractometer, *see* Diffractometer, x-ray
X-ray emission, chemical influences in, 53
  discussion, 45-52
X-ray emission electron microprobe, *see* Electron microprobe

X-ray emission energies, table, 521-523
X-ray emission lines, characteristic, chemical influences on, 53
  origin of, 179-186
  relationship between absorption edges and, 49-51
  revealed by Barkla's absorption measurements, 19-22
  see also Characteristic x-ray emission spectra
X-ray emission spectra, characteristic, *see* Characteristic x-ray emission spectra
  and selection rules, 193
X-ray emission spectra of the elements, tables, 513-521
X-ray emission spectrography, a look at the electron microprobe, 452-456
  best guarantee of results, 348, 349
  bromine, as internal standard for the determination of tungsten, 419-421
  bulk-solid samples, 442, 443
  choice of internal standards, 427-430
  comparative method in, 417-427
  counting strategy, 347, 348
  difficulties, and kind of sample, 435-452
  see also Absorption effects; Enhancement effects; Interelement effects; Spectral interference
    attending sample preparation, 414-435
    caused by unwanted contributions to intensity, 377-379
    traceable to equipment, 375-377
  excitation of analytical lines, 358-374
  general discussion of the history and development of, 355-358
  importance of sample preparation, 414-416
  large-scale applications, 493-497
  localized determinations, 451, 452
  "matrix" defined, 386
  methods, of prior concentration of traces, for determinations by, 321-324
    used extensively in the Fluo-X-Spec Analytical Laboratory, 487-489
  miscellaneous samples, 446, 447
  need of pulse-height selection for reducing background, 384, 385
  peaks encountered, 378
  powdered samples, 443-446
  principal difficulties and common

INDEX 565

remedies, 374-414
process control of flowing mineral-slurry streams in Cu and Zn production, table, 494, 495
qualitative introduction to interelement effects, 386-393
references to determinations of various elements by, 535-549
regression treatments of interelement effects, 351-354, 393-399, 403-405
reliability of, 328-354
remedies, and compensations for difficulties attending sample preparation, 414-435
 for background difficulties, 379-386
 sample preparation by dilution, 435-441
  by fusion, 437-440
  with liquids, 436, 437
  with solids and briquetting, 440, 441
 selected applications of broad significance, 487-493
  qualitative and semiquantitative determinations, 487-490
 selected determinations on complex materials, − thucolite; rock analyses; rare earths; portland cement, 490-493
 use, of compensatory standards for monitoring, 421-425
  in determination of composition of alloy and oxide films, 319-321
  in examination of genuine and counterfeit bank notes, 309-312
  in film thickness determinations, 290-298
  of internal standards in, 419-421
  of ion-exchange papers as solid diluents, 441
  of scattered background as compensatory standard, 424, 425
  of "spiking," 425, 426
X-ray energy-level diagram for uranium, 184, 185
X-ray energy levels of atoms, 179-186, 192
X-ray excitation by radioactive sources, 359, 371-374
 useful sources cited, table, 372
X-ray excitation of x-rays, 45, 46, 51, 52, 174-179, 358, 363-370
 compared with electron excitation, 370, 371
 efficiency of, 174, 176-179
"X-ray fluorescence," see X-ray emission spectrography
"X-ray fluorescent spectroscopy", see X-ray emission spectrography
X-ray generation, 4
X-ray instruments, comparison and nomenclature of various, 223-225
X-ray intensity, see Intensity
X-ray lines, 179-197
 relative intensity of, table, 524
X-ray methods of analysis, comparative, statistics, 397, 413, 434, 439
 use of Coolidge tubes in, 12
X-ray photometer, see Photometer, x-ray
X-ray photometers, commercial, 162-168
X-ray production, efficiency of, 16
X-ray production by electron excitation, 12-17
X-ray refraction, 41-44
X-ray research, Nobel Prize awards for, 4
X-ray resolution, see Resolution
X-ray safety, 501-503
X-ray spectra, comparison with optical spectra, 194-197
 fine structure of, 195, 196
X-ray spectrograph, losses in, 225-227; see also Spectrograph, x-ray
X-ray spectrography, see X-ray emission spectrography
X-ray spectrometer, see Spectrometers, x-ray
"X-ray spectrometry"−and still other names, see X-ray emission spectrography
X-ray spectrophotometer, see Spectrophotometer, x-ray
X-ray spectrum, continuous, see Continuous x-ray spectrum
 simplicity of, 2
X-ray story, introductory remarks, 1, 2
X-ray targets, see Targets
X-ray tube, impurities in, 175, 176
 self-rectified, 7, 8
X-ray tube for diffraction, choice of, 245
X-ray tube power supplies, 7
X-ray tubes, for excitation of ultrasoft analytical lines, 368, 369; see Coolidge tubes; Gas (or ion) tubes

X-ray wavelength, *see* Wavelength
X-ray wavelength-intensity distributions, 13-17
X-rays, and the Bohr theory of atomic structure, 186-188
  covenient classification of, 54
  discovery by Roentgen, 3, 4
  discussion of certain properties of, 2
  effects on samples exposed to, 53, 54
  "hard" and "soft," 4
  origin, 1-17
  and the periodic table, 188-193
  Roentgen's discovery of, 3, 4
  secondary, produced by Roentgen, 20
  significance in atomic structure, 1, 2
  significance in quantum theory, 1, 2
  and visible light, effects of, compared, 69
X-rays and electrons, interplay of, 11-17
X units, conversion to angstroms (Å), 14

Zinc, determinations of, after collection on ion-exchange membranes, 322-324
ZRD-Dow Search-Match program, 268